Y0-BTC-502

Radiotracer Techniques and Applications

VOLUME 1

edited by

E. Anthony Evans

Organic Department
The Radiochemical Centre, Limited
Amersham, Buckinghamshire
United Kingdom

and

Mitsuo Muramatsu

Department of Chemistry
Faculty of Science
Tokyo Metropolitan University
Setagaya-ku
Tokyo, Japan

MARCEL DEKKER, INC.   New York and Basel

Library of Congress Cataloging in Publication Data
Main entry under title:

Radiotracer techniques and applications.

1. Radioactive tracers. I. Evans, Eustace Anthony. II. Muramatsu, Mitsuo.
QD607.R32 543'.088 76-20000
ISBN 0-8247-6496-X

MARCEL DEKKER, INC.

270 Madison Avenue, New York, New York 10016

Current printing (last digit):
10 9 8 7 6 5 4 3 2 1

PRINTED IN THE UNITED STATES OF AMERICA

PREFACE

This book is aimed at identifying problems and their solutions in the numerous applications of radiotracer techniques in chemical and biological research. Contributors were selected who are expert in the various scientific disciplines, and who are very familiar with the problems pertaining to radiotracer methodology. Much of the information contained within these two volumes provides a basic understanding of the problems and knowledge in the applications of radiotracer techniques.

The book is divided into two volumes, with twenty-three chapters covering three principal sections of scientific interest. The fundamental techniques in the design of tracer experiments are discussed in chapters 1 through 7 of this volume. These include the selection of radionuclides, handling and health physics aspects of radiotracer uses, measurements of radioactivity, with special emphasis on biological samples, the philosophy of selecting methods for the preparation of radiotracer compounds, the essential factors to consider and methods for the analysis and quality control of radiotracers, and the difficulties in controlling self-decomposition of radiochemicals in storage.

Chapters 8 through 14, also in this volume, discuss problems in the application of radiotracer techniques for studies familiar to chemists, namely the special behavior of radionuclides at maximum isotopic abundance, isotope effects, exchange processes, solution properties, diffusion and interfacial phenomena, and important environmental studies with radiotracers.

Volume 2, chapters 15 through 23, will survey problems and pitfalls in many applications of radiotracer techniques to biological research. These include applications to studies of biosynthesis, and radiotracer techniques in drug metabolism and in cytology. The assay of enzymes and enzymatic assays using the high sensitivity of radiotracer techniques can often lead to a better understanding of mechanisms in the biochemistry of man, and subsequently to new methods of diagnosis, prognosis, and treatment of diseases. An understanding of the metabolism of chemicals used in agriculture is very important, especially for the safe protection of our food supplies and of our environment, areas which are readily investigated with radiotracers. Similarly, radiotracers help to further our understanding of marine biology. Biological research with radiotracers helps to set down the foundations upon which to build procedures for routine clinical diagnosis for disease control, and the final three chapters are of special interest to clinical investigators. These chapters deal with the assay of drugs and hormones by competitive protein binding (radioimmunoassay), the uses of not only inorganic ions of radionuclides but also of radiochemicals in diagnostic medicine, and a discussion of the philosophy and difficulties in developing radioactive drugs for the radiotherapeutic treatment of cancer.

The nomenclature which has been adopted for this text is referred to as the "square-brackets-preceding" system recommended by the American and British Chemical Societies and by most biochemical journals. The symbol for the isotope is placed in square brackets directly attached to the front of the chemical name, or to that part of the chemical name to which the label refers.

Some readers will already be aware that the familiar units of radioactivity and radiation dose, namely the *curie* and the *rad*, respectively, may well disappear in time as the intention is to introduce the International System of Units (SI units) of the becquerel (Bq) and the gray (Gy) for activity and for absorbed dose, respectively, as agreed by the General Conference on Weights and Measures and ratified by the Council of the European Community in 1972. The becquerel is equal to 1 sec^{-1} and the gray to 1 J kg^{-1}. The use of these units is already legally approved in several countries. However, as there is a large international trade in radioactive materials, and also a well familiarized use of the curie units in medical practice for applications of radiopharmaceuticals, it is clearly most desirable that all countries should agree to change to SI units at the same time. With these problems in mind therefore, one suspects that it will be many years before the "old" familiar units finally disappear from use. (See p. xiii for the SI Units for radioactivity.)

We trust that many scientists will find the wealth of knowledge discussed in this text useful in helping them to develop further valid applications of radiotracer techniques. Finally, we should like to express our appreciation for the excellent cooperation of the scientists whose contributions to this book have made its publication possible.

E. Anthony Evans
Mitsuo Muramatsu

February, 1977

CONTRIBUTORS TO VOLUME 1

CHARLES H. BOVINGTON, Chemistry Department, Rutherford Laboratories, Royal Military College of Science, Shrivenham, Swindon, Wiltshire, England

M. W. CARTER, Georgia Institute of Technology, Atlanta, Georgia

JOHN R. CATCH, Controller of Technical Development, The Radiochemical Centre, Limited, Amersham, Buckinghamshire, England

BRIAN DACRE, Chemistry Department, Rutherford Laboratories, Royal Military College of Science, Shrivenham, Swindon, Wiltshire, England

E. ANTHONY EVANS, Organic Department, The Radiochemical Centre, Limited, Amersham, Buckinghamshire, England

PETER J. F. GRIFFITHS, Department of Chemistry, Institute of Science and Technology, University of Wales, Cardiff, Wales

FRANCIS J. JOHNSTON, Department of Chemistry, University of Georgia, Athens, Georgia

J. R. JONES, Chemistry Department, University of Surrey, Guildford, England

FRANTIŠEK KEPÁK, Department of Radioactive Wastes, Nuclear Research Institute, Řež, Czechoslovakia

YUTAKA KOBAYASHI,* Worcester Foundation for Experimental Biology, Shrewsbury, Massachusetts

DAVID V. MAUDSLEY, Worcester Foundation for Experimental Biology, Shrewsbury, Massachusetts

ICHIRO MIYANAGA, Division of Health Physics and Safety, Japan Atomic Energy Research Institute, Tokai-Mura, Japan

---

*Current Affiliation, New England Nuclear Corporation, Boston, Massachusetts.

A. A. MOGHISSI, Georgia Institute of Technology, Atlanta, Georgia

YUKIO MURAKAMI, Department of Chemistry, Faculty of Science, Tokyo Metropolitan University, Tokyo, Japan

MITSUO MURAMATSU, Department of Chemistry, Faculty of Science, Tokyo Metropolitan University, Tokyo, Japan

GEOFFREY S. PARK, Department of Chemistry, Institute of Science and Technology, University of Wales, Cardiff, Wales

G. SHEPPARD, Quality Control Department, The Radiochemical Centre, Limited, Amersham, Buckinghamshire, England

JOHN A. SPINK, Division of Tribophysics, CSIRO, University of Melbourne, Parkville, Victoria, Australia

YASUO SUZUKI, Department of Industrial Chemistry, Faculty of Engineering, Meiji University, Kawasaki, Japan

RITCHIE THOMSON, Quality Control Department, The Radiochemical Centre, Limited, Amersham, Buckinghamshire, England

YOSHIKI WADACHI, Division of Health Physics and Safety, Japan Atomic Energy Research Institute, Tokai-Mura, Japan

DAVID E. YATES,* Colloid and Surface Chemistry Group, Department of Physical Chemistry, University of Melbourne, Parkville, Victoria, Australia

---

*Current affiliation: School of Chemistry, University of Bristol, Bristol, England

CONTENTS

Cumulative indexes will appear in Volume 2

CONTENTS OF VOLUME 2

INTERNATIONAL SYSTEM OF UNITS (SI UNITS) FOR RADIOACTIVITY

The SI unit for radioactivity is the becquerel (Bq) equal to 1 disintegration per second. The table below gives the conversion values for becquerels to curies.

1 becquerel	1 Bq	27.03 picocuries
1 kilobecquerel (1 kBq)	10^3 Bq	27.03 nanocuries
1 megabecquerel (1 MBq)	10^6 Bq	27.03 microcuries
1 gigabecquerel (1 GBq)	10^9 Bq	27.03 millicuries
1 terabecquerel (1 TBq)	10^{12} Bq	27.03 curies
1 petabecquerel (1 PBq)	10^{15} Bq	27.03 kilocuries
1 exabecquerel (1 EBq)	10^{18} Bq	27.03 megacuries
	1 microcurie = 37 kBq	

Chapter 1

THE DESIGN OF RADIOTRACER EXPERIMENTS

Mitsuo Muramatsu

Department of Chemistry
Faculty of Science
Tokyo Metropolitan University
Tokyo, Japan

E. Anthony Evans

Organic Department
The Radiochemical Centre, Limited
Amersham, Buckinghamshire
England

It is an unrewarding exercise to apportion excessive credit to any single scientific study for its part in the progress of knowledge in the chemical and biological sciences. Nevertheless, certain aspects of scientific achievement are unequivocally indebted to radiotracer experiments for direct information about the fate of specified species -- atoms, molecules, ions, fragments, aggregates, etc. -- in terms of their mass, which is a primary measure for the extent of chemical and biological processes. Such information can, in principle, refute or definitively confirm knowledge previously established by conventional techniques. This is indeed the case when the radiotracer techniques appropriate to the experimental conditions under which the radioactivity R of a radionuclide is absolutely proportional to the mass or number n of the specified species thereby concerned. To establish the constancy of

$$\frac{R}{n} = k \tag{1}$$

variously contradictory requirements from different aspects must be simultaneously satisfied. The constant k is conventionally defined as

$$k = \eta\lambda q\sigma \tag{2}$$

with

$$\eta = \frac{R}{-dN'/dt} \tag{3}$$

for the detection coefficient (or overall counting efficiency) of N' radioactive atoms at time t,

$$\lambda = \frac{-dN'/dt}{N'} \tag{4}$$

for their decay constant,

$$q = \frac{N'}{n'} \tag{5}$$

to specify the chemical form of n' molecules labeled with N' radioactive atoms, and

$$\sigma = \frac{n'}{n} \tag{6}$$

for the ratio of the number of radiolabeled, n', to that of the total (labeled plus unlabeled), n, molecules. The last term σ (moleculues/molecule or atoms/atom) or, more conveniently, the product of $\lambda q\sigma$ (atoms/time/molecule, dpm/mol, or Ci/mol) is called specific activity. This is an important factor in designing the radiotracer experiments.

In many cases, n for a chemically or biologically treated specimen is determined by comparing its R value with the radioactivity R_s for a known number n_s (or mass) of the same molecules in an untreated standard sample, so that the relationship

$$n = \frac{n_s R}{R_s} \tag{7}$$

holds. It must be borne in mind, however, that Eq. (7) can be rationalized only when k $(= \eta\lambda q\sigma)$ for the unknown specimen equals to that for the standard. For this purpose, we usually choose such conditions as

$$\frac{\eta}{\eta_s} = \frac{\lambda}{\lambda_s} = \frac{q}{q_s} = \frac{\sigma}{\sigma_s} = 1 \tag{8}$$

where the subscript s refers to the standard sample.

Fulfillment of the first condition, $\eta = \eta_s$, depends primarily on the choice of radionuclide (Chap. 2) and the method used for measurement of radioactivity (Chap. 4). The former determines the type and energy of radiation to be detected, while the latter is concerned with how adequately and efficiently it is quantified. In general expression,

$$\eta = \eta_g \cdot \eta_a \cdot \eta_b \cdot \eta_{sa} \cdot \eta_{sc} \cdot \eta_c \tag{9}$$

for the detection coefficient η as a function of the sample-geometric factor η_g, the attenuation factor η_a, the back-scattering factor η_b, the self-absorption and -scattering factors, η_{sa} and η_{sc}, respectively, and the counting efficiency η_c, each factor would be made equal between the two samples; otherwise appropriate correction must be made for the difference. The correction methods for differences in η between different samples are discussed in the cases of dose assessment in radiation monitoring (Chap. 3), liquid scintillation and other measurement techniques for various specimens (Chaps. 4 and 6), estimation of radioisotopes incorporated in human body for diagnosis (Chap. 22), and sample preparation for environmental studies (Chap. 14). Autoradiographic techniques (Chap. 4) are also applicable, especially in determining the distribution of radiolabeled compounds adsorbed heterogeneously at solid/solution interfaces (Chap. 13), as well as those incorporated in animal tissues and cells (Chaps. 16, 17, and 20).

The attenuation of radiation in matter is skillfully utilized for determination of diffusion constants (Chap. 12) and adsorbed amounts (Chap. 13) of various compounds labeled with soft β-emitters such as ^{14}C, ^{35}S, and ^{3}H. The radioactivity thereby observed is contributed mostly by the radiolabeled molecules within the relevant medium. In Chapter 23, which deals with radioactive drugs for cancer therapy, the author describes ambitious experiments in which cancerous cells are irradiated with a tritiated compound incorporated into such cells.

The specific activity $\lambda q \sigma$ must be of a value large enough to assure a high sensitivity required from experimental aspects (Chaps. 10-21) and small enough to make self-radiolysis (Chap. 7) of negligible importance under the experimental conditions and to carry out the experiment safely (Chap. 3). From this viewpoint, much thought is needed in preparation of radiotracer compounds (Chap. 5) compatible with requirements from aspects of quality control (Chap. 6). With subscripts m and i for the main and impurity components, respectively, a radiotracer compound must satisfy a simultaneous requirement on large values of the chemical purity

$$\omega_c = \frac{n_m}{n_m + n_i} \qquad (10)$$

to assure that the chemical or biological phenomena are associated predominantly with n_m main molecules, of the radiochemical purity

$$\omega_{rc} = \frac{q_m \sigma_m n_m}{q_m \sigma_m n_m + q_i \sigma_i n_i} \qquad (11)$$

to establish that all of the radioisotopic atoms are exactly located to n_m' $(= \sigma_m n_m)$ main molecules, and of the radionuclidic purity

$$\omega_{rn} = \frac{\eta_m \lambda_m q_m \sigma_m n_m}{\eta_m \lambda_m q_m \sigma_m n_m + \eta_i \lambda_i q_i \sigma_i n_i} \qquad (12)$$

to confirm that the radioactivity R $(= R_m + R_i)$ is contributed solely by N_m' $(= q_m \sigma_m n_m)$ radioactive atoms that are radiochemically (Chap. 2) different from N_i' $(= q_i \sigma_i n_i)$ atoms. As behavior of n_m molecules is not always the same as that of n_i molecules, ω_{rc} and ω_{rn} values of a chemically or biologically treated specimen may largely differ from those of an untreated sample (Chaps. 15 and 19; see also Chaps. 6 and 7). The difference depends on the mode and extent of such a treatment of the "labeled" material and on its response to such a treatment. In Chapter 6, the authors detail valuable discussions on these problems and their possible solutions. In harmony, Chapter 5 states in a unique manner the principles for selecting the labeling method suitable for the radiotracer experiment. Thus, the reader will find an important expression, "for what?" (or its equivalent), in many areas throughout these chapters.

One might say that the condition $\lambda = \lambda_s$ is automatically satisfied. Equation (4), however, is based on a statistical treatment for which sufficiently high count rate R must be assured. This leads eventually to a requirement that the specific activity must be high enough to establish the validity of the treatment. Most of the commercially available samples of labeled compounds are of an order of magnitude of 10^{-1} to 10^{2} Ci/mmol for their specific activity, the value being sufficiently large for most of the radiotracer experiments. Sometimes there is a need for a higher specific activity in order to achieve even higher sensitivity. Typical examples are the solubility (ppb or less in order of magnitude) measurement of sparingly soluble compounds (Chap. 11), determination of unimolecularly adsorbed amounts (10^{-10} mol/cm^{2}) at various interfaces (Chap. 13), and protein-binding assays for determination of extremely small amounts (say, 10^{-12} mol) of antigens, haptens, hormones, vitamins, and drugs (Chap. 21). Essentially, the maximum value of the specific activity is determined by λ. Such an upper limit for ^{14}C- or ^{11}C-labeled compound, for example, is 62.4 and 9.18×10^{9} mCi/m atom of carbon, respectively, for $q = 1$ and $\sigma = 1$. Apparently, the shorter the half-life, the more conveniently the radiotracer experiment can be designed, although too short-lived radionuclide is rather difficult to handle and often impossible to convert into a useful chemical form. Chapter 22 discusses the advantages and disadvantages of these radionuclides for diagnosis of human body. Chapter 23 discusses how the intracellular irradiation of cancerous tissue is achieved with a tritiated compound of $\lambda q\sigma = 116$ Ci/mmol at maximum. It is a unique advantage of the radiotracer method that the sensitivity alterable by isotopic dilution can a priori be established by virtue of the second law of thermodynamics. Sometimes, however, extra care is needed in treating carrier-free radionuclide in an extremely dilute solution. Behavior of the radionuclide in such a dilute solution is not always explicable by extrapolation from that in higher concentrations. Chapter 8 (and in part Chap. 3) shows how it deviates from the theoretical presumption based on the thermodynamical laws.

The condition, $q = q_s$, implies something more than it says mathematically. In observation of kinetic processes without perturbation, $\lambda q\sigma$ must be low enough to assure that the experimental conditions are not disturbed by self-radiolytic products (Chap. 7). This is the absolute condition for the experiments of the isotopic exchange (Chap. 10) and self-diffusion (Chap. 12), in which experimental details must be arranged so that the spontaneous processes due purely to entropy increase can be observed without dissipation of enthalpy (or total energy). Indeed, these interesting processes could not have been studied if there were no isotopic tracers. On the other hand, q is generally different from q_s when a radiotracer experiment involves chemical reaction(s) at the expense of bond-breaking energy (Chaps. 9, 15, and 16); one is interested in the numerical value of q/q_s. In this case the mechanisms of a chemical or biological reaction are studied by tracing the fate of an isotopically labeled atom (Chaps. 15-20). Such labeling should be made at a nonlabile position, otherwise correction must be made for intra- and/or intermolecular transfer of the specified atom due to isotopic exchange (Chap. 10). Needless to say, the deviation of q/q_s from unity involves an isotope effect, from which one can expect to obtain information about the "nature" of the chemical bond involved in its rate-determining step (Chap. 9). Currently, most of these experiments do not require large $\lambda q \sigma$ values.

Many radiotracer experimenters tacitly accept the condition of $\sigma = \sigma_s$; behavior of radiolabeled species is identical to that of unlabeled, both in equilibrium and in dynamic states. This is almost true inasmuch as we deal with heavy species involved in a molecular process without bond rupture. Such a process may be exemplified by ionic transport in solution (Chap. 11), diffusion in inert medium (Chap. 12), or physical adsorption (Chap. 13). In any case, the fate of labeled molecules or ions is negligibly affected by the primary isotope effect, i.e., the effect attributable to the difference in the kinetic energies of the colligative units (in sense of statistical kinetics). In the case of chemical reactions, however, the fate of a light atom or radical depends also on the secondary isotope effect (differences between labeled and unlabeled species in their vibration frequencies at the process-governing point) including the tunneling effect (mass number-dependable readiness of a chemical process without surmounting the energy barrier). These eventually appear as the ratio of the kinetic rate constants, $k_{labeled}/k_{unlabeled}$, for the chemical process (Chaps. 9, 10, 15, and 16).

Much care must be taken to carry out the radiotracer experiments safely. Chapter 3 deals with safe handling of radiotracer compounds, radiation monitoring techniques, radioactive contamination and decontamination, and waste management for safe disposal. Similar topics are discussed briefly in Chapter 22, especially for safe use of radioisotopes for diagnostic and clinical studies. From any of these aspects, the $\lambda q \sigma$ value for radiotracer experiments would be chosen so low as to meet the national regulations based on the international recommendations. In ICRP recommendations, for example, a labeled compound can be disposed into air only when its concentration does not exceed the maximum permissible concentration in radioactivity/volume. As its vapor pressure (mass/volume) is constant at a given temperature, the "degree of hazard" of such a compound is determined only by its $\lambda q\sigma$ value in radioactivity/mass.

It is obvious in principle that the radiotracer experiment must be designed from two main aspects; one must seek the conditions for efficient and reproducible measurements of radioactivity, and one must choose an appropriate specific activity, $\lambda q\sigma$, which should be within the range of the upper and lower limits. What can we do if these limits cross each other? Partial improvement can be made by auxiliary adjustments such as the use of a low-background counter, change of the sample storage conditions, and modification of experimental details. If these are found to be ineffective, what should we do? The best way could possibly be to call off the experimental program.

Throughout the following chapters, 37 experienced radiotracer specialists describe up-to-date techniques in individual research areas and, in many cases, state rigorously and critically what the current problems are and how one can expect to overcome them.

Chapter 2

SELECTION AND PROPERTIES OF RADIONUCLIDES

Yukio Murakami

Department of Chemistry
Faculty of Science
Tokyo Metropolitan University
Tokyo, Japan

I. INTRODUCTION

As stated in the preceding chapter, selection of radionuclides is of primary importance in designing radiotracer experiments. Essentially, it depends not only on the radiochemical nature of the nuclides thereby concerned but also on the experimental requirements which are too diversified to be summarized in one chapter. Nevertheless, all these conditions are conjugated with one another and it may be worthwhile to look at the fundamental principle of selection of radionuclides from the aspects of their radiochemical properties, including their decay modes, type of radiation, half-life, radiotoxicity, etc. in relation to their production method. For example, currently useful radioiodine includes ^{123}I, ^{124}I, ^{125}I, ^{131}I, and ^{132}I, of which the cheapest, ^{131}I, has a half-life ideally too long (8.05 days) for human administration. In this respect, ^{132}I (2.33 hr) can be substituted for ^{131}I [1]. The disadvantage can be overcome by adoption of a milking or "cow" system (see Section II D): ^{132}Te (78 hr) $\rightarrow$ $^{132}I \rightarrow {}^{132}Xe$ (stable) [2]. The short half-life can now be an advantage of this nuclide, assuring high sensitivities (Chap. 1) and making it possible to carry out the experiments successively. Both of these β-γ emitters, ^{131}I and ^{132}I, give large exposure dose values to the human body. On the other hand, ^{123}I, an electron capture (EC) nuclide [3], is characterized by a small exposure dose [4] and ready detection of its 159 keV γ-rays, though its half-life (13.3 hr) is rather short. In this respect, ^{125}I (60.2 days) is useful, in spite of its disadvantage in determining its low-energy (35 keV) γ-rays with a low emission rate (7%) [5] for routine experiments such as the protein-binding assays (Chap. 21) and cytological studies (Chap. 17). Incidentally, ^{125}I is not included in the ICRP radiotoxicity table (Table 1 of Chap. 3). However, it should have been included in a medium toxicity group [6].

The physical and chemical properties of radionuclides are considered to be exactly the same as those of nonradioactive nuclides within a deviation due to the isotope effects (Chap. 9). Sometimes we encounter peculiar behavior of carrier-free radionuclides (Chap. 8) when we deal with their solutions at extremely low concentrations. In the present chapter, statements are made of the properties of radionuclides to be understood from radiochemical aspects. It could be of some importance to discuss how such properties can influence the selection of radionuclides to be made compatible, as much as possible, with the requirement of radiotracer experiments.

II. PROPERTIES OF RADIONUCLIDES

A. Useful Radionuclides and Their Decay Modes

Radionuclides identified so far include 58 natural and approximately 1300 artificial ones. Most of them, however, are not always useful as radiotracers, mainly because of difficulties in preparation, undesirable half-life time, and/or inadequacy of radiation for quantification. Some 220 useful radionuclides are listed in Table 1 with their chemical characteristics as reflected in the periodic table. These radionuclides are produced mostly in the nuclear reactor by such reactions as (n,γ), (n,p), and (n,α). Some short-lived nuclides such as ^{18}F and ^{52}Fe can be produced routinely by cyclotron bombardment [7-10], and some others, such as ^{99m}Tc, ^{90}Y, and ^{140}La, are prepared by milking commercially available "cow" systems (see Table 5) [11,12]. In any case, production procedure must be followed more or less by chemical treatment, for which one always needs information about the chemical nature of the individual radionuclide and its environmental (Table 1) elements as possible radionuclidic impurities (Section III B).

In Table 1, no mass numbers are attached to some elements, e.g., He, Li, B, and Ne, as none of them has radionuclide of appropriate half-life. One has to use stable isotopes for tracing the fate of these atoms or pertinent compounds. This condition is similar to the cases of nitrogen and oxygen, for which stable isotopes, ^{15}N and ^{18}O, respectively, are currently employed at the expense of tedious measurement procedures and high cost for mass spectroscopy. Such difficulties are not involved in the radiotracer experiments if these nuclides are substituted for the cyclotron-producible, short-lived ^{13}N and ^{15}O, respectively [13]. Chapter 1 briefly discusses the advantages and disadvantages of

TABLE 1

Useful Radionuclides on the Periodic Table

Row	IA	IIA	IIIB	IVB	VB	VIB	VIIB	VIII			IB	IIB	IIIA	IVA	VA	VIA	VIIA	0
1	3 H																	He
2	Li	7 Be											B	11 14 C	13 N	15 O	18 F	Ne
3	22 24 Na	28 Mg											28 Al	31 Si	32 33 P	35 38 S	36 38 Cl	37 41 43 Ar
4	40 42 43 K	45 47 Ca	44 46 47 Sc	44 Ti	48 49 V	51 Cr	52 54 56 Mn	52 55 59 Fe	56 57 58 60 Co	59 63 65 Ni	61 62 64 Cu	62 65 69+69m Zn	67 68 70 72 Ga	68 71 77 Ge	72 73 74 76 77 As	72 75 79 Se	77 80m 82 Br	79 81m 85 85m Kr
5	81 82m 84 86 Rb	85 87m 89 90 Sr	87 88 90 91 Y	95 97 Zr	90 94 95 Nb	93m 99 Mo	99m 99 Tc	97 103 105 106 Ru	99 105 Rh	103 109 Pd	105 110m 111 Ag	107 109 115m 115 117 Cd	111 113m 114m 116m In	113 119m 121 Sn	117 122 124 125 Sb	125m 127+127m 129m 132 Te	123 124 125 131 132 I	125 127 133 Xe
6	129 130 131 132 134 137 Cs	131 133 140 Ba	140 La*	175 181 Hf	182 Ta	181 185 187 188 W	183 186 188 Re	185 191 Os	192 194 Ir	193 197 Pt	195 198 199 Au	194 197 197m 203 Hg	204 Tl	203 212 210 Pb	206 207 210 Bi	208 210 Po	At	222 Rn
7	Fr	226 228 Ra	227 Ac**															
* Lanthanoids				139 141 143 144 Ce	142 143 Pr	147 149 Nd	145 147 Pm	145 153 Sm	152 152m 154 Eu	153 Gd	160 161 Tb	157 159 Dy	166 Ho	169 171 Er	170 171 Tm	169+175 Yb		177 Lu
** Actinoids				228 230 231 Th	233 Pa	232 233 237 U	237 239 Np	238 239 241 Pu	241 Am	Cm	Bk	252 Cf	Es	Fm	Md		No	Lr

IA: Alkaline metal group
IIA: Alkaline earth group
IB-VIIB and VIII: Transition Metals
IIIB: Scandium group
VIIB: Manganese group
IB: Copper group
IIIA: Aluminum group
IVA: Carbon group
IVB: Titanium group
VIII: (left to right) Iron, Palladium and platinum groups
IIB: Zinc group
VA: Nitrogen group
VIA: Oxygen group
VB: Vanadium group
VIB: Chromium group
VIIA: Halogen group
0: Rare gas group

such short-lived nuclides for radiotracer experiments. Chapter 22 presents some examples of short-lived radionuclides for diagnostic application.

Any of these radionuclides decays, with or without intermediate stage, to the daughter nuclide in the ground state, accompanying an energy dissipation in the form of various kinds of radiations such as α ($^{4}He^{2+}$), β^- (negatron), β^+ (positron), γ (electromagnetic wave), and/or characteristic or continuous x-rays. In the figure, an energy difference is given as a distance between two horizontal lines representing nuclear energy levels. These levels are correlated with the energy dissipation accompanying an increase, ↘ (β^-), or a decrease, ↙ (α, β^+, or EC), of atomic number. The vertical arrow, ↓, denotes the isomeric transition (IT) from excited to ground state, while the symbol ↙ represents β^+ decay followed by a positron conversion into two annihilation γ-rays of 0.511 MeV each in the opposite direction as a result of its interaction with surrounding materials. Corresponding to (a)-(g) in the figure, brief explanations are given as follows:

(a) ^{35}S (half-life, 87.6 days), a pure β^--emitter, emits β^--rays of 0.167 MeV.

(b) ^{137}Cs (30.0 years) emits β^--rays of 1.17 MeV (6.5%) and 0.51 MeV (93.5%). The former is associated with formation of ^{137}Ba in the ground state, whereas the latter is concerned with formation of ^{137m}Ba which emits γ-rays of 0.66 MeV.

(c) ^{13}N (10.0 min) emits 1.20 MeV β^+ which annihilates into two γ-rays of 0.511 MeV each.

(d) An EC nuclide, ^{54}Mn (303 days), emits γ-rays of 0.84 MeV due to a nuclear transition from the excited to the ground state. Another radiation involves 5.989 keV KX-rays characteristic to its daughter nuclide, ^{54}Cr. It depends on the counting conditions which one of these radiations is selected for radioactivity determination.

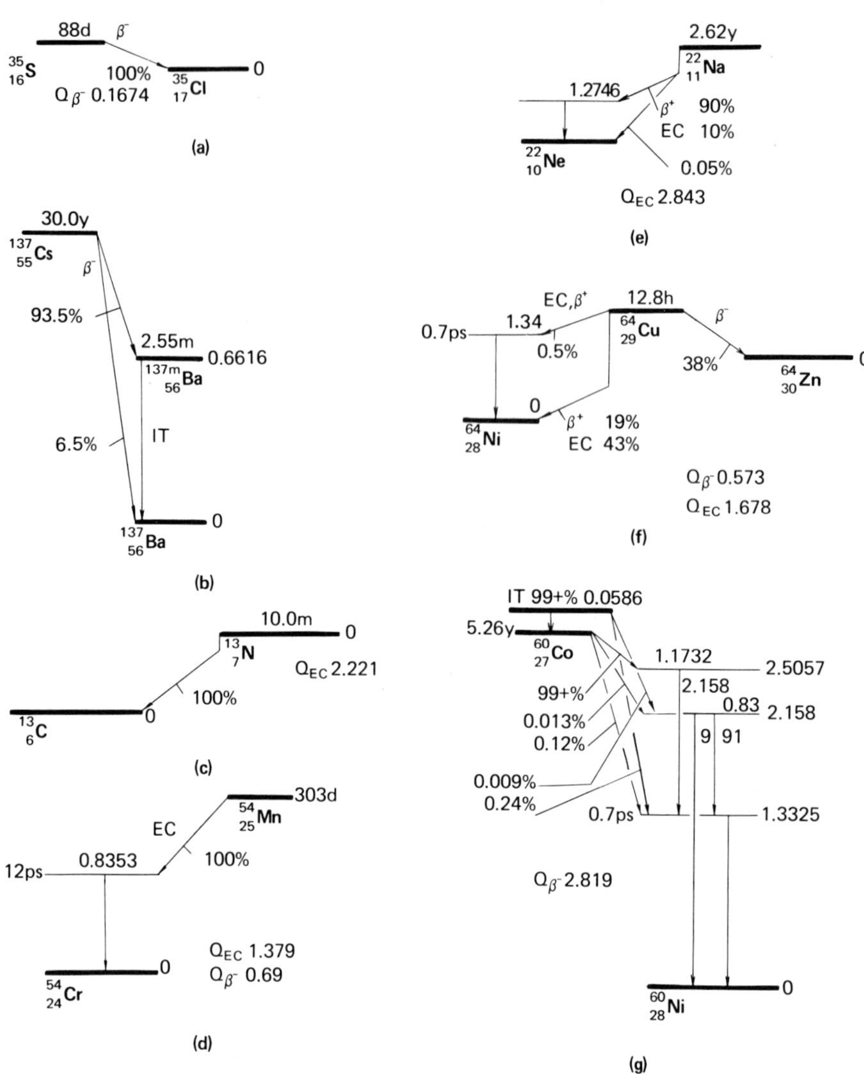

FIG. 1. Examples of decay schemes.

(e) A decay energy of 2.84 MeV for ^{22}Na (2.60 years) is contributed by a β^+ (90%) energy of 0.54 MeV, two annihilation γ energies of 0.511 MeV each, and an energy of 1.28 MeV γ-rays for the nuclear transition. The remaining portion (10%) is contributed by an EC process.

(f) The decay of ^{64}Cu is contributed by EC (43%), β^+ (19%), and β^- (38%) emissions, resulting in the formation of ^{64}Ni in excited and ground states and ^{64}Zn, respectively. Again, experimental conditions determine which one of these radiations would be utilized for a radiotracer experiment.

(g) A portion of >99% of the (n,γ)-produced ^{60m}Co (10.47 min) yields ^{60}Co (5.26 years) which emits the three β^--rays followed by the corresponding γ emissions. However, ^{60}Co practically available for radiotracer experiments gives the cascade radiation including, mainly, 1.17 (= 2.5057 - 1.3325) and 1.33 MeV γ-rays.

In selecting radionuclides, radiotracer experimenters should consult with such decay schemes in pertinent literature [14,15].

B. Type of Radiation

Radiation detectable for quantification of radionuclides in tracer experiments include β^--, β^+-, α-, γ- and x-rays. Of these, the first two give continuous spectra, whereas the remainder, except the continuous x-rays due to bremsstrahlung, are of discrete energies characteristic of the decay product. It is thus advantageous to select such a characteristic radiation for determination of a specified nuclide. In the case of an EC nuclide, for example, the KX- or LX-rays characteristic of its daughter nuclide is so soft (5.5, 31.8, and 80.7 keV for ^{51}Cr, ^{123}I, and ^{197}Hg, respectively) that the effective determination can hardly be achieved even with a Ge(Li) detector of high resolution. It might be better to make use of a γ radiation due to the isomeric transition from an excited to the ground state. When such a γ radiation is of low energy, it is frequently accompanied by internal conversion into an electron (e^-) of discrete energy. Thus, ^{51}Cr, an EC nuclide, would be determined by counting of 0.320 MeV γ-rays (9%) or of 0.315 MeV conversion electrons. This situation is similar to the case of β^+ radiation, which is usually accompanied by two annihilation γ-rays (see Scheme C in the figure). Thus, ^{22}Na (Scheme e) would be quantitized by determination of either 1.28 MeV (100%) or 0.511 MeV (180%) γ radiation rather than of its characteristic x-rays.

Little perplexities are involved in counting β and γ activities. Table 2 summarizes the useful β and γ emitters classified from their radiation energies and the half-lives. It is obvious from Table 2A that such β^- emitters as ^{33}P and ^{45}Ca (in the second column of Table 2) are much inferior to ^{32}P and ^{47}Ca (seventh column), respectively, in that the self-absorption correction [16] is necessitated for the samples of the former nuclides. Such examples can be found in other combinations of the β emitters in Table 2A, as well as of the γ emitters in Table 2B. This is similar to the aspect of half-life which is discussed in the next section.

Leaving details in counting techniques to Chapters 3 (Section III), 4, and 6, Table 3 lists practically useful β emitters, making it easy to envisage how the counting samples should be prepared for the most reliable and efficient counting. In regard to bremsstrahlung, i.e., conversion of β energies into continuous x-rays wherein β particles are decelerated in the nuclear electric field, the conversion ratio is increased with increases of the energy of original β-rays and of the atomic number of the interacting material. Thus, such a hard β^- emitter as ^{32}P (1.70 MeV) or ^{90}Y (2.27 MeV) would be stored in a container lined doubly with C/H/O-composing plastics such as Perspex for stopping β-rays and with lead against the continuous x-rays which may have generated in the inner shield.

Another aspect in the selection of radionuclides is the safe handling of them. In Table 4 main γ emitters are classified from a viewpoint of the specific γ-ray constant, which is defined as the dose rate in R/hr at 1 cm distance from 1 mCi source. The radionuclides included in groups I and II always need great caution for radiation protection.

C. Half-life

One of the selection criteria is obviously an optimum choice of half-life τ, which should be long enough to make the decay correction meaningful or unnecessary and, at the same time, short enough to make waste disposal easy, as well as to carry out the tracer experiments safely (Chap. 3). Practically, it should be made compatible with the duration time d for an experiment with consideration that the decay correction is 3.5, 6.6, or 10% for $d/\tau = 0.04$, 0.10, or 0.15, respectively. If, for example, an experiment for tracing the fate of iron nuclide were to be performed over a period of many months, one would avoid the use of ^{59}Fe ($\tau = 45.6$ days, a β^--γ emitter) and adopt ^{55}Fe (2.60 years, EC) for which one has to measure 6.5 keV KX-rays arising from its daughter nuclide, ^{55}Mn, or continuous bremsstrahlung to 0.23 MeV (0.004% though). On the other hand, the former can be used even for corrosion studies of iron in contact with aqueous solutions of $^{59}Fe^{3+}$ in nitrogen atmosphere [18]. However, none of them should be administered to human beings for medical diagnosis for which cyclotron-producible ^{52}Fe (8.2 hr; β^+-EC) [19] is better suited [20,21]. Such examples are reviewed in various chapters of this book.

The situation becomes slightly complicated when one deals with an isotopic mixture of different half-lives such as ^{32}P ($\tau = 14.28$ days) and ^{33}P (25.5 days) or ^{198}Au (2.697 days) and ^{199}Au (3.15 days). A mixture of two radionuclides of different decay constants, λ_1 and λ_2, can be analyzed usually by semilog

TABLE 2A

Classification of β-emitting Nuclides*

** half life \ Emax	<0.3 MeV	0.3 - 0.5 MeV	0.5 - 0.75 MeV	0.75 - 1.0 MeV	1.0 - 1.5 MeV	>1.5 MeV
< 1h				^{11}C ^{69}Zn	^{13}N ^{116m}In	^{15}O ^{67}Ge ^{70}Ga ^{38}Cl
1 - 3h			^{18}F		^{43}Ar ^{149}Nd ^{31}Si	^{132}I ^{117}Cd ^{65}Ni ^{56}Mn
3 - 9h		^{107}Cd ^{28}Mg	^{52}Fe ^{127}Te ^{212}Pb ^{64}Cu	^{85m}Kr ^{43}K	^{44}Sc ^{105}Ru ^{171}Er ^{24}Na	
9h - 1d			^{187}W		^{187}W	^{152m}Eu ^{77}Ge ^{42}K ^{142}Pr
1 - 3d		^{231}Th ^{121}Sn ^{82}Br ^{239}Np	^{105}Rh ^{77}As	^{153}Sm ^{198}Au	^{151}Pm ^{143}Ce ^{140}La ^{149}Pm ^{115}Cd ^{99}Mo	^{76}As ^{166}Ho ^{140}La ^{122}Sb ^{90}Y
3 - 10d	^{132}Te	^{199}Au ^{175}Yb ^{133}Xe ^{177}Lu ^{169}Er	^{47}Sc ^{52}Mn ^{47}Ca ^{132}Cs ^{131}I		^{186}Re ^{210}Bi ^{111}Ag	^{124}I ^{47}Ca
10 - 30d	^{191}Os ^{33}P ^{233}Pa			^{147}Nd ^{143}Pr	^{140}Ba ^{74}As	^{32}P ^{86}Rb
30 - 100d	^{95}Nb ^{103}Ru ^{203}Hg ^{35}S	^{181}Hf ^{59}Fe ^{95}Zr ^{58}Co ^{185}W ^{46}Sc	^{141}Ce ^{142}Ir	^{160}Tb	^{89}Sr ^{56}Co ^{123}Sn	^{115m}Cd ^{91}Y ^{124}Sb
100d - 1y	^{45}Ca ^{110m}Ag ^{144}Ce	^{144}Ce	^{182}Ta ^{110m}Ag	^{88}Y	^{170}Tm	
1 - 100y	^{147}Pm ^{3}H ^{241}Pu ^{210}Pb ^{227}Ac ^{63}Ni	^{60}Co	^{134}Cs ^{22}Na ^{125}Sb ^{85}Kr ^{90}Sr ^{137}Cs	^{204}Tl ^{154}Eu	^{152}Eu	
$>10^2$y	^{14}C ^{99}Tc					

*Compiled by the author from the data in Ref. [14].

**Categorized by the time of temporary storage for waste disposal; h = hours, d = days, y = years.

TABLE 2B

Classification of γ-emitting Nuclides*

Energy / ** half life	< 0.3 MeV	0.3 - 0.5 MeV	0.5 - 0.75 MeV	0.75 - 1.0 MeV	1.0 - 1.5 MeV	>1.5 Mev
< 1h			^{15}O ^{13}N ^{11}C			^{38}Cl
1 - 3h	^{116m}In ^{117}Cd	^{87m}Sr ^{113m}In	^{68}Ga ^{18}F ^{132}I	^{132}I ^{56}Mn	^{41}Ar ^{65}Ni	^{56}Mn
3 - 9h	^{80m}Br ^{85m}Kr ^{99m}Tc ^{171}Er ^{52}Fe	^{85m}Kr ^{171}Er ^{157}Dy	^{44}Sc ^{52}Fe ^{107}Cd		^{44}Sc	
9h - 1d	^{123}I ^{197}Pt ^{28}Mg ^{187}W ^{197m}Hg ^{231}Th	^{77}Ge ^{69m}Zn ^{43}K ^{105}Rh	^{77}Ge ^{43}K ^{72}Ga ^{97}Zr ^{187}W	^{72}Ga ^{28}Mg	^{24}Na ^{28}Mg	^{42}K ^{24}Na ^{142}Pr
1 - 3d	^{151}Pm ^{143}Ce ^{153}Sm ^{239}Np ^{197}Hg ^{97}Ru ^{111}In ^{67}Ga	^{76}As ^{151}Pm ^{140}La ^{115}Cd ^{198}Au	^{151}Pm ^{82}Br ^{115}Cd ^{198}Au ^{99}Mo ^{122}Sb	^{82}Br ^{149}Pm ^{115}Cd		^{140}La
3 - 10d	^{175}Yb ^{133}Xe ^{132}Te ^{47}Sc ^{237}U	^{175}Yb ^{206}Bi ^{111}Ag ^{131}I	^{124}I ^{206}Bi ^{132}Cs	^{124}I ^{52}Mn	^{47}Ca ^{52}Mn ^{48}V	
10 - 30d	^{131}Ba ^{140}Ba ^{191}Os ^{103}Pd	^{131}Ba ^{140}Ba ^{233}Pa ^{51}Cr	^{147}Nd ^{48}V ^{74}As	^{48}V		
30 - 100d	^{169}Yb ^{141}Ce ^{105}Ag ^{181}Hf ^{203}Hg ^{114m}In ^{183}Re ^{160}Tb	^{169}Yb ^{103}Ru ^{105}Ag ^{181}Hf ^{175}Hf ^{7}Be ^{192}Ir	^{124}Sb ^{85}Sr ^{95}Zr ^{58}Co ^{192}Ir ^{185}Os	^{59}Fe ^{91}Y ^{56}Co ^{46}Sc ^{182}Ta ^{110m}Ag	^{59}Fe ^{91}Y ^{46}Sc	^{124}Sb ^{56}Co
100d - 1y	^{113}Sn ^{182}Ta ^{75}Se ^{159}Dy ^{195}Au ^{153}Gd ^{57}Co ^{144}Ce ^{49}V	^{113}Sn ^{75}Se ^{144}Ce	^{127m}Te ^{110m}Ag	^{88}Y ^{182}Ta ^{110m}Ag ^{144}Ce ^{54}Mn	^{65}Zn	^{88}Y ^{110m}Ag
1 - 100y	^{109}Cd ^{55}Fe ^{125}Sb ^{133}Ba ^{152}Eu ^{154}Eu ^{227}Ac ^{44}Ti	^{125}Sb ^{133}Ba ^{134}Cs ^{152}Eu ^{154}Eu	^{152}Eu ^{154}Eu	^{134}Cs ^{22}Na ^{152}Eu ^{154}Eu ^{207}Bi	^{60}Co	
> 100y						

*Compiled by the author from Data in Ref. [14].

**Categorized by the time of temporary storage for waste disposal.

TABLE 3

Maximum Energy, Average Energy and Maximum Range of β rays*

Nuclide	H. L.**	Emax	$E_{average}$		Max. Range
^{3}H	12.35 y	0.0186 MeV	0.0057 MeV		0.23 mg/cm²
^{14}C	5730y	0.156	0.045		20
^{35}S	87.4 d	0.167	0.0488		35
^{147}Pm	2.62y	0.224	0.070		49
^{45}Ca	165d	0.252	0.075		60
^{185}W	75d	0.429	0.14		121
^{59}Fe	45.6 d	0.46 0.255	0.150 0.085	0.120	137
^{64}Cu	12.80h	0.578 (β^-) 0.689 (β^+)	0.175 0.265	0.205	200
^{131}I	8.06 d	0.606	0.19		210
^{143}Pr	13.59d	0.939	0.31		390
^{210}Bi	5.013d	1.160	0.330		508
^{24}Na	14.96 h	1.389	0.540		621
^{32}P	14.28 d	1.710	0.695		810
^{234m}Pa	1.157m	2.232	0.865		1105
^{90}Y	64.0 h	2.27	0.93		1065

*Compiled by the author mainly from Ref. [14]. **Half life: y = years; d = days; h = hours; m = minutes.

TABLE 4

Classification of γ-emitters based on their Specific γ-ray Constant*

Group	Typical nuclide		Other nuclides (in decreasing order of dose rate)
	Nuclide	Dose rate R/h**	
I	^{24}Na	19	> ^{52}Mn > ^{56}Co > ^{28}Mg
	^{60}Co	13.2	> ^{82}Br > ^{110m}Ag > ^{140}Ba > ^{22}Na > ^{132}I > ^{140}La > ^{46}Sc
II	^{38}Cl	7.27	> ^{124}Sb > ^{66}Ga > ^{134}Cs > ^{56}Mn > ^{226}Ra > ^{59}Fe
	^{58}Co	5.60	> ^{11}C > ^{13}N > ^{18}F > ^{152}Eu > ^{47}Ca > ^{63}K > ^{54}Mn > ^{95}Zr
III	^{198}Au	2.44	> ^{65}Zn > ^{76}As > ^{131}I > ^{75}Se
	^{42}K	1.4	> ^{106}Rh > ^{203}Hg > ^{64}Cu
IV	^{86}Rb	0.6	> ^{123}I > ^{125}I > ^{197}Hg
	^{51}Cr	0.19	> ^{7}Be > ^{111}Ag

*Compiled by the author from the data in Ref. [17].

plotting of the total radioactivity A against the elapsed time t. The half-lives of both nuclides can be determined from the decay curve. It must be kept in mind, however, that the errors associated with the long-lived nuclide are always superimposed on those due purely to the short-lived one. Such a difficulty can be avoided to some extent by constructing the $\log A \sim t$ relationship for at least three times as long as the half-life for the short-lived nuclide. In some cases, it is effective to plot $A[\exp(\lambda_2 t)]$ against $\exp[(\lambda_2 - \lambda_1)t]$ in a linear relationship so that one can estimate the two radioactivities at $t = 0$ as the slope of the straight line and its intercept with the ordinate [22-24]. Even more tedious procedures are required for an isotopic mixture of mother and daughter nuclides, both being radioactive. This will be discussed in the following section.

D. Successive Decay and Radioactive Equilibrium

When a decay product is radioactive, it decays to emit its own radiation. With subscripts 1 and 2 for the parent and daughter radionuclides, respectively, the numbers of radioactive atoms, N_i's ($i = 1$ and 2), at time t are expressed as

$$N_1 = N_1^0 \exp(-\lambda_1 t) \tag{1}$$

and

$$N_2 = N_2^0 \exp(-\lambda_2 t) + \frac{\lambda_1}{\lambda_2 - \lambda_1} N_1^0[\exp(-\lambda_1 t) - \exp(-\lambda_2 t)] \tag{2}$$

where λ_i ($i = 1$ or 2) is the decay constant and the superscript 0 denotes $t = 0$.

If $N_2^0 = 0$, N_2 gives a maximum value at

$$t_{max} = \frac{2.303}{\lambda_2 - \lambda_1} \log \frac{\lambda_2}{\lambda_1} \tag{3}$$

Under the condition that $\lambda_2 > \lambda_1$, the observed total radioactivity A

$$A = A_1 + A_2 = \eta_1\lambda_1N_1 + \eta_2\lambda_2N_2 \tag{4}$$

with η_i for the respective detection coefficient, becomes

$$A_\infty = \left(\eta_1 + \frac{\eta_2\lambda_2}{\lambda_2 - \lambda_1}\right)\lambda_1N_1^0 \exp(-\lambda_1t) \tag{5}$$

at a time long enough to disregard the $\exp(-\lambda_2t)$-containing terms in Eq. (2). Equation (5) indicates that A_∞ decreases with t as if there were only the parent nuclide (amount, $N_1^0 \exp(-\lambda_1t)$) decaying with her own decay constant but emitting a mixture of her own radiation and her daughter's. This means that the daughter growth is equilibrated with her decay. Such an equilibrium is called transient equilibrium, wherein the relationship

$$\frac{N_2}{N_1} = \frac{\lambda_1}{\lambda_2 - \lambda_1} \tag{6}$$

holds. If, further, $\lambda_2 >> \lambda_1$, we obtain

$$\lambda_1N_1 = \lambda_2N_2 \tag{7}$$

This is called the secular equilibrium. In general, both of these are called radioactive equilibria.

The daughter nuclide under an equilibrium is separated by milking from its cow system, which is suited for obtaining short-lived radionuclides in the carrier-free state within the amount given by Eq. (6) or (7). The separation is achieved usually by such experimental means as solvent extraction, coprecipitation, ion exchange, and any other method that is appropriate to individual combination of the pair radionuclides. For example, a carrier-free sample of ^{140}La can be obtained by coprecipitation with Fe^{3+} added (by NH_4OH) to a mixture of ^{140}Ba (τ_1 = 12.8 days) and ^{140}La (τ_2 = 40.2 hr) under the transient equilibrium. Separation procedure could be made simpler when a parent nuclide in solid state produces gaseous daughter nuclide. Typical examples are gaseous ^{222}Rn originating from solid ^{226}Ra, and ^{127}Xe as the decay product of ^{127}Cs (EC) which is produced by $(\alpha,4n)$ reaction of ^{127}I. In any case, the decay curve for a cow system after milking is always upward convex because of the consumption, followed by growth, of the daughter nuclide. Such milking systems are given in Table 5.

In radiotracer experiments, we frequently deal with parent/daughter systems under nonequilibrated states wherein N_2^0/N_1^0 values vary greatly with the experimental conditions. In such a case, $A \sim t$ relationship must be carefully analyzed for obtaining correct N_1^0 and N_2^0 values, which are actually necessitated in the experiment [25]. One more remark should be given to the selective determination of A_1 and A_2. Unlike the former, the latter is a sum of N_1^0- and N_2^0-containing terms with different proportionality constants [see Eqs. (1), (2) and (4)]: It informs us N_2 $(= A_2/\eta_2\lambda_2)$ but not N_2^0.

III. RADIONUCLIDIC PURITY

A. Specific Activity and Carrier-free State

The maximum specific activity s (Ci/g) of a radionuclide of mass number M is given by

$$s = \frac{(0.6931/\tau)(6.023 \times 10^{23}/M)}{3.7 \times 10^{10}} = \frac{1.128 \times 10^{13}}{\tau\ M} \tag{8}$$

with τ in seconds.* Thus, ^{32}P (τ = 14.3 days) in the carrier-

---

*The equation is applicable to each of the dual decays of a radionuclide. For example, carrier-free ^{252}Cf has two specific activities: 0.536 Ci/mg for its α decay (96.9%, τ = 2.646 years) and 0.01669 Ci/mg for its spontaneous fission (3.1%, 85 years) [26].

free state has a specific activity of 2.85×10^5 Ci/g. The carrier-free state can be expected when it is produced by a nuclear reaction such as $^{32}S(n,p)^{32}P$, but not $^{31}P(n,\gamma)^{32}P$, because ^{32}P is chemically separable from unreacted ^{32}S but not from ^{31}P. Carbon-14, which is produced by $^{14}N(n,p)^{14}C$ reaction can be made at almost 100% isotopic abundance with a specific activity of 4.46 Ci/g for the s value based on Eq. (8) using $\tau = 5730$ years and $M = 14$, provided care is taken to avoid dilution during processing. However, for practical use ^{14}C is often supplied with a specific activity of $1 \sim 4.23$ Ci/g. On the other hand, tritium produced by $^6Li(n,\alpha)^3H$ reaction is commercially available with almost 99% of the s value given by Eq. (8). From a practical standpoint, the carrier-free radionuclides are not always routinely available.

When a target nuclide of abundance ratio θ of element (atomic weight M) is irradiated with a neutron flux of f $(n{\cdot}cm^{-2}sec^{-1})$ for time t (sec), the resulting (n,γ)-product gives a specific activity s (Ci/g) of

$$s = \frac{6.02 \times 10^{23} f(\sigma \times 10^{-24})\theta[1 - \exp(-0.6931\, t/\tau)]}{3.7 \times 10^{10} \times M} \tag{9}$$

where σ is the activation cross section (barn) of that target nuclide. The largest s value at $t = \infty$ is given by

$$s_{max} = \frac{1.630 \times 10^{-11} f\sigma\theta}{M} \tag{10}$$

Radionuclide obtained by the milking process (Section II D) is in the carrier-free state, giving a high specific activity based on Eq. (8) in comparison with s_{max} based on Eq. (10). For example, a specific activity of ^{90}Y based on the milking of the $^{90}Sr \rightarrow {}^{90}Y \rightarrow {}^{90}Zr$ chain is 5.44×10^5 Ci/g compared to 23.8 Ci/g for the s_{max} value of the same nuclide produced by $^{89}Y(n,\gamma)^{90}Y$ reaction ($\sigma = 1.3$ barn and $\theta = 1.00$) using a neutron flux of 10^{13} $n{\cdot}cm^{-2}sec^{-1}$. A similar situation exists in the case of selection of arsenic radionuclides. A specific activity of >100 Ci/g is obtainable with ^{77}As as the β^- decay product of ^{77}Ge produced

TABLE 5

Example of Cow Systems

Parent		Daughter			
Nuclide[a]	H.L.[b]	Nuclide	H.L.	Substitute for	Remarks
^{28}Mg	21.2h	^{28}Al	2.31m	(n,γ) produced ^{28}Al or ^{26}Al (7.4×10^5y)	high γ energy, C.A.[c]
^{38}S	2.87h	^{38}Cl	37.29m	(n,γ) produced ^{38}Cl and ^{36}Cl (3.08×10^5y)	C.F., useful for repeated administration
^{44}Ti	48y	^{44}Sc	3.92h	^{46}Sc (83.9d)	β^+ emitter
^{62}Zn	9.13h	^{62}Cu	9.76m	(n,γ) produced ^{64}Cu (12.80h)	useful for repeated administration
^{68}Ge	275d	^{68}Ga	68.3m	(n,γ) produced ^{72}Ga (14.12h)	β^+ emitter, suited for bone scintigraphy, C.A.
^{72}Se	8.4d	^{72}As	26h	(n,γ) produced ^{76}As (2.64h)	β^+ emitter, β^+ scanning
^{87}Y	80h	^{87m}Sr	2.83h	^{90}Sr (27.7y)	0.388 MeV γ, suited for bone scintigraphy, C.A.
^{90}Sr	27.7y	^{90}Y	64.0h	(n,γ) produced ^{90}Y	high β^- energy (2.25 MeV), C.A.
^{99}Mo	66.7h	^{99m}Tc	6.049h		a low γ energy (0.140 MeV), C.A.

^{111}Ag	7.5d	^{111m}Cd	48.61m	^{109}Cd (453d) ^{115m}Cd (43d)	low γ energy (0.247 MeV), suited for scintigraphy
^{113}Sn	115d	^{113m}In	99.8m	(n,γ) produced ^{114m}In (50.0d)	a low γ energy (0.390 MeV)
^{132}Te	77.7h	^{132}I	2.26h	^{131}I (8.05d)	C.A., remarkable reduction exposure dose for thyroid therapy
^{194}Hg	1.9y	^{194}Au	39.5h	(n,γ) produced ^{198}Au (2.697d)	EC
^{137}Cs	30y	^{137m}Ba	2.6m	(n,γ)-produced ^{131}Ba and ^{133}Ba	C.A.

[a] ^{38}S, ^{44}Ti, ^{62}Zn, ^{68}Ge, ^{72}Se, ^{87}Y, and ^{194}Hg are cyclotron-producible.

[b] Half-life: y = years; d = days; h = hours.

[c] Commerically available.

by $^{76}Ge(n,\gamma)^{77}Ge$ reaction [27], whereas the specific activities of commercially available samples of ^{76}As produced by $^{75}As(n,\gamma)^{76}As$ are usually in the range of 0.5-2 Ci/g. For 1 mCi of radioactive arsenic, therefore, the mass of the former sample would be only <0.01 mg, compared to 2-0.5 mg for the latter. It is possible that the usable limit of arsenic from chemical toxicity might be somewhere between these values.

Some remark should be given of so-called carrier-free radionuclides separated from the fission product. It is almost impossible to separate ^{90}Sr from stable ^{86}Sr and ^{88}Sr as the destination of the decay chains of mass numbers 86 and 88, respectively, formed in the mixed fission products. Similarly, ^{137}Cs obtained from the product contains more or less nonradioactive ^{133}Cs originating from ^{133}Sb. Readers should be aware that these carrier-containing ^{90}Sr and ^{137}Cs (also such nuclides as ^{95}Zr, ^{99}Mo, ^{106}Ru, ^{125}Sb, and ^{155}Eu) are called "carrier-free."

Carrier-free radionuclide can also be obtained by the Szilard-Chalmers method [28,29], i.e., recoiling out of radionuclides produced by nuclear bombardment. Carrier-free ^{64}Cu, for example, can be prepared by neutron bombardment on water-soluble Cu phthalocyanin in a nuclear reactor. An acidic extract of the irradiated sample is treated with a cation exchanger for chemical purification. Upon irradiation with a neutron flux of 10^{12} or 10^{13} $n \cdot cm^{-2} sec^{-1}$ for 72 or 17 hr, respectively, the specific activity of the final product is reported to be 50 Ci/g [30], compared to 8.05 Ci/g for ^{64}Cu based on Eq. (10) using $f = 10^{13}$ $n \cdot cm^{-2} sec^{-1}$, $\sigma = 4.5$ barns and $\theta = 0.691$ for ^{63}Cu target.

B. Radionuclidic Impurities

Inasmuch as the radionuclidic purity ω_{rn} is defined as Eq. (12) of Chapter 1, it is concerned with the radiation itself of the nuclide. Consequently, it depends on the choice of radiation for quantitization of the main nuclide. Thus, a sample of ^{59}Fe (a β^--γ emitter) containing ^{55}Fe (EC) as a radionuclidic impurity can be considered to be $\omega_{rn} = 100\%$ if the β^- radiation is exclusively determined. This does not mean, however, that the sample is radionuclidically equivalent to a pure sample of ^{59}Fe.

It is stated in Section III A that ^{32}P (τ = 14.3 days; β^-, 1.70 MeV) obtained by ^{31}P(n,γ)^{32}P cannot be in the carrier-free state. In addition, it is usually accompanied by ^{33}P (25.5 days; β^-, 0.248 MeV) due to the successive neutron capture giving rise to the lowering of its radionuclidic purity. Naturally, the radionuclidic purity of ^{32}P decreases with time elapsed for storage, though the soft β^- radiation of ^{33}P can be discriminated upon counting. This is similar to the case of ^{198}Au (2.697 days) produced by (n,γ) reaction which is followed by a successive neutron capture yielding ^{199}Au (3.15 days) as an impurity. Contrary to these cases from a nuclear-chemical aspect, a possible (n,2n) side reaction of ^{55}Mn(n,γ)^{56}Mn results in the formation of ^{54}Mn (303 days) as an impurity of ^{56}Mn (2.576 days). Note that the impurity emits 0.835 MeV γ-rays, making it difficult to spectrometrically separate it from 0.847 MeV γ radiation due to ^{56}Mn. An even more serious problem may be involved in neutron irradiation of target, if it contains chemical impurities of large cross sections. For example, ^{60}Co (5.24 years) produced by (n,γ) reaction of ^{59}Co implies a possibility to contain such radionuclidic impurities as ^{58}Ni(n,p)^{58}Co (71.3 days), ^{45}Sc(n,γ)^{46}Sc (83.9 days), ^{47}Ti(n,p)^{47}Sc (3.43 days), ^{51}V(n,α)^{48}Sc (1.83 days), and/or ^{50}V(n,α)^{47}Sc. The radionuclidic purity of its sample must be raised when cooling time is chosen appropriately for decaying-out of the impurities. Similar care must be paid to the use of an isotopic mixture in radioactive equilibrium such as:

^{90}Sr (τ = 27.7 years) + ^{89}Sr (51 days)
^{106}Ru (1.00 year) + ^{103}Ru (39.8 days)
^{113m}Cd (5 years) + ^{115}Cd (2.21 days)
^{125}Sb (2.0 years) + ^{126}Sb (9 hr)
^{137}Cs (30.0 years) + ^{136}Cs (12.9 days) or
^{127m}Te (105 days) + ^{125m}Te (56 days) + ^{129m}Te (33.5 days)

all being obtainable from the fission product.

REFERENCES

1. M. Green, J. J. Pinajian, and P. S. Baker, Rept. ORNL-IIC-4, Jan., 1969.
2. L. G. Stang, Jr., BNL-864 (T-347), 31, (1964).
3. H. B. Hupf, J. S. Eldridge, and J. E. Beaver, Int. J. Appl. Rad. Isotopes, 19:345 (1968).
4. W. A. Myers, H. O. Anger, and J. F. Lamb, Radiopharmaceuticals and Labelled Compounds, Vol. 1, IAEA, Vienna, 1973, p. 249.
5. J. S. Eldridge and P. Crowther, Nucleonics, 22:56 (1964).
6. P. M. Daniel, M. M. Gale, and O. E. Pratt, Nature, 196:1065 (1962).
7. J. C. Clark, C. M. E. Mathews, D. J. Silvester, and D. D. Vonberg, Nucleonics, 25:54 (1967).
8. Cyclotron-produced nuclides, in Radiopharmaceuticals and Labelled Compounds, Vol. 2, IAEA, Vienna, 1974, pp. 197-341.
9. M. M. Ter-Pogossian and H. Wagner, Jr., Nucleonics, 24:50 (1966).
10. K. Svoboda, Production of radioisotopes using accelerator, in Radioisotope Production and Quality Control, IAEA, Vienna, 1971, p. 747.
11. Nicole Falconi, Isotope generators, in Radioisotope Production and Quality Control, IAEA, Vienna, 1971, p. 659.
12. R. Henry, J. Nucl. Biol. Med., 15:105 (1971).
13. J. C. Clark and P. D. Buckingham, Short-lived Radioactive Gases for Clinical Use, Butterworths, London, 1975.
14. C. M. Lederer, J. M. Hollander, and I. Perlman, Table of Isotopes, 6th ed., Wiley, New York, 1967.
15. K. Way (ed.), Nuclear Data A and B, 6 issues/year, Academic Press, New York.
16. P. Massini, Science, 133:877 (1961).
17. W. Marth, Atompraxis, 12:392 (1966).
18. K. H. Lieser and A. Krueger, Z. Physik. Chem., 68:185 (1969).
19. F. Akiha, T. Aburai, T. Nozaki, and Y. Murakami, Radiochim. Acta, 18:108 (1972).
20. R. A. Fawwaz, H. S. Winchel, M. Pollycove, T. Sargent, and H. O. Anger, J. Nucl. Med., 7:569 (1966).
21. D. van Dyke, C. Shkurkin, D. Price, Y. Yano, and H. O. Anger,

Blood, 30:364 (1967).

22. E. C. Freiling and L. R. Bunney, Nucleonics, 14:112 (1956).

23. D. H. Perkel, Nucleonics, 15:103 (1957).

24. F. Lagoutine, Y. Legallic, and J. Legrand, Int. J. Appl. Rad. Isotopes, 19:475 (1965).

25. M. Muramatsu, in Surface and Colloid Science Vol. 6, (E. Matijevic, ed.), Wiley-Interscience, New York, 1973, p. 110.

26. W. Schirmer and N. Wächter, Actinides Rev., 1:125 (1968).

27. W. Gebauer, Z. Anal. Chem., 185:339 (1962).

28. G. Stocklin, Chemie heisser Atome, Chemische Reaktionen als Folge Kernprozessen, Verlag Chemie, Weinheim, 1969.

29. G. Harbottle and M. Hilman, Szilard-Chalmers processes for isotope production, in Radioisotope Production and Quality Control, IAEA, Vienna, 1971, pp. 617-632.

30. H. Ebihara, Radiochim. Acta, 6:120 (1966).

Chapter 3

SAFETY ASPECTS OF RADIOTRACER EXPERIMENTS

Mitsuo Muramatsu

Department of Chemistry
Faculty of Science
Tokyo Metropolitan University
Tokyo, Japan

Yasuo Suzuki

Department of Industrial Chemistry
Faculty of Engineering
Meiji University
Kawasaki, Japan

Ichiro Miyanaga and Yoshiki Wadachi

Division of Health Physics and Safety
Japan Atomic Energy Research Institute
Tokai-Mura, Japan

I. INTRODUCTION

All living organisms, including man, have been exposed throughout their lifespans to ionizing radiation from natural sources such as cosmic rays and radionuclides present in the earth's crust and atmosphere. The discovery of natural radioisotopes led to the knowledge that exposure to radiation has damaging effects on the living organism. The importance of this aspect has intensified with the increase in the number of artificial radionuclides produced by various methods (see Chap. 2). The majority of the radionuclides in Table 1 of Chapter 2 are currently used as radiotracers for studying a variety of chemical and biological processes. Thus, thorough precaution must be exercised in the handling of such radiotracer compounds in order to maintain safety standards in each laboratory.

The concept of safety standards has been markedly influenced over the past 15 years by the ever-increasing use of ^{14}C- and ^{3}H-labeled compounds in chemical and biological studies. In the 1950s, health physicists were interested in evaluation of external exposure to γ- and hard β-rays, as well as in the removal of such simple ionic contaminants as $^{60}Co^{2+}$, $^{137}Cs^{+}$, and $^{131}I^{-}$. At present, great importance is placed on the assessment of intracorporeal contamination resulting from those organic compounds, which may have been evaporated or dispersed into the aerial phase (aerosols). Similarly, the decontaminabilities of ^{14}C and ^{3}H depend on the physicochemical properties of their compounds rather than on their nuclear characteristics. The situation becomes even more complicated when we consider the soft nature of their β-rays which are characterized by large values for the ionization/energy ratio (Section V C 2) and the ability to produce "spurs" in matter (see Chap. 7) including the human body.

All these facts suggest that the health-physical problems associated with soft β emitters (especially ^{3}H) must be interpreted in terms of the molecular processes as physicochemical phenomena rather than of the electromagnetic processes arising from nuclear disintegration. This is exactly as was predicted in 1959 [79]; the danger posed by γ or hard β emitters such as ^{60}Co or ^{90}Sr is governed by such radiological quantities as those having the units of roentgen, rad, rem, etc., whereas the danger posed by ^{14}C or ^{3}H is determined primarily by such physicochemical quantities as vapor pressure, solubility, adsorptivity, and colloidal stability of the compound in which it has been incorporated. These physicochemical properties are of potential importance in assessment of the derived working limit (DWL) which should be correlated eventually with the maximum permissible dose (MPD) (see Section III A). In keeping with all these points, the present chapter deals with (1) safe handling of radiotracer compounds, (2) monitoring of radioactivity for radiological protection, (3) the nature of radioactive contamination and decontamination and related subjects, and (4) management of waste disposal, which is a matter of prime importance in relation to environmental protection.

II. SAFE HANDLING OF RADIOTRACER COMPOUNDS

A. General Considerations

In handling radiotracer compounds, general laboratory procedure is even more rigidly controlled than in standard chemical and biological experimentation. In addition, there is a need for extreme care in preventing any contamination with radioactive materials. In some cases where large amounts of radionuclides are employed, personnel must be adequately protected from radiation. Upon selecting a radionuclide appropriate to the experimental condition, the relative toxicity shown in Table 1 should also be taken into consideration. It is quite possible that some contamination takes place during a radiotracer experiment, even though the most skillful chemist performs it. A "cold run" should

TABLE 1

Classification of Radionuclides According to Relative Radiotoxicity Per Unit Activity*

Group 1 (High toxicity)								
^{210}Pb	^{210}Po	^{223}Ra	^{226}Ra	^{228}Ra	^{227}Ac	^{227}Th	^{228}Th	^{230}Th
^{231}Pa	^{230}U	^{232}U	^{233}U	^{234}U	^{237}Np	^{238}Pu	^{239}Pu	^{240}Pu
^{241}Pu	^{242}Pu	^{241}Am	^{243}Am	^{242}Cm	^{243}Cm	^{244}Cm	^{245}Cm	^{246}Cm
^{249}Cf	^{250}Cf	^{252}Cf						
Group 2 (Medium toxicity -- upper subgroup A)								
^{22}Na	^{36}Cl	^{45}Ca	^{46}Sc	^{54}Mn	^{56}Co	^{60}Co	^{89}Sr	^{90}Sr
^{91}Y	^{95}Zr	^{106}Ru	^{110m}Ag	^{115m}Cd	^{114m}In	^{124}Sb	^{125}Sb	^{127m}Te
^{129m}Te	^{124}I	^{126}I	^{131}I	^{133}I	^{134}Cs	^{137}Cs	^{140}Ba	^{144}Ce
^{152}Eu(12.7 y)		^{154}Eu	^{160}Tb	^{170}Tm	^{181}Hf	^{182}Ta	^{192}Ir	^{204}Tl
^{207}Bi	^{210}Bi	^{211}At	^{212}Pb	^{224}Ra	^{228}Ac	^{230}Pa	^{234}Th	^{236}U
^{249}Bk								
Group 3 (Medium toxicity -- lower subgroup B)								
^{7}Be	^{14}C	^{18}F	^{24}Na	^{38}Cl	^{31}Si	^{32}P	^{35}S	^{41}Ar
^{42}K	^{43}K	^{47}Ca	^{47}Sc	^{48}Sc	^{48}V	^{51}Cr	^{52}Mn	^{56}Mn
^{52}Fe	^{55}Fe	^{59}Fe	^{57}Co	^{58}Co	^{63}Ni	^{65}Ni	^{64}Cu	^{65}Zn
^{69m}Zn	^{72}Ga	^{73}As	^{74}As	^{76}As	^{77}As	^{75}Se	^{82}Br	^{85m}Kr
^{87}Kr	^{86}Rb	^{85}Sr	^{91}Sr	^{90}Y	^{92}Y	^{93}Y	^{97}Zr	^{93m}Nb
^{95}Nb	^{99}Mo	^{96}Tc	^{97m}Tc	^{97}Tc	^{99}Tc	^{97}Ru	^{103}Ru	^{105}Ru
^{105}Rh	^{103}Pd	^{109}Pd	^{105}Ag	^{111}Ag	^{109}Cd	^{115}Cd	^{115m}In	^{113}Sn
^{125}Sn	^{122}Sb	^{125m}Te	^{127}Te	^{129}Te	^{131m}Te	^{132}Te	^{130}I	^{132}I
^{134}I	^{135}I	^{135}Xe	^{131}Cs	^{136}Cs	^{131}Ba	^{140}La	^{141}Ce	^{143}Ce

TABLE 1 (continued)

^{142}Pr	^{143}Pr	^{147}Nd	^{149}Nd	^{147}Pm	^{149}Pm	^{151}Sm	^{153}Sm	^{152m}Eu(9.3h)
^{155}Eu	^{153}Gd	^{159}Gd	^{165}Dy	^{166}Dy	^{166}Ho	^{169}Er	^{171}Er	^{171}Tm
^{175}Yb	^{177}Lu	^{181}W	^{185}W	^{187}W	^{183}Re	^{186}Re	^{188}Re	^{185}Os
^{191}Os	^{193}Os	^{190}Ir	^{194}Ir	^{191}Pt	^{193}Pt	^{197}Pt	^{196}Au	^{198}Au
^{199}Au	^{197}Hg	^{197m}Hg	^{203}Hg	^{200}Tl	^{201}Tl	^{202}Tl	^{203}Pb	^{206}Bi
^{212}Bi	^{220}Rn	^{222}Rn	^{231}Th	^{233}Pa	^{239}Np			
Group 4 (Low toxicity)								
^{3}H	^{15}O	^{37}Ar	^{58m}Co	^{59}Ni	^{69}Zn	^{71}Ge	^{85}Kr	^{85m}Sr
^{87}Rb	^{91m}Y	^{93}Zr	^{97}Nb	^{96m}Tc	^{99m}Tc	^{103m}Rh	^{113m}In	^{129}I
^{131m}Xe	^{133}Xe	^{134m}Cs	^{135}Cs	^{147}Sm	^{187}Re	^{191m}Os	^{193m}Pt	^{197m}Pt
^{232}Th	Th(natural)		^{235}U	^{238}U	U(natural)			

*Reproduced with minor changes from Ref. 190 by courtesy of Pergamon Press, Inc.

normally be done before entering into the entire course of the radiotracer experiment. A "stand-in," as well as a cheaper analog, which are either radioactive or nonradioactive, may also be useful.

B. Apparatus and Tools for Safe Handling

1. Gloves. For most purposes, thin surgical gloves, or those specified as "for radioisotope handling," are conveniently used, although they sufficiently protect only from α- and very soft β-particle radiation. Thin, polyethylene or polypropylene gloves are cheaper, provide less grip, allow less fingertip sensitivity, and have less mechanical durability than rubber gloves, but the latter are more resistant to organic solvents. Discomfort due to sweating on hands may be prevented somewhat by putting on thin cotton gloves underneath the rubber ones, although the doubly worn gloves considerably reduce fingertip sensitivity.

Extreme care is required for wet and slippery gloves; even those with a molded rugged pattern on fingers and palm do not provide as good a grip as bare hands, and yield only poor grip when wet. Unfortunately, there have been many accidents caused by improper use of rubber gloves.

The outside of gloves should be regarded as contaminated: the apparatus and ware are classified into those which may have been contaminated and those which can be handled with unprotected hands, though there may be some borderline items. Upon turning a switch with gloves, for example, a sheet of paper tissue should be used to avoid a direct contact. Similarly, counting samples, planchets, and surveying or monitoring equipment should not be operated with gloves, unless the gloves are not contaminated. Before putting on gloves, it is desirable to apply talcum powder to the hands for easy wearing. Gloves may be used repeatedly for the same experiments, if they are not contaminated or spoiled. Used gloves should be kept inside out on a paper tissue for drying, or right side out, leaving cuff folded back [1].

2. Laboratory coats. Laboratory coats for radiotracer laboratories should be made distinguishable from those for the "cold" laboratories by coloring or marking. The coats should not be mixed together, particularly in washers (Section IV C 3).

3. Tissues and sheets. Paper tissues are conveniently used for both preventing and removing contamination as they are generally cheap, disposable, and available in various types and sizes. However, inadequate or excessive use of tissues should be avoided: simultaneous use of a glove and a tissue for a pipette filler often reduces fingertip sensitivity too much to operate the filler properly. Protective bench papers, prepared to cover the laboratory tables or bench tops, are effective for minimizing the spread of radioactive spills. Polyethylene-backed cellulose paper may be used effectively, particularly on the bench top of the fume hood, where most of the unsealed radioactive materials are handled. Sheet papers on the bench top may occasionally cause another problem: small and lightweight ware, such as polyethylene beakers and bottles placed on the paper, are often tipped over.

The floor should be coated with a sheet of plastic such as polyvinyl chloride, or more safely in case of fire, polyvinyl acetate.

4. *Trays*. Contaminated glassware such as beakers, bottles, or separatory funnels should be segregated in an absorbent-lined plastic tray from the clean items. The use of such a tray under a well-ventilated fume hood appreciably reduces spreading contamination.

5. *Plastic ware*. Unbreakable laboratory ware such as beakers, bottles, cylinders, flasks, funnels, pipettes, etc., are mostly disposable and useful for radiotracer experiments, although they are not quite as heat-resistant as glass. Chemical resistance and the physical properties, including adsorptive characteristics, of plastics have to be carefully inspected before use.

6. *Pipette-filling assemblies*. In radiotracer laboratories, smoking, drinking, eating, and applying cosmetics must be prohibited. The mouth should not come into contact with any appliances, including pipettes, which should be handled with the aid of pipette fillers [1,2]. Dropping pipettes, including graduated ones may be conveniently used for transferring and adding a solution, though they have poor accuracy and control. A bulb of sufficient capacity and of proper rigidity is desirable. Incidentally, glassblowing should be done in a separate room where there are no radioactive materials.

7. *Stirring rods*. Stirring rods prepared from glass tubing or rod are generally used. Use of "policemen" for scraping precipitates adhering to the vessel should be restricted, since

they are apt to leave some radioactive materials in the capillary space. Careless withdrawal of a rod from the beaker containing radioactive materials may cause contamination.

8. Filtration assemblies. Filtration for preparation of counting samples may be carried out by the use of a separable or demountable filter assembly composed of filter paper, fritted glass disk, and chimney [1]. To prevent creeping of the precipitate, the chimney wall should be coated with a hydrophobic film or washed inside with ethanol. For holding a chimney on a fritted glass bar, a single-action clamp such as a Millipore clamp is recommended for safe operation.

9. Washbottles. Mouth-blowing of washbottles should not be used in radiotracer laboratories; instead, polyethylene, squeeze-type washbottles may be employed. When an intake of fumes or gas through the spout is not desirable, a bottle having a separate screw cap with air vent may be favorably used.

10. Hot plates. Thermostatted hot plates wherein electric circuits are tightly sealed are useful for most radiotracer experiments. Household electric skillets or hot plates may also be used in the laboratory; those with Teflon coating are preferable for easier decontamination.

11. Face protectors. Equipment for face protection may include safety spectacles, goggles, and an eye-and-face wash which is instantly operated for immediate rinsing of both reagents and radioactive materials. A squeeze-type eyewash bottle may also be kept at hand.

12. Waste pails and buckets. Transferring solid wastes from one pail or bucket to another should be avoided, especially when powdered items are being handled. Highly contaminated wastes should be stored in a separate pail and transferred to temporary waste storage as soon as possible.

Mixing of two or more kinds of liquid waste may yield dangerous materials. For example, cyanide solutions, whether containing ^{14}C or not, should not be mixed with acids.

C. Practice of Safe Handling

In general, good experimental scientists can also do well in radiotracer experiments. Should it be found that a particular person tends to repeat his errors in the matter of contamination, he may have to undergo further training in laboratory safety.

In designing a radiotracer experiment (see Chap. 1), safety should be taken into account. Some modification of the conventional procedure and ware may be required accordingly. The ready convertibility of ^{14}C- and ^{3}H-labeled substances to volatile compounds yields outstanding difficulties both in handling them and in disposing of wastes. A good glove box with moderate ventilation should be used for handling ^{14}C- or ^{3}H-labeled compounds in volatile state.

Radioactivity is always potentially hazardous and frequent use of small or large amounts of radioactive materials often makes us oblivious to the dangers of high-level radioactivity. Use of radioactive materials should not be considered as a privilege; the hazards affect us directly, and may also affect the public. In conclusion, it must be emphasized that there should be a well-designed fire precaution and fire fighting system in the radiotracer laboratory. In some laboratories, precaution against great earthquakes must be taken.

III. MONITORING FOR RADIATION PROTECTION

A. Derived Working Limit (DWL) and Investigation Level (IL)

The objective of radiation monitoring is to control the exposure of personnel to external radiation and internal contamination under the relevant dose limits and dose commitment recommended by the International Commission on Radiological Protection (ICRP) [4-6]. As it is impossible to measure directly the organ doses actually received by individuals, secondary standards must be derived for interpretation of the results obtained for the external dose rates and radioactivity levels of contaminants at various surfaces, in air, in drinking water, in excreted urine, and so on, all being susceptible to the conventional measurements. These secondary standards, often called derived working limits (DWLs) [3], are estimated from the relevant dose limits with reasonable assumptions of the time of a person's occupancy of a radiating area (for external radiation) or on the modes of intake of radionuclides by inhalation or ingestion and their subsequent metabolism in the body (for internal contamination). As an example of the DWL, Ref. 6 gives a voluminous table listing the maximum permissible concentrations of individual radionuclides in air and in drinking water for a 40 hr/week working person (see, e.g. Table 2 in Section IV), as well as the maximum permissible concentrations in air and water for a 168-hr week and the critical organ.

For effective monitoring, it is convenient to set checking levels well below the DWLs. These levels, called investigation levels (ILs) [3], varying with the type of operation and conditions of workplaces are given in such a way that more precise investigation is needed only when the surface contamination or external dose rate in a tracer laboratory exceeds these levels.

B. Area Monitoring

1. External Radiation

External radiations in a radiotracer laboratory may be β- and γ-rays of energies up to a few MeV (see Table 2, Chap. 2). In reality, radiation field has a wide energy distribution because of the scattering with and the absorption by surrounding materials.

a. γ Radiation. An ionization chamber is a single and rugged device for a reliable determination of the radiation dose rate when a proper consideration is taken of the wall thickness to assure electronic equilibrium [7,8]. Various types of survey meters are available with a full scale range from a few mR/hr to more than 1000 R/hr. Simple and more sensitive instruments include a Geiger-Müller (GM) and an inorganic scintillation counter principally useful for detection or scanning of radiations. The response of the GM [9] and inorganic scintillation counters [10] for tissue dose can be improved over a limited range of photon energy by addition of an appropriate filter shield. Investigation was carried out to give detector tissue-equivalent atomic number by an appropriate combination of different plastic scintillators [11].

In recent years, a technique to adjust the response of an NaI(Tℓ) scintillation counter to γ radiation independent of its energy was developed by calculating a conversion function from the γ response spectrum of the detector to the absorbed dose in tissue and applying its inverse function to bias-voltage modulation in electronic circuits [12-15]. This technique indicates a possible approach to the practical radiation dosimetry which meets the requirements of tissue equivalence and electronic equilibrium for the detector.

b. <u>β Radiation</u>. β Radiation is usually detected by the use of an ionization chamber or a GM counter with a thin window. Discrimination of β radiation from γ can be done by a subtraction method in which β radiation is cut with an additional shield window. A thin plastic scintillation counter is able to detect only β radiation even in the presence of γ radiation. Because of the low penetrating power and a wide distribution of β-ray energies, β dose rate changes substantially with distance, and the dose received by the skin layer ($\sim$7 mg/cm^2) is of primary importance. Measurement of the absorbed dose at or above the surface of β-emitting source is extremely difficult with a conventional survey meter [16]. In a routine procedure, a standard source calibrated with an extrapolation chamber is useful. Another practical method for evaluation of dose is a calculation based on the estimated amount of radioactivity at the surface. For this purpose, accurate depth-dose data from sources calculated by the transport theory are available [17,18].

2. Surface Contaminations

Surface contaminations are classified into fixed and transferable contaminations. The former can be an external radiation source, while the latter causes an internal contamination due to resuspension of the contaminants into air. Considerable effort has been made to determine the resuspension factor [19-21], from which the surface contamination for MPC_a values gives 10^{-3}-10^{-4} $\mu Ci/cm^2$ for β emitters and 10^{-4}-10^{-5} $\mu Ci/cm^2$ for α [22-24]. Experience shows strong evidence for no significant intake of radioactivity due to air contamination, if the level of surface contamination is below the above limits.

The extent of transferable contamination is usually estimated with the smear or adhesive paper samples. The smear technique is applicable to ^3H contamination on the surface by the use of a glycerol-soaked filter paper [25]. This technique, however, provides no information about the presence and amount of air contamination. Sensitivity of the method can be made very high by low background counting technique. A new method [26] for collecting transferable contaminants was devised by employing air impingement to disperse contaminants from a surface and simultaneously collect them with a conventional filter sampler. This method is suited for elucidating the mechanisms of resuspension of the contaminants (though not practical).

A thin GM, proportional or scintillation counter is sometimes useful for measuring total surface contamination -- fixed and transferable -- due to β emitters including ^{14}C and ^{35}S, but not ^3H [27,28]. For detecting α contamination, a proportional counter with an extremely thin window ($\sim$1 mg/cm^2) or a ZnS scintillation counter is useful. Detection limits of these instruments reach as low as 1/10 $\sim$ 1/100 of the derived working limits (Section III A) of surface contamination. In small laboratories, information about surface contamination is obtained by surveying mops, dust (collected with vacuum cleaner), shoes, and clothing.

3. Air Monitoring

A constant inhalation of contaminated air of the MPC_a during the year (8 hr/working day, 5 days/week, and 50 weeks/year) or the maximum permissible annual intake (MPAI) of 2000 MPC_a-hr corresponds to the maximum permissible dose commitments. In laboratory scale, the common procedures of good housekeeping provide adequate freedom from air contamination. Air contamination may arise in particulate and gaseous forms. Radioactivities of particulate usually collected by drawing air through onto a filter paper does not always represent the air breathed by personnel. To solve this problem, a personal air sampler [29] would be used, which is carried by an individual with a sampling head close to the nose. When the "dominant" or "hot" particle is in airborne particulates, a large error may result from the possible discrepancy of the number of such inhaled particles from that collected in the air sample [30,31]. Particle size of the contaminants should also be taken into consideration in the estimation of internal does, because the location and the amount of deposition of inhaled particles in the lung, and consequently the dose received by the lung, would vary with their size distribution [32]. In order to solve this problem, a "size selection dust sampler" [33] has been devised in which the size distribution is made, by appropriate centrifugation, almost identical to the theoretical relationship between particle size and pulmonary deposition. Measurement of α-, β-, or γ-rays emitted from the collected particulates can be carried out by conventional methods. In the case of α emitter, correction for its self-absorption in the filter paper [34] and for the effect of natural activities is usually necessary.

Gases are sampled in a chamber or monitored with a continuous flow ionization chamber. Appropriate chemical or physical reactions can be utilized for sampling particular gases. For example, CO_2 is sampled in basic solutions [35], and so is SO_2, with a filter paper impregnated with activated charcoal [36].

Tritiated water vapor is sampled in a cold trap, a water bubbler or silica gel [37,38]. Sampling technique of airborne iodine is rather complicated, because it includes I_2, HI, HIO, and organic forms such as CH_3I and higher molecular weight compounds. A device for collection and classification of airborne iodine [39,40] has been developed, which consists of a particle filter, silver screens, activated charcoal-loaded filter, and a cartridge of activated charcoal.

C. Individual Monitoring

1. External Exposure

For many years, photographic film and pocket ionization chambers have been widely used for individual monitoring [41]. The advantage of the film badge is its capability of distinguishing different radiations by the use of multifilters, making it possible to provide manifold information obtainable from visual pattern and to file the record for medicolegal consideration. It has, however, undesirable characteristics such as fading of the latent image due to the changes in temperature and humidity, energy dependence, and directional dependence of the blackening density.

Recently a solid-state dosimeter using modern photoluminescent glass or thermolunimescent materials has been developed, and many laboratories have replaced the film badge with these in routine monitoring [42-45]. Their characteristics are superior to those of film in precision, sensitivity, and linearity between radiation dose and its response [46,47].

The most widely used glass is silver-activated metaphosphate glass, which enables one to measure 10 mR with an error of about ±10%. Contamination of the glass by dust and grease reduces the accuracy in low dose measurements due to spurious luminescence. Washing with neutral detergent and methanol is, therefore, essential for the accurate measurement of low dose. Currently available thermoluminescent materials are CaF_2:Mn, $CaSO_4$:Mn [48], Mg_2SiO_4:Tb [49], LiF [50], $Li_2B_4O_7$:Mn [51], and BeO [52]. All these materials make it possible to measure γ radiation of less than 10 mR with an error of about ±5 to 10%. The possibility of measuring a radiation as low as 10 μR is reported by using $CaSO_4$:Mn [48]. The lower limit of the range is determined by spurious triboluminescence. CaF_2:Mn and $CaSO_4$:Mn undergo fading at room temperature, but this is improved by using Tm or Dy as an activator [53]. Materials such as CaF_2:Mn, $CaSO_4$:Mn, and Mg_2SiO_4 characterized by their intermediate values of atomic number are energy-dependent in their response to tissue dose and this is compensated for by the use of a metal filter as in the case of photographic film. Fortunately, the effective atomic number of LiF, $Li_2B_4O_7$, or BeO is almost equivalent to that of human tissue.

Solid state dosimeters also respond to β radiations, and their relative response to γ radiation largely depends on the physical shape of the detector [54,55]. Thin teflon discs impregnated with pulverulent LiF are available for finger dosimetry of β radiation [56,57]. This is applicable to the measurement of dose from surface contamination.

Both glass and thermoluminescent dosimeters (TLD) are reusable by annealing. TLD can be made quite small (0.001 to 0.1 g) and are available in various shapes such as rods, discs, and ribbons. A simple readout operation makes it possible to fully automate the TLD system, although problems arise from wet procedures needed in the film and glass dosimetries. It has been developed to make an extremely sensitive dosimeter as an application of thermally stimulated exo-electron (TSEE) emission from some thermoluminescent materials [58-60].

2. Internal Contamination

Two methods are currently available for assessment of body or organ contamination: these are the estimation based on radiochemical analysis of excreta or other biological samples and the direct measurement of body or organ content of radioactivity from outside the body. The former is applied to α and β emitters, while the latter is employed for γ emitters and nuclides emitting high energy β radiations, such as ^{32}P and ^{90}Sr, and consequently radiating bremsstrahlung in tissue.

a. Bioassay. For assessment of internal contamination, excreta and other biological samples should be chosen appropriately to the nuclide, its chemical form, metabolic pattern, and routes of contamination.

Urine is the most useful sample for "soluble" or "transferable" compounds of nuclides, i.e., the compounds which readily transfer from one site (gut, lung, and wound) into extracellular fluid and then other organs. To correlate the urine analysis with the body content it is desirable to collect a full 24-hr urine sample. Otherwise, a full day's excretion of urine should be estimated by comparing the creatinine content in the sample with the daily creatinine excretion which is relatively constant under normal conditions [61]. Mathematical formulation on the dose calculation from urine analysis is given with metabolic data for various nuclides [62]. In a radiotracer laboratory, urine may be sampled only when intake of radionuclide is anticipated from an accidental spill or finding of abnormal surface or air contamination. Before urine sampling, analysis of nasal swabs should be made for the detection of possible inhalation. This analysis does not give a quantitative estimation of internal contamination, but provides information of the significance of internal contamination and the identification of the responsible nuclides. Breath may also be analyzed for the estimation of content of some nuclides such as ^{14}C (in the form of exhaled $^{14}CO_2$) [35], ^{3}H (in exhaled water vapor) [37] and ^{226}Ra (in exhaled ^{222}Rn) in the body [63].

The sensitivity of the radiochemical method for urine analysis permits the detection of the amount of radioactivity in the body corresponding to 1/20 of the MPD recommended by the ICRP [4-6] for intake after more than 2-3 months. For example, 1.5 mCi of ^{3}H gives the dose of 0.25 rem (1/20 MPD) to the whole body, and the excretion of ^{3}H in urine is 0.5 μCi/liter 60 days after intake, while the detection limit of conventional liquid scintillation counting is 0.1 μCi/liter [62].

b. Whole-body counting. (See also Chapter 22.) The whole-body counter [64,65] consisting of one or more large NaI(Tl) (15-23 cm in diameter) or plastic scintillators is placed in a heavily shielded room for decreasing background radiation. The crystalline scintillator in combination with a multichannel analyzer for identification of nuclides is characterized by high sensitivity. A single shielded crystal with an open "shadow shield" [66,67] for the body is sufficient for routine counting of the personnel. A large-area proportional counter or 3-5 mm-thick NaI(Tl) crystal is used for high-energy β or low-energy x-ray emitters such as ^{90}Sr and ^{239}Pu retained in lung [68,69]. It is desirable for the estimation of the body content to consult the results thereby obtained with those based on the bioassay.

D. Other Measurements for Radiation Protection

1. Liquid Scintillation Counting

Ionization chamber and proportional counter can be employed for the measurement of tritium. The former is restricted by interference of external radiation and other radioactive gases such as ^{41}Ar and ^{85}Kr, as well as difficulty in decontaminating the chamber. The proportional counter offers the most sensitive method at present, which has an advantage over the former in its energy discrimination performance. The problem is how to make the tritiated contaminant in gaseous form.

Tritiated samples in routine monitoring are usually in the form of tritiated water (H^3HO). Recent developments in scintillater systems and techniques of sample preparation and low background counting have made it possible to apply this technique to various types of samples [70-72]. For example, the Y-value (defined as the minimum detectable activity at a confidence level of one sigma of statistical error for 1-min counting time) is obtained by about 1 pCi/cm^3 for homogeneous solution [73], 0.7

pCi/cm^3 using the detergent Triton N-101 [74], and 0.2 pCi/cm^3 with a large volume vial (250 ml) and three photomultipliers [75]. This technique is, of course, applicable to the measurement of other soft β emitters such as ^{14}C and ^{35}S.

2. γ Spectrometry

Spectrometry for identification of nuclides is often required in routine monitoring. The most commonly used spectrometers are NaI(Tl) scintillator and Ge(Li) semiconductor coupled with a multichannel pulse height analyzer. NaI(Tl) detector is preferred for its high sensitivity, whereas Ge(Li) detector is suited for detailed analysis of mixed nuclides. One disadvantage of the Ge(Li) detector is the need, due to its narrow bandgap, for continuous cooling with liquid nitrogen ($T \lesssim 100\ ^\circ K$). Widening of the bandgap may be achieved by the use of GaAs, CdTe, or HgI_2 which seems to be a strong tool in the use of the crystal counter at room temperature [76-78].

IV. RADIOACTIVE CONTAMINATION AND DECONTAMINATION

A. Nature of Radioactive Contamination and Decontamination

Radioactive contamination (RC) and decontamination (RD) are essentially different from conventional soiling and cleaning of solid surfaces, a difference characterized by the following aspects [79]:

1. RC can be caused even by a trace amount of contaminant which may exceed the maximum permissible level of surface contamination. In the case of carrier-free radionuclide solution, for example, a maximum permissible concentration in radioactivity/volume gives an extremely small amount of that in mass/volume, as exemplified in Table 2. Little is known about the mechanism of interfacial processes involved in such an extremely low concentration as 10^{-16}-10^{-13} g/cm^3 (see the last column of Table 2). At the same time, RC at such a high level as 10^{-5}-10^{-3} µCi/cm^3 (see the

TABLE 2

Maximum Permissible Concentration of Carrier-Free Radionuclide in Soluble State

Radionuclide	Specific activity* (Ci/g)	Maximum permissible concentration in water**	
		μCi/cm^3	g/cm^3
^{45}Ca	1.91×10^{4}	3×10^{-4}	1.6×10^{-14}
^{60}Co	1.14×10^{3}	1×10^{-3}	8.8×10^{-13}
^{144}Ce	3.18×10^{3}	3×10^{-4}	9.4×10^{-14}
^{131}I	1.23×10^{5}	6×10^{-5}	4.9×10^{-16}
^{32}P	2.88×10^{5}	5×10^{-4}	1.7×10^{-15}

*Based on data in Ref. 80.

**A DWL for occupational exposure (40 hr/week) to radionuclide in soluble state [6].

third column) may be of some importance in radiological safety. It has been pointed out [81-83] that the mode of removal of radioactive oils and fats from various cloths at an extremely low level is considerably different from that at a higher level for conventional detergency. A similar conclusion is revealed in the experimental results on the removal of metal ions in minute amounts from the solid surfaces of semiconductors for integrated circuits [84-86].

2. Of various kinds of isotopic species involved in RC, only the radioactive species plays an important role. For example, a drastic decrease of surface radioactivity (e.g., by ^{60}Co) can be observed by soaking a radioactively soiled specimen in a concentrated solution of nonradioactive, isotopic species (e.g., Co^{2+}) for the exchange equilibrium. Such a decrease may be accompanied by an equal or an excessive amount of the nonradioactive species (Co) for substitution. From the aspect of

conventional detergency, this process must be considered as an additional contamination rather than decontamination.

3. Since radionuclides for tracer experiments are in many cases in the form of ionic or simple organic species, the pertinent contaminants are essentially different in the mode of interaction with solid surfaces from conventional soils, which are mostly a mixture of oils, fats, proteins, carbohydrates, silicates, and other compounds of complicated structure [87-90]. Thus, ordinary detergents are not always effective for removal of such an ionic contaminant as $^{45}Ca^{2+}$, $^{60}Co^{2+}$, $^{144}Ce^{3+}$, $^{131}I^{-}$, $^{35}SO_4{}^{2-}$, or $^{32}PO_4{}^{3-}$ (see Section IV B 2 b for details).

4. Differing from nonradioactive contamination, the soiling extent is given in radioactivity as a function of time with which it is either monotonously decreased (in most cases) or changed in some complicated manner [91] for a radionuclide involved in a series of successive decays (see Chap. 2, Section II D). As an extreme case, RC due to a radionuclide of short half-life can best be "removed" by leaving the subject alone.

The mode and extent of RC and, accordingly, the method of RD, largely depend on permeability, polarity, and other physicochemical properties of solid surfaces, as well as on the dissolved state of contaminants in their solutions. Nonpermeable surfaces such as metal and plastics are characterized by quick saturation at a low level of RC [79,92,93], whereas slow processes involving penetration and diffusion of contaminant into the solid interior (see Chaps. 10 and 12 for details) are observed specifically with such permeable material as wood, fiber, hair, skin, and other gelatinous material [92,94,95]. Facility of RD is determined by the extent of adherence, physical adsorption, chemisorption, and formation of interfacial compound. The RD due to the first two (surface-physical contamination, *a*) can readily be removed by wiping and rinsing in an adequate solvent, whereas RD of the persistently contaminated surface (surface-chemical contamination, *b*) requires an adequate choice of chemical decontaminants such as EDTA [79,96,97], citrate [96,97], polyphosphate [98], TiO_2- [99] or bentonite- [100] paste, and/or nonradioactive compound for isotopic substitution [101,102]. The choice, therefore, depends on the type and mode of RC, as well as on the properties of the decontaminant.

Even more persistent contamination is the case of RC of a permeable surface, across which the radiolabeled species penetrates into the interior phase (mobile interior contamination, *c*). This kind of RC is removable only after a very long time. Rinsing-out, however, becomes ineffective when the isotope species has been bound to form immobile compounds with the matrix of diffusible material (fixed contamination, *d*) [79,92,103]. In the case of living skin, contamination is transferred further to some organs through circulatory systems (incorporated contamination, *e*) [104-107]. This is the most hazardous situation. RC in reality is comprised of more than one of these modes, *a* - *d*, or *a* - *e*. The contribution of each component to the observed radioactivity depends on the physicochemical properties of the contaminant and the solid material, as well as on the physiological conditions of the living system. In addition, it is largely governed by the time elapsed from the first contact of the contaminant with the solid surface. It must be noted that, when one leaves a contaminant at a solid surface, the mode of RC is changed in the direction of $a \rightarrow b \rightarrow c \rightarrow d$ and, for living systems, up to *e*. Experimental results show that the process of *e* takes place at 30 min after the first contact of $^{90}Sr^{2+}(^{90}Y)$ or $^{32}PO_4{}^{3-}$ with the skin of a living rabbit [104]. Obviously, RC of living skin must be removed as soon as possible, preferably within no more than 30 min after contact [104].

The mode and extent of RC are affected also by the dissolved state of contaminants. There is certain aspect of RC, in that the binary mixture of a radioactive solute *A and a solvent B contains various kinds of chemical species such as

$$n{*}A \rightleftarrows {*}A_n \quad \text{by molecular (or ionic) association} \tag{1}$$

$$*A + B \rightleftarrows {*}A{\cdot}B \quad \text{by solvation} \tag{2}$$

$$*A + B \rightleftarrows {*}A{\cdot}C + (B - C) \quad \text{by solvolysis} \tag{3}$$

where C is a fragmental product (OH^-, for example) of the dissociation of B (H_2O). Consequently, RC is determined by the type and amount of *A, $*A_n$, *A·B, and *A·C. For example, RC due to aqueous $^{60}CoCl_2$ depends on the amounts of $^{60}Co(H_2O)_6{}^{2+}$, $^{60}Co(H_2O)_5(OH)^+$, $^{60}Co(H_2O)_4(OH)_2$, and oligomers of these species in the solution, as well as on the nature of solid surfaces (anionic, cationic, or nonionic), to which the individual contaminant species bind in a variety of interaction manners [108]. The extent of the reactions (1)-(3) can be estimated or presumed [109] from information about the stability constant, the degree of solvation, and the solvolysis constant, all available in the pertinent literature [110-112]. Some of these data can be experimentally obtained by appropriate use of radiotracer techniques, to which Chapter 11 is devoted.

RC due to carrier-free radionuclides seems to be rather complicated in their characteristic behavior at an extremely low concentration. Behavior of such a nuclide at a concentration as low as 10^{-6} mol/liter or less is not always explicable as an extrapolation from that at a higher concentration. It has not yet been established whether the peculiar behavior can be attributed to the formation of so-called "radiocolloidal particles." Even so, it is a matter of conjecture as to what are the main constituents and/or components of such radiocolloidal "particles." Nevertheless, there are some phenomena explicable by plausible assumption of radiocolloidal particles (see Chap. 8 for details). Adsorption, for example, of carrier-free radionuclide at solid/solution interfaces is markedly enhanced when the solution contains, apparently, radiocolloidal particles. Such a strong adsorption is involved in the mechanisms of RC of pig skin surfaces by ^{90}Y [103], ^{125}Sb, ^{147}Pm, ^{210}Po [113], and ^{239}Pu [114].

Using intestinal protein films, Tashiro [115,116] found that such a radiocolloidal contaminant as ^{147}Pm or ^{239}Pu chloride cannot be removed by rinsing with detergent solution. From Eq. (1) one may write

$$*A_{mo} \rightleftarrows *A_{rc} \tag{4}$$

for the equilibrium between molecular, $*A_{mo}$, and radiocolloidal, $*A_{rc}$, species of a radioactive solute $*A$. Because of the high adsorptivity of colloidal particles, one can assume that surface-chemical contamination due to the reaction

$$*A_{rc} + \text{solid} \rightleftarrows *A_{rc}\cdot\text{solid} \tag{5}$$

is much more than that based on the reaction

$$*A_{mo} + \text{solid} \rightleftarrows *A_{mo}\cdot\text{solid}. \tag{6}$$

The fundamental aspect of effective RD is therefore to provide a condition for shifting the equilibrium (4) or (5) toward the left direction [79,92,108] (see Section IV B 2 a).

Extent of RD is expressed conventionally by

$$\text{decontamination factor (DF)} = \frac{R_1}{R_2} \tag{7}$$

$$\begin{aligned}\%\ \text{removal} &= 100\,\frac{R_1 - R_2}{R_1}\\ &= 100\,\frac{\text{DF} - 1}{\text{DF}}\end{aligned} \tag{8}$$

or by the decontamination index, DI, defined as

$$\text{DI} = \log \frac{R_1}{R_2} \tag{9}$$

for the radioactivity before, R_1, and after, R_2, the decontamination procedures. Equation (9) is suited for discussing the case of $R_1 >> R_2$. Such a marked decrease of radioactivity is observed practically when R_1 is contributed solely by mode *a* (surface-physical contamination) which is characterized by the readiest decontamination.

B. Properties and Choice of Radioactive Decontaminants

1. Surface-Physical Contamination

Because of weak attraction between a contaminant and a solid surface, any kind of contaminant should in principle be removed by wiping with or rinsing in a solvent capable of dissolving it. The readiness of RD depends on the rate of solution rather than solubility. Vaporizability of the solvent would be so low as to facilitate the decontamination procedures and high enough to permit quick evaporation for final waste disposal to be made in the solid form. For most ionic or dipolar contaminants, water is best suited for its ready availability.

2. Surface-Chemical Contamination

RC of this sort can hardly be removed by simple treatment with a single liquid. This is indeed the case of ionic contamination which is caused solely by an ionic and/or ion-dipolar attraction between a radioactive ion and a solid material. Adequate choice of decontaminant for the best RD is thus necessitated.

a. Single liquids. For reasons similar to the case of surface-physical contamination, water is the decontaminant that should be first attempted for removing ionic contaminants from solid surfaces. In this case, the readiness of RD is governed primarily by the acidity of water. A cationic contaminant is in general removed more with increased acidity, as shown in Fig. 1, probably because of the increased amount of $[H^+]$ available for ion exchange with the radioactive ions at the surface [93,98]. It may be, however, that the extraordinary increment of DI value at $pH < 4$ reflects corrosive reaction at the metal surfaces [79]. On the other hand, an increase in solution pH does not always result in an increased removal of an anionic contaminant [93].

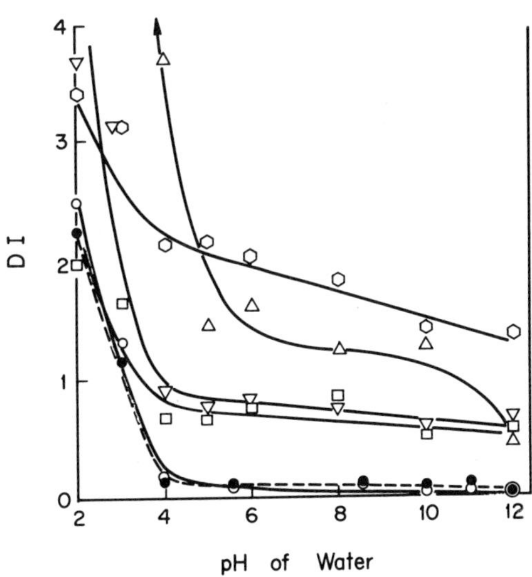

FIG. 1. pH dependence of removal of $^{60}Co^{2+}$ (open symbols) and $^{144}Ce^{3+}$ (filled circles) from the surfaces of nickel (○ and ●), copper (△), lead (▽), iron (□), and stainless steel (⬡) by rinsing in water at 30 ± 1°C. pH was controlled with HCl or NaOH. [Reproduced by permission from Ref. 137 (circles) and Ref. 98 (remainder).]

To remove nonionic and nonpolar contaminants such as ^{14}C-labeled fats or oils and tritiated esters, one needs the appropriate choice of a nonpolar solvent. Again, conditions for such a solvent is similar to the case of surface-physical contamination. Decontaminants for polar-nonpolar (=amphipathic) compounds would be chosen among a variety of solvents -- from typically polar (like water) to perfectly nonpolar (such as cyclohexane and CCl_4) -- depending on the balance of the hydrophilic (=oleophobic) and hydrophobic (=oleophilic) moieties of the contaminant molecule. [^{14}C]Stearic acid, for example, can be removed from various surfaces by repeated treatment with a nonpolar solvent such as benzene or petroleum ether. On the other hand, sodium [^{14}C]dodecylsulfate at nonpolar surfaces should be decontaminated by repeated treatment of the soiled surface with water.

It is noteworthy that the removal of RC is enhanced by the aid of ultrasonic irradiation [117-119] or by a small amount of detergent added to the decontamination solvent [120]. This clearly indicates an important role for solvent penetrability into the microcracks where RC has taken place.

b. Detergent solutions. Parallel to the enhanced penetrability mentioned above, solutions of many surfactants exhibit a marked detergency effect at oil(or fat)/solid interfaces. Thermodynamically, detergency is a physicochemical process accompanying a large amount of free energy dissipation $\Delta\gamma$ (<0) given by

$$\Delta\gamma = \gamma_{sl} + \gamma_{ol} - \gamma_{os} \tag{10}$$

where γ_{sl}, γ_{ol}, and γ_{os} are the free energies at solid/solution, oil(or fat)/solution, and oil(or fat)/solid interfaces, respectively. In this respect, a good detergent is not always a good penetrant, which is characterized by a large value of spreading coefficient η as the difference between γ_{sl} and the free energy γ_{as} at air/solid interface,

$$\eta = \gamma_{sl} - \gamma_{as} \tag{11}$$

A surfactant, on the other hand, is defined as a substance that lowers the air/solution-interfacial free energy γ_{al}, which is conventionally called surface tension. Thus, a mere decrease of γ_{al} as evidenced by, e.g., foam formation, is not a direct indication that the surfactant will function well as a detergent or as a penetrant. Although decreases in γ_{sl}, γ_{ol}, and γ_{al} are commonly caused by adsorption of an amphipathic compound at the interfaces concerned (see Chap. 13 for details), criteria for a good detergent must be based on the cleaning test, for which the radiotracer method is best suited [81-83,87-90,121-128].

Detergent molecules at a specific concentration, called the critical micelle concentration (cmc), associate to form its micelles surrounded by molecularly dispersed detergent solution. Because of their hydrophilicity at the micelle/solution interface and oleophilicity in the micellar interior, foreign oil dissolves stably in the micellar phase (solubilization) [129]. In addition, oil droplets existing in a sufficiently concentrated solution of detergent are stabilized by adsorption of molecules at the drop/ solution interface (emulsification) [130]. The stabilization contributes to maintaining emulsion of the greasy material in the solution, once it is detached from the solid surface, and thus prevents its redeposition on the solid substrate [131]. The same is true for the suspension of particulate soil such as carbon black or clay. Using artificial particulate comprised of neutron-irradiated kaolin, ^{14}C- and ^{3}H-labeled oils, Gordon et al. [126, 127] reported that the considerable fraction of the clay particles remaining at fabric surface is contributed by soil redeposition rather than its retention.

Perhaps, Ashcraft [87] was the first investigator who clearly showed a parallelism (see Fig. 2) between radiologically determined percent removal of particulate soil from solid surfaces and optically determined percent reflectance which had been conventionally adopted as a criterion in the cleaning test [132].

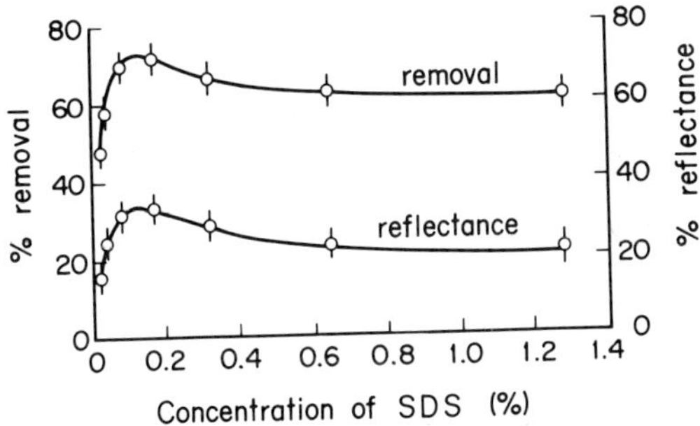

FIG. 2. Effects of concentration of sodium dodecylsulfate (SDS) upon percent removal of [^{14}C] carbon black from cotton swatch and percent reflectance of the specimen after washing [87]. (Reproduced by permission of the American Society for Testing and Materials from STP 215, copyright 1958.)

The maximum detergency at a concentration beyond the cmc was later corrected by many investigators [81,82,124], who reported onset of detergency at or near cmc, followed by gradual saturation or continuous increase of soil removal in the micellar region. This aspect is revealed in Fig. 3, in which fairly good parallelisms are seen in the relationship between the percent removal and the amount of a water-insoluble dye solubilized in the solution of nonionic detergent. Regarding the type of detergent, the following order of increasing power of detergency is generally suggested by many investigators [81-83,124,125]

cationic < anionic < nonionic

although details vary with many factors such as type of soiling material, its dispersity at the interface, soil/detergent ratio, solution pH, nature of solid substrate, temperature, and washing time [81-83,87-90,121-128]. Table 3 shows a typical example.

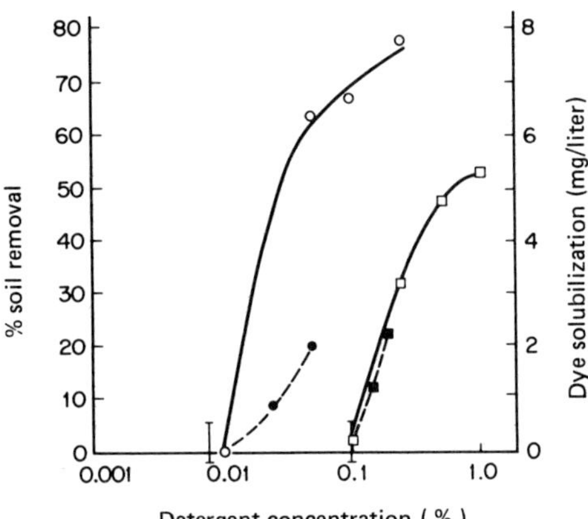

FIG. 3. Effects of concentration of decaethyleneglycol decyl ether (squares) and of decaethyleneglycol tridecyl ether (circles) on percent removal of [^{14}C] triolein from frosted glass (open symbols) and on solubilization of 1-o-tolylazo-2-naphthol in their solutions (filled). The vertical bars represent the respective cmc values. (Reproduced by permission from Ref. 82.)

TABLE 3

Percent Removal of Individual Components Contained in a 7-Component Artificial Soil* from Cotton Swatch by Washing with Detergent Solution**

Surfactant	Hydrocarbon oil	Tri-stearin	Stearic acid	Oleic acid	Octa-decanol	Cholesterol
Dodecaethyleneglycol ether of straight-chained C_{12-15} alcohol	51	67	70	60	64	47
Tridecaethyleneglycol ether of branch-chained C_{11-15} alcohol	50	63	68	59	62	46
Alkyl (C_{14-15}) sulfate	41	57	73	63	70	49
Alkylaryl sulfonate	37	43	67	52	52	43

*Composed of hydrocarbon oil (25 wt%), tristearin (10%), arachis oil (20%), stearic acid (15%), oleic acid (15%), octadecanol (8%), and cholesterol (7%); labeled with ^{3}H for either one of the first two components and ^{14}C for one of the last four.

**Reproduced by permission from Ref. 83.

The importance of cmc is involved also in the decontamination of radioactive metal ions from solid surfaces. In Fig. 4 the rise in percent removal approximating concentration curves near cmc is explicable either by adsorption of radioactive metal cations at the aqueous interface of negatively charged micelles [135] or by solubilization of the metal soap which has been formed in the nonmicellar phase [79,134]. Whatever it is, cationic contamination can be removed by treatment with the micellar solution of an anionic detergent. It is thus reasonable that neither $^{60}Co^{2+}$ nor $^{32}PO_4^{3-}$ can be removed from a cotton

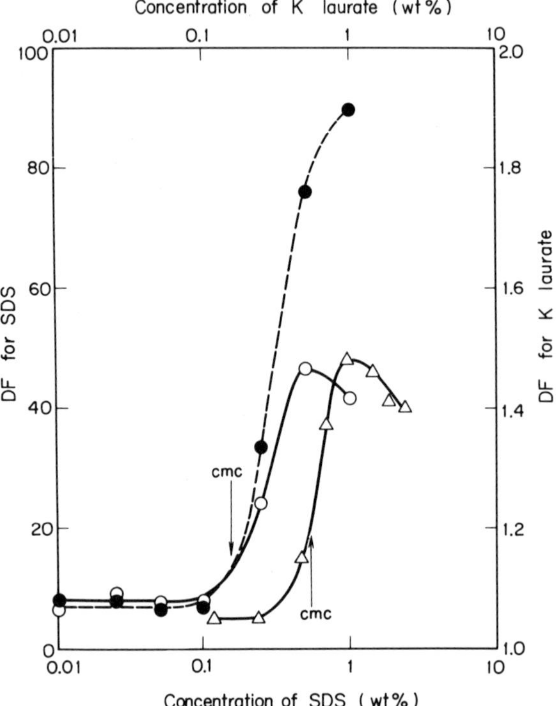

FIG. 4. Removal of $^{60}Co^{2+}$ (open circles) and $^{45}Ca^{2+}$ (filled circles) with SDS from cotton cloth [133] and of $^{91}Y^{3+}$ with potassium laurate from aluminum plate (triangles) [134]. (Reproduced by permission from the copyright holders.)

surface by washing with micellar solutions of nonionic detergents such as Tween 80 and polyethyleneglycol nonylphenyl ether [136]. Interestingly, dodecylbenzyldimethylammonium chloride exhibits a significant RD effect for removal of $^{45}Ca^{2+}$ and $^{60}Co^{2+}$ [133] rather than $^{32}PO_4{}^{3-}$ [136] from cotton surfaces. Takada and his collaborators [136] assumed that the mechanisms would involve complex formation between the metalic radionuclides and the NH-containing surfactant. Establishment of such an aspect will require further investigation.

In summary then, RD of radioactive oil, fat, and other organic compounds is best achieved with nonionic detergent, whereas RC of radioactive metal ions should be removed with anionic detergent. In any case, it is necessary that the detergent concentration be higher than its cmc. Addition of builders such as NaCl and Na_2SO_4 reinforces the RD effect [97,138] as well as in the case of the conventional detergency [132].

c. Aqueous sequestrants (chelating agents). Ionic contamination due to heavy metal radionuclides is effectively removed by treatment of the soiled surfaces with an aqueous solution of such a sequestrant as ethylenediaminetetracetic acid (EDTA), citrate [96,97,139], polyphosphate [98,139], or nitrilotriacetate [140]. The mechanisms in general involve the following steps:

$$\text{surface}\cdot\text{metal} \rightleftarrows \text{surface} + (\text{metal cation}) \tag{12}$$

$$(\text{metal cation}) + (\text{sequestrant anion}) \rightleftarrows \text{metal}\cdot\text{sequestrant} \tag{13}$$

RD conditions should thus be chosen for the rightward shifting of the equilibrium reactions (12) and (13), though these are sometimes contradictory to each other [79]. An increase in the solution pH, for instance, will cause an increased number of sequestrant ions available for the complex formation by Eq. (13), but lead to the lack of hydrogen ions for substitution for the metal ion to be detached from the solid surface. The rather complicated feature of pH dependence of decontamination effect [139] can thus result. Needless to say, the stability constant

[110,111] governs Eq. (13) and accordingly the overall RD effect, if Eq. (12) is negligibly contributed [141]. If, however, the rate-determining step is involved in reaction (12), inefficacy of RD of some heavy metals likely occurs in some cases.

Some reports [96,142-144] deal with some multiplicative effects of the simultaneous use of sequestrant and anionic detergent. Practically, a combined use of these decontaminants is recommended in laboratory manuals for radioactive decontamination (see Section IV C 1).

d. Decontamination paste. The RD of solid surfaces is achieved by scrubbing them with aqueous paste or gel of such reactive powders as titanium oxide [99], bentonite, kaolin [100], or sodium alginate [145,146]. The mechanisms are believed to involve the steps similar to Eqs. (12) and (13), with substitution of the powder acting as an adsorbent for the sequestrant in these equations [79]. In addition, the RD effect is significantly augmented by abrasive action, as well as by corrosion due to an excessive acidity which is necessitated for rightward shifting of an Eq. (12)-like expression. The adsorbent therefore would have a wide coverage of pH range for effective decontamination without serious damage to solid surfaces. In the case of TiO_2 powder, for example, anatase type in comparison with rutile is characterized not only by a high RD effect but also a wide coverage of optimum pH region [147].

e. Isotopic solutions. A radionuclide *R at solid surface exchanges with an isotopic species R in its solution. The equilibrium constant given by

$$\frac{(\text{specific activity})_{\text{solution}}}{(\text{specific activity})_{\text{surface}}}$$

(see Chap. 10) or by the ratio of the kinetic rate constants (Chap. 9) is very close to unity or within the order of magnitude of 10^{-1}-10^{1}, being much smaller than

$$\frac{(\text{number of R atoms})_{\text{solution}}}{(\text{number of *R atoms})_{\text{surface}}}$$

For this reason, *R is replaced eventually by R from the surface with an exchange rate governed largely by the concentration of R (see Chaps. 10 and 13 for details).

The isotopic exchange method is applicable, in principle, to the removal of any kind of RC; The principle is based on the exchange equilibrium but not necessarily on the rate. Thus, ionic contamination is readily removable, as exemplified in Fig. 5 [148], whereas a similar but less distinct effect has been reported for the isotopic exchange of nonradioactive octadecylamine with [1-^{14}C]octadecylamine at steel and glass surfaces [149].

From a practical aspect, the use of an isotopic exchange method is limited to cleaning of nonpenetrable surfaces. It is applicable to penetrable surfaces only when the exchange rate is high enough to finish the entire procedure before serious damage of solid material.

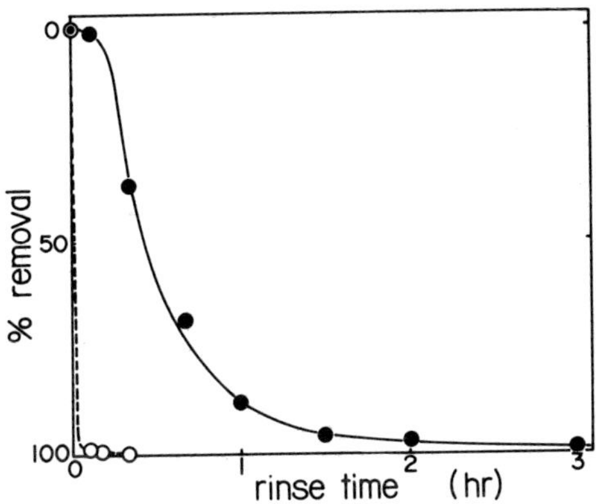

FIG. 5. Removal of $^{60}Co^{2+}$ from surface of sulfonated polystyrene by washing with distilled water (filled circles) and with nonradioactive cobaltous chloride (0.1 mol/liter) (open). (Reproduced by permission from Ref. 148.)

f. Acid, alkaline, and oxidative mixtures. Many papers deal with the use of mixtures of strong acids such as nitric, sulfuric, hydrochloric, and acetic acids for cleaning contaminated surfaces in nuclear engineering [150-152]. (Citric and tartaric acids seem to belong to the category of sequestrant.) Their action includes chemical dissolution of the soiling radionuclides and deprival of the soiled material in the surface layer. To reinforce the action, addition of such oxidative reagents as hydrogen peroxide, chromic acid, and permanganate is sometimes effective. For painted surfaces, solutions of such caustics as sodium or potassium hydroxide with or without the oxidative reagent is useful.

These decontaminants are probably the most powerful tool for RD and the most destructive to the surface. Studies on these decontaminants have thus been made for nuclear plants but not for radiotracer laboratories. For example, one of the most serious problems is how to remove RC due to the fission product (especially Nb and Ru) [153], which is seldom needed by radiotracer experimenters. These decontaminants therefore should be considered as the last resort for cleaning glass ware, painted walls, and floors in radiotracer laboratories.

3. Interior Contamination

Interior contamination -- whether mobile or immobile -- is removed either by rinsing the soiled material in an adequate solvent for a long time or by depriving its surface. The former is applicable, though less practical, to gel or jelly substances, whereas the latter is achieved by the use of either chemical reagents (Section IV B 2 f) or such removers as emery cloth and steel wool [154]. To facilitate this procedure, the use of strippable paint such as volcanized latex [155] or copolymers of polyvinyl chloride (83-85%) and vinyl acetate [156] may be effective.

C. Laboratory Manual of Radioactive Decontamination

1. Equipment and Tools

In principle, the first RD procedures should be mild, by wiping the contaminated surface with a solution of detergent -- nonionic or cationic for organic contamination and anionic for cationic. Efficient RD is achieved by ultrasonic irradiation with relatively low frequency such as 40 kHz [117], 35-40 kHz [142], 20-25 kHz [118], or 16.1 kHz [157]. The minimum power necessary for RD is reported to be 0.37 W/cm^2 for 1:10 in solid/solution volume ratio, 0.75 W/cm^2 for 1:4, and 1.5 W/cm^2 for 1:2 [118]. Simultaneous use of a sequestrant is, as stated in Section IV B 2 c, advantageous in removing RC due to heavy metal cations. For removal of such contaminants, Blythe [142], for example, recommends the following mixed decontaminant:

Surfactant (alkylaryl sulfonate)	20-22 wt%
Sodium tripolyphosphate	40 wt%
Sodium carboxymethyl cellulose	3 wt%
EDTA	1 wt%
Filler (sodium carbonate or sulfate) to	100 wt%

An isotopic exchange method (Section IV B 2 e) is especially applicable to cleaning of laboratory glass ware, which is considerably cleaned by soaking in an isotopically identical solution overnight or little longer. For details of other techniques, the reader should refer to the pertinent literature [150-154,158-160].

2. Working Area

For a large area, the use of a steam cleaner [154,159,160] could be advantageous in preventing the inhalation of radioactive dust. Any kind of decontaminant can be used for this purpose. In laboratory scale, it is sufficient to employ the methods described in Section IV C 1. A small area can be cleaned also by application of the decontamination paste (Section IV B 2 d).

3. Protective Clothes

Decontamination of protective clothes would be performed routinely by the use of an automatic laundering machine under the conditions of a wash-load ratio (weight of washing solution/clothes weight) of at least 10:1 for preventing cross-contamination [161]. Again, the combined use of synthetic detergent and sequestrant is effective for decontamination of protective clothes soiled with radioactive heavy metals [143,144,161,162].

4. Human Skin

Experimental results both in vivo [104-107,163] and in vitro [103, 113,114,164] indicate that ionic contaminants quickly penetrate through animal and human skin layers. Many papers [150,152,165-167] prescribe the standard methods for complete removal of RC at skin surfaces. Additional precautions should be taken for quick completion of the procedures, minimum spreading of the contaminant, and avoidance of heavy stimulation in the local area. Any of the prescribed RD standards always start with scrubbing in running water, followed by brushing with the solution of a light-duty detergent. In case this is still insufficient, the detergent should be replaced with a heavy-duty one [150,152]. All these procedures must be completed within 30 min after the RC. If none of them is effective, the paste method (Section IV B 2 d) may be useful in an additional 10-min period for removal of persistent contaminant [168]. Precisely classified methods seem to be rather impractical, apart from their physiological importance. In fact, it is most important to complete the whole procedure as quickly as possible.

V. WASTE MANAGEMENT IN CHEMICAL AND BIOLOGICAL LABORATORIES

In principle, radioactive wastes should be treated and disposed under rules and regulations in accordance with the recommendations of ICRP [6]. Separation of radioactive wastes from nonradioactive ones and classification into several categories are to be done first. However, the recommendations by IAEA [169] and USASI [170] concern gross radioactivity (in Ci) and/or MPC (in Ci/ml or R/hr) and lack a more important consideration, readiness of the subsequent procedures to concentrate the wastes for containment or to dilute them for safe disposal [171]. The readiness depends on the type and chemical form of the radionuclide (see Chap. 2) and its specific activity. Fortunately, most radiotracer experimenters use a relatively small amount of gross radioactivity and, therefore, they are responsible merely for safe condensation of the wastes for temporary storage and subsequent transportation. In the following sections, discussion will be made of how to achieve the condensation procedures, of the problems involved in them, and of possible solutions of these points.

A. Condensation of Liquid Wastes

1. Evaporation

Heating is a simple but useful procedure for condensation of nonvolatile radioactive wastes dissolved or dispersed in nonradioactive solvent(s). Bubbling can be avoided by addition of a small amount (0.001-0.01%) of silicone oil [172,173] or by IR irradiation below boiling point of the liquid at pH $\lesssim$ 9 [174]. Difficulties may arise when the radioactivity is contributed by volatile constituents such as ruthenium tetroxide, iodine, water, and some organic compounds, or when the liquid contains explosive material. For a thermally unstable solid, the solution can be concentrated at room temperature by evaporation from the surface of the filter paper in which it has been soaked. This is especially effective for a mixture of radioactive organic compound and a small volume of nonradioactive volatile solvent.

2. Ion Exchange

Application of this method is limited to simple ionic contaminant in aqueous waste with a total solid content of less than 1 g/liter [175,176]. Usually, a monobed of anion or cation exchanger or a combination of such counterbeds is used for condensation of ionic radionuclide. Fairly satisfactory results have been reported with the use of a mixed bed (column of cation and anion exchangers) [177,178], which, however, may not be convenient for reuse. Dowex A-1 containing iminodiacetate radicals has been reported to be effective in selective chelation for collection of radioactive cations [179]. In any case, much care must be taken in pH control for preventing the formation of nonexchangeable species such as hydroxides.

3. Chemical Reaction

Precipitation by addition of an adequate reagent is one of the surest ways to convert radioactive solute into solid form. For example, ${}^{35}SO_4{}^{2-}$ in aqueous solution is best collected as $Ba^{35}SO_4$. Solubility is thus the primary factor that determines the effectiveness of the collection. Additional importance is involved in the readiness of the multistage collection comprised by successive cycles of precipitation, filtration and isotopic dilution of the supernatant. If a binary mixture of a radioactive solute (speccific activity, σ cpm/g) and mg nonradioactive solvent is reacted to produce a precipitate and a saturated solution (solubility, s g solute/g solvent) to be diluted isotopically with A g nonradioactive solute, the radioactivity R_n of the supernatant after nth stage is given roughly by

$$R_n = \frac{\sigma(sm)^{n+1}}{(sm + A)^n} \tag{14}$$

or

$$\frac{R_n}{R} = \left(\frac{sm}{sm + A}\right)^n = \left(\frac{sm}{A}\right)^n \tag{15}$$

because $sm \ll A$. For example, radioactivity of $^{35}SO_4^{2-}$ in 100 g water is decreased to $R_n/R_0 = 2.6 \times 10^{-5}$, 6.6×10^{-10}, 1.7×10^{-14}, etc. for $n = 1, 2, 3$, etc., respectively, by alternating addition of Ba^{2+} and SO_4^{2-} on the basis of $s = 1.1 \times 10^{-5}$ mol/liter at 25°C [180] and $A = 10$ g.

Sometimes one can utilize coprecipitation for removal of ionic wastes in aqueous solutions. Because of reasons similar to the above example, multistage coprecipitation gives a considerably large effect [181,182]. The host precipitates useful for removal of heavy metal ions include aluminum and ferric hydroxides and calcium phosphate [183-185]. It has been reported that alkali metal ions coprecipitate with ferrocyanides of transition metals [186,187]. In fact, almost 100% of $^{137}Cs^+$ was removed by coprecipitation with $Ni_2[Fe(CN)_6]$ [187], as illustrated in Fig. 6. The chemical sludges resulting from the coprecipitation are usually swollen by a great extent of hydration. It is desirable, therefore, to reduce sludge volumes to save space, time, and effort in transportation.

B. Packaging of Wastes

For safe transportation of radioactive wastes from a laboratory to a central facility for further treatment to dispose, much care has to be paid in accordance with the national regulations based on the suggestions of IAEA [188]. The containers suited for liquid wastes are polyethylene bottles for aqueous waste and stainless steel bottles for wastes containing organic solvent. Solid wastes which can be contained in a cardboard box, plastic bag, or steel

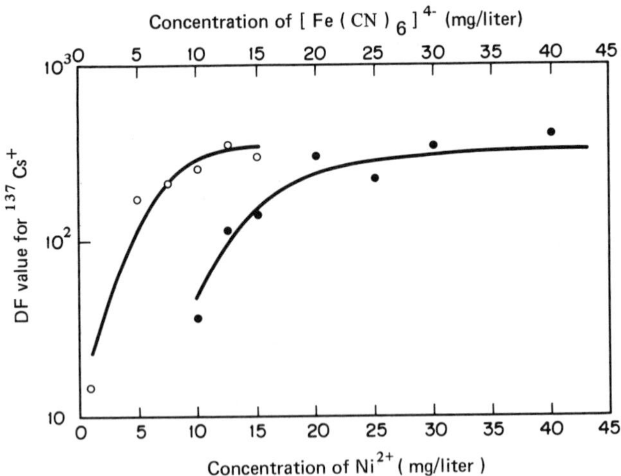

FIG. 6. DF for aqueous $^{137}Cs^+$ as functions of $[Ni^{2+}]$ at a constant value of $[Fe(CN)_6]^{4-}$ = 20 mg/liter (open circles) and of $[Fe(CN)_6]^{4-}$ at a constant value of $[Ni^{2+}]$ = 7.5 mg/liter (filled circles). (Reproduced from Ref. 187 by permission.)

drum should be segregated according to whether they are combustible and/or whether they are compressible [175,189]. For a higher level waste, the container is required to shield with concrete, steel, or lead sheet to reduce the external radiation. The sludge obtained by evaporation or chemical treatment should be dried by addition of vermiculite or other absorbent [175]. The contaminated animal carcass dipped in aqueous formaldehyde is contained in a wide-mouthed bottle, which is placed in the can with vermiculite or another absorbent for filling the gap [175].

A tag attached to each of the waste containers should state the information including the radionuclide(s) present in the waste, approximate quantity (mCi) or concentration (μCi/ml), external radiation level (mR/hr), physically and chemically dangerous properties of the waste, packing data, and other remarks which may be helpful to the subsequent waste treatment.

C. Disposal of Wastes

For sewage disposal of a small quantity of low specific activity sample from laboratory, the recommendation of ICRP [190] can be a good guide to procedures for liquid, solid and gaseous wastes. In the present subsection, sewage disposal of liquid waste containing short-lived radionuclide such as ^{131}I and ^{32}P and disposal of the waste containing ^{14}C and ^{3}H are described.

1. Sewage Disposal

The sewage disposal of short-lived radionuclides can be done if they are soluble in water and their amounts are so small that the final concentrations are far below the (MPC) values. The flushing time should be long enough to fulfill the condition required by the national regulations. A constant-drip discharge bottle (see Fig. 7) may be useful for maintaining relatively uniform discharge of the radioactive liquid over a period of several hours [191].

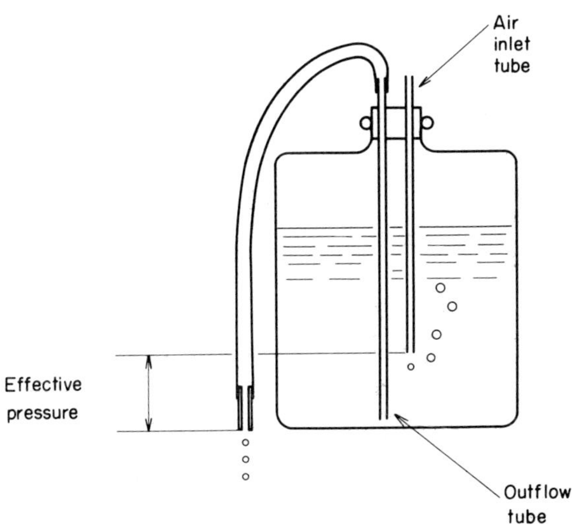

FIG. 7. Gallon bottle setup for constant pressure drip discharge. (Reproduced from Ref. 191 by permission.)

2. Disposal of ^{14}C- and ^{3}H-Labeled Compounds

Probably because of the soft nature of the β-rays, ^{14}C and ^{3}H belong to the lower radiotoxity groups in the classification by ICRP (see Table 1). However, great care must be taken upon treatment of wastes containing these isotopes, because their labeled compounds are more or less volatile and the air-ionization/energy ratios for their β-rays are by far larger than those due to the other isotopes [192]. The degree of "hazard" associated with a ^{14}C-labeled compound is determined primarily by its vapor pressure and specific activity. For example, palmitic acid which is usually considered as nonvolatile gives 9.8×10^{-6} mmHg for its vapor pressure at 46.6°C [193], the value corresponding to only 4.9×10^{-11} mol/liter for the gaseous density but 1088 dpm/liter for its radioactive concentration if its specific activity is 10 mCi/mmol.

^{14}C-labeled compounds should not be mixed with strong acid, which may cause generation of gaseous $^{14}CO_2$. The National Bureau of Standards [194] recommends disposal of ^{14}C-labeled compounds after isotopic dilution with the same nonradioactive compounds so as to give a specific activity of 1 μCi/10 g stable carbon.

The situation is much the same, or even more serious, in the case of tritiated compounds. Organic substance labeled at the nonlabile C-^{3}H position can be disposed in the manner similar to ^{14}C-labeled material. Labile ^{3}H atom, however, must be converted into the C-^{3}H form in order to avoid possible exchange with atmospheric water; this is not very easy though (see Chap. 5). Such a compound, as it is, may be necessarily contained in a perfectly sealed box or bottle. An alternative way is to convert tritiated compound into $^{3}H_2O$ and discharge by dilution with water. Gaseous tritium in an amount of laboratory scale can thus be catalytically oxidized [195,196] to isotopically mix with tap water for sewage disposal. The discharge bottle like that in Fig. 7 greatly helps this procedure.

REFERENCES

1. R. A. Faires and B. H. Parks, Radioisotope Laboratory Techniques, 3rd ed., Butterworths, London, 1973.
2. Safe Handling of Radioisotopes, IAEA Safety Series No. 1, IAEA, Vienna, 1958.
3. ICRP Publication No. 12, Pergamon Press, London, 1968, p. 3.
4. ICRP Publication No. 9, Pergamon Press, Oxford, 1966.
5. ICRP Publication No. 22, Pergamon Press, Oxford, 1973.
6. ICRP Publication No. 2, Pergamon Press, Oxford, 1959.
7. ICRU Report 20, Washington, D. C., 1971, p. 10.
8. F. W. Spiers, Radiation units and theory of ionization dosimetry, in Radiation Dosimetry (G. J. Hine and G. L. Brownell, eds.), Academic Press, New York, 1956, p. 1.
9. E. B. Wagner and G. S. Hurst, Health Phys., 5:20 (1961).
10. F. M. Miller, UCRL-10986, University of California, 1963.
11. W. J. Ramm, Scintillation detectors, in Radiation Dosimetry (G. J. Hine and G. L. Brownell, eds.), Academic Press, New York, 1956, p. 246.
12. S. Moriuchi and I. Miyanaga, Health Phys., 12:541 (1966).
13. S. Moriuchi and I. Miyanaga, Health Phys., 21:1481 (1966).
14. I. Miyanaga and S. Moriuchi, NSJ-Tr 137, Japan Atomic Energy Research Institute, 1968.
15. G. S. Hurst and R. H. Ritchie, Health Phys., 8:117 (1962).
16. D. Herman, Nukleonik, 8:320 (1966).
17. W. G. Cross, AECL-2793, Chalk River, Canada, 1967.
18. M. J. Berger, Health Phys., 26:1 (1974).
19. H. J. Dunster, Health Phys., 8:4 (1962).
20. I. S. Jones and S. F. Pond, Some experiments to determine the resuspension factor of plutonium from various surfaces, in Surface Contamination (B. R. Fish, ed.), Pergamon Press, London, 1967, p. 83.
21. R. T. Brunskill, The relationship between surface and airborne contamination, in Surface Contamination (B. R. Fish ed.), Pergamon Press, London, 1967, p. 93.
22. H. J. Dunster, The concept of derived working limits for surface contamination, in Surface Contamination (B. R. Fish ed.), Pergamon Press, London, 1967, p. 139.

23. H. Bratz and M. Eisenbud, The establishment of limits for radioactive surface contamination, in Surface Contamination (B. R. Fish ed.), Pergamon Press, London, 1967, p. 163.

24. Safe Handling of Radioisotopes, IAEA Safety Series No. 1, IAEA, Vienna, 1958, p. 95.

25. J. D. Eakins and W. P. Hutchinson, AERE-R 5988, Atomic Energy Research Establishment, Harwell, 1969.

26. G. W. Royster, Jr., and B. R. Fish, Techniques for assessing "removable" surface contamination, in Surface Contamination (B. R. Fish ed.), Pergamon Press, London, 1967, p. 201.

27. Radiation Protection Procedures, IAEA Safety Series No. 38, IAEA, Vienna, 1973.

28. NBS Handbook No. 92, U. S. Department of Commerce, Washington, D.C., 1964, p. 55.

29. R. J. Sherwood, Am. Ind. Hyg., 27:98 (1966).

30. R. J. Sherwood and D. M. S. Greenhalgh, Ann. Occup. Hyg., 2:127 (1960).

31. W. A. Langmead and D. T. O'Connor, in Proc. 1st Int. Congr. Radiation Protection (S. W. Snyder ed.), Vol. 2, Pergamon Press, London, 1968, p. 1167.

32. ICRP Task Group on Lung Dynamics, Health Phys., 12:173 (1966).

33. D. C. Stevens and J. Stephenson, Aerosp. Sci., 3:15 (1972).

34. Y. Yoshida, K. Kitano, M. Murata, and S. Moriyasu, Proc. 1st Int. Congr. Radiation Protection (W. S. Snyder ed.), Vol. 2, Pergamon Press, London, 1968, p. 1561.

35. J. R. Kennally, Health Phys., 16:813 (1969).

36. S. Fukuda and Y. Yoshida, in Peaceful Uses of Atomic Energy, IAEA, Vienna, 1972, p. 259.

37. W. D. Chiswell and G. H. C. Dancer, Health Phys., 17:331 (1969).

38. R. V. Osborne, AECL-2699, Chalk River, Canada, 1967.

39. W. J. Megaw and F. G. May, J. Nucl. Eng. Parts A/B, 16:427 (1962).

40. F. D. McCormack, BNWL-1145, Battelle Memorial Institute, Pacific Northwest Laboratory, 1969.

41. K. Becker, Photographic Film Dosimetry, Focal Press, London, 1966.

42. E. Piesch, in Radiation Dose Measurements, Their Purpose, Interpretation and Required Accuracy in Radiological Protection, ENEA, Stockholm, 1967, p. 151.

43. R. Maushart and E. Piesch, in Proc. 1st Int. Congr. Radiation Protection, Vol. 2, Pergamon Press, London, 1968, p. 803.

44. F. H. Attix, Health Phys., 22:287 (1972).

45. L. F. Kocher, R. L. Kathren and G. W. R. Enders, Health Phys., 25:567 (1973).

46. K. Becker, Health Phys., 14:17 (1968).

47. J. F. Fowler and F. H. Attix, Solid state integrating dosimetry, in Radiation Dosimetry (F. H. Attix and W. C. Roesh, eds.), Vol. 2, Academic Press, New York, 1966, p. 241.

48. G. Bjärngard, Rev. Sci. Instr., 33:1129 (1962).

49. T. Hashizume et al., Adv. Phys. Biol. Radiation Detectors, IAEA, Vienna, 1971, p. 91.

50. J. F. Fowler and F. H. Attix, Solid state integrating dosimetry, in Radiation Dosimetry (F. H. Attix and W. C. Roesh, eds.), Vol. 2, Academic Press, New York, 1966, p. 278.

51. J. J. Thompson, Health Phys., 22:399 (1972).

52. G. Scarpa, G. Benincasa, and L. Ceravolo, in Proc. 3rd Int. Conf. on Luminescent Dosimetry, Risö Report No. 249, 1971, p. 427.

53. T. Yamashita, N. Nada, H. Onishi and S. Kitamura, Health Phys., 21:295 (1971).

54. R. Yokota, S. Nakajima and H. Sakai, Health Phys., 5:219 (1961).

55. I. Miyanaga and H. Yamamoto, Health Phys., 9:965 (1963).

56. I. A. Bernstein, B. E. Bjärngard and D. Jones, Health Phys., 14:33 (1968).

57. B. E. Bjärngard and D. Jones, in Proc. 1st Int. Congr. Radiation Protection, Vol. 1, Pergamon Press, London, 1968, p. 473.

58. J. Kramer, in Proc. 2nd Int. Conf. on Luminescent Dosimetry, USAEC, CONF-680920, 1968, p. 180.

59. K. Becker, IAEA Atomic Energy Review, 8:173 (1970).

60. J. Kramer, in Proc. 3rd Int. Conf. on Luminescent Dosimetry, Risö Report No. 249, 1971, p. 622.

61. S. Jackson, Health Phys., 12:843 (1966).

62. ICRP Publication No. 10, Pergamon Press, Oxford, 1968.

63. J. Vennart, G. Maycock, B. E. Godfrey, and B. L. Davies, Measurement of radium in radium luminizers, in Assessment of Radioactivity in Man, Vol. 2, IAEA, Vienna, 1964.

64. Whole-Body Counting, IAEA Series, IAEA, Vienna, 1962.

65. Directory of Whole-Body Radioactivity Monitor, 1970 ed., IAEA, Vienna, 1970.

66. H. E. Palmer and W. C. Roesh, Health Phys., 11:1213 (1965).

67. J. N. Brady and F. Swanberg, Health Phys., 11:1221 (1965).

68. G. R. Laurer and M. Eisenbud, In vivo measurements of nuclides emitting soft penetrating radiations, in Diagnosis and Treatment of Deposited Radionuclides (H. A. Kornberg and W. D. Norwood, eds.), Excerpta Medica, New York, 1968, p. 189.

69. K. H. Linden and R. C. McCall, Low-Energy-Photon Detectors for Whole-Body Counting, IAEA, Vienna, 1962, p. 145.

70. O. R. J. Budniz, Health Phys., 26:165 (1974).

71. Organic Scintillators and Liquid Scintillation Counting (D. L. Horrocks and C. T. Peng, eds.), Academic Press, New York, 1971.

72. D. L. Horrocks, in Proc. 1971 Tritium Symp. (A. A. Moghissi and M. W. Carter eds.), Messinger Graphics, Phoenix, 1973, p. 143.

73. A. A. Moghissi, H. L. Kelley, J. E. Regner and M. W. Carter, Int. J. Appl. Radiat. Isotopes, 20:145 (1969).

74. R. Lindeman and A. A. Moghissi, Int. J. Appl. Radiat. Isotopes, 21:319 (1970).

75. A. A. Moghissi, Adv. Chem. Ser., 93:419 (1970).

76. J. E. Eberhardt, R. D. Ryan, and A. J. Tavendale, Nucl. Instr. Methods, 94:463 (1971).

77. R. O. Bell and F. V. Wald, IEEE Trans. Nucl. Sci. NS-19, No. 3, 334 (1972).

78. H. L. Malm, IEEE Trans. Nucl. Sci. NS-19, No. 3, 263 (1972).

79. M. Muramatsu and T. Sasaki, Isotopes Radiation, 2:451 (1959).

80. A Basic Toxicity Classification of Radionuclides, IAEA Tech. Rep. Ser. 15, IAEA, Vienna, 1963.

81. K. Takada, Y. Wadachi, and M. Muramatsu, Radioisotopes, 19: 253 (1970).

82. M. E. Ginn and J. G. Harris, J. Am. Oil Chemists Assoc., 38:605 (1961).

83. B. E. Gordon, J. Roddewig and W. T. Shebs, J. Am. Oil Chemists Assoc., 44:289 (1967).

84. A. B. Kuper, Surface Sci., 13:172 (1969).

85. W. Kern, RCA Rev., June, 207, 234 (1970).

86. W. Kern, Solid State Tech., Feb., 39 (1972).

87. E. A. Ashcraft, ASTM Spec. Tech. Publ., 215:30 (1958).

88. V. Mahadevan, J. Am. Oil Chemists Assoc., 40:372 (1963).

89. J. W. Hensley and C. G. Inks, ASTM Spec. Tech. Publ., 268:27 (1963).

90. J. W. Hensley, J. Am. Oil Chemists Assoc., 42:993 (1963).

91. M. Muramatsu, Radioactive tracers in surface and colloid science, in Surface and Colloid Science, Vol. 6 (E. Matijevic ed.), Wiley-Interscience, New York, 1973, p. 112.

92. M. Muramatsu and T. Sasaki, Radioisotopes, 7:42 (1958).

93. Y. Wadachi, JAERI-1165, Japan Atomic Energy Research Institute, 1968.

94. M. Mongrieff-Yeates and H. J. White, Jr., The interaction of cellulose with simple salt solutions and with dyebaths, in Chemical Dynamics, (J. O. Hirschfelder and D. Henderson, eds.), Wiley-Interscience, New York, 1971, p. 685.

95. p. 152 of Ref. 91.

96. F. D. Snell, G. Segura, S. Stigman, and C. T. Snell, Soap Sanitary Chemicals, 1:2 (1953, Oct.); Chem. Ind., 1270 (1953).

97. A. P. Talboys and E. C. Spratt, NYO-4990, USAEC, 1954.

98. Y. Tajima and Y. Wadachi, J. Chem. Soc. Japan, Pure Chem. Sec., 81:891 (1960).

99. H. A. Kunkel, Strahlentherapie Sonderbaende, 95:326 (1954).

100. H. Unger, Isotopenpraxis, 7:189 (1971).

101. J. W. Shepard and J. P. Ryan, J. Phys. Chem., 63:1729 (1959).

102. J. P. Ryan, R. J. Kunz and J. W. Shepard, J. Phys. Chem., 64:525 (1960).

103. Y. Wadachi, S. Tashiro, Y. Tajima and T. Hiyama, J. At. Energy Soc. Japan, 5:938 (1963).

104. Y. Wadachi, S. Tashiro, Y. Tajima, and T. Hiyama, J. At. Energy Soc. Japan, 5:994 (1963).

105. T. Kanayasu and Y. Yoshizawa, Proc. 6th Japan. Conf. Radioisotopes, Tokyo, 1964, A/H-9.

106. J. E. Wahlberg, Acta Dermatol. Venereol., 45:397 (1965).

107. M. A. Khodyreva, Med. Radiol., 10:42 (1965).

108. M. Muramatsu and T. Sasaki, Radioisotopes, 7:36 (1958).

109. J. N. Butler, Ionic Equilibria, A Mathematical Approach,

Addison-Wesley, Reading, Mass., 1964.

110. J. Bjerrum, G. Schwarzenbach, and L. G. Sillen, Stability Constants, Spec. Publ. No. 6, Chemical Society, London, 1956.

111. L. G. Sillen and A. E. Martell, Stability Constants of Metal-Ion Complexes, Spec. Publ. No. 17, Chemical Society, London, 1964.

112. H. S. Harned and B. B. Owen, Physical Chemistry of Electrolytic Solutions, 3rd ed., Reinhold, New York, 1957, p. 546.

113. S. Tashiro, Y. Wadachi, and M. Muramatsu, Radioisotopes, 15:224 (1966).

114. S. Tashiro, Y. Wadachi, and M. Muramatsu, J. Nucl. Sci. Technol., 5:160 (1968).

115. S. Tashiro, J. Nucl. Sci. Technol., 8:513 (1971).

116. S. Tashiro, J. Nucl. Sci. Technol., 9:344 (1972).

117. R. L. Rod, Nucleonics, 16:104 (1958).

118. T. F. D'Muhala, Ultrasonic cleaning, in Decontamination of Nuclear Reactors and Equipment (J. A. Ayres, ed.), Ronald Press, New York, 1970, pp. 400-438.

119. B. Carlin, Ultrasonics, 2nd ed., McGraw-Hill, New York, 1960, p. 281.

120. T. Sasaki, M. Muramatsu, H. Hotta, and Y. Wadachi, Radioisotopes, 7:47 (1958).

121. J. W. Hensley, M. G. Kramer, R. D. Ring, and H. R. Suter, J. Am. Oil Chemists Assoc., 32:138 (1955).

122. R. E. Wagg and C. J. Britt, J. Textile Inst., 53:T 203 (1962).

123. M. C. Bourne and W. G. Jennings, J. Am. Oil Chemists Assoc., 40:517, 523 (1963).

124. B. A. Scott, J. Appl. Chem., 13:133 (1963).

125. T. Fort, Jr., H. R. Billica, and T. H. Grindstaff, J. Am. Oil Chemists Assoc., 45:354 (1968).

126. B. E. Gordon and E. L. Bastin, J. Am. Oil Chemists Assoc., 45:754 (1968).

127. B. E. Gordon and W. T. Shebs, J. Am. Oil Chemists Assoc., 46:537 (1969).

128. J. C. Illman, B. M. Finger, W. T. Shebs, and T. B. Albin, J. Am. Oil Chemists Assoc., 47:379 (1970).

129. M. E. L. McBain and E. Hutchinson, Solubilization and Related Phenomena, Academic Press, New York, 1955.

130. P. Becher, Emulsion, 2nd ed., Reinhold, New York, 1965.

131. A. W. Adamson, Physical Chemistry of Surfaces, 2nd ed., Wiley-Interscience, New York, 1967, p. 495.

132. K. Durham, Surface Activity and Detergency, McMillan, New York, 1961, p. 216.

133. Y. Wadachi, K. Takada, Y. Yamaoka, and S. Noguchi, J. Nucl. Sci. Technol., 2:104 (1965).

134. R. C. Chandler and W. E. Shelberg, J. Colloid Sci., 10:393 (1955).

135. P. C. Tompkins, O. B. Bizzel and C. D. Watson, Nucleonics, 7:42 (1950).

136. K. Takada, Y. Wadachi, and Y. Yamaoka, J. At. Energy Soc. Japan, 6:575 (1964).

137. Y. Wadachi, S. Iwaya, and C. Machida, Unpublished.

138. K. Takada, Y. Wadachi, Y. Yamaoka, S. Noguchi, and M. Muramatsu, Radioisotopes, 14:487 (1965).

139. H. Hotta, Y. Wadachi and H. Fukuda, J. Chem. Soc. Japan, Pure Chem. Sec., 80:37 (1959).

140. S. Chaberek and A. E. Martell, Organic Sequestering Agents, Wiley, New York, 1959, p. 401.

141. F. Basolo and R. G. Pearson, Mechanisms of Inorganic Reactions, Wiley, New York, 1958, pp. 105-107.

142. H. J. Blythe, AERE-R 4307, Atomic Energy Research Establishment, Harwell, 1963.

143. F. Reiff, K. Schuster, H. Spoor, and M. Stoeppler, EUR-3276·e, European Atomic Energy Community-EURATOM, 1967.

144. B. W. Ariss and C. R. Thomas, in Proc. 1st Int. Symp. Decontamination of Nuclear Installations (H. J. Blythe, A. Catherall, A. Cook, and H. Wells eds.), Cambridge at the University Press, 1967, pp. 8-10.

145. A. Cook, AHSB(S)-R-32, UKAEA, 1962.

146. Y. Tajima, Y. Wadachi, S. Tashiro, T. Hiyama, and K. Tobita, in Proc. 5th Japan. Conf. Radioisotopes, Tokyo, 1963, A/b-6.

147. H. Hotta and H. Nakamura, J. Chem. Soc. Japan, Pure Chem. Sec., 81:1 (1960).

148. T. Seimiya and T. Sasaki, Bull. Chem. Soc. Japan, 35:1567 (1962).

149. A. Block and B. B. Simms, J. Colloid Interface Sci., 25:514 (1967).

150. P. C. Tompkins, Surface contamination and decontamination, in Radiation Hygiene Handbook (H. Blatz, ed.), McGraw-Hill, New York, 1959, 18-19-20.

151. H. J. van der Westhuizen, J. W. van Averbeke, and G. L. F. E. Bruyninckx, PEL-187, Atomic Energy Board, Pretoria, Union of South Africa, 1969.

152. E. J. Vallario, Evaluation of radiation emergencies and accidents -- Selected criteria and data, IAEA Tech. Rept. Ser. 152, IAEA, Vienna, 1974.

153. A. B. Meservey, Progr. Nucl. Energy, Ser. IV, 4:377-398 (1961).

154. H. J. Blythe, Decontamination of buidings and laboratories, in Decontamination of Nuclear Reactors and Equipment (J. A. Ayres, ed.), Ronald Press, New York, 1970, pp. 670-713.

155. C. G. Venis, Health Phys., 11:1091 (1965).

156. O. A. Bernaola and A. Filevich, Health Phys., 19:685 (1970).

157. J. M. White, AECL-1427, Chalk River, Canada, 1961.

158. A. J. Hill, Jr., Specialized equipment, in Decontamination of Nuclear Reactors and Equipment (J. A. Ayres, ed.), Ronald Press, New York, 1970, pp. 376-399.

159. C. J. MacFarlane and R. J. Beal, AECL-1465, Chalk River, Canada, 1962.

160. J. Neil and A. M. Marko, AECL-1531, Chalk River, Canada 1962.

161. H. Yasunaka, Y. Ashikagaya, Y. Wadachi and S. Iwaya, Hoken Butsuri (Japan), 3:348 (1968).

162. J. L. Norwood, HW-38218 Rev., Hanford Atomic Products Operation, 1955.

163. G. H. Dancer, A. Morgan and W. P. Hutchinson, Health Phys., 11:1055 (1965).

164. J. E. Wahlberg, Acta Dermatol. Venereol., 45:415 (1965).

165. W. J. Bair and V. H. Smith, Progr. Nucl. Energy Ser. XII, 2, Part I:176-183 (1969).

166. NBS Handbook No. 48, U. S. Department of Commerce, 1951, pp. 5-7.

167. J. H. Schulte, Arch. Environ. Health, 13:96 (1966).

168. H. Yasunaka and Y. Wadachi, Hoken Butsuri (Japan), 8:25 (1973).

169. Standardization of Radioactive Waste Categories, IAEA Tech. Rept. Ser. 101, IAEA, Vienna, 1970.

170. Proposed Definition of Radioactive Waste Categories, United States of America Standards Institute (USASI), 1967.

171. F. Gera, Health Phys., 27:113 (1974).

172. Design and Operation of Evaporators for Radioactive Wastes, IAEA Tech. Rept. Ser. 87, IAEA, Vienna, 1968.

173. W. Bähr, W. Hempelmann, H. Krause and O. Nentwich, in Management of Low- and Intermediate-Level Radioactive Wastes, Proc. Symp. Aix-en-Provence, Sep. 1970, IAEA, Vienna, 1970, pp. 461-484.

174. J. Hirling and O. Pavlik, in Management of Low- and Intermediate-Level Radioactive Wastes, Proc. Symp. Aix-en-Provence, Sep. 1970, IAEA, Vienna, 1970, pp. 773-785.

175. The Management of Radioisotope Wastes Produced by Radioisotope Users, IAEA Safety Ser. 19, IAEA, Vienna, 1966.

176. Operation and Control of Ion-Exchange Processes for Treatment of Radioactive Wastes, IAEA Tech. Rept. Ser. 78, IAEA, Vienna, 1967.

177. C. P. Straub, W. J. Lacy and R. J. Morton, in Proc. Int. Conf. Peaceful Uses of Atomic Energy, Geneva, 1955, United Nations, New York, Vol. 9, 1956, pp. 24-27.

178. C. B. Amphlett, Treatment and Disposal of Radioactive Wastes, Pergamon Press, London, 1961, p. 156.

179. H. Matsuzuru and Y. Wadachi, J. Nucl. Sci. Technol., 10:551 (1973).

180. C. H. Bovington and A. L. Jones, Trans. Faraday Soc., 66:764 (1970).

181. V. L. Zolotavin, A. A. Konstantinovich, V. N. Sanatina, V. V. Pushkarev and V. S. Petrov, Soviet Radiochem., 13:167 (1971) (translated from Radiokhimiya, 13:164 (1971).

182. H. Matsuzuru, Y. Wadachi and T. Hashino, Radioisotopes, 21:333 (1972).

183. Report of the Joint Program of Studies on the Decontamination of Radioactive Wastes, ORNL-2557, Oak Ridge National Laboratory, 1959.

184. C. P. Straub, Low-Level Radioactive Wastes, Division of Technical Information, USAEC, 1964.

185. C. W. Christenson, LA-DC-10007, Los Alamos Scientific Laboratory, 1968.

186. N. van de Voorde and K. Peeters, in Management of Low- and Intermediate-Level Radioactive Wastes, Proc. Symp. Aix-en-Provence, Sep. 1970, IAEA, Vienna, pp. 669-688.

187. P. Pottier and P. Chauvet, CEA-N-1241, Centre d'Etudes Nucléaires de Fontenary-aux-Roses, 1970, pp. 90-93.

188. Regulations for the Safe Transport of Radioactive Materials, IAEA Safety Ser. 6, revised ed., IAEA, Vienna, 1973.

189. The Management of Radioactive Wastes Produced by Radioisotope Users, IAEA Safety Ser. 12, IAEA, Vienna, 1965.

190. ICRP Publication No. 5, Pergamon Press, Oxford, 1964.

191. NBS Handbook No. 49, U. S. Department of Commerce, 1951.

192. G. Friedlander and J. W. Kennedy, Nuclear and Radiochemistry, Wiley, New York, 1955, p. 196.

193. M. Davies and V. E. Malpass, J. Chem. Soc., 1048 (1961).

194. NBS Handbook No. 53, U. S. Department of Commerce, 1953, p. 1.

195. I. Heertje, G. K. Koch, and M. R. M. Geheniau, J. Labelled Compounds, 8:95 (1972).

196. M. Muramatsu, J. At. Energy Soc. Japan, 14:39 (1972).

Chapter 4

DETERMINATION OF RADIOACTIVITY IN BIOLOGICAL MATERIAL

Yutaka Kobayashi* and David V. Maudsley

Worcester Foundation for Experimental Biology
Shrewsbury, Massachusetts

I. INTRODUCTION

The two major counting techniques in use in most laboratories today are those utilizing liquid or crystal scintillation counters. This contrasts sharply with the situation twenty years ago when radiation detectors such as Geiger counters were based mainly on gas ionization techniques. The primary advantage of liquid

*New England Nuclear Corporation, Boston, Massachusetts.

scintillation counting over other procedures is that it facilitates the measurement of tritium, which is now the most widely used isotope in the biological sciences. A discussion of liquid scintillation counting, however, is inextricably linked with the problems of sample preparation and both are given emphasis in this chapter.

The increase in the use of crystal (sodium iodide) or gamma counting arises from the emergence of radioimmunoassay as the predominant technique in the quantitation of substances of biological significance. These procedures utilize proteins which specifically bind the compound of interest and commonly use radioiodinated reagents which are γ emitters. This popularity of gamma counting is likely to continue with the increasing availability of ^{125}I which has several important advantages over ^{131}I. A primary advantage of gamma counting over liquid scintillation counting is that it does not require any special sample preparation.

Other methods of detecting radioactivity of great interest to chemists and biologists are radiochromatography and autoradiography. With the increasing use of tritium, radiochromatography has fallen somewhat into decline, since self-absorption losses often render these techniques of limited use. Significant improvements in detector efficiency appear to be necessary if radiochromatography is to be sustained as a viable technique. Autoradiography, on the other hand, has made significant progress in the past five years; first, in the application of light microscopic autoradiography to diffusible substances, and second, in the resolution, sensitivity, and quantitation of autoradiograms at the electron microscope level (see also Chaps. 16 and 20).

II. GAS IONIZATION COUNTERS

When a charged particle travels through a gas, the electrostatic interaction between the particle and the gas results in ionization of the interacting molecules. Ionization produces an ion pair consisting of a heavy positive ion and a negative electron. The degree of ionization is dependent on two factors: the charge of the ionizing particle, which determines the magnitude of the interaction, and the velocity of the ionizing particle, which determines the duration of this interaction. It has been found that the ionization per unit pathlength traveled by the ionizing particle, known as the specific ionization, is roughly proportional to its charge squared and inversely to its velocity squared.

A comparison of the range of typical specific ionization values of α and β particles and γ radiation are given in Table 1. α Particles are easily detected by gas ionization because of their high mass, high charge, and slow speed, whereas γ

TABLE 1

Some Nuclear Emanations

Name	Symbol	Mass units	Charge	Number of ion pairs/cm
Alpha	α	4.002777	+2	5,000 to 80,000
Beta	β	0.000548	-1	50 to 500
Gamma	γ	hc/λ*	0	1 to 50

*h = Planck's constant, c = speed of light, λ = wavelength.

radiation is not easily detected because it has almost no mass, is not charged, and travels at relatively high velocity.

The various forms of radiation detectors based on the ionization of gas molecules are illustrated in Fig. 1. Consider a closed vessel containing an ionizable gas. The surface of the vessel is an electrical conductor. This serves as the negatively charged electrode (cathode). A positive electrode (anode), insulated from the cathode, is placed in the center. The vessel is connected in series to a device to measure current flow (ammeter) and a variable direct current power supply. A radioactive source emitting particles with sufficient energy to penetrate the walls of the vessel is placed near the vessel to initiate ionization of the gas contained therein. The ionized gas molecules can either recombine or be collected at the oppositely charged electrode. As the voltage imposed between the cathode and the anode is raised, some of the ionized gas particles are attracted to the oppositely charged electrodes causing a current to flow in the system (region 1 of Fig. 2). As the voltage is increased further, a steady rise in current flow is observed until a plateau is reached (region 2). This region represents the point where quantitative collection of the ionized particles occurs, that is, every single

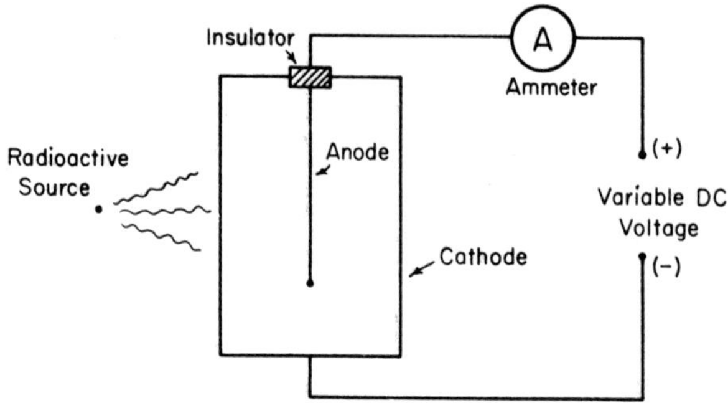

FIG. 1. A model ionization chamber connected to a current meter and a variable DC source for determining the ionization characteristics of a gas by radiation.

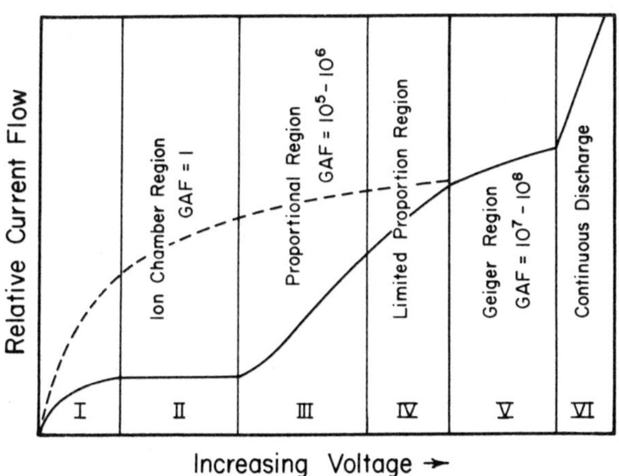

FIG. 2. Relationship of ionization to voltage derived from the model shown in Fig. 1.

ionized particle formed is collected at the electrodes. This is called the ion chamber region and ion chamber detectors can differentiate α from either β particles or γ radiation because of the large difference in their respective specific ionization factors. As the voltage is further increased, the current flow begins to rise again (region 3) because the voltage applied is sufficiently great to accelerate the ionized gas particles to a point where they can cause secondary ionization by collision with nonionized gas molecules. The number of secondary ionizations initiated by a single ion is called the gas amplification factor and is between 10^5 and 10^6 in region 3. Region 3 is known as the proportional region because the current produced is proportional to the voltage applied. Proportional counters can differentiate α from either β or γ particles. As the voltage is increased further, a region of limited proportionality is traversed (region 4) which is not used in radiation detection devices. If the voltage is raised further (region 5) the Geiger-Müller region is reached. Here, the gas amplification factor is so high (10^7 to 10^8) that α, β, or γ particles cannot be differentiated.

Counters operated in this region are very stable and relatively trouble-free compared to proportional or ion-chamber counters. The major drawback is that these instruments cannot accommodate very active samples, that is, those containing more than a few thousand counts per minute without a correction for coincident count losses. In other words, Geiger-Müller counters are "slow" counters. If the voltage is further increased (region 6) counting becomes impossible because the gas is continuously ionized.

In the Geiger-Müller region where the amplification factor is large, a radioactive particle initiates an avalanche of electrons which, because of their small mass, are completely collected at the anode wire in about 10^{-6} sec or less. At the same time, the heavier and slower moving cloud of the positive anions must be discharged at the cathode before a second pulse can be counted. When the electrons are collected, the positive ions, because of their greater mass, are still moving relatively slowly toward the cathode, and this creates a positive space charge and reduces the electric field intensity around the anode wire. This stops the avalanche ionization. In about 10^{-4} sec the positive space charge will reach the cathode surface. The speed of discharge of the positive ion cloud determines the dead time of the detector. When the positive ion cloud is neutralized at the cathode, the process results in the formation of excited neutral gas molecules which quickly return to the ground state with the release of photons. These photons will have sufficient energy to release an electron from the cathode and, in the Geiger-Müller region, a single electron is sufficient to initiate a second avalanche, and thereby a single discharge can become self-perpetuating. The occurrence of a repetitive discharge can be prevented by electrical or chemical means, the latter being the preferred method in modern Geiger-Müller counting.

III. CRYSTAL COUNTING

γ-Rays (and x-rays) are absorbed by ionization by a unique process different from that for α or β particles. Like light and other electromagnetic radiation, γ-rays possess properties not fully explained by wave theories and appear at times to consist of discontinuous packets of energy or "particles". These packets or units (quanta) of energy are called photons. γ Photons produce no direct ionization by collision along their path as they are devoid of mass, but are absorbed by one of three mechanisms: the photoelectric effect, the Compton effect, and by pair production.

In the photoelectric effect, each γ photon retains all its energy until it interacts with a planetary electron of an atom in the absorbing medium. In this process, the photon gives up all its energy to the electron which is ejected at high speed and the photon ceases to exist. The ejected electron, called a photoelectron, dissipates its energy by ionization of other atoms in the same manner as a β particle.

In the Compton effect, the incident γ photon with an energy, $h\nu$, interacts with a planetary electron of an atom in the absorbing medium. In this process, the photon gives up a part of its energy to the planetary electron which is ejected and a photon of lesser energy, $h\nu'$, is "scattered" in a manner such that both energy and momentum are conserved. The ejected electron is known as a recoil electron and is also called a Compton electron. The Compton electron, as in the photoelectric effect, dissipates its energy in a similar manner to β particles. The scattered photon is further absorbed by either the photoelectric or Compton process.

In pair production, some of the incident photon energy is converted to mass according to Einstein's equation (1):

$$E = mc^2 \tag{1}$$

where E is the energy in ergs, m is the mass in grams and c is the velocity of light in cm/sec. The incident γ photon is annihilated in an unknown manner in the nuclear field of an atom of the absorbing medium and two particles, an electron and a positron, are produced. It can be calculated that a minimum of 1.02 MeV of energy are required for this event and the energy not transformed into mass is imparted as kinetic energy to the two particles, equally. The electron dissipates its energy, as already described. The positron, however, has a very short existence. As soon as it slows down, it is neutralized by an electron in the absorbing medium and this annihilation process results in a pair of γ photons, each of 0.51 MeV energy, which are ultimately absorbed by the photoelectric or Compton effect.

The γ photons, being without mass, are very penetrating and are best absorbed by materials of high electron density such as lead. Atoms with high atomic Z numbers are associated directly with high electron density. In terms of detection, certain inorganic salts absorb γ photons efficiently and emit light photons proportionate in intensity to the energy of the absorbed γ photons. Thallium-activated sodium iodide, for example, has been found to be superior in this regard because of its relatively high density (sp gr 3.67) due to the high Z number of the iodine atom, its high light yield per unit of energy absorbed, and the transparency of the crystal to the light emitted.

A solid crystal detector assembly for γ photons consists of a "canned" thallium-activated sodium iodide crystal mounted on the face of a photomultiplier tube (Fig. 3). The "canned" crystal is a solid cylindrical piece of thallium-activated sodium iodide, the top and sides of which are covered with a thin aluminum skin to protect it from light and moisture because sodium iodide is hygroscopic. To improve reflectivity, the sodium iodide crystal is sealed with a glass plate and is placed in direct contact with

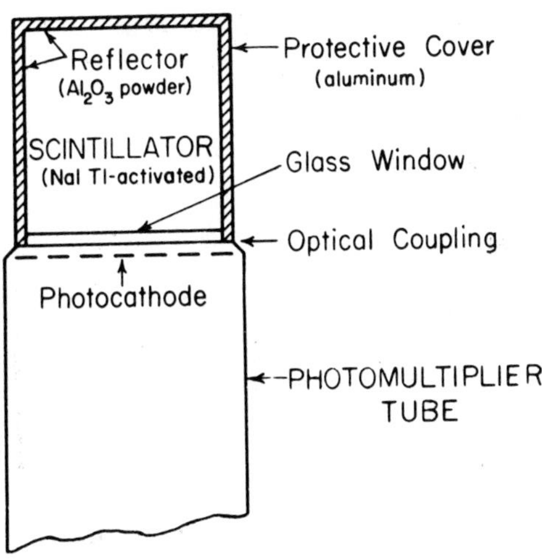

FIG. 3. A typical arrangement of a sodium iodide crystal detector and photomultiplier tube.

the face of the photomultiplier tube and optically coupled with grease. The entire assembly is light tight. The γ photons easily penetrate the thin aluminum skin of the crystal and are absorbed with high efficiency by the crystal. The crystal emits light photons of proportionate energy to the incident γ photons and the photomultiplier tube, in turn, converts the energy of the light photons to an electrical pulse. The proportionate nature of the various energy conversion processes, i.e., from the emission of the γ photons through to the production of an electrical pulse and the nature of γ photon absorption, is such that the γ-emitting radioisotope may be both identified and quantitated through an analysis of the energy spectrum.

The usual crystal gamma counter is designed to detect efficiently both the photoelectric and Compton effects. However, the efficiency of detection decreases with increasing energy of the γ photon. With the size of the sodium iodide crystal used in most commercial gamma counters the photoelectric effect predominates at low photon energies, say, below 400 keV, and the

Compton effect predominates around 1 MeV. Between these energies, both effects occur with near equal frequency. Pair-production is poorly detected because the size of the crystal used is relatively small.

The decay scheme (see Chap. 2, Section II A) of a typical γ emitter, ^{51}Cr, which is used clinically for the determination of red blood cell half-life, is shown in Fig. 4. This radioactive chromium isotope decays primarily by electron capture and has a half-life of 27.8 days. By electron capture, the atomic number of the atom is decreased by 1 and thereby becomes an isotope of vanadium. The frequency of the occurrence of decay by electron capture to the ground state of vanadium is 91% and results in the subsequent emission of a weak x-ray of 5 keV, which is usually not measurable because the x-ray from the sample is absorbed before it can penetrate into the sodium iodide crystal. In 9% of the time, ^{51}Cr decays by electron capture to an excited nuclear state of vanadium and immediately decays to the stable ground state by the emission of a 320 keV γ-ray which can be easily detected.

If the ^{51}Cr energy spectrum is observed with the aid of a multichannel analyzer, the spectrum shown in Fig. 5 results. A sharp peak called the photopeak is observed at 320 keV and is the result of the dissipation of the γ photon energy by the photoelectric effect. However, not all the energy is dissipated by this process, and therefore a contiguous series of broader and

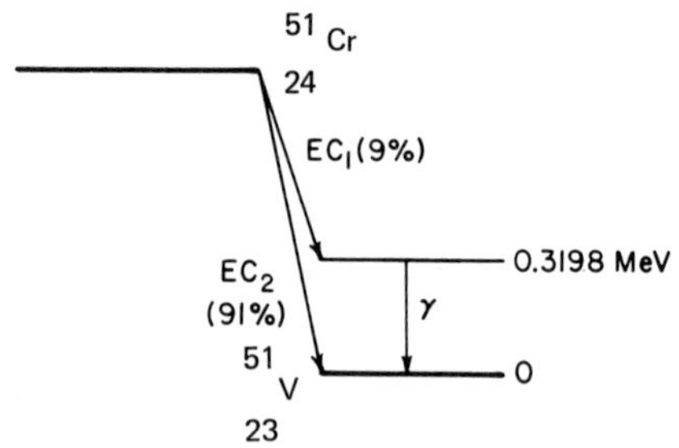

FIG. 4. Radioactive decay scheme of chromium-51.

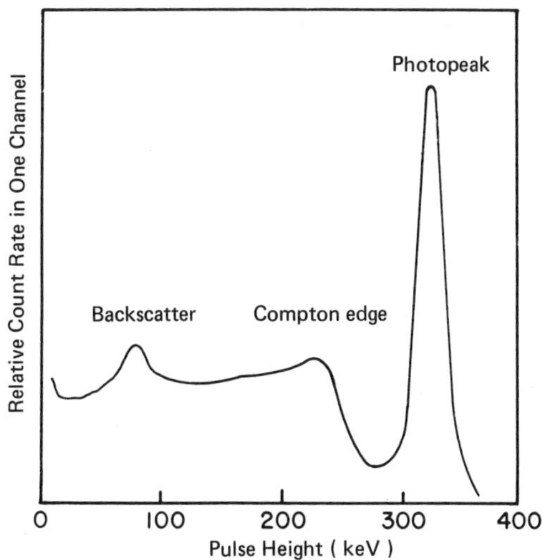

FIG. 5. Energy spectrum of chromium-51.

less-defined peaks occur at lower energies due to the dissipation of the photon energy by the Compton effect. The opposing side of the trough from the photopeak is called the Compton edge. The diffuse peaks of lower energy than those in the Compton region are due to the γ-rays which escape from the crystal during the Compton effect and are scattered back into the crystal from surfaces surrounding it. The scattered photons would be of much lower energy having dissipated part of it by undergoing the Compton effect and inelastic scattering. γ-Emitting isotopes are identified by their characteristic photopeak(s).

The performance of gamma counters is usually compared on the basis of their ability to resolve the 662 keV photopeak of ^{137}Cs. The resolution of the detection system is a measure of the spread of a photopeak, which is defined as the full width of the peak (in keV) at one-half the maximum peak height divided by the maximum pulse height (in keV) of that photopeak multiplied by 100. In the example shown in Fig. 6, the resolution is (694 - 630)(100)/662 = 9.7%. Under best conditions with selected

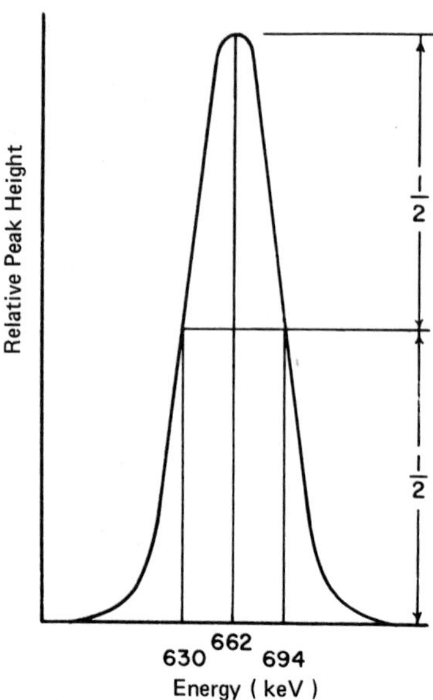

FIG. 6. The photopeak of cesium-137 used to determine the resolution of the gamma detection system. The full width of the peak at one-half the peak height is 64 keV. Resolution is equal to

$$\left(\frac{\Delta E_{1/2 \text{ peak height}}}{E_{\text{photopeak}}}\right)(100) = \left(\frac{64 \text{ keV}}{662 \text{ keV}}\right)(100) = 9.7\%$$

photomultiplier tubes, a resolution of 7% can be achieved [1]. However, the average well-crystal counters exhibit poorer resolution because of inferior optics yielding resolution figures of about 12%. In general, the resolution of photopeaks can be shown to improve in a near-linear fashion with increasing energy of the γ photon if the resolution is plotted against the log of the γ photon energy.

IV. LIQUID SCINTILLATION COUNTING

In crystal counting, the radiation must first penetrate the "canned" crystal scintillator to produce photons which are subsequently detected by a photomultiplier tube. In liquid scintillation counting, both the radioactive source and the scintillator are in solution. This intimate mixing of the radioactive sample with the scintillator results in high detection efficiency for weak β emitters and, as in crystal counting, photons emitted by the scintillator are detected by the photomultiplier tube. The energy transfer process: β decay $\rightarrow$ solvent $\rightarrow$ scintillator $\rightarrow$ photomultiplier tube, which transforms radiant energy originating from radioactive decay to electrical energy by the photomultiplier tube is a near-linear process. This ability to relate the radiant energy absorbed to the size of the electrical pulse emerging from the photomultiplier tube is the basis for scintillation counters being used as proportional counters to differentiate between different radioisotopes. These electrical pulses are then amplified, analyzed for the proper pulse height, and the acceptable pulses are counted.

For low-energy β emitters, the resulting electrical impulses are small and may be obscured by the background noise arising from the thermal emission of electrons from the photocathode of the photomultiplier tube. Two procedures were originally used to combat this problem. The first was to reduce the noise by cooling the photomultiplier tubes. However, this is not so vital for the more recently introduced types of photomultiplier tubes and the advantages of cooled or controlled temperature counters over ambient temperature units have been the subject of much intensive debate. These discussions have ranged from problems of instrument drift to the effect of temperature on sample stability, but controlled temperature units are strongly recommended for double isotope studies where only low levels of activity are expected. The second procedure used to reduce the background is the coincident technique. Two photomultiplier tubes are used with a coincidence unit which only passes signals which arrive simultaneously from both tubes. Since noise pulses in the photomultiplier tubes are emitted in a random manner, they are rejected by the coincidence unit.

The β particles resulting from the disintegration of an isotope such as tritium have a continuous distribution of energy from zero up to a maximum which is characteristic of that particular isotope. The amplitude or height of the pulse emerging from the photomultiplier tube is, as has already been stated, proportional to the energy of the absorbed radiation, and therefore a given isotope gives a characteristic pulse-height distribution. A pulse-height analyzer is simply a device which can be set to accept pulses within a preselected range of pulse heights and reject all others. In the early coincident counters, pulse-height analysis was achieved using the output of one photomultiplier tube, while the other photomultiplier tube was used to monitor coincident pulses. It soon became apparent, however, that the output from the second photomultiplier tube could be used for analysis as well as for monitoring coincidence. This concept, known as pulse summation, is now standard in commercial instruments and it effectively doubles the signal input to the pulse-height analyzer without increasing the background. Its major practical contribution is a markedly improved separation of ^{14}C from tritium. A block diagram of a single-channel unit with pulse summations is shown in Fig. 7. Units with up to three independent channels are available.

One of the most important attributes of a liquid scintillation counter is the ability to determine two isotopes simultaneously in the same sample. For this technique to be

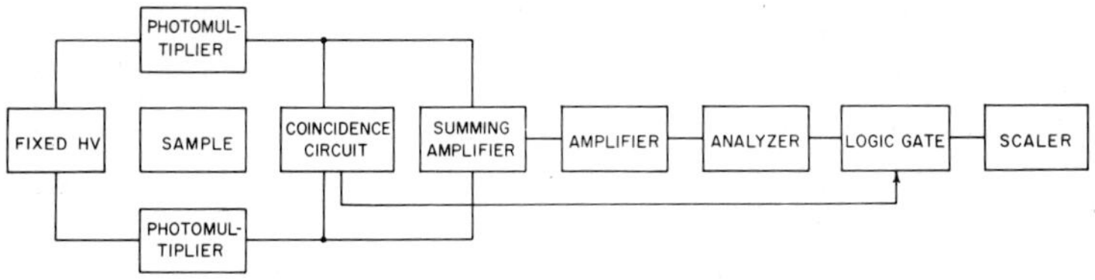

FIG. 7. Block diagram of a single-channel liquid scintillation counter with pulse summation.

successful, the β spectra of the two isotopes must be sufficiently different to be amenable to separation by pulse-height analysis. This is the case for the isotope pairs most commonly used, such as ^3H and ^{14}C, ^3H and ^{35}S, ^3H and ^{32}P, and ^{14}C and ^{32}P. Where the spectra of the two isotopes are closely matched as, for example, with ^{14}C and ^{35}S, chemical separation of the isotopes is necessary, and they must be counted individually. Figure 8 shows the spectra of ^{14}C and tritium and illustrates two important aspects of double isotope counting. First, the higher energy isotope can, if required, be counted without any interference by the lower energy isotope. Second, the lower energy isotope cannot be counted without also counting some of the superimposed ^{14}C. The counts of the lower energy isotope must always be corrected for

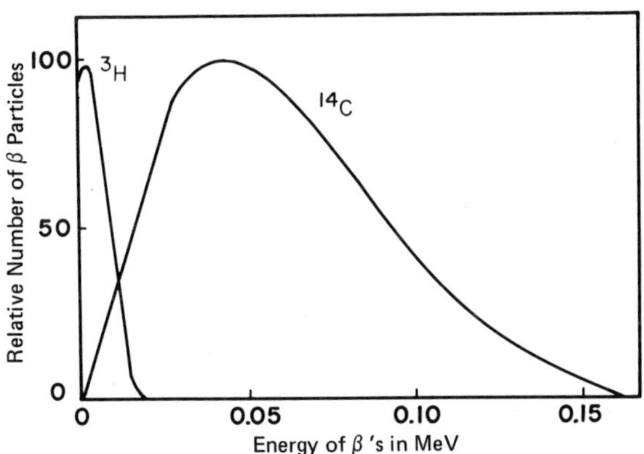

FIG. 8. β Spectra of carbon-14 and tritium.

the contribution of the higher energy isotope. The only real problem in double isotope counting is arriving at the best conditions for counting the lower energy isotope of the mixed pair.

The solvents of choice for liquid scintillation counting are those which dissolve the organic scintillator and efficiently transfer the energy from the radioactive sample to the scintillator. The best solvents are the alkyl benzenes, some of which are listed in Table 2. The range of biological material which can be accommodated by the best solvents such as toluene or xylene, however, is very limited and in recent years, considerable effort has been expended in developing methods for rendering biological samples suitable for counting [2].

A scintillator for liquid scintillation counting should exhibit the following characteristics: (1) a high efficiency of light production when activated by radiation; (2) the emitted light should be at a wavelength that corresponds to the area of maximum sensitivity of the photomultiplier tubes; (3) it should be both soluble and stable under conditions imposed by the nature of the sample and the working temperature of the counter; and (4) it should be readily available and of low cost. An extensive survey of hundreds of compounds as possible scintillators led to the conclusion that compounds containing conjugated rings combined in a linear fashion (*p*-terphenyl) rather than a fused ring system (anthracene) were good scintillators [4-8]. Of the many ring structures studied, oxazole and 1,3,4-oxadiazole have formed the basis of the most useful scintillators some of which are listed in Table 3. The improved sensitivity in the region of 3900 Å of the more recent photomultiplier tubes has made the use of secondary scintillators superfluous [2].

The quantitation of the amount of radioactivity present in a sample in absolute units (dpm) depends on the knowledge of the sample counting efficiency and is calculated from the following mathematical relationship (2):

TABLE 2

Relative Counting Efficiencies of
Some Solvents Used in Liquid Scintillation Counting [3]

Solvent	Relative Efficiency*
p-Xylene	112
Xylene (mixed isomers)	107
Toluene	100
Ethylbenzene	96
Benzene	85
Anisole	83
p-Cymene	80
Fluorobenzene	65
Benzyl alcohol	45
Hexane	30
Cyclohexane	27
1,4-Dioxane	20
Acetone	10
Pyridine	0
Ethyl alcohol	0
Ethylene glycol	0
Acetic acid	0

*With 3 g 2,5-diphenyloxazole per liter of solvent.

$$\text{dpm} = \frac{\text{cpm}}{\text{efficiency}} \tag{2}$$

where dpm = disintegrations per minute; cpm = observed net counts per minute; and efficiency = counting efficiency (100% = 1.00). In liquid scintillation counting, the sample counting efficiency is usually determined using one of three methods: the internal standard method, the sample channels ratio method, or the external standard method.

The internal standard method consists of adding a known amount of a nonquenching standard to the sample after it has been

TABLE 3

Primary Scintillators

Scintillator	Abbreviation	Solubility in toluene (g/liter) 0°C	20°C	Conc.* (g/liter)	Maximum (Å) fluorescence	Relative pulse height
p-Terphenyl	TP	4.0	8.6	8	3600	1.00
2,5-Diphenyl oxazole	PPO	238	414	4	3650	1.01
2-phenyl-5-(4-biphenylyl)-1,3,4-oxadiazole	PBD	10	21	9	3650	1.28
2-(4-*t*-butylphenyl)-5-(4-biphenylyl)-1,3,4-oxadiazole	Butyl-PBD	57	119	7	3650	1.28

*Typical concentration in toluene for unquenched samples.

counted and recounted. The additional counts due to the added standard are used to compute the counting efficiency. The sample channels ratio method for efficiency determination is based on the fact that the β spectrum is always shifted when quenching occurs. By counting two different portions of the spectrum simultaneously in a two-channel counter, the ratio of the counts appearing in the two channels will change in one direction as quenching is increased. A plot of the ratio change vs counting efficiency can be constructed by preparing a series of variably quenched standards. The accuracy of the ratio determination is a function of the amount of radioactivity contained in the experimental sample. Low activity samples are not well suited for the sample channels ratio method. The external standard method was conceived to accommodate any type of sample. In this method a highly active gamma source is used to irradiate the experimental sample to generate Compton electrons within the sample. The ratio of the counts generated by the Compton electrons are related to counting efficiency as in the sample channels ratio method.

V. SAMPLE PREPARATION FOR LIQUID SCINTILLATION COUNTING

The primary considerations in choosing a method for preparing samples for scintillation counting are:

1. The chemical and physical characteristics of the sample
2. The nature of the isotope
3. The anticipated level of radioactivity

The characteristics of the sample determine the type of counting solution required and whether conversion to a form more suitable for counting is required. With respect to the nature of the isotope, tritiated samples, for example, require more attention than samples labeled with ^{14}C because tritium is more sensitive to quenching. Finally, the level of radioactivity can influence the method chosen for preparing the sample. For example, samples which contain high levels of activity may be prepared by the most economical method in terms of time and convenience. For samples which are very low in activity, however, these factors may be secondary to a technique which produces little or no quenching.

Whatever method is chosen should produce a series of uniform, stable, and homogeneous samples such that the counting efficiencies can be determined either by the channels ratio method or by external standardization. If the samples require internal standardization, further manipulations and additional counting time are necessary.

Techniques for sample preparation for materials that cannot be directly incorporated into a toluene solution for counting, fall into three main categories: emulsion counting, solubilization techniques, and combustion procedures. Emulsion counting is used for aqueous samples and solubilization procedures are used for substances, such as proteins, which are normally insoluble in water or the base organic solvents used in counting solutions. Combustion techniques are used whenever they are more convenient as for material which is difficult to treat by any other procedure. Counting on a support medium is of diminishing importance because of the difficulty in accurately determining counting efficiency.

A. Emulsion Counting

The great attraction of emulsion counting lies in its capacity for accommodating large volumes of aqueous samples which can be counted at reasonable efficiencies. However, the evidence is not at all clear that this is a simple technique in which the aqueous sample is simply shaken up with the emulsifier and counted.

The most widely used emulsion system is based on a mixture of toluene and Triton X-100 (2:1 v/v). The success of this system has prompted the development of several commercial products such as Instagel, Aquasol, Ready-Solv 1, Multisol, Handifluor, and PCS. These products have been formulated as either toluene- or xylene-based systems which can accommodate up to 50% water. The physical characteristics of the emulsion change as the aqueous concentration is increased and a typical phase diagram is illustrated in Fig. 9. The phases are temperature-dependent, and in a Triton X-100 system the counting efficiency increases linearly about 10% when the temperature is reduced from 17°C to 4°C [9]. Lower temperatures do not increase the efficiency any further. These workers also found that emulsions gave reproducible results only if they were first heated to 40°C, allowed to cool without shaking, and kept at 4°C for 2 to 4 hr. Heating may exclude trapped oxygen and cooling without shaking presumably reduces the entrapment of air within the sample.

The effect of added water on counting efficiency is exemplified in Fig. 10 [10]. As the concentration of the aqueous phase increases the counting efficiency for tritium decreases since the water is acting as a quenching agent. When the first phase change occurs, there is a sharp fall in counting efficiency because the tritium stays in the aqueous phase, whereas the scintillator remains dissolved in the organic phase. As more water is added an emulsion is formed which consists of submicrometer

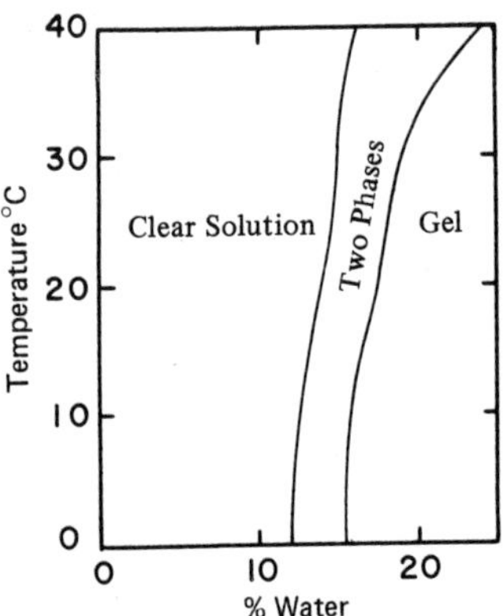

FIG. 9. Phase diagram of a commercial emulsifier, Aquasol. The physical and optical characteristics of the solution change with increasing aqueous content ranging from a clear liquid to an opaque, stiff gel. Between these two states, there is a narrow range of water content at which phase separation occurs. The addition of water will eliminate this phasing and result in the formation of a homogeneous gel of long-term stability. (Courtesy of New England Nuclear Corporation.)

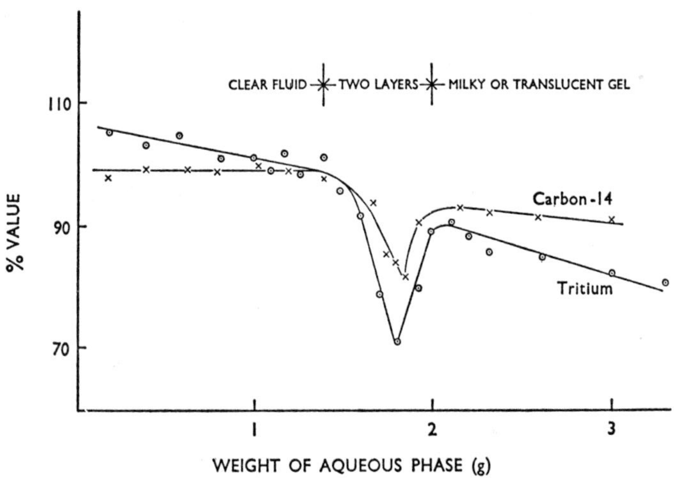

FIG. 10. Emulsion counting of D-[U-^{14}C]glucose and tritiated water using toluene/Triton X-100, 2:1 (v/v) [10].

water particles dispersed in the organic phase. Since the size of the water particles is less than the mean free path of the average tritium β particles there is an effective interaction between the β particles and the scintillator with a subsequent increase in counting efficiency. The opacity of the emulsion also helps to prevent internal light trapping. If the aqueous concentration is increased still further the size of the dispersed water particles increases. The counting efficiency, therefore, is reduced due to the increase in self-absorption. In contrast, ^{14}C samples can be counted at constant efficiency in the "solution" phase and near constant efficiency in the emulsion phase due to the more energetic nature of ^{14}C β particles.

For the determination of counting efficiency, either internal standardization or channels ratio methods can be used although the conditions are more stringent than for homogeneous solutions. The external standard cannot be used to monitor counting efficiency because the Compton electrons produced by the gamma source within the organic phase are not attenuated by the dispersed water particles.

Typical applications of emulsion counting are illustrated by a commercial product, Aquasol (Table 4).

B. Solubilization Techniques

Many biological samples are normally insoluble in toluene and have to be chemically treated so that they can be incorporated into a counting solution. The most versatile of the solubilizing agents are Hyamine, NCS, Biosolv, Soluene, and Protosol. They are all strong quaternary ammonium bases and are all used in a similar manner. The sample, e.g. protein, is simply digested with the base for a few hours, usually by warming. When the sample has dissolved it is cooled and the counting solution added.

Hyamine was the first solubilizing agent to be introduced and was used initially as a trapping agent for $^{14}CO_2$. There are, however, better agents now available.

A common advantage of these agents is their versatility, for they permit samples as diverse as tissues, proteins, blood, amino acids, plasma, and urine, as well as aqueous samples to be incorporated directly into a toluene-based counting cocktail. The disadvantages are that (1) they are all quenching agents to varying degrees, (2) many samples require additional treatment with benzoyl peroxide to decolorize the samples, (3) they cannot be used when butyl-PBD is the primary scintillator because they yield a colored solution (PPO, therefore, should be employed when these agents are used), and (4) many samples prepared by these agents will exhibit chemiluminescence, i.e., spurious high count rates which are independent of sample activity. This can be eliminated by dark adaptation of the samples and by neutralizing the sample with acid. Chemiluminescence is greater when impure dioxane solutions are used and decay is prolonged at ambient temperatures.

C. Combustion Techniques

When dealing with radioactive biological samples, the majority of sample preparation problems are concerned with the estimation of tritium and/or ^{14}C. The existence of a wide spectrum of types and classes of organic compounds has precluded the discovery of a single universal procedure for preparing an acceptable counting solution of these organic compounds for analysis by liquid scintillation counting. However, a common feature of all organic matter is that it can be burned to carbon dioxide and water. By selectively trapping the water and carbon dioxide, uniform samples can be prepared which can all be counted at the same efficiency for each isotope. A method for combusting organic compounds within a sealed flask was first developed by Schöniger [11] and applied to radiolabeled biological compounds by Kalberer and Rutschmann [12] and Kelly et al. [13]. A number of other combustion techniques have been used to prepare samples for liquid scintillation counting. These include wet oxidation with Pirie's reagent [14], an oxygen train [15], and oxidation in a sealed tube

TABLE 4

Performance Data for a Commercial Emulsifier, Aquasol*

Sample type	Sample volume	Aquasol volume	% of Sample	^3H Counting efficiency (% range)	Figure of merit**	Physical characteristics Ambient	Physical characteristics 5°C
	0.0 ml	15 ml	0.0	30-50	0	Clear	Clear
H_2O	0.5 ml	15 ml	3.2	25-45	106	Clear	Clear
H_2O	3.5 ml	15 ml	18.9	20-40	575	Cloudy, gel	Cloudy, gel
H_2O	5.0 ml	15 ml	25.0	20-40	700	Hazy gel	Hazy gel
0.5 M HCl	0.5 ml	15 ml	3.2	15-35	99	Slight haze	Clear
0.5 M HCl	1.5 ml	15 ml	9.1	20-45	241	Clear	Clear
0.5 M HCl	3.5 ml	15 ml	18.9	20-40	493	Stiff, cloudy gel	Stiff, cloudy gel
0.9 M NaCl	0.5 ml	15 ml	3.2	15-30	105	Cloudy	Clear
0.9 M NaCl	1.5 ml	15 ml	9.1	25-40	260	Clear	Clear
0.9 M NaCl	3.5 ml	15 ml	18.9	20-40	567	Cloudy	Cloudy gel

0.5 M NaOH	0.5 ml	15 ml	3.2	20-35	96	Cloudy	Hazy
0.5 M NaOH	1.5 ml	15 ml	9.1	25-40	242	Clear	Clear
0.5 M NaOH	3.5 ml	15 ml	18.9	20-35	516	Clear	Cloudy gel
30% sucrose	1.5 ml	15 ml	9.1	20-35	270	Cloudy	Clear
30% sucrose	3.0 ml	15 ml	16.7	20-35	448	Clear	Clear
30% sucrose	6.0 ml	15 ml	28.6	15-30	712	Clear	Cloudy gel
				^{14}C Counting efficiency (% range)			
H_2O	0.5 ml	15 ml	3.2	70-90	281	Clear	Clear
H_2O	5.0 ml	15 ml	25.0	70-90	2133	Stiff, hazy gel	Stiff, hazy gel

*Data from New England Nuclear Corporation, Boston, Mass.

**Figure of merit = (percent added sample)(efficiency). 6 samples at each point. Internal standard = 1.19 × 10^5 dpm.

[16]. Combustion techniques, however, as applied to liquid scintillation counting, are tedious and not well suited for handling large numbers of samples. The combustion requires a separate flask for each sample and the quantitative trapping of the water and carbon dioxide requires great care. Also, the hazardous nature of the combustion process and the abovementioned drawbacks discouraged potential users of this procedure.

The situation has been changed by the introduction of semiautomatic combustion apparatus which demonstrated that it was possible to perform combustions rapidly, quantitatively and, most important, safely. A recent version of such an instrument is illustrated in the flow diagram in Fig. 11. The oxidizer operates at atmospheric pressure and the water and carbon dioxide are collected at ambient temperature. The operator simply sets the dispensers to indicate whether tritium and/or ^{14}C are to be analyzed and the combustion timing clock is adjusted to allow adequate combustion time for the largest anticipated sample. To operate this system, two empty sample vials are positioned under the two condensers and the specimen to be burned it placed into the combustion basket; the "start" button is pressed. The combustion then proceeds automatically until completion. The sample is burned and the combustion flask purged with steam to remove all combustion products under a continuous flow of nitrogen. The water is condensed, trapped in the tritium exchange column, and quantitatively removed from the exchange column with a special scintillator solution by countercurrent flow and collected in the tritium vial. The carbon dioxide gas is insoluble in the scintillation solution and is swept into the carbon reaction column where it is trapped by Carbosorb, an organic amine which forms a toluene-soluble carbamate with carbon dioxide by countercurrent flow. An exchange column ensures the quantitative recovery of carbon dioxide. The adsorbed carbon dioxide is then washed into the ^{14}C vial by the carbon dioxide scintillator solution. The specifications claim recoveries of 99 ± 1% for either

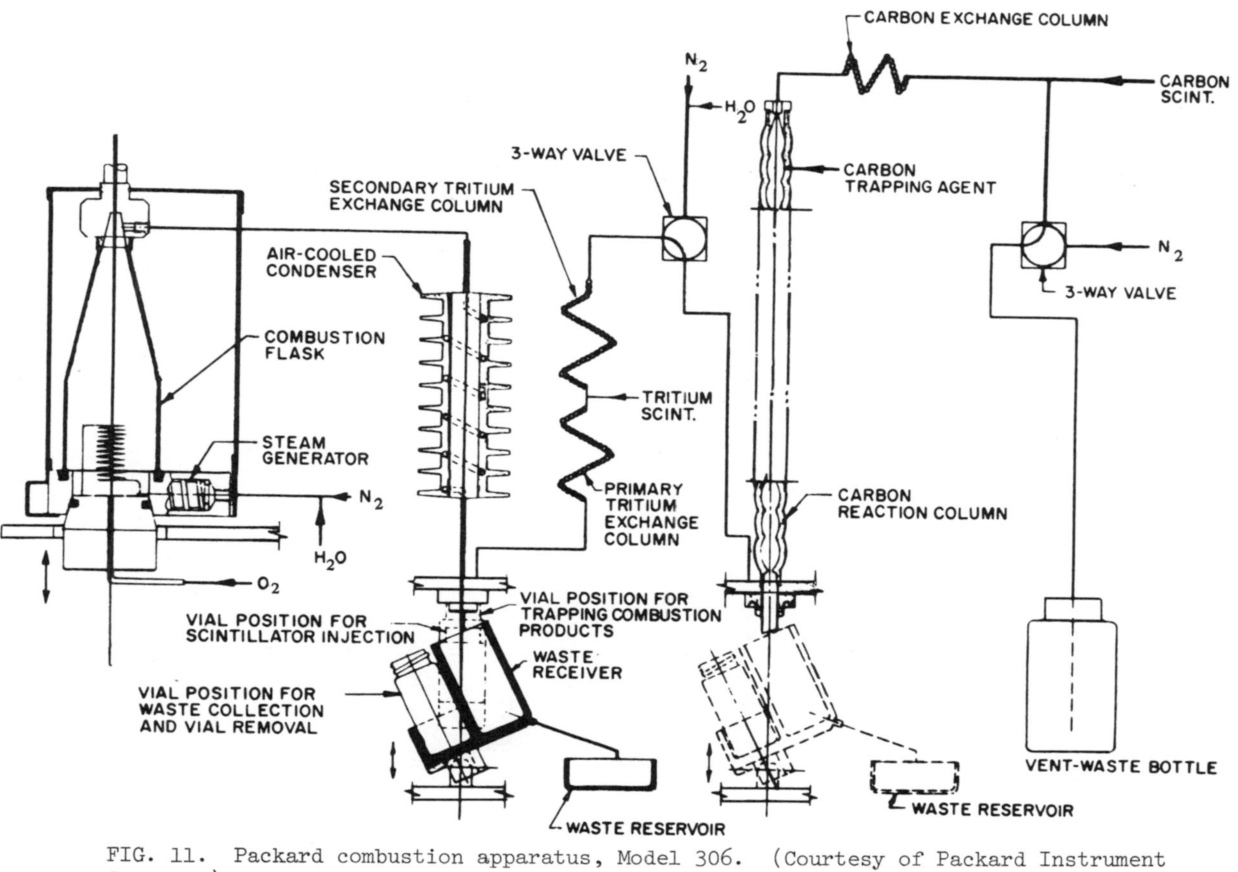

FIG. 11. Packard combustion apparatus, Model 306. (Courtesy of Packard Instrument Company.)

isotope and a spillover of ^{14}C in the tritium vial of less than 0.02% and tritium in the ^{14}C vial of less than 0.001%. The memory is less than 0.05% for either isotope. The sample throughput ranges from 20 to 60/hr for a single labeled sample depending on the combustion time used.

VI. CERENKOV COUNTING

Cerenkov light is generated when a charged particle travels through a medium faster than the speed of light through that medium. The emitted light is directional and cone-shaped with a half-angle, ϕ. Cosine ϕ is defined as follows (3) [17]:

$$\text{Cos } \phi = \frac{1}{\beta n} \tag{3}$$

where

$$\beta = \frac{\text{velocity of particle}}{\text{speed of light}} = \frac{v}{c}$$

n = refractive index of medium

The threshold electron energy, E_{min}, for Cerenkov radiation can be computed from the relationship (4) [18]:

$$E_{min}\ (\text{keV}) = 511 \left[\frac{n}{\sqrt{n^2 - 1}}\right]^{-1} \tag{4}$$

In water, which has a refractive index of 1.332, the minimum electron energy for Cerenkov radiation is 263 keV. If it is assumed that the medium is nondispersive, there is a continuous spectrum of radiation emitted with the greatest number of photons being produced in the ultraviolet and extending into the visible but negligible in the infrared. Cerenkov radiation is distinct from either fluorescence or bremsstrahlung and is characterized by the short duration of the light pulse of less than 1 nsec.

Cerenkov counting is currently useful for energetic β emitters. Parker and Elrick [17] have experimentally determined the Cerenkov counting efficiencies of a number of β-emitting radioisotopes in a 10-ml solution (Table 5). The wavelength shifter used in Table 5 was the water-soluble salt of 2-naphthylamine-6,8-disulfonic acid (100 mg/liter). If the wavelength shifter had been used with the RCA tube, higher efficiencies than those given would have been observed.

As a means of analysis, Cerenkov counting offers many advantages. It is a simple technique because the sample can be counted directly without the addition of any scintillator. The sample size is limited only by the capacity and geometry of the sample counting chamber and is not subject to chemical quenching. The sample is not contaminated by extraneous chemicals and the solvent can be almost any colorless liquid.

The disadvantages of the method are that the counting efficiencies are relatively low due, in part, to the directional nature of the Cerenkov radiation, and it is sensitive to color quenching. However, poor counting efficiencies can be improved by the addition of a wavelength shifter and color quenching can, in most instances, be removed by chemical treatment.

The most useful application in the biological sciences is the measurement of ^{32}P which can be counted directly at greater than 40% in a counter equipped with bialkali photomultiplier tubes (Table 5). Higher counting efficiencies can be realized by the addition of a wavelength shifter and by the use of plastic vials. The latter help scatter the light more efficiently and also reduce the background.

TABLE 5

Experimentally Determined Cerenkov Counting Efficiencies*

		Counting efficiency (% of disintegration)		
		S11 photocathode		RCA 4501-V3 photocathode
Nuclide	E_{max} (MeV)	Aqueous sample	Wave-length shifter	Aqueous sample
^{204}Tl	0.77 (98%)	1.3	2.6	--
^{137}Cs	0.51 (92%), 1.17 (8%)	2.1	--	--
^{36}Cl	0.71 (98.3%)	2.3	4.7	5.3
^{198}Au	0.96 (99%)	5.4	--	--
^{47}Ca	0.66 (83%), 1.94 (17%)	7.5	14.8	--
^{40}K	1.32 (89%)	14	31	34.3
^{24}Na	1.39 (100%)	18	40	--
^{86}Rb	0.68 (8.5%), 1.77 (91.5%)	23	46	--
^{32}P	1.71 (100%)	25	50	43.3
^{144}Ce-^{144}Pr	2.98 (97.7%)	54	75	59.2
^{42}K	2.0 (19%), 3.6 (82%)	60	85	76.3
^{106}Ru-^{106}Rh	2.0 (3%), 2.4 (12%) 3.1 (12%), 3.6 (70%)	62	85	75.3

*Comparisons are given with and without the water soluble sodium/potassium salt of 2-naphthylamine-6,8-disulfonic acid (100 mg/liter) using now obsolete photomultipliers. The values given for more modern phototubes are in the absence of wavelengths shifter; had shifter been added, even higher efficiencies might be expected [17].

VII. RADIOCHROMATOGRAPHY

Chromatography is one of the most important techniques for the separation of chemical compounds from biological material. In conjunction with the use of radiolabeled substances, it has been widely applied to the mapping of metabolic pathways and in the quantitative estimations of biological molecules. The major variants of the technique are paper, thin-layer, column, and gas-liquid chromatography. The problems of measuring the radioactivity in aqueous samples are discussed elsewhere in this chapter and emphasis in this section will lie on the methods for detecting radioactivity in thin-layer and paper chromatograms (see also Chap. 6).

After an initial period of rapid growth, the utility of radiochromatography has abated a little. There are at least two reasons for this. First, there is an increasing number of examples coming to light where several chromatographic systems have failed to separate closely related compounds or detect impurities in radiochemical substances. None of these techniques, therefore, should be relied upon exclusively as a means of separation and, least of all, identification. Second, the preferential use of tritium in many biochemical investigations requires extremely sensitive methods for radiotracer detection. The range of a β particle from tritium is so short that over 95% of the radiation present in a thin-layer chromatogram may be lost within the support matrix. Further developments in radiochromatography, therefore, require a breakthrough in detection techniques.

Most of the detection mechanisms applicable to radiochromatography use x-ray film, a β-particle detector, or a luminescence detector. In many of these procedures the samples are left uncontaminated and can be subjected to further processing.

A. Autoradiographic Detection Methods

Labeled compounds can be visualized by blackening of photographic emulsion and autoradiography is a useful adjunct to paper and thin-layer chromatography. It is also used for detecting radioactivity in polyacrylamide gels. The procedure is applicable only to those isotopes having sufficient energy for the radiation to reach the film emulsion. For ^{14}C, ^{35}S, and ^{32}P there is little difficulty, but tritium presents special problems because of the short range of the radiation emission.

The procedure is the same for both paper and thin-layer chromatograms. A holder is required to provide light-tight contact between the x-ray film and the chromatogram. After a suitable period of time the film is removed and developed according to standard techniques. Guidelines for the approximate times of exposure for different isotopes in the preparation of autoradiograms for thin-layer chromatograms have been published [19]. When a moderate vacuum is applied there is a reduction factor of about 2 for the exposure time for tritium, indicating that there is absorption of the β particles in the thin air gap between the chromatogram and the photographic emulsion [20].

A modification of this procedure is scintillation autoradiography or fluorography. This procedure involves interaction of the β particle with an organic scintillator. The resultant fluorescence produces an image on x-ray film. Fluorography was first carried out using *p*-terphenyl as the organic scintillator, but anthracene has also been used. Exposure at low temperatures is advantageous. In the method developed by Lüthi and Waser [21] the thin-layer adsorbant (silica gel) is mixed with anthracene in a suitable ratio, and then the radioactive compounds separated on the mixture. Anthracene is only slightly soluble in most solvents used in chromatography and it possesses a high fluorescence efficiency. The fluorograms are exposed at -70°C. Originally the lowered temperature was thought to enhance the scintillator efficiency, but it has subsequently been shown that the increased light sensitivity of the photographic emulsion at the lower temperature is the reason for the increased sensitivity [22].

An alternative procedure to mixing the scintillator with the chromatographic material is to spray a scintillator solution onto the chromatogram, evaporate the scintillator solvent and then apply the chromatogram to an x-ray film. The sensitivity of the method depends on the amount of scintillator which can be applied to the chromatogram. The scintillators, such as the phenyloxazoles which are easily soluble, are, therefore, the most suitable for this purpose. Exposure of the chromatogram is carried out at dry-ice temperature (-80°C) and overnight exposure is sufficient to detect 0.1 μCi of tritium.

A comparison of an autoradiogram and fluorogram indicated that for tritium the fluorogram may be as much as 100 times more sensitive than the autoradiogram [23]. It is, however, limited to molecules which do not exhibit any chemiluminescence at low temperatures.

B. Beta-Particle Detection Techniques

A general requirement is that the detector aperture should be as close to the sample as possible in order to avoid the absorption of low-energy particles before they reach the detector. Most of the radiochromatogram sources currently available use windowless gas flow GM tubes. The chromatogram is automatically drawn under the detector and the pulses produced by a radioactive spot are analyzed by a rate meter and the count rate is recorded on a paper chart. As the chart recorder moves at the same speed as the chromatogram, the radioactive spots can be accurately located by aligning the two together. The amount of radioactivity is indicated by the area appearing under the curve.

Approximate sensitivities for some radioisotopes are shown in Table 6. The sensitivity is particularly low for tritium. For detection on paper chromatograms it is possible to count at 4π geometry by using two detectors, one on each side of the chromatogram.

C. Scintillation Detection Techniques

When the position of the compound of interest on the chromatogram is known, it can be quantitatively removed by elution with a suitable solvent. If necessary, the solvent is evaporated off and the sample can then be counted by liquid scintillation counting. In the case of a thin-layer chromatogram the appropriate zone is scraped into a mound and collected on a fritted disk in a glass aspirator attached to a vacuum system [24]. The aspirator is then inverted and the eluting solvent poured over the adsorbent and aliquots of the eluate taken for counting. To ensure that the sample has been completely eluted the residual adsorbent can be combusted to determine whether any radioactivity remains.

A special technique (zonal scanning) has been described by Snyder [25]. An automatic scraper takes out narrow successive sections of the plate, the activity is counted by liquid scintillation counting and a histogram of the distribution of

TABLE 6

Typical Efficiencies in a Gas Flow Strip Scanner for a Thin-Layer Chromatogram 2π Counting Efficiency

Isotope	Percent efficiency with 2π geometry
Tritium	1.5
Carbon-14	8
Sulfur-35	8
Phosphorus-32	15
Iodine-131	10

radioactivity along the chromatograms is established. The method is exact and sensitive, but time-consuming and expensive. When accurate quantitation is required the sample on the support medium should be either completely soluble or completely insoluble in the scintillation fluid. Partial elution from the supporting medium gives inconsistent results. Breakdown of the molecule may occur on the chromatogram and the crucial time is the period during which the adsorbent is dry and exposed to air. The time taken to locate and scrape the zone and elute the sample should not be prolonged.

D. Gas-Liquid and Column Chromatography

Radioactivity in the effluent of a gas chromatograph column has been monitored using flow through ionization chambers [26], proportional counters [27], and scintillation counters [28]. Combustion procedures have also been developed for quantitatively converting the organic materials in the column effluent into carbon dioxide and water. Measurement may be made either using fractions collected intermittently or by continuously assaying the chromatographic effluent. Since measuring the radioactivity during the course of an analysis is so much more convenient, fractionating the effluent is generally reserved for those experiments in which the level of radioactivity is too low to permit the measurement to be carried out in any other way. Furthermore, the continuous monitoring method frequently enables a distinction to be made between trace components of high specific activity and major components of low activity with similar retention times. The choice among the variety of methods is based on the chemical nature of the compounds to be assayed, how much radioactivity can be injected, and the specific activity of the compound being studied. The article by Karmen [28] should be consulted for the relevant technical details and merits of the various procedures.

In view of their sensitivity for measuring weak β emitters, it is not surprising that liquid scintillation counters have been used extensively for the continuous assay of chromatographic effluents. Originally, Schram and Lombaert [29] used a cell made of a grooved Plexiglas center section sandwiched between two sheets of plastic scintillator. Steinberg [30], however, found that anthracene crystals gave higher counting efficiencies than a plastic scintillator, and anthracene-packed cells are still widely used. Many continuous flow cells are now shaped like counting vials and can easily be inserted into a conventional counter. There are three main types of flow cells:

1. External counting cells in which the aqueous fluid is circulated through a cell of plastic scintillator
2. Cells packed with suspended scintillators or crystals. Anthracene belongs to this type
3. Homogeneous counting systems where the effluent is mixed with liquid scintillator before passing through the cell

The third procedure gives the highest efficiencies for tritium. Hunt [31] described the use of such a cell for recording tritium and ^{14}C. The initial cell consisted of a coil of polyethylene tubing, but subsequent modification [32] substituted Teflon for the polyethylene tubing. The cell is contained in a modified plastic counting vial and the effluent from the column is mixed with counting solution before passing into the flow cell (Fig. 12).

VIII. AUTORADIOGRAPHY

In biological autoradiography, labeled substances in the organism, tissue or cell, are made visible by preparing thin sections and exposing them to a suitable photographic film. After developing, the silver grains can be visualized and the distribution of the isotope correlated with the underlying biological structures. Autoradiography has been combined with both light and electron microscopy and, although a variety of different isotopes have been used, tritium is favored because the low energy and penetrating

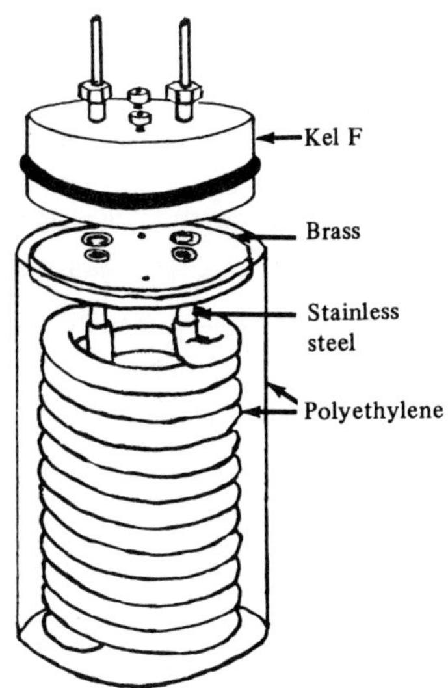

FIG. 12. Plastic flow cell. The cell is constructed from PE 280 polyethylene or Teflon tubing. The outer casing is made by cutting off the top of a 20-ml scintillation counting vial [31, 32].

power of its β particles permits the greatest resolution. It is also available in high specific radioactivities and is readily introduced into many different types of organic molecules during synthesis (see Chap. 5).

A. Light Microscope Autoradiography

Whole-body autoradiography is widely used in drug metabolism studies. The labeled compound is administered to a number of animals, usually mice, and at preselected times the animals are rapidly frozen by immersion into cooled hexane. Microtome sections (5 to 20 μm) are then taken through the whole body at different levels at -15°C. To facilitate the handling of sections, Scotch tape is applied to the surface. The sections are dried by sublimation and pressed against x-ray film or dry mounted onto nuclear emulsions. The latter procedure is being preferred because the nuclear emulsions developed originally for electron microscope autoradiography have a higher resolution. This procedure, originally developed by Ullberg [33], has undergone various technical modifications but remains essentially as outlined above.

The primary objective of autoradiography is to determine precisely the localization of the labeled molecule. The resolving power of whole-body autoradiography is small, however, since thick sections are used and it is mainly used in establishing the differences in the distribution of the isotope among different tissues.

There have been many attempts to correlate whole-body autoradiography with more detailed cellular morphology using light microscopy [34]. For labeled compounds which are firmly bound to cellular structures there are few problems, but for substances which are not bound or are freely soluble the conventional histological methods for fixing the tissues are generally unsuccessful in autoradiography because of the dangers of translocation. The localization of estradiol in the uterus probably represents the classic example of the confusion that can arise through inadequate technique. This steroid has been shown to be present mainly in the cytoplasm [35] or in the nucleus [36] of the endometrial cells or evenly distributed. It was either absent [37] or present in the glandular lumina [36]. The major causes of this confusion were artifacts arising in the preparation of the autoradiogram. These may be introduced at various stages during the preparation of the tissue and liquid fixatives, dehydrating agents, organic solvents, and embedding media are all potential sources of diffusion artifacts. The abrupt cessation of metabolic processes obligatory in high-resolution autoradiography is rarely achieved by liquid fixation and freezing methods with liquid nitrogen-cooled hydrocarbons are preferred.

In the method devised by Stumpf and Roth [38], the freshly excised tissue attached to a tissue holder is plunged into liquified propane at about -180°C. The tissue is cut in a low-temperature cryostat between -30° and -60°C depending on the thickness of the section required, the thin ones being cut at the lower temperature. The sections are then transferred to a vial in a tissue carrier and freeze-dried for 24 hr in a cryosorption pump in a vacuum between 10^{-5} and 10^{-6} mmHg. The tissue is then allowed to come to room temperature and the vacuum broken. The freeze-dried sections, still in the vials, are stored in a desiccator at room temperature. Mounting is carried out at room temperature under conditions of low relative humidity (20-40%). The sections are placed on a small piece of Teflon and then pressed firmly onto a slide coated with a photographic emulsion. The slides are kept at -5°C under low humidity and eventually developed.

B. Electron Microscope Autoradiography

With the introduction of tritium, autoradiographic resolutions of the order of 1 μm became possible in the light microscope, but further improvements require an extension of the technique to the level of the electron microscope. The basic principles are the same. The main difference is that the specimen and the photographic emulsion have to be very thin compared with those used in light microscopy and longer exposure times are necessary. The application of electron microscopy autoradiography to diffusable substances has not so far proved very successful [39].

The specimen is generally a thin section of fixed tissue embedded in a polymerized acylic or epoxide medium which is in close contact with the emulsion during exposure and photographic processing. One of the major technical problems is in the application of the photographic emulsion to the tissue section. A uniform emulsion layer is of prime importance, otherwise it is impossible to relate the distribution of developed grains to radioactive sources with any certainty. Various methods have been tried, including dipping the grid in liquid emulsion [40], applying the emulsion with a brush or a pipette [41], or as a thin membrane of gelled emulsion which is produced in a wire loop [42, 43].

The theoretical and practical aspects of sensitivity as they apply to electron microscope autoradiography have been discussed in several publications [44,45]. Two independent errors contribute to limit the resolution; these are the errors caused by the photographic factors, that is, effects attributable to processing the emulsion and errors caused by geometric factors arising from the distance between the radioactive source and the silver halide crystals in the emulsion. Geometric errors predominate in electron microscope autoradiography.

When a β particle interacts with a silver bromide crystal the latent image produced can form anywhere in the crystal and not necessarily at the point of impact. The maximum distance between the path of the particle and the latent image is, therefore, the diameter of the undeveloped crystal. The smaller the crystal in the emulsion the smaller the photographic error and the higher the resolution.

When the image is formed it consists of silver atoms which must be enlarged and this is accomplished during development. The resulting silver grains are variable in shape and size depending on the type of emulsion and on the method of development. An error during development arises, therefore, from the uncertainty regarding the spatial relationship between the developed silver grain and the latent image. The total photographic error is dependent on the size of the silver halide crystals and also on the developed grain. Of these two factors, the size of the silver halide crystal is the major one in determining resolution and little is gained by making the developed grain much smaller than the undeveloped crystal.

The geometric error is concerned with the interaction between the silver halide crystals and the source of the emitted radiation. The relevant factors include the energy of the ionizing particles, the thickness of the material, and the angle at which the particles leave the source. Ionizing particles are emitted in all directions and will impinge upon a number of silver halide crystals in the overlying emulsion. This results in a spread of the photographic error and the problem is how this distribution relates to the original source. According to Salpeter [46] the thickness of the section plus the intermediate layer has greater influence on the geometric error than the thickness of the emulsion.

The main disadvantage of electron microscope autoradiography is that it is a relatively insensitive method for detecting radioactivity. Since the thickness of the section is reduced to a minimum to achieve the highest resolution, only a small fraction of the radioactivity in the sample is measured.

Efficiency in an autoradiogram can be defined as the number of silver grains produced by a radioactive disintegration [44]. It is dependent upon a number of factors, particularly the sensitivity of the emulsion and the energy of the isotope. For thin sections (less than 1000 Å) the radioactivity detected is directly proportional to the thickness of the section. For quantitative comparisons it is therefore essential that sections be of uniform thickness.

REFERENCES

1. G. J. Hine, in Instrumentation in Nuclear Medicine (G. J. Hine, ed.), Academic Press, New York, 1967, p. 95.

2. Y. Kobayashi and D. V. Maudsley, in Liquid Scintillation Counting--Recent Advances (P. Stanley and B. Scoggins, eds.), Academic Press, New York, 1974.

3. F. N. Hayes, B. S. Rogers, and P. C. Sanders, Nucleonics, 13:46 (1955).

4. H. Kallmann and M. Furst, Nucleonics, 8:32 (1951).

5. M. Furst, H. Kallmann, and F. H. Brown, Nucleonics, 13:58 (1955).

6. F. N. Hayes, D. G. Ott, Y. N. Kerr, and B. S. Rogers, Nucleonics, 13:38 (1955).

7. D. G. Ott, F. N. Hayes, E. Hansbury, and Y. N. Kerr, J. Am. Chem. Soc., 79:5448 (1957).

8. D. G. Ott, Y. N. Kerr, F. N. Hayes, and E. Hansbury, J. Org. Chem., 25:872 (1960).

9. P. H. Williams and T. Florkowski, in Radioactive Dating and Methods of Low Level Counting, IAEA, Vienna, 1967, p. 703.

10. J. C. Turner, Int. J. Appl. Radiation Isotopes, 19:557 (1968).

11. W. Schöniger, Microchim. Acta, 1:123 (1955).

12. F. Kalberer and J. Rutschmann, Helv. Chim. Acta, 44:1956 (1961).

13. R. G. Kelly, E. A. Peets, S. Gordon, and D. A. Buyske, Anal. Biochem., 2:267 (1961).

14. H. Jeffay and J. Alvarez, Anal. Biochem., 2:506 (1961).

15. E. A. Peets, J. R. Florini, and D. A. Buyske, Anal. Chem., 32:1465 (1960).

16. H. I. Jacobson, G. N. Gupta, C. Fernandez, S. Hendrix, and E. V. Jensen, Arch. Biochem. Biophys., 86:89 (1960).

17. R. P. Parker and R. H. Elrick, in The Current Status of Liquid Scintillation Counting (E. D. Bransome, Jr., ed.), Grune and Stratton, New York, 1970, p. 110.

18. H. H. Ross in Ref. 17, p. 123.

19. R. M. Fink, C. E. Dent, and K. Fink, Nature, 160:801 (1947).

20. S. Prydz, Anal. Chem. 45:2317 (1973).

21. U. Lüthi and P. G. Waser, Nature, 205:1190 (1965).

22. J. F. Koren, T. B. Melo, and S. Prydz, J. Chromatog., 46:129 (1970).

23. U. Lüthi and P. G. Waser, Atomlight, No. 50, New England Nuclear Corp., Boston, 1966.

24. B. Goldrick and J. Hirsch, J. Lipid Res., 4:482 (1963).

25. F. Snyder in Ref. 17, p. 248.

26. L. H. Mason, H. J. Dutton, and L. R. Blair, J. Chromatog., 2:322 (1959).

27. R. Wolfgang and F. S. Rowland, Anal. Chem., 30:903 (1958).

28. A. Karmen, Packard Technical Bulletin No. 14, Packard Instrument Co., Inc., Downers Grove, Ill. (1965).

29. E. Schram and R. Lombaert, Biochem. J., 66:21P (1957).

30. D. Steinberg, Nature, 182:740 (1958).

31. J. A. Hunt, Anal. Biochem., 23:289 (1968).

32. J. A. Hunt, Biochem J., 116:199 (1970).

33. S. Ullberg, Acta Radiol. Suppl., 118 (1954).

34. W. E. Stumpf and L. J. Roth, Isotopes in Experimental Pharmacology, University of Chicago Press, Chicago, Ill., 1965, p. 133.

35. B. G. Mobbs, J. Endocrinol., 27:129 (1963).

36. W. E. Stumpf and L. J. Roth, J. Histochem. Cytochem., 14:274 (1966).

37. J. C. DePaepe, Nature, 185:264 (1960).

38. W. E. Stumpf and L. J. Roth, Adv. Tracer Meth., 4:113 (1968).

39. J. Jacob, Int. Rev. Cytol., 30:91 (1971).

40. J. L. Liquier-Milward, Nature, 177:619 (1956).

41. R. T. O'Brien and L. A. George, Nature, 183:1461 (1959).

42. R. P. Van Tubergen, J. Biophys. Biochem. Cytol., 9:219 (1961).

43. L. G. Caro and R. P. Van Tubergen, J. Cell Biol., 15:173 (1962).

44. L. Backmann and M. M. Salpeter, J. Cell Biol., 33:299 (1967).

45. B. M. Kopriwa, J. Histochem. Cytochem., 15:501 (1967).

46. M. M. Salpeter, Meth. Cell Physiol., 2:229 (1966).

Chapter 5

THE PREPARATION OF RADIOTRACER COMPOUNDS

John R. Catch

Controller of Technical Development
The Radiochemical Centre, Limited
Amersham, Buckinghamshire
England

I. INTRODUCTION

Radioactive tracer compounds are mostly used in two fields of work which interact but which are in practice fairly distinct from each other: research in the life sciences and clinical diagnosis.

This chapter is written primarily with the first of these fields in mind.

The distinction between the two fields, which is reflected in the present-day isotope industry, was not always so clear as it is now. Diagnostic uses were once highly experimental, as research uses still are; the initiative in innovation was wholly with the clinical user, who would call on the isotope supplier to provide a labeled chemical, and neither was bothered with regulatory controls. The scale of use was fairly small, but even so the very good record of safety in nuclear medicine does credit to all concerned.

Growing recognition of the usefulness of tracer methods in diagnosis, both for in vivo scintigraphy and--even more so--for in vitro assays, has been one major cause of a change in outlook. There is a large and rapidly growing volume of routine uses in nuclear medicine and clinical biochemistry which call for reliable, readily available, simply used, and reasonably cheap reagents or "kits." The medical user is commonly a clinician with no particular interest in isotopes as such, any more than in the technicalities of ultrasonics or electrocardiography or thin-layer chromatography, and the laboratory operator may be a junior technician.

The second major influence on isotopes in medicine, which follows naturally from the above, has been the extension of drug legislation to cover radioactive materials.

The production of a reagent or kit, thoroughly tested in all likely circumstances, reliable, simple, and legally approved is much more expensive and time consuming than supply of a research chemical. Isotope suppliers naturally concentrate on major rather than minor foreseeable needs and contribute increasingly to actual innovation, of principle as well as of detail. They are less able to provide for a wide range of speculative uses, and are (in short) conforming more and more to the pattern of pharmaceutical industry as a whole. There is some loss in this for the enterprising individual in nuclear medicine, who cannot so easily try out a new idea, but it is probably more than offset by the raising of standards and extension of clinical use.

The standards of quality in the supply of labeled research chemicals have certainly improved over the years, but they remain, and must by their nature remain, experimental materials--materials for experiment. The almost infinite variety of research uses of labeled compounds very severely limits the amount of standardization and evaluation in use which is practicable. The present writer has discussed this more fully elsewhere [1,2]. The research worker is on the frontier of scientific knowledge, whereas the clinical user is "behind the lines" in occupied country. A research project using one tracer compound commonly suggests further experiments needing half-a-dozen others, so that the demand for more, different, compounds increases exponentially. There must always be a limit to what can be economically manufactured and supplied, so that the researcher must often prepare labeled compounds for himself, as an ancillary activity to his investigation which sometimes proves to be the most laborious part of it.

There have sometimes been proposals for sharing such "Do-It-Yourself" tracer compounds, or for overcoming in other ways the gaps in the range of commercially available compounds. Sharing is practical on a small scale on a basis of personal communication. Any more extended scheme faces big problems of administration and finance. Some special funded exercises, such as the supply of labeled steroids by the National Institutes of Health in the United States and by the Medical Research Council in Britain, have been valuable. Other proposals or attempts at cooperative distribution have failed, for the basic reason that tracer compounds are inextricably bound up with the research for which they are wanted, so that cooperation in tracer compounds is a very big step toward cooperation in research programs. It is unrealistic to expect such cooperation, which might also be held to compromise the freedom of scientific exploration, on any but a small scale. Apart from this, the predictable problems of correspondence, records, storage, quality control, and finance appear insuperable to anyone with experience in the supply of radioactive tracer compounds.

This chapter does not attempt to deal with "hot-atom" or recoil labeling, interesting though such work is, except in the rare case of its producing a useful tracer compound. Examples are selected to illustrate general principles, recent progress, novel achievements, and problems. A very large quantity of good work is therefore passed by because it applies well-known or predictable methods. The selection of examples inevitably reflects personal judgment. There is a special problem in biosynthetic work in distinguishing between a "preparative" method (producing a useful labeled compound) and a method demonstrating the biosynthetic route, and the distinction has to be a matter of opinion; there are no simple criteria for selection. The word "useful" is inexact; "useful for what purpose?" is the true question, to which there can be no definitive general answer.

II. OBJECTIVES AND CONSTRAINTS

These include choice of the nuclide, which includes choice of isotopic or nonisotopic labeling, position of labeling, specific activity, and criteria of purity. They all depend primarily on the experiment to be made. There is usually some freedom of choice in nuclide and position of labeling, which is fortunate. Basic information for this discussion will be found in Chapters 1-4, and related material will be found in many places in later chapters.

A. Choice of Nuclide

Attempts to use easy and quick nonisotopic labeling in the 1940s and 1950s, both in biochemical and in physicochemical tracing, were often unfortunate and discredited this approach. It survived for protein labeling (with ^{131}I or ^{125}I), for which there was and is little alternative, and interest is now growing again in the use with "modified" molecules for biochemical assay procedures [3] which is very promising.

It remains true that, for tracer investigations and metabolism, strictly isotopic labeling is always safer and often obligatory. Even with larger molecules such as proteins, nonisotopic labeling with iodine has its problems.

At the risk of stating the obvious it may be noted at this point that measurements of radioactivity follow the labeled *atom(s)* of the experiment. Any conclusion about its molecular context requires other, usually chemical, evidence.

Beyond this the questions are fairly obvious: cost, ease of labeling, sensitivity needed, the evidence sought, and (rarely) the duration of the experiment: rarely, because the most widely eligible nuclides are ^{14}C and ^{3}H, and it is not often that one has any option to use ^{32}P, ^{33}P, ^{35}S, ^{75}Se, ^{131}I, or ^{125}I as alternatives. Carbon-11 deserves mention if only because of the tours de force it has prompted [4], but its potential is very severely restricted by the half-life of 20 min.

Cost is of two kinds: purchase of nuclide or intermediate, and "own time." A college laboratory will pay more attention to the first, because it has to be authorized in cash; and "own time" will often have its own educational value in (for example) synthesizing a labeled compound which might have been bought. If a realistic cost is attributed to "own time" it usually pays to buy labeled compounds, unless there is an unusually large and recurrent need.

A short-lived nuclide such as ^{32}P is relatively cheap in itself, but in a long experimental program the cost of repetitive synthesis soon overshadows the basic cost of the nuclide.

In the common choice between tritium and ^{14}C, cost, ease of labeling, and sensitivity all favor tritium. Tritium is also technically better for autoradiography at the cellular and subcellular levels. But it is a tracer for hydrogen (though sometimes subject to large isotope effects) and not for carbon, and when it is used as an accessory tracer for carbon it is necessary to think carefully about the stability of carbon-hydrogen bonds involved.

The situation may be summarized something like this: For tracing unknown carbon skeleton transformations, ^{14}C is always safer. With enough knowledge of the stability of tritium in the planned experiment, tritium is likely to be first choice. Halogen isotopes, except as strictly isotopic tracers, are useful only for very large molecules (and then with circumspection) and in specific uses for which they can be carefully checked. ^{125}I is much the most useful of them. Isotopes of phosphorus, and ^{36}Cl, are rarely useful except as isotopic tracers. ^{35}S is rather different in that it might in principle be useful as an accessory tracer for an intact carbon structure, although it is unlikely to have any advantage over ^{14}C. ^{75}Se, with its convenient half-life (118.5d) and γ energy, and its history of use in selenomethionine as a pancreas scanning agent, seems likely to attract interest in coming years as a label for intact organic molecules.

B. Position of Labeling

This is determined by existing information (e.g., on metabolism) and by the information sought in the experiment. If the problem is a very specific one there may be no choice at all. If there will be no metabolic transformation the choice may be very wide (for isotopes other than ^{14}C and ^{3}H there is usually no choice, at least in small molecules).

The selection of one of several acceptable alternatives will depend on the ease of the synthesis (in relation to the quantity and specific activity required) and this may depend on ready availability of a labeled or unlabeled precursor.

It will depend also on technical problems associated with the scale of working. It would not be very useful in the limited space available, nor is it necessary, to discuss the "technology" of isotopic syntheses; the subject is well documented [6-12].

Cholesterol (I) will illustrate the complexities of isotopic isomerism and the limitations in ^{14}C-labeling of a familiar substance.

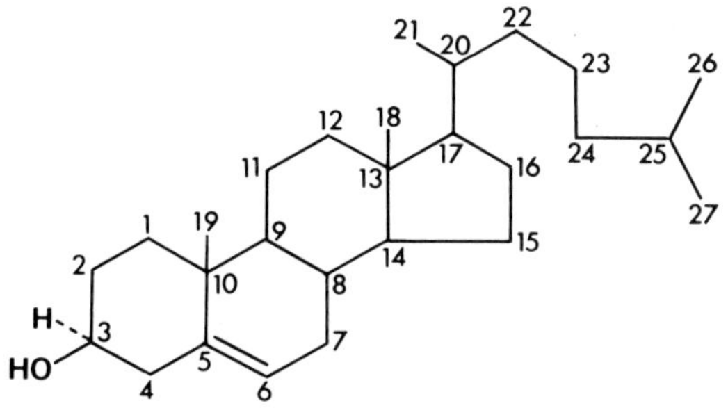

(I)

It has been labeled specifically in positions 3 [13], 4 [14], 20 [15], 21 [16], 23 [17], and 26 [18]. There has probably been no great need to label any of the 20 other positions, but some of them would be difficult, and it seems that position 4 has met most needs for labeling in the nucleus.

C. Specific Activity

The task should not be made unnecessarily difficult by setting this higher than the experiment needs, and needs vary enormously, from 2.8×10^6Ci/g-atom (^{125}I for extreme sensitivity in assay systems) to 7×10^{-6}Ci/g-atom (radiation-labeled [^{14}C]isonicotinic hydrazide [21]); a range of nearly 10^{12}.

In favorable circumstances, using "clean" reactions and microanalytic procedures, work on the microgram scale is possible. Conventional synthesis, particularly extended multistage synthesis, begins to become difficult and inefficient below 1-2 mmol--say 100-200 mg of a fairly simple compound. A Friedel-Crafts reaction with [^{14}C]acetyl chloride, or an aromatic sulfonation with $^{35}SO_3$, can only be done on a smaller scale by a tour de force requiring more specialized experience than a nonspecialist will be willing to acquire.

The practical problem is summarized in the elementary equation:

$$\text{Molar specific activity (Ci/mol)} = \frac{\text{Total activity (Ci)}}{\text{Quantity (mol)}} \tag{1}$$

Small chemical scale makes some operations easier (e.g., thin-layer chromatography for preparative separations) and others more difficult (crystallization). It can make it very difficult to attain or even determine chemical (as distinct from radiochemical) purity. It may not be possible to assure specific activity with certainty. Fatty acids uniformly labeled with ^{14}C may for example be prepared from green algae (see Section III B 1) at specific activities in excess of 1 Ci/mmol, and in good radiochemical purity, but in weights so small that they are uncertain (±10%) by direct measurement. Spectroscopic measurements of mass are often uncertain because of impurities (from solvents, paper, adsorbents, resins, etc.).

Dilution with pure carrier is one of the stratagems of labeling, either to facilitate a reaction or to help purify a product in good yield. It must be remembered that any impurities which do remain may have a much higher specific activity than the diluted product. Labeled compounds at high specific activity have often a lower chemical purity than high-grade unlabeled reagents (because rigorous chemical purification is impossible or too wasteful); reactions are often quite sensitive to chemical purity, and dilution may greatly improve yields, which may otherwise be much lower than in "dry" runs with chemically pure reagents. Potassium cyanide (cf. Ref. 19), diethylmalonate, and (of course) all materials sensitive to traces of water or oxygen, are examples of compounds which give this kind of trouble.

Molar specific activities are more convenient and informative, for nearly all purposes, than specific activities (as in the older sense) referred to mass.

D. Criteria of Purity

Experimental requirements vary greatly in kind and degree [1]. Strange as it may seem, research workers do not always recognize that the adjective "pure" has no absolute meaning, nor do they always consider what demands an experiment may make for purity of a tracer compound.

The preparative route may occasionally be significant in this respect. [^{14}C]Benzene prepared by polymerization of [^{14}C]acetylene is liable to contain traces of [^{14}C]styrene [20]: that prepared by ring expansion of [*methyl*-^{14}C]pentane [21] will not.

Thought should always be given to the value, limits of error and possible artifacts in methods of purification and analysis (see, for example, Chap. 6). The writer has known more than one occasion when a user, with the best of intentions, "purified" a tracer compound by an insufficiently tested method, leaving it much less pure than when he started.

III. CARBON-14 LABELING

Carbon-14 is extracted from irradiated targets of aluminum nitride or beryllium nitride. The yield is only 2.5 mCi per annum per gram nitrogen at a flux of 10^{13}n cm^{-2} sec^{-1}. Annual world requirements are now measured in hundreds of curies, so that long irradiations of large quantities of target are necessary. The isotopic abundance of the carbon-14 is dependent mainly on the carbon-12 content of the target. By keeping this to a low value, and by long irradiations, abundances of 90-95% are now readily attainable. At 100% abundance 14 mg of elementary carbon-14 is equivalent to 62.4 mCi.

The most convenient form for primary extraction is carbon dioxide and this is the starting point for all syntheses. Attempts to use direct syntheses from "hot atom" reactions have proved of very limited use because only low specific activities and small quantities can be produced in this way. The approach is therefore by chemical synthesis (which includes use of more or less purified enzymes) or biosynthesis in vivo--"isotope farming."

The need to start from carbon dioxide is the most fundamental controlling factor in ^{14}C-labeling but a wide range of ^{14}C-labeled synthetic intermediates may now be purchased commercially, and there is an extensive and useful literature of ^{14}C preparations [7,8,10,11].

The research worker preparing a special compound for his own use should nevertheless be aware of the various approaches, even if he only completes a synthesis from a purchased intermediate. It will help to choose the best route, it will show the possibilities for alternative labeling, and may give information on likely trace impurities.

The two most important factors in practice are the small scale of working and the rather high cost of the isotope. One practical point may be noted: the need to check a synthesis throughout with *representative materials*. Yields at successive stages may be checked with pure reagents, but when the whole synthesis is carried through from carbon dioxide the overall yield is likely to be lower, sometimes much lower, since yields at some stages may be greatly affected by small amounts of impurity. In small-scale total synthesis it is not always possible to purify thoroughly at intermediate stages.

A. Carbon-14 Synthesis

Synthesis with carbon-14 is systematic; there are 12 possible primary stages:

1. Carboxylation of an organometallic compound to give a carboxyl-labeled acid

2. Reduction to methanol
3. Reduction to cyanide
4. Reduction to carbide, for acetylene preparation
5. Reduction to formate
6. Reduction to cyanamide
7. Reduction to carbon monoxide
8. Conversion to urea via ammonium carbonate
9. Reduction to methane
10. Reduction to formaldehyde
11. Reduction to elemental carbon
12. Conversion to an alkyl carbonate via silver carbonate

Of these only the first six are of much importance. The manner in which these are exploited is exemplified in Table 1. For further examples see *The Radiochemical Manual* [22]. Such tables represent only a fraction of known isotopic syntheses.

When ^{14}C-labeling was novel, and the nuclide and its simpler chemical forms were considerably more expensive than they are now, there was always a challenge to develop novel preparative routes for economy of ^{14}C. This challenge is still there for the commercial manufacturer, but for the likely reader of this book (see Section I) it will often be cheaper in the event to use a well-explored existing synthetic route, even if it seems wasteful of ^{14}C. It takes a great deal of time to verify a novel synthesis with a fairly expensive nuclide to be used, and time is not free.

The important basic conversions of $^{14}CO_2$ were largely worked out in the late 1940s and early 1950s, and there have not been many further advances. A recent patent [23] describes the preparation of hydrogen [^{14}C]cyanide from [^{14}C]methane and ammonia at high temperatures, using in fact a well-known reaction in large-scale manufacture. It appears to be a material improvement on older methods. Trimerization of [^{14}C]acetylene using a Ziegler catalyst [24] has superseded other methods, such as Turner's synthesis [25]. But few readers of this text will wish to involve themselves in these primary conversions of ^{14}C, which are a rather specialized technology.

TABLE 1

Carbon-14 Compounds--Syntheses via Carboxyl-Labeled Acids*

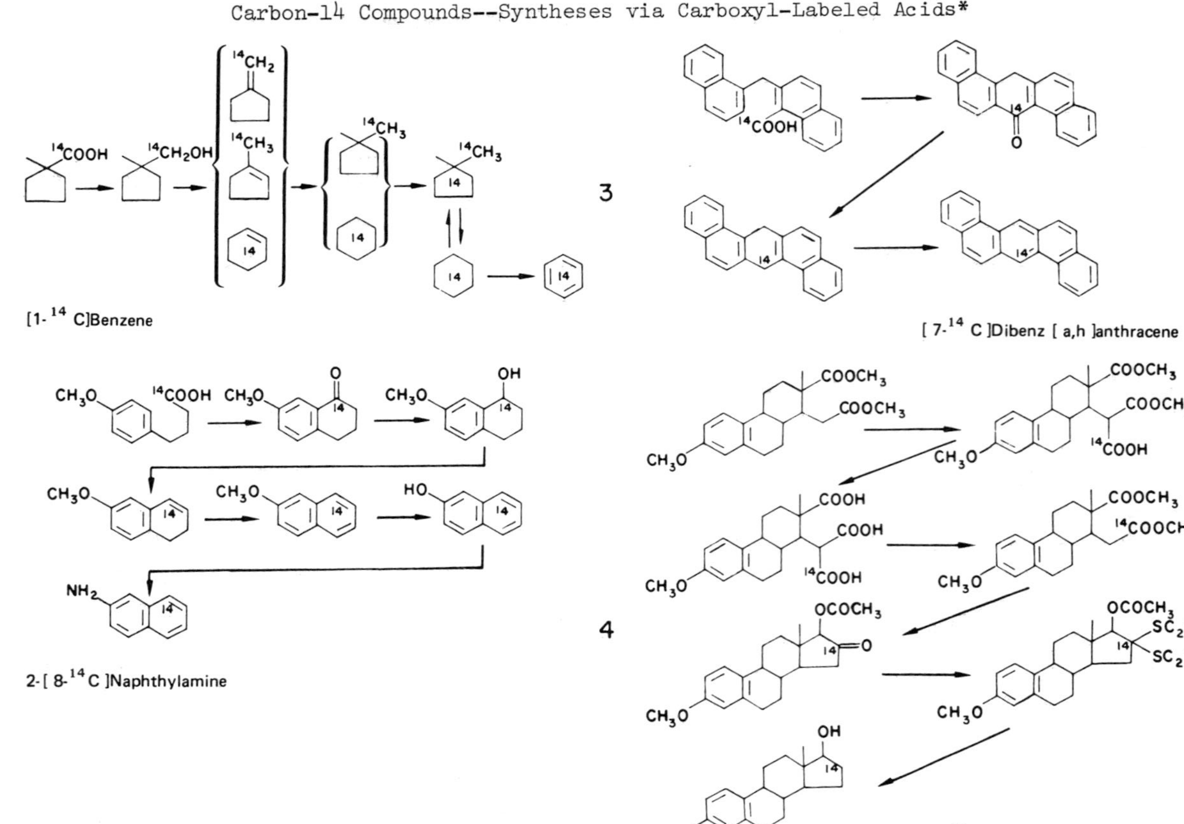

5 2-[Methyl-^{14}C]-1,4-naphthoquinone

6 2-Methyl-1,4-[4-^{14}C] naphthoquinone

7 2-Methyl-1,4-[8-^{14}C] naphthoquinone

8 3-Phenyl [3-^{14}C] alanine

9 [β-^{14}C] Styrine

10 $(CH_3)_2CH^{14}COOH \longrightarrow (CH_3)_2CH^{14}CH_2OH \longrightarrow [(CH_3)_2CH^{14}CH_2-]_2Cd$

[24-^{14}C] Cholesterol

*Reproduced by permission from <u>Radiochemical Manual</u> (1966).

The advantages of synthesis, as compared with biosynthesis, are specificity of labeling, control over the products, and easier purification. Synthetic labeling with ^{14}C is, with rare exceptions, *completely* specific. Unforeseen rearrangements are very exceptional; a synthesis proceeds to greater "order" and selection by purification. Any conclusion that a synthetic ^{14}C compound is not specifically labeled must be examined carefully; it most likely means a misinterpretation of the evidence (see Chap. 6).

Synthesis can be directed for one product rather than another, and yields can often be improved by careful study of the reaction conditions. There is little such control over a living organism. Synthetic products are easier to purify than products of "isotope farming", which have to be separated from very complex mixtures.

Disadvantages of synthesis include difficulties with stereoisomers and limitation of specific activity. Resolution of optically active compounds on a small scale is troublesome and becomes very inefficient when two or three active centers are involved, unless stereospecific enzymes can be used. Multiple labeling or uniform labeling is rarely practicable, and this limits molar specific activities.

The limitations of ^{14}C synthesis are otherwise those of organic synthesis in general. Proteins and other natural macromolecules are difficult to label usefully. This is particularly so if they occur in small quantities (protein hormones, etc.); and isotopic labeling of human proteins with ^{14}C is not possible. The rapidly increasing knowledge of the mechanism of such syntheses in nature, perhaps allied with expertise in cultures of specific cells, may conceivably overcome this difficulty in time. Recent advances have allowed of the synthesis of extended polypeptide chains with controlled sequences and configurations, but corresponding labeled syntheses at useful specific activities are not exactly easy.

The specific labeling of ring structures (other than pyrimidines and purines) is another problem in which no general advance has been made, but this may be partly because there is no great need. Specific tritium labeling can often be used with confidence for problems which formerly would have demanded difficult ^{14}C syntheses.

B. Carbon-14 Biosynthesis

Biosynthesis is effective and important for labeling a rather limited range of common intermediary metabolites: the common amino acids, carbohydrates and nucleosides, which are readily available in their natural configurations by growing green algae on $^{14}CO_2$ or by photosynthesis in detached plant leaves. This approach, properly used, can give high specific activities (up to 90-95% isotopic abundance in all carbon atoms) and good yields. Few readers of this text will be concerned with work in this limited field; the commercial supply is so highly developed that it is scarcely worthwhile for any individual to make his own amino acids by photosynthesis, particularly since he is unlikely to need more than a few of the wide range inescapably produced.

Beyond this point the limitations of biosynthesis quickly become apparent. Whatever one can do by selective breeding, mutation, environmental control, enzyme blocking, and so on, living organisms taken as a whole are not very amenable to control of their biochemical products; and they are sensitive to radiation.

The labeling of bilirubin from [2-^{14}C]glycine [26] is a satisfactory although not a very simple biosynthesis, and has a hopeful basis in that incorporation of eight labeled atoms from [2-^{14}C]glycine into the hemoglobin of duck red cells is quite efficient and one might hope for fairly efficient conversion of the hemoglobin in the dog to bilirubin. The yield was >50% of the glycine used, and the specific activity about 0.3 mCi/mmol. A direct conversion of [2-^{14}C]glycine gave a yield of perhaps 0.5% [30].

Other examples are less successful. Certain plant products, such as alkaloids and cardiac glycosides, have been labeled in small quantities and at rather low specific activities by growing the plants in $^{14}CO_2$. This aspect of "isotope farming" has been disappointing, the most serious problem being the sensitivity of plants to radiation; pronounced abnormalities appear at a level of 1 mCi/g carbon (12 mCi/g-atom). If it were not for this, the second major difficulty, which is the very small conversion of carbon dioxide into the desired product, might possibly have been overcome by using and recycling large amounts of ^{14}C. The yield of digitoxin, for example, appears to have been about 0.02% of the $^{14}CO_2$ used [27].

Good yields of sucrose, up to 70% of $^{14}CO_2$, are obtainable in detached leaves [28] even with ^{14}C at >90% isotopic abundance because the time of exposure is short (hours) and no reproductive growth is required.

The microscopic green alga *Chlorella* is very resistant to radiation damage and can be grown through several reproductive cycles at >90% isotopic abundance, although the cells are abnormally enlarged and the growth rate progressively retarded. This growth is remarkable, for there will be many thousands of disintegrations in each cell during each cycle, even though much of the energy of the β particles is dissipated outside the cell, in the aqueous medium. A still better, blue-green, alga has recently been identified in *Anacystis nidulans* [29,29a], which is even more tolerant of self-irradiation than is *Chlorella*, and has the specific advantage of giving a practicable yield of deoxyribonucleic acid. Attempts to use *E. coli* for this purpose, with a substrate of [U-^{14}C]glucose at high specific activity, were unsuccessful, apparently because of sensitivity to radiation [31]. The commonest sugars and amino acids are thus labeled readily enough, but corresponding approaches for uncommon ones do not in general exist.

1. Carbon Dioxide Substrate

[^{14}C]Carbon dioxide is the cheapest substrate but implies a photosynthetic organism; other CO_2 assimilation reactions are either too inefficient or have practical difficulties. One would not, for example, choose to grow a [^{14}C]methane-reducing bacterium with large amounts of ^{14}C without great care to avoid risk of fire or explosion. Except for tobacco or *Canna* leaves [28] and green algae [29,29a] already referred to, the approach is not much used; it is rarely possible to find an organism which will give a reasonable conversion of ^{14}C into the compound desired.

Leaf and algal photosyntheses, properly carried out, give (as already mentioned) very high specific activities, as a consequence of labeling in all carbon positions: and this labeling may be made uniform within the limits of experimental detection. This has at times been questioned (for a remarkable recent example see Ref. 32), but the evidence of ninhydrin decarboxylation of amino acids [33], degradative oxidation of L-[U-^{14}C]lysine [33a], and verified degradation of D-glucose [34] is conclusive; and it is supported by the direct physical evidence of mass spectrometry [35].

The common sugars and amino acids are of course very useful as starting materials for other labeled compounds, using chemical or biological methods.

From the practical point of view leaf and algal photosyntheses are quite simple, except that the reliable growth of large quantities of ^{14}C-algae is not quite straightforward either for apparatus or for conditions of growth. The greater part of the work and skill is in separating and purifying the products; the difficulty in determining the chemical purity and hence the true specific activity has been mentioned above (Section III C). At The Radiochemical Centre cultures of [^{14}C]*Anacystis* are grown only once or twice in a year, and may take less than a week; but the subsequent separations go on continuously throughout the year.

2. Organic Substrate

Use of a suitable organic substrate gives a much wider choice of biosynthetic organisms. The conversions of [^{14}C]glucuronic acid into the corresponding 2-keto-3-deoxy-D-glucotate, by a mutant of *E. coli*, is a good example of such a preparation (conversion 20%) [36]. It is interesting that the author found that the yield fell to 10-15% on raising the specific activity of the culture to 0.1 mCi/ml, and this was attributed to a radiation effect on the cells.

The preparation of *trans*-11-[1-^{14}C]octadecenoic acid from [1-^{14}C]linoleic acid in 60% yield, by microorganisms from sheep rumen, is another good recent example [37].

A conspectus of earlier examples will be found in Chapter 4 of Ref. 11. The caution there expressed about reported yields may be repeated. Some publications on biosynthetic labeling still appear with unsatisfactory basic information. An example is the reported yield and specific activity for [^{14}C]aflatoxin B_1 from [1,2-^{14}C]acetate [38]. The specific activity for the product must be in error, but a correct value cannot be deduced from the given data.

The risk of nonspecific labeling probably discourages some uses of biosynthesis. [3,4-^{14}C]Glucose, in which from 4-25% of the activity may "wander" into carbon-1, -2, -5, and -6 from a specifically labeled substrate [39,40], is a classical example. Carbon-14 labeling of this kind, neither completely specific nor uniform, but "general" (G) is of very limited value. It is not even easy to determine exactly what the pattern of labeling really is; as Turner's work showed [34] it needs much careful work to ensure that degradation reactions are properly interpreted. The reaction

$$\begin{array}{lc}
(1) & \mathrm{CHO} \\
 & | \\
(2) & \mathrm{H\ C\ OH} \\
 & | \\
(3) & \mathrm{HO\ C\ H} \\
 & | \\
(4) & \mathrm{H\ C\ OH} \\
 & | \\
(5) & \mathrm{H\ C\ OH} \\
 & | \\
(6) & \mathrm{CH_2OH}
\end{array}
\qquad
\begin{array}{ll}
(1) & \mathrm{CO_2} \\
 & + \\
(2) & \mathrm{CH_3} \\
 & | \\
(3) & \mathrm{CH_2OH} \\
 & + \\
(4) & \mathrm{COOH} \\
 & | \\
(5) & \mathrm{CHOH} \\
 & | \\
(6) & \mathrm{CH_2OH}
\end{array}
\tag{2}$$

for the fermentation of D-glucose by L. mesenteroides undoubtedly represents truth, but not the whole truth.

Any expectation that a more detailed knowledge of biosynthetic routes will make efficient biosyntheses possible from specific labeled precursors may be optimistic. The synthetic chemist's concept of a specific route with efficient conversions is, after all, not characteristic of natural processes.

IV. TRITIUM LABELING

It is usually much easier to introduce tritium into an organic molecule than to introduce a carbon isotope. Unfortunately, the tritium often comes out again relatively easily. Nevertheless, the value of tritium as an auxiliary tracer for carbon is well established and has brought about a pronounced raising of standards in the preparation and use of tritium compounds. Rational synthetic methods predominate; nonspecific exchange reactions become less and less important, radiation-induced syntheses are rarely used, and biosynthesis is much less important than with ^{14}C.

Chemists approaching tritium labeling for the first time may be disconcerted by lack of stoichiometry for yields and specific activities. Much is now known about the stability of tritium in organic molecules but not enough to predict certainly what will happen in a new reaction. It is rather unusual for theoretical specific activities to be reached, although they are occasionally exceeded (implying of course additional substitution elsewhere than was foreseen), and this very fact calls for caution in interpreting an apparent specific activity.

The exchange of tritium, more or less, with many solvents, and its cheapness, cause the use to be very wasteful. For exchange labeling in particular the radioactive yields are very small, and nearly all the isotope eventually has to be disposed of as waste.

Uniform labeling with tritium is hardly ever possible because hydrogen atoms vary so much in stability; conversely, completely specific labeling is not always easy. Only purely chemical (not catalytic) methods may be relied upon with confidence.

A. Synthetic Labeling

The primary labeling processes with tritium are:

1. Exchange with heterogeneous catalysis
2. Exchange with homogeneous catalysis
3. Exchange catalyzed by radiation
4. Substitution by chemical reduction
5. Addition (hydrogenation)

These, and many other aspects of work with tritium compounds, are considered in great detail in Evans' book [5].

1. Exchange with Heterogeneous Catalysis

This is often the easiest and quickest method of tritium labeling. The unlabeled organic substance is heated, with, for example, tritiated water or acetic acid and a catalyst such as palladium or platinum. After disposal of the excess solvent, labile tritium is removed by repeated equilibration with water or other appropriate solvents and the product purified by suitable methods. Chemically catalyzed exchange can give quite high specific activities (1-20 Ci/mmol). It results in labeling which is general but rarely uniform. The resulting pattern of labeling may be determined by chemical degradation [41], or by NMR spectroscopy [42], but it is not exactly easy, certain, or quick, and it is not very often done. The method is not, of course, applicable to compounds which are unstable under the conditions used, and it is not often possible to obtain a theoretical equilibrium concentration without excessive breakdown of the starting material. Purification of the products is not usually too difficult, and the method, although rather unpredictable, has been generally useful. The erratic results of these exchange reactions are suspected to be due to variation in the activity of the catalyst used, and greater reliability could no doubt be attained by more thorough investigation. Tritium is, however, cheap in elemental form, and general exchange is less and less used, so there is no great urge to improve these methods. However, numerous compounds can be labeled by catalyzed exchange in solution with gaseous tritium at room temperature [5]. Labeling by this technique is often specific [42a].

2. Exchange with Homogeneous Catalysis

Hydrogen atoms in "labile" positions undergo rapid exchange and equilibration with tritium oxide. Examples are 2-naphthoic acid and malonic acid.

$$C_{10}H_7COOH + T_2O \rightleftharpoons C_{10}H_7COOT + THO \qquad (3)$$

$$CH_2(COOH)_2 + 3T_2O \rightleftharpoons CHT(COOT)_2 + 3THO$$

The products are not of much interest in themselves, because the label will exchange off again equally readily in aqueous solution; but by reaction of the substituted naphthoic acid with diazomethane, followed by hydrolysis, tritium-labeled methanol is readily obtained. Similarly, by decarboxylation of the labeled malonic acid

$$C_{10}H_7COOT + CH_2N_2 \longrightarrow C_{10}H_7COOCH_2T + N_2 \xrightarrow{NaOH} CH_2TOH \qquad (4)$$

followed by removal of the labile tritium, one obtains acetic acid labeled with tritium in the methyl group. Exchanges of this kind are necessarily possible only in particular cases, but, as in the examples quoted, are useful as practical methods.

Hydrogen exchanges catalyzed by acid reagents, although extensively studied for their theoretical interest, are not much used in practical tritium labeling. This no doubt is because the conditions required for a high degree of substitution are rather drastic, so that relatively few compounds will survive them; and for those which will, for example, the simpler aromatic compounds, more efficient alternative methods of labeling are already available.

We may note here studies of hydrogen exchanges using homogeneous catalysis by noble metals in solution [49], but these have little practical effect as yet for tritium labeling.

3. Exchange Catalyzed by Radiation

Radiation-induced labeling as described by Wilzbach and others [43,44] has often appealed to the beginner in tracer work because of its superficial simplicity. In its simplest form--there are elaborations of the method--the finely divided organic compound is exposed to elemental tritium. Hydrogen exchange occurs in some measure with most compounds, but is complicated by side reactions such as additions to unsaturated centers and, above all, by extensive radiation decomposition. This decomposition seriously limits the specific activities attainable, and unfortunately is most marked with large and sensitive molecules for which the method would otherwise be particularly valuable. As a consequence, a very complex mixture of labeled compounds results, and special care is necessary in purification [45]. On really rigorous purification and analysis it will often be found that the desired product is present only in very small concentration compared with labeled impurities. Observations published at the Brussels Conference in 1963 [46], two of them dealing with lysozyme, show interesting differences of opinion on the success of the Wilzbach labeling of protein materials, and suggest that much care is necessary in using it. The success of labeling by the Wilzbach method or its variants is very unpredictable, although it can occasionally score a pronounced success as in the example of atropine [47]. The method is, however, at least relatively simple to try, although it is seldom that of choice.

Labeling by recoil tritons, e.g., by neutron irradiation of a mixture of an organic substance with a lithium salt, may also be mentioned under this heading. As a means of preparing labeled compounds it is less useful than the Wilzbach method, and is at present of academic interest only.

4. Substitution by Chemical Reduction

Reduction of a halogen compound by a metal, such as zinc, in the presence of tritium oxide or tritium labeled acetic acid, of the general type

$$RX + M^{2+} + T_2O \rightarrow RT + MOTX \qquad (5)$$

will produce a compound with a tritium atom in place of the original halogen.

Reductions of this kind, although sometimes used for labeling, require a large excess of tritium in the water, acetic acid, or other comparable solvent used. This large excess is avoided by converting the halide RX into an organometallic compound such as a Grignard reagent, and reacting this with an equivalent quantity of tritium oxide,

$$RMgX + T_2O \rightarrow RT + MgOTX \tag{6}$$

The method obviously lends itself to high specific activities and work on a small scale. A comparable method of chemical reduction uses sodium borotritide for reduction of carbonyl compounds

$$4R_2CO + NaBT_4 + 2H_2O \rightarrow 4R_2CTOH + NaBO_2 \tag{7}$$

while reductions with tritiated lithium borohydride are often used for labeling carbohydrates [48]. These chemical reductions can usually be relied upon to give completely specific labeling, although isotope effects and exchange reactions sometimes reduce the specific activity of the product as it would be predicted from the simple equation.

Another variant is reduction using elemental tritium and a catalyst such as platinum or palladium:

$$RX + T_2 \xrightarrow{\text{Pd or Pt}} RT + TX \tag{8}$$

This is often a very efficient method, but it should not be assumed that the labeling by this means is completely specific, since hydrogen migration can occur in the presence of catalysts. As in labeling by addition of tritium, care is needed in the selection of solvents and other conditions such as pH. Elemental tritium in the presence of platinum or palladium exchanges rapidly with water, ethanol, acetic acid, and many other solvents that are used for catalytic hydrogenations or reductions. Ethers (particularly dioxane), and esters have proved particularly useful as solvents, since they exchange less readily. High specific activities may be achieved by replacing more than one halogen atom.

5. Addition (Hydrogenation)

Addition to unsaturated centers (particularly carbon-carbon double and triple bonds) catalyzed by platinum or palladium is a generally useful and efficient method of tritium labeling. There is always some risk that exchange or hydrogen migration will occur at the same time, and it cannot be assumed that tritium will be located only at places indicated by simple addition to the double or triple bond. Another source of trouble is tritium exchange with solvent, which has been referred to in the preceding paragraph.

As with multiple halogen substitutions it is possible to get very high specific activities by adding tritium to more than one unsaturated center. At the time of writing this work seems not to be documented in the scientific literature but is established in commercial supply. Prostaglandin-E_2 is, for example, labeled in (nominally) seven positions, 5,6,8,11,12,14,15, by enzymatic synthesis from [5,6,8,9,11,12,14,15-^{3}H(n)]arachnidonic acid; [1,2,4,5,6,7-^{3}H(n)]5α-dihydrotestosterone is obtained by reduction of 1,4,6-androstatriene-17β-ol-3-one [50]. The specific activities achieved in practice, although very high, are still rather less than the theoretical maximum, as is usual.

6. SUMMARY

The value of the various methods (Section IV A 1-5) may be summarized briefly; for more complete discussion see Ref. 5:

1. Although easy, the method is not always reliable or reproducible and may not give very high specific activities. Its use is probably declining.
2. This method is only of limited use.
3. Radiation-induced labeling has been widely studied, but difficulties of purification and low specific activities are serious handicaps.

4. These methods, especially halogen-tritium replacement, are increasingly important. They are clean, give very high specific activities and more or less specific labeling.
5. Hydrogen addition is also very valuable for the high specific activities it can give.

B. Biosynthesis with Tritium

Biosynthetic labeling with tritium, except by efficient specific enzyme syntheses, has proved of little use. Problems of radiation dose limit the specific activity which can be used (Porter and Watson, 1954) with tritiated water as a substrate. Even if this were not so, any labeling comparable with the growing of ^{14}C-algae would involve very large quantities of tritium oxide if high specific activities were to be achieved. As in ^{14}C work, there is rather more scope for biosyntheses from ^{3}H-labeled organic substrates; but exchange and synthetic approaches are relatively easy with tritium so that biosynthetic methods have been less studied than those for ^{14}C.

V. LABELING WITH ISOTOPES OF HALOGENS

Halogen isotopes are mostly used as nonisotopic tracers, and interest concentrates on ^{125}I and large molecules. Other iodine isotopes (even ^{131}I) are too short-lived to be convenient; the energetic γ radiation of ^{131}I is a nuisance, and the higher specific activity of ^{125}I a real advantage for many important uses (see p. 10). ^{77}Br, ^{80}Br, ^{82}Br, and ^{36}Cl are of too limited importance (because of half-lives and cost) to merit more than a passing mention.

The size of the iodine atom, and its chemical character, severely limit the use as a nonisotopic tracer. In large molecules such as proteins the difficulty of size can be overcome; and it can be avoided in specific applications, as with labeling antigens for immunoassay [3,51] if the distortion of the molecule does not affect a highly specific reaction site.

The basic reaction used is generally either exchange or addition (9).

$$(\mathrm{H} \rightleftharpoons {}^{131}\mathrm{I}) \quad \text{or addition} \quad (\mathrm{C{=}C} \longrightarrow \mathrm{IC{-}CI}) \qquad (9)$$

Exchange is the most used, addition less so, and conventional replacements by way of diazonium salts only rarely. This is because interest is mainly in iodine-labeling of proteins and polypeptides using a replaceable hydrogen in a tyrosine residue.

A. General Problems

These have been well reviewed in detail (Ref. 51, passim; [52]).

One is the permissible degree of substitution with iodine, since a protein will often have more than one tyrosine residue. The old "rule of thumb" that the substitution should not average more than one iodine atom per protein molecule is only partly true; other important factors are the protein concerned, the experimental use, and the method of iodination used.

Another general problem, arising from the minute chemical scale often used (μg level) is the effect of impurities. These may be reactive, and take up the iodine to give irrelevant labeled impurities; or they may damage a sensitive protein; or they may interfere in some obscure way with the iodination reaction.

This irritating phenomenon, "failure to iodinate," occurs in every laboratory which does not concern itself only with the most robust methods (such as chloramine-T, used in generous excess). It has been attributed at different times to deficiency of reducing agent, excess of reducing agent, selenium or tellurium compounds derived from target material, radiolytic products such as chlorate or iodate, dust in the reagents, and maybe other causes. It has so far defied exact investigation and satisfactory explanation. Scrupulous attention to purity and cleanliness of reagents and vessels reduces but does not eliminate the problem.

B. Choice of Method

Many reagents and methods have been used, including:

1. Elemental iodine is now little used. Readily volatile and fat-soluble, it is particularly difficult to control personal exposures (although this problem arises in some degree with all methods).
2. Chloramine-T

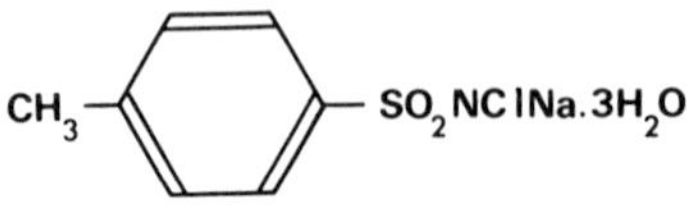

$$CH_3-C_6H_4-SO_2NClNa.3H_2O \qquad (10)$$

has provided the classical oxidizing agent since its use for this purpose was first published [53]. It is the most generally reliable method, giving high incorporation of activity, particularly if the protein will tolerate an excess of chloramine-T. It is, however, sometimes too vigorous.
3. Iodine monochloride was developed by McFarlane from 1956 onward [54] with considerable success and is still widely used as a "mild" and well-controlled reagent.
4. Electrolysis [55]
5. Lactoperoxidase [56]
6. Chlorine/Hypochlorite [57]
7. (N-succinimidyl-3-(4-hydroxy-5[^{125}I]iodophenyl)propionate; Bolton and Hunter reagent [57a].

It is difficult to give any further guidance on the choice of method for any particular protein. Unsolved problems remain, particularly in respect of stability of the labeled protein; no method of iodination tried at The Radiochemical Centre will produce a stable iodinated human fibrinogen or human growth factor, although for some other proteins the method of iodination is a major factor in stability.

C. Double Labeling

Although plasma proteins are rarely labeled strictly isotopically, this has been done with ^{14}C in classical work by McFarlane [54, 58]. If such a material is used in double-labeling experiments it is important not to iodinate the [^{14}C]protein, i.e., to work with uniodinated [^{14}C]protein mixed with iodinated [^{12}C]protein; otherwise any degradation in the iodination reaction will not be distinguished. This is one of the few examples of an experiment in which there is a real distinction between intramolecular and intermolecular double labeling.

VI. LABELING WITH OTHER NUCLIDES

Those deserving some comment are ^{32}P, ^{33}P, ^{35}S, ^{75}Se, ^{57}Co, ^{58}Co, ^{197}Hg, ^{99m}Tc, ^{90}Y, ^{169}Yb, and ^{111}In, although most of these are of specific rather than general interest. ^{15}O, ^{18}F, and ^{13}N are unluckily too short-lived to allow of their synthesis into any but the simplest labeled compounds.

A. Phosphorus Labeling

^{32}P is mostly used, being cheap and readily available, despite its rather short half-life (14.3 days). ^{33}P is hardly longer-lived enough (25 days) to support the greater cost in wide use.

Attention was concentrated in the early days of radioactive tracer work on chemical synthesis, typical objectives being labeled nerve poisons of the kind extensively used as insecticides. The chemistry of such work is conveniently summarized in Tables 2 and 3.

This kind of work with ^{32}P has discouraging features. The half-life means that it has to be repetitive, and speed and

TABLE 2
Phosphorus-32 Compounds--Routes to Intermediates

$^{32}P(n,\gamma)^{32}P$
Red phosphorus

(S)

(C)

$Al^{32}P$

(H_2SO_4)

$^{32}P_2S_5$
Phosphorus pentasulfide-P32

$^{32}P_2O_5$
Phosphorus pentoxide-P32

$^{32}PH_3$
Phosphine-P32

$^{32}S(n,p)^{32}P$
Elemental sulfur

High temperature reduction

$^{32}PO_4'''$
Carrier-free phosphate

^{32}P
Elemental
phosphorus-32

(Cl_2)

$^{32}PCl_3$
Phosphorus trichloride-P32

(PCl_5) ·

($S+AlCl_3$)

(Sb)

(O_2 or $KClO_3$)

Isotope dilution

$^{32}PSCl_3$
Thiophosphoryl chloride-P32

$^{32}POCl_3$
Phosphorus oxychloride-P32
(I)

(PCl_5)

$H_3{}^{32}PO_4$
Phosphoric acid-P32

(H_2S)

(Cl_2)

(H_2O)

$^{32}PCl_5$
Phosphorus pentachloride-P32

Nitric acid oxidation

TABLE 3

Phosphorus-32 Compounds--Syntheses from Phosphorus Trichloride

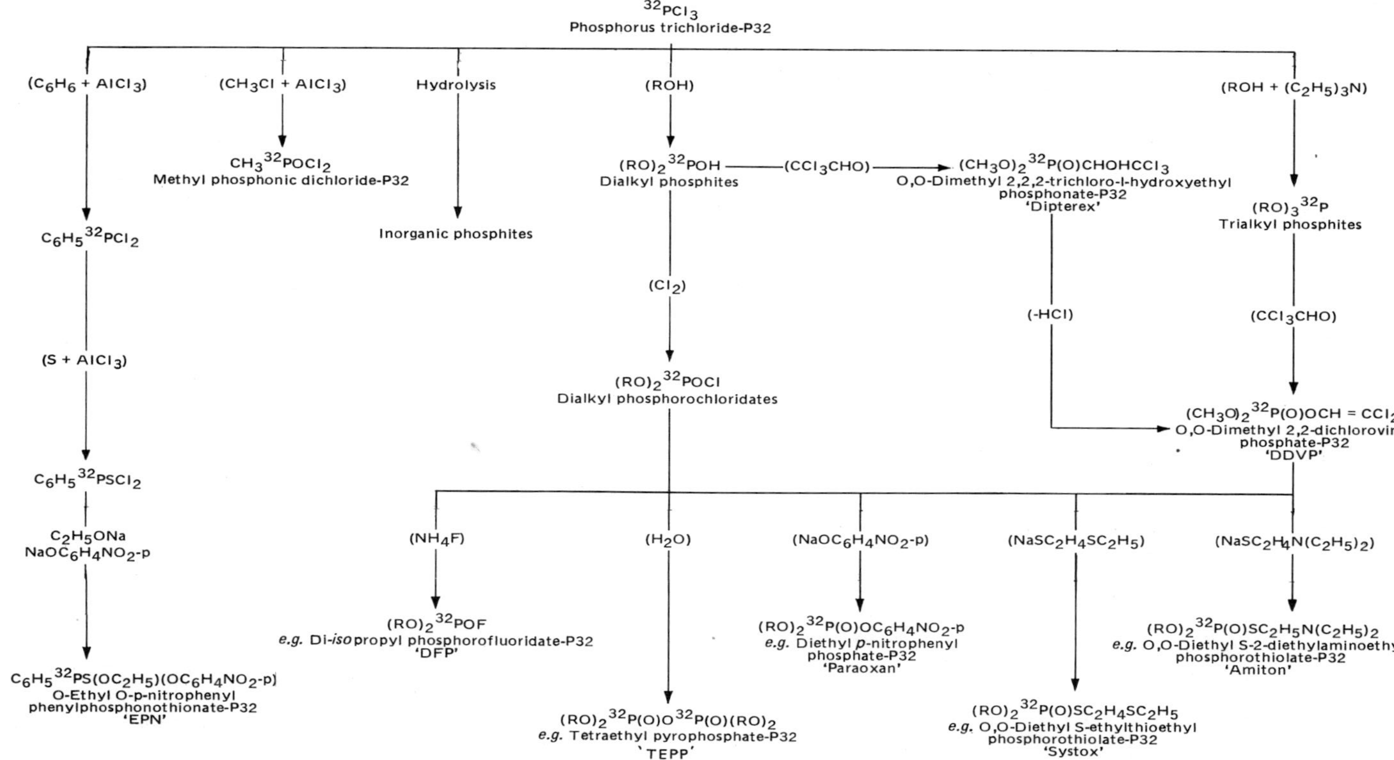

certainty are more important than the best possible yield; but the energetic β radiation is a real problem if quantities become at all large. The characteristic end-products of Table 3 are, moreover, extremely toxic; it is at least an advantage of the radioactive forms that they can be monitored. Except with quite small amounts of activity, ^{32}P synthesis of this kind is confined mainly to a few specialist laboratories.

Growing interest in nucleotides and above all in DNA and protein synthesis has led to a change in emphasis [59]. Various chemical methods are available for the preparation of ^{32}P nucleotides. Two of the most widely used methods are phosphorylation of the nucleoside with [^{32}P]cyanoethylphosphate [60,61] or with [^{32}P]orthophosphoric acid using trichloroacetonitrile as condensing agent [62]. When unprotected ribo- and deoxyribonucleosides are used, all hydroxyl groups are susceptible to phosphorylation.

Enzymatic syntheses, now extensively used in the preparation of ^{14}C-labeled nucleotides, are also useful for ^{32}P and tritium compounds. The use of enzymatic methods to prepare labeled nucleotides has been considered to be inadvisable when more than one enzyme step is involved in the conversion of substrate to product [60]. Historically, the limitations were those of poor yield and reduced specific activities. However, all specifically labeled ^{14}C nucleotides are prepared at The Radiochemical Centre by multienzyme syntheses from the labeled purine or pyrimidine base with yields of 50% or more and without any significant drop in molar specific activity. Impurities arising from preparations of this type are usually not troublesome as the various labeled or unlabeled intermediates differ sufficiently from one another to separate well in most chromatographic systems. The routes are not, however, so readily applied to phosphorus labeling, since the required labeled phosphorus donor (such as PRPP) may itself be difficult to label.

"Isotope farming" of *E. coli* on [^{32}P]orthophosphate has been used for ^{32}P-nucleotides [63]. There is, of course, a formidable problem, with the short half-life, in separating pure components, particularly since it is not easy to separate RNA and DNA quickly with good recoveries.

B. Sulfur and Selenium Labeling

These two nuclides may be considered together since their characteristics have much in common.

1. Sulfur-35

This has some importance as an isotopic tracer, there being an important and extensive organic chemistry and biochemistry for this element, and much interest in the sulfur-containing amino acids cystine and methionine, drugs such as thiouracil and similar anti-thyroid substances, and a fairly wide range of detergents, sulfonamides, anti tubercular drugs, etc. containing sulfur.

The labeling of the common sulfur amino acids was achieved quite early by biosynthesis from $[^{35}S]$sulfate (e.g., Ref. 64). Synthetic methods were, however, long preferred, being upon the whole more convenient to work with the fairly short half-life of 87 days. The first approach to $[^{35}S]$methionine [Eq. (11)]:

$$CH_3SH + CH_2{=}CHCHO \rightarrow CH_3SCH_2CH_2CHO \quad \text{etc.} \tag{11}$$

was superseded by syntheses of the type (12)

$$BzSH + X\,(CH_2)_{1-2}CH\begin{matrix} \nearrow NHR_1 \\ \searrow COOR_2 \end{matrix} \tag{12}$$

which have the advantage that stereospecific synthesis is possible. Such methods are used also for $[^{35}S]$cystine. More recently the biosynthetic method has again come into use for the sake of the very high specific activities it can provide. General approaches to ^{35}S labeling are conveniently summarized in Tables 4, 5, and 6.

Sulfur can replace carbon in many organic molecules without undue distortion and might be used as a nonisotopic tracer in

TABLE 4

Sulfur-35 Compounds--Routes to Intermediates

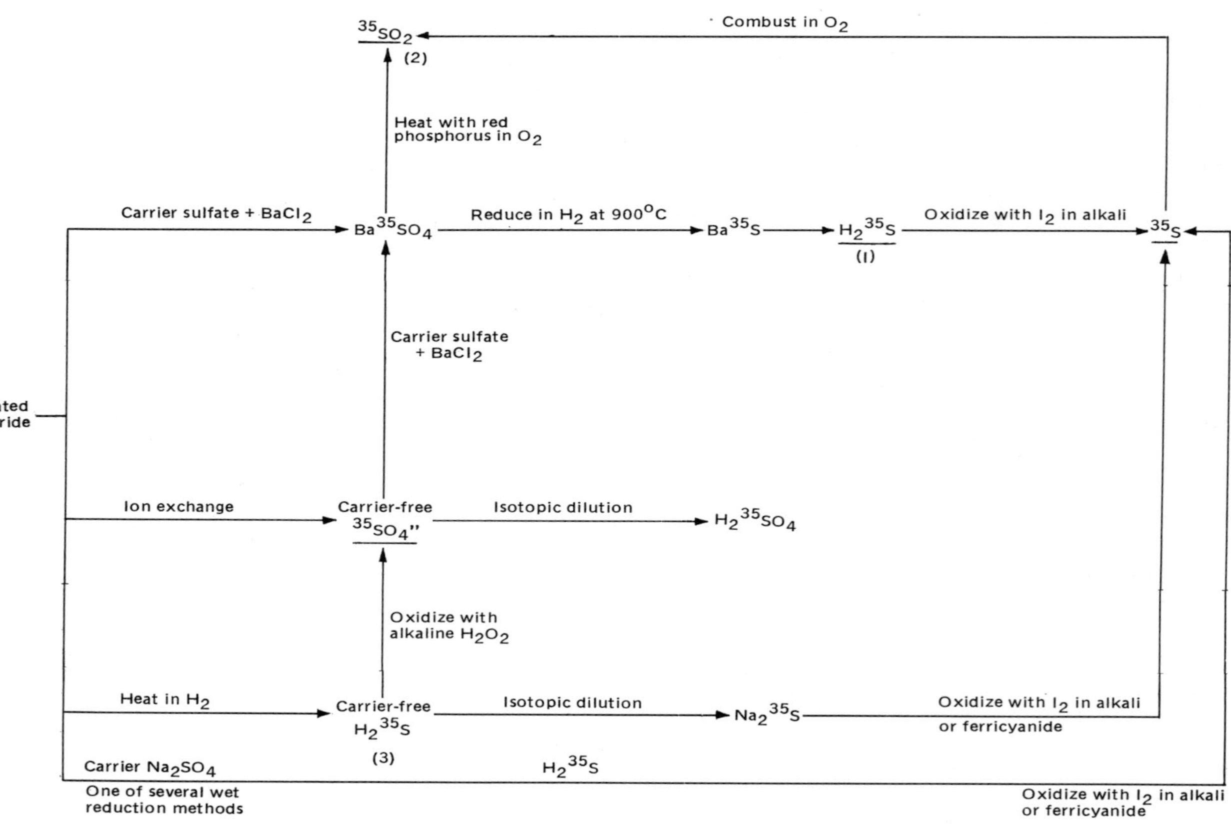

TABLE 5

Sulfur-35 Compounds--Syntheses via Sulfur Dioxide

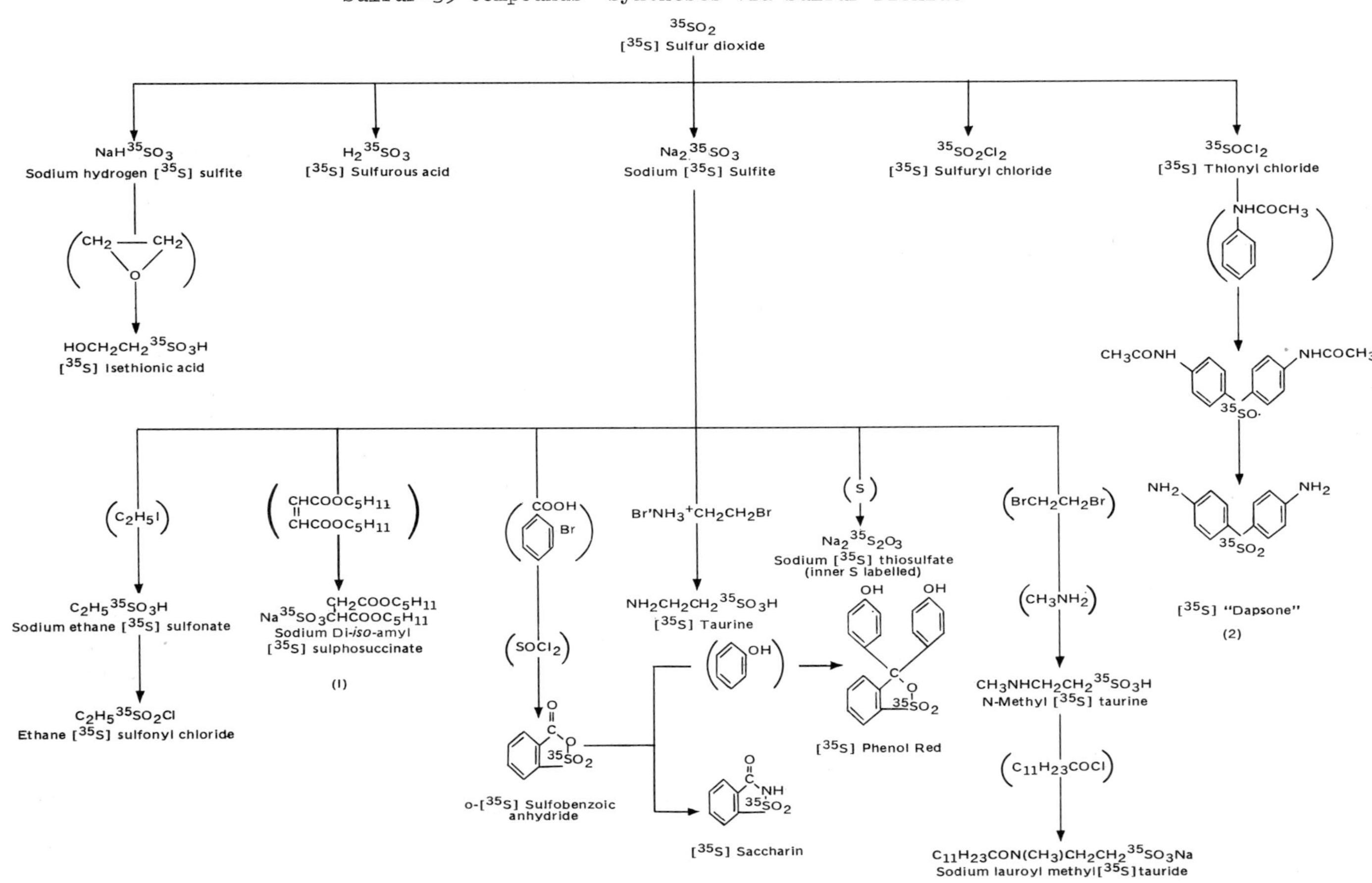

TABLE 6
Sulfur-35 Compounds--Routes from Benzyl Mercaptan to DL-Methionine

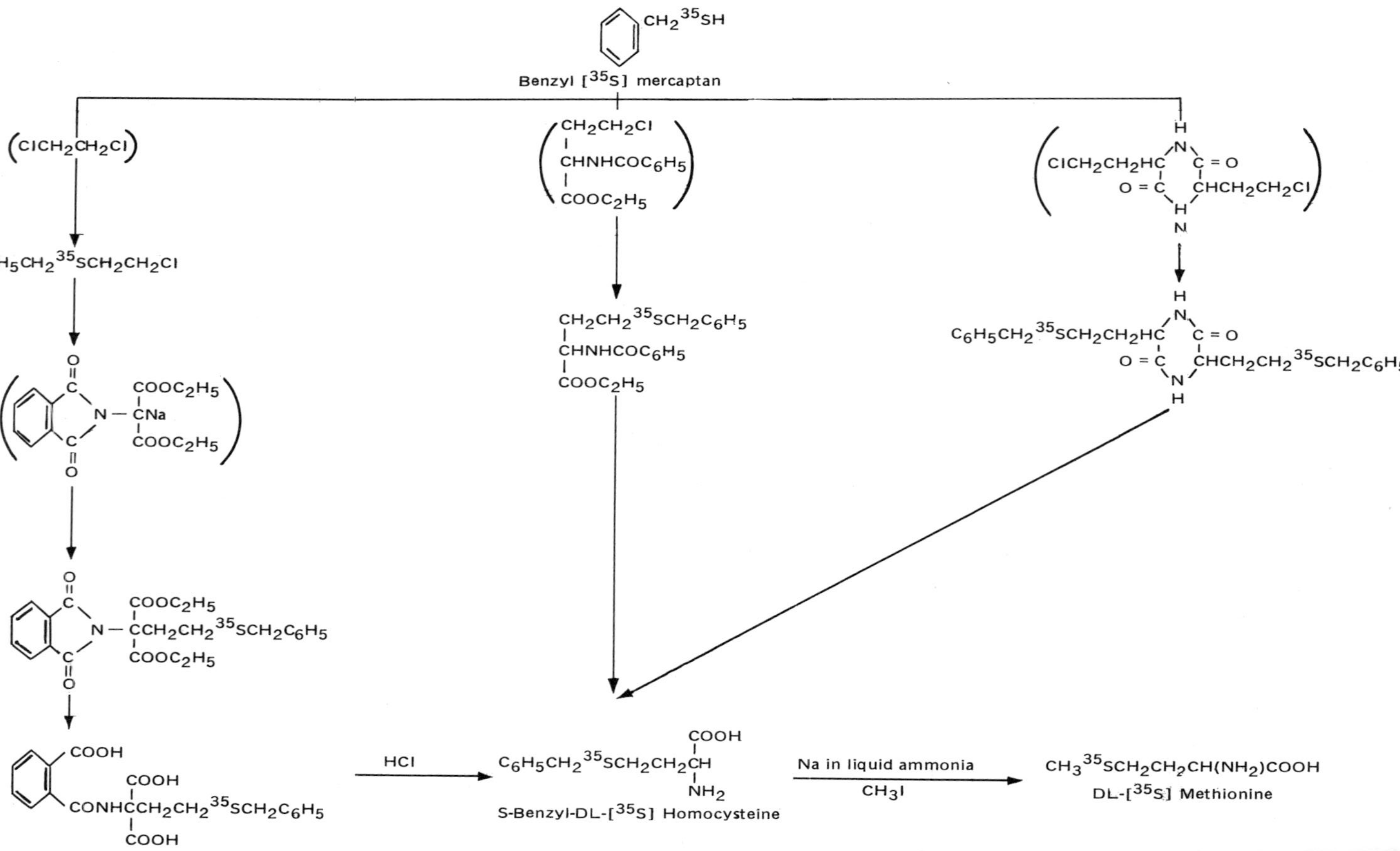

applications for which the higher specific activity (theoretical maximum 1.2×10^{10} mCi/mmol) might be a material advantage over ^{14}C, or even ^{3}H, e.g., in radioimmunoassay. It has not been so used, mainly no doubt because ^{125}I has met the need with the advantage (for this purpose) of a γ label.

2. Selenium-75

This is the only one of several selenium isotopes to have a useful half-life (118.5 days) and the γ emission (decay by electron capture) and absence of associated β radiation are valuable features. Like sulfur, it has some scope as a nonisotopic label, apart from its specific but limited use as an isotopic tracer for selenium.

Until recently interest in ^{75}Se has been almost wholly in scintigraphy of the pancreas. The close biochemical similarity of L-methionine and its analog L-selenomethionine had been observed in 1957 [65] and L-[^{75}Se]selenomethionine was first described in pancreas scanning in 1962 [66,67], having been selected as an accessible γ-labeled analog of a material amino acid. It is widely used, and although not ideal because of the high uptake by liver it has not yet been improved on.

One may safely predict that interest in ^{75}Se as a nonisotopic tracer will grow. Work has been published with analogues of urea and uracil [68] but other uses have been few and sometimes rather esoteric, as in labeling snake venoms [69]. A practical difficulty is the need for γ-shielding of preparative work on any considerable radioactive scale. Selenium-75 compounds can be used in saturation analysis as alternatives to ^{125}I compounds (see Chapter 21). An example is the use of 2-methyl[^{75}Se]-selenoprednisolone for the assay of cortisol in urine and in plasma [74,75].

C. Other Nuclides

Most of these are of highly specific interest in scintigraphic methods. The most versatile is ^{99m}Tc, which has had great attention because of its very favorable radiation characteristics and half-life. It has been used to label albumin aggregates and a wide range of colloidal preparations for scintigraphy of lung, liver, kidney, etc. Its wider use in biochemistry as a nonisotopic tracer seems unlikely.

The biochemical uses of the cobalt isotopes are almost wholly in studies of cyanocobalamins, and are mainly in clinical methods for determining vitamin B_{12} uptake. The mercury isotopes ^{197}Hg and ^{203}Hg have again been mainly of clinical interest for kidney scanning, being used as labeled forms of known diuretics such as chlormerodrin.

Yttrium-90 colloids, such as the colloidal silicate, have some use in radiotherapy. Ytterbium-169 chelates have been patented [70] for use in (for example) brain scanning and study of kidney function [71].

More recently indium-111 has attracted attention, as the complex with the antitumor agent bleomycin, for scintigraphy of tumor of soft tissue [72,73]. Its standing remains uncertain at the time of writing.

These and similar tracers are more properly discussed in texts on nuclear medicine (see Chap. 22). They represent an area of work which has been largely empirical. The reasons are that, first, without practicable γ-emitting isotopes of carbon, hydrogen, nitrogen, and oxygen, it is difficult to use existing knowledge of uptake and turnover of organic compounds for a logical search for better scintigraphic agents, and second, there is very little information available for the γ emitters and their compounds which are available.

REFERENCES

1. J. R. Catch, Purity and Analysis of Labelled Compounds, The Radiochemical Centre, Review 8, Amersham, England, March 1968.
2. J. R. Catch, Labelled Compounds: the User and the Supplier, 164th National Meeting of the American Chemical Society, New York 1972; J. Label. Compounds, 9:737 (1973).
3. E.g., British Patents 1323015 and 1323239 to Becton, Dickinson (1973).
4. D. R. Christman, R. M. Hoyte, and A. P. Wolf, J. Nucl. Med., 11:474 (1970).
5. E. A. Evans, Tritium and its Compounds 2nd ed., Butterworths, London, 1974, pp. 238-439.
6. A. Murray and D. L. Williams, Organic Syntheses with Isotopes, Part II. Organic compounds labelled with isotopes of the halogens, hydrogen, nitrogen, oxygen, phosphorus, and sulfur. Wiley-Interscience, New York, 1958, 950 pp.
7. A. Murray and D. L. Williams, Organic Syntheses with Isotopes, Part I. Compounds of isotopic carbon. Wiley-Interscience, New York, 1958, 1146 pp.
8. V. F. Raaen, G. A. Ropp, and H. P. Raaen, Carbon-14, McGraw-Hill, New York, 1968, 338 pp.
9. E. A. Evans, Tritium and Its Compounds, 2nd ed., Butterworths, London, 1974, 822 pp.
10. M. Bubner and L. Schmidt, Die Synthese Kohlenstoff-14-markierter organischer Verbindungen, Georg Thieme, Leipzig, 1966, 181 pp.
11. J. R. Catch, Carbon-14 Compounds, Butterworths, London, 1961, 128 pp.
12. J. R. Catch, Preparation of Radioactive Tracers: Synthetic Organic Tracers, International Encyclopaedia of Pharmacology and Therapeutics, Section 78, Vol. 1, pp. 97-130; Radionuclides in Pharmacology, Oxford, Pergamon, Oxford, 1972.
13. E. Schwenk, M. Gut, and J. Belisle, Arch. Biochem. Biophys., 31:456 (1951).
14. W. G. Dauben and J. F. Eastham, J. Am. Chem. Soc., 73:4463 (1951).

15. A. M. Porto and E. G. Gros, J. Label. Compounds, 6:369 (1970).

16. P. Kurath and M. Capezzuto, J. Am. Chem. Soc., 78:3527 (1956).

17. C. Katsaros, Univ. Michigan, Ann Arbor, Michigan: Pub. No. 7044: Diss. Abs., 14:236 (1954).

18. W. G. Dauben and H. L. Bradlow, J. Am. Chem. Soc., 72:4248 (1950).

19. H. Rutner and R. Rapun, J. Label. Compounds, 91:71 (1973).

20. J. W. Breitenbach, G. Billek, E. Faltlhansl, and E. Weber, Monats. Chem., 92:1100 (1961).

21. A. P. Wolf, Symposium on Preparation and Biomedical Application of Labelled Molecules, Venice 1964 (J. Sirchis, ed.), European Atomic Energy Community, Brussels, p. 423. (ERU 2200e).

22. The Radiochemical Manual, pp. 268-275, The Radiochemical Centre Ltd, Amersham, 1966.

23. Brit. Pat. No. 1141483, Process and Equipment for Preparing ^{14}C-Labelled Alkali Metal Cyanides (1969).

24. S. Ikeda and A. Tamaki, Radioisotopes, 12:368 (1963).

25. H. S. Turner, R. J. Warne, J. Chem. Soc., 789 (1953).

26. L. E. Custer, T. Abei, B. R. Chipman, F. L. Iber, J. Lab. Clin. Med., 64:

27. G. T. Okita, F. E. Kelsey, E. J. Walaszek, and E. M. K. Geiling, J. Pharmacol. Exp. Ther., 110:244 (1954).

28. E. W. Putman and W. Z. Hassid, J. Biol. Chem., 196:749 (1952).

29. N. G. Carr and K. G. Oldham (to The Radiochemical Centre) British Patent 1342098 (1973).

29a. K. C. Tovey, G. H. Spiller, K. G. Oldham, N. Lucas, and N. G. Carr, Biochem. J., 142:47 (1974).

30. J. D. Ostrow, L. Hammaker, and R. Schmid, J. Clin. Invest., 40:1442 (1961).

31. K. G. Oldham, unpublished work at The Radiochemical Centre, Amersham, 1970.

32. A. W. Wood, M. E. McCrea, and J. E. Seegmiller, Anal. Biochem., 48:581 (1972).

33. K. H. Hallowes, F. P. W. Winteringham, and W. J. LeQuesne, Nature, 181:336 (1958).

33a. I. J. Christensen, P. Olsen Larsen, and B. L. Møller, Anal. Biochem., 60:531 (1974).

34. J. C. Turner, J. Label. Compounds, 3:217 (1967).

35. W. R. Waterfield, unpublished work at The Radiochemical Centre, Amersham, 1970.

36. J. Pouyssegur, J. Label. Compounds, 9(1), 3 (1973).

37. W. W. Christie, M. L. Hunter, and C. G. Harfott, J. Label. Compounds, 9:483 (1973).

38. M. S. Mabee, J. R. Chipley, and K. L. Applegate, J. Label. Compounds, 9:277 (1973).

39. J. C. Bevington, E. J. Bourne, and C. N. Turton, Chem. Ind., 1390 (1953).

40. E. J. Bourne and H. Weigel, Chem. Ind., 132 (1954).

41. H. Simon and H. G. Floss, Bestimmung der Isotopenverteilung in Markierten Verbindungen, Springer Verlag, Berlin, 1967, 2477 pp.

42. J. Bloxsidge, J. A. Elvidge, J. R. Jones, and E. A. Evans, Org. Magnet. Resonance, 3:127 (1971).

42a. E. A. Evans, H. C. Sheppard, J. C. Turner, and D. C. Warrell, J. Label. Compounds, 10:569 (1974).

43. K. E. Wilzbach, in Symposium on the Detection and use of Tritium in the Physical and Biological Sciences, IAEA, Vienna, 1962, pp. 2, 3.

44. M. Wenzel and P. E. Schulze, Tritium-markierung, Walter De Gruyter, Berlin, 1962, 176 pp.

45. P. H. Jellinck and D. G. Smyth, Nature, 182:46 (1958).

46. European Atomic Energy Community, Proc. Conf. Methods of Preparing and Storing Marked Molecules, Brussels, 1963, 1st ed., European Atomic Energy Community, Brussels, (EUR 1625e), (1964).

47. E. A. Evans, Chem. Ind., 2097 (1961).

48. H. S. Isbell, H. L. Frush, and J. D. Moyer, J. Res. Nat. Bur. Std., 64A:359 (1960).

49. A. F. Thomas, Deuterium Labelling in Organic Chemistry, Appleton-Century-Crofts, New York, 1971, 518 pp.

50. See the Batch Analysis Sheets of The Radiochemical Centre, product codes TRK.431 and TRK.443.

51. U. Rosa and R. Malvano, in Radiopharmaceuticals and Labelled Compounds, Vol. II, Vienna, IAEA, 1973, p. 100.

52. L. Donato, G. Milhaud, and J. Sirchis, (eds.), Labelled Proteins in Tracer Studies, European Atomic Energy Community, Brussels, Oct. 1966, (EUR 2950).

53. W. M. Hunter and F. C. Greenwood, Nature, 194:495 (1962).

54. A. S. McFarlane, Nature, 182:495 (1962).

55. F. Pennisi and U. Rosa, J. Nucl. Biol. Med., 13:64 (1969).

56. J. I. Thorell and B. G. Johansson, Biochim. Biophys. Acta, 251:363 (1971).

57. S. S. Lynch and M. R. Redshaw, J. Endocrinol., 60:527 (1974).

57a. A. E. Boulton and W. M. Hunter, Biochem. J., 133:529 (1973).

58. A. S. McFarlane, Biochemical J., 62:527 (1974).

59. R. Monks, K. G. Oldham and K. C. Tovey, Labelled Nucleotides in Biochemistry, Radiochemical Centre Review No. 12, 1971, The Radiochemical Centre Ltd., Amersham, England.

60. H. J. Gray, in Methods of Cancer Research, (H. Busch, ed.), Vol. 3, Academic Press, New York, 1967, p. 245.

61. G. M. Tener, J. Am. Chem. Soc., 83:159 (1961).

62. R. H. Symons, Biochim. Biophys. Acta, 190:548 (1969).

63. R. B. Hurlbert and N. B. Furlong, in Meth. Enzymol., 12:193 (1967).

64. D. B. Cowie, E. T. Bolton, and M. K. Sands, Arch. Biochem. Biophys., 35:140 (1952).

65. D. B. Cowie and G. N. Cohen, Biochim. Biophys. Acta, 26:252 (1957).

66. M. Blau and M. A. Bender, Radiology, 78:974 (1962).

67. M. Blau, Biochim. Biophys. Acta, 49:389 (1961).

68. E. A. Carr and B. J. Walker, Arch. Exp. Pathol. Pharmakol., 248:287 (1964).

69. D. Lebez, F. Gubensek, and Z. Maretic, Toxicon, 5:263 (1968).

70. Brit. Pat. 1273446, to Minnesota Mining and Manufacturing Corp. (1972).

71. F. Hosain, R. C. Reba, and H. N. Wagner, Radiology, 93:1135 (1969).

72. M. L. Thakur, Int. J. Appl. Radiat. Isotopes, 24:357 (1973).

73. M. V. Merrick, S. W. Gunasekera, J. P. Lavender, A. D. Nunn, M. L. Thakur, and E. D. Williams, in IAEA Symposium on Medical Radioisotope Scintigraphy, Vol. 2, Monte Carlo, Oct. 1972, IAEA Vienna, 1973, p. 721.

74. V. E. M. Chambers, R. Tudor, and A. L. M. Riley, J. Steroid Biochem., 5:298 (1974) (Abstract only).

75. V. E. M. Chambers, J. S. Glover, and R. Tudor, Steroid Radioimmunoassay (E. H. D. Cameron, S. G. Hillier, and K. Griffiths, eds.), Proc. 5th Tenovus Workshop, Cardiff, April, 1974, Alpha, Omega, Cardiff, 1975, p. 177.

Chapter 6

QUALITY CONTROL AND ANALYSIS OF RADIOTRACER COMPOUNDS

G. Sheppard and Ritchie Thomson

Quality Control Department
The Radiochemical Centre, Limited
Amersham, Buckinghamshire
England

I. INTRODUCTION AND QUALITY CHARACTERISTICS

A. Introduction

The importance of maintaining a high standard of quality for radiotracers has been recognized from the early days of work with them, in order that the research worker or diagnostician using them can have confidence in his results. In general, the tracer principle demands that a radioactive compound behaves in exactly the same way (except for predictable isotope effects) in the system it is tracing as the corresponding nonradioactive compound under study. This assumption will be invalid if the nature of the radioactive material is in doubt, that is, if it consists of more than one species or has the wrong identity, or if its behavioral characteristics are not known. In many cases it is also necessary to know accurately the quantity of radioactivity being used.

In this chapter the subject of quality control and analysis is treated both from the viewpoint of the user who is concerned to know the limitations of quality control methods and their effect on his application of the radiotracer and those research workers who need to synthesize and analyze their own radiochemicals. There is a special emphasis on the problems and errors possible in quality control and analysis. The principles underlying quality control are outlined and, although a detailed description of analytical techniques is outside the scope of this work, those particularly applicable to radiotracers are described, with brief critiques of their limitations. The chapter is concluded with a selection of examples showing the effects of impurities on the use of radiotracers.

There is a continuous learning process in the quality control of radiochemicals as new applications are found and criteria of quality become more critical and demanding. In this way the research worker becomes more aware of the problems and pitfalls in quality control and analysis and is better able to determine the suitability of radioactive material for his application; also commercial suppliers can develop more discerning procedures and analytical techniques and improve the specification of their products.

B. Quality Characteristics

The quality characteristics for radiotracers are of two kinds, _variable_, that is, those which can be measured on a continuous scale and _attributive_, that is, "go or no-go" parameters. Among the former are the following parameters:

Radionuclidic purity--the proportion of radioactivity in the form of the specified nuclide (in the case of short half-life nuclides, at a specified date)

Radiochemical purity--the proportion of radioactivity in the specified chemical form, which may include isomeric or stereoisomeric form, and may also specify biological or immunological properties

Chemical purity--the proportion by mass in the specified chemical form (this term normally refers to nonradioactive impurities; radiochemical impurities, of course, also constitute chemical impurities but are usually of very small mass)

Radioactive concentration--the amount of radioactivity per unit quantity of material, usually expressed in curies (Ci) or millicuries (mCi) per gram or milliliter

Total activity--the total quantity of radioactivity (in curies) present; frequently measured as the product of total volume or mass and radioactive concentration

Specific activity--the ratio of radioactivity to mass of all the atoms or molecules containing isotopes of the specified radionuclide, usually expressed in curies or millicuries per millimole

Compatibility of components and final product performance--used particularly for radiotracers supplied in kit form, measuring such parameters as discrimination, sensitivity, reproducibility both for components and for kits in the final packaged form

Attributes which may be considered are:

Appearance--physical state, color, freedom from particulate matter, particle size for colloids, etc.

Sterility--conformity to standard criteria for sterility, such as a certified sterilization record and absence of detectable contamination by a standard microbiological test

Freedom from pyrogens--conformity to standard tests for apyrogenicity

Position and pattern of labeling--conformity to the specified position or distribution of labeling of the radioactive atoms present in the molecule (in certain cases compounds with the label in different positions from those specified may be considered as radiochemical impurities)

II. THE NEED FOR QUALITY CONTROL

Quality control as a concept developed naturally in large-scale manufacturing, but its basic principles can also be applied to small-scale chemical syntheses by the individual research worker. It usually entails both the measurement of quality and the prevention of quality defects. In particular, the minimization of the presence of unexpected and unwanted by-products from a preparation greatly simplifies the analysis of the final product and reduces the wastage of valuable effort and starting materials. Quality control also involves the idea that quality does not arise naturally at the end of a process but will only be achieved by critical attention to the need for quality throughout the whole process.

The quality and reliability of the final product initiates with the selection of the correct starting materials, continues with the meticulous performance of the preparation and concludes with a searching assessment of the quality of the product and of its suitability for use. Quality-mindedness means that the possible consequences to the final product of any changes in the preparation are properly considered and that in any set of control procedures there should be provision for feedback so that the lessons learned, usually from failures of quality, can be introduced into the preparation at the right place.

Thus the reliability of the preparation is improved and the risk that the final product is unsatisfactory is minimized. To "build in" quality in this way implies as full an understanding as possible of the preparation, the impurities which may arise from it and the procedures necessary to measure and control them. The overall quality control schedule may be idealized as shown in Fig. 1; in this the preparation of a radioactive compound is divided into four areas, but it should be appreciated that there is a considerable degree of interaction between them and they cannot be separated fully in principle or in practice.

A. Control of Starting Materials

Identification of the starting material distinguishes those cases of misdispensing, wrong labeling, or mixup of the starting material by its supplier, whether it is radioactive or inactive. The formation of by-products in the preparation of labeled compounds can often be related to impurities in the starting material which are labeled in the course of the synthesis; for example, in the preparation of stereospecifically labeled [^{3}H]mevalonic acid lactones by the Cornforth method [1], the starting material is an olefin, whose geometrical isomerism determines the stereochemistry and the stereospecificity of labeling of the product, and hence the ratio of cis-trans isomers present in it has to be known.

The purity of carrier materials used for the dilution of finished products is particularly important; for example, the purity of adenosine 5'-triphosphate (ATP), used as a bulk phosphorylating agent in an enzymatic reaction, is less critical than when it is used as a carrier material to dilute the specific activity of [^{3}H]ATP since the specific activity of the final

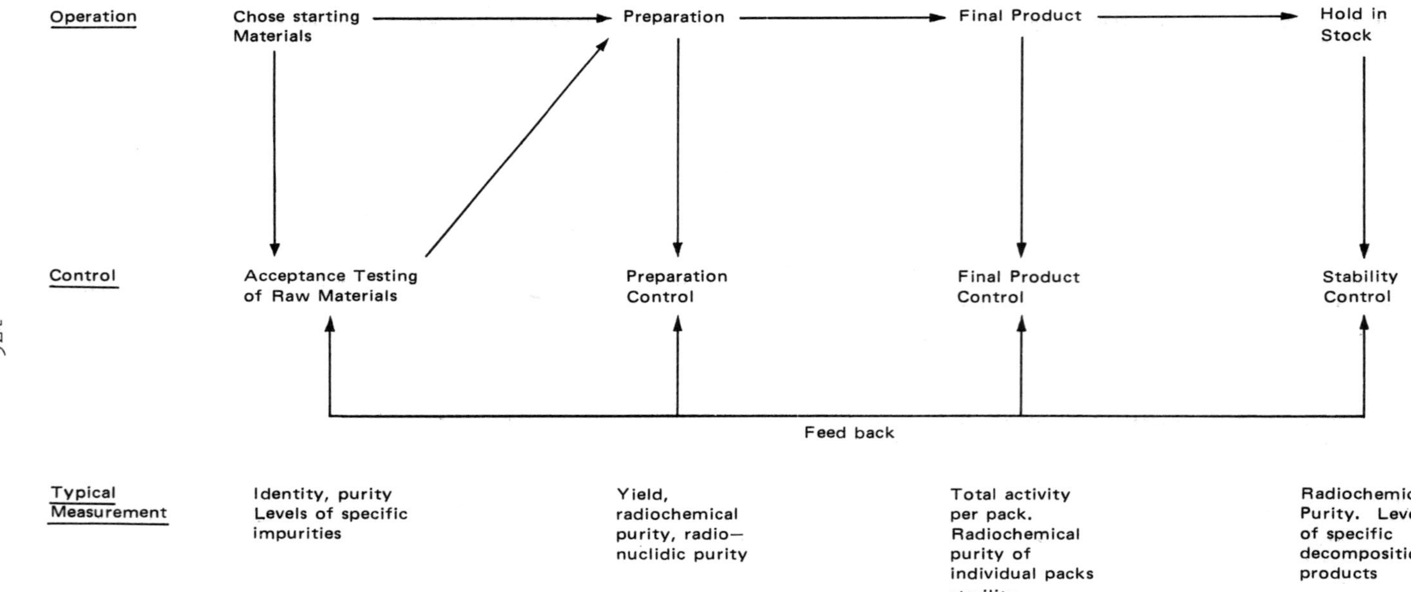

FIG. 1. Generalized production and quality control flow scheme.

diluted product depends on the purity of the material used. To many research workers, the production of a compound at high molar specific activity is of great importance. Specific activity may be reduced at any stage of a preparation from starting material onward; thus in the production of sugars uniformly labeled with carbon-14 by photosynthesis of detached leaves in an atmosphere containing [^{14}C]carbon dioxide, endogenous sugar residues in the leaf will reduce the specific activity of the sugars. The leaf, the starting material, has to be treated carefully by starvation in the dark to minimize this dilution effect.

The purity of target materials used to produce radionuclides, either by neutron irradiation in a nuclear reactor or by bombardment in a cyclotron, has also to be considered. The presence of impurities, even in amounts which are insignificant by normal criteria for purity, can give rise to radionuclidic impurities which form a significant proportion of the radioactivity of the radioisotope required. Impurities which do not form interfering radioactive species but are isotopic with the desired radionuclide nevertheless lead to an unwanted reduction of its specific activity. A recent review [2] gives details of suitable target materials for production of high quality radionuclides.

The quality of a starting material can only be assessed sometimes by "compatibility testing," that is, by testing its efficiency in the preparation or in the final product in conjunction with other components which have been tested and shown to be up to specification. Its performance can then be examined by statistical techniques, such as analysis of variance, to test the significance of any departures from normal, and to assess the quality and the compatibility of the components under test. Thus, the charcoal adsorbent used in certain adenosine 3',5'-cyclic monophosphate assay procedures, using a competitive binding protine and [^{3}H]cyclic AMP may only be examined by measuring its efficiency in absorbing free cyclic AMP in the presence of radioactive material.

Packaging material may also be considered as a potential source of variation in the quality or usability of a radiotracer. For example, all containers and closures used for radiotracers should be tested for inertness, since they may affect the stability of the material [3] or react irreversibly with it. This applies especially in those cases where a user wishes to subdivide a bulk solution of a radiochemical into smaller packages; inert plastic or neutral glass containers are preferable. The effect of impurities in the solvent on the compound's stability has also to be considered [4]. The properties of all packaging used for radiopharmaceuticals have to be closely specified--in particular their efficiency in preventing leakage-in (bacterial contamination), or leakage-out (external contamination with radioactivity) and their compatibility with the radioactive material, especially their effect on its stability under all conditions (including terminal autoclaving). Normally all raw materials and packaging used for radiopharmaceuticals are quarantined when received and have to be shown to be satisfactory before use as an essential part of good pharmaceutical manufacturing practice.

B. Control of Preparation

1. General Considerations

The control of the preparation may be separated into two general areas, the control of the chemical reactions to form the required product and the controls and precautions against external adventitious contamination. The former control is familiar and is similar to that encountered in nonradioactive synthesis. In preparations of radiotracer compounds (see Chap. 5) the extremely small masses, half-lives, and radiological considerations often preclude full analysis of intermediates or the final product and control of the synthesis becomes even more critical to ensure the formation of the right product of the right purity. The small masses and volumes used often mean that the chemical reactions are more easily affected by variations in temperature, local concentrations, inefficient mixing, etc.; in addition, two important properties of the product, its specific activity, and the position and extent of labeling are affected by the conditions of the preparation. Thus, in the iodination of many proteins for use in radioimmunoassay the degree of substitution and damage to the protein depend critically on the vigor of the reagents and the procedure used; in particular the quantity of chloramine-T, reaction time, and temperature are critical [5].

In-process controls are frequently used to monitor the progress of the synthesis and to determine the purity and identity of the product at intermediate stages. Those employed in radioactive synthesis include:

1. Radiochemical separations of radionuclides; due to the specificity of chemical reactions, e.g., ion exchange, precipitation, radionuclidic impurities present in the major radionuclide may be removed from the mainstream of the reaction.
2. Physical or chemical fractionation; distillation, solvent extraction, crystallization, etc., often fix the identity of the product at successive stages of the reaction.
3. Chromatographic separations regulate the purity and identity of the product, especially when the technique is highly specific for individual compounds--for example, the use of ion-exchange chromatography to separate the individual ^{14}C-labeled amino acids from algal protein hydolysate.

Determinations of the purity of intermediates and the yield of the radioactivity and mass at intermediate stages are used to indicate the successful progress of the synthesis, for example, to confirm that all the product has been collected into the correct phase of a solvent extraction.

2. Control of Position of Labeling

For radiotracers three types of labeling may be identified:

Specific--when the positions labeled are defined, which may also include the stereochemical configuration of the label for complex organic molecules

Uniform--statistical uniformity of labeling with the radionuclide for all atoms of the element through the molecule; at high specific activity multiply-labeled molecules are likely to be formed

General--when more than one atom of the radioelement may occur at different positions in the molecule, but the position and distribution of labeling are not specified

Complex molecules, such as proteins which may be labeled in several different positions by iodination, give rise to complex patterns of labeling depending on the residues or atoms labeled. These patterns may vary from batch to batch and it is difficult to specify the distribution of labeling; however, this is frequently not an important criterion of quality for iodinated proteins used in such techniques as radioimmunoassay. For a foreign label (i.e., nonisotopic labeling) the performance under the conditions of use is the important criterion.

The position of labeling may depend on the following variables:

1. The labeled precursor used: for example, $^{14}CO_2$ used to prepare uniformly labeled materials by photosynthesis, sodium borotritide to introduce a tritium atom by reduction of carbonyl function, and tritium gas for the reduction of double bonds or for exchange with hydrogen atoms
2. The reagents used: for example, the catalyst used for reaction of Δ^1-steroids with tritium gas influences the stereochemistry of the 1 and 2 tritium atoms introduced [6,7]. Also, the reactivity of the reagents used for the iodination of proteins affects the degree of substitution with iodine atoms [8]

3. The reaction conditions: for example, the time of exposure to $^{14}CO_2$ controls the uniformity of labeling of sugars prepared by photosynthesis with detached leaves; the elevation of the temperature, or prolongation of the time of an exchange with tritium gas increases the probability of less accessible positions in the molecule being labeled

In the case of rational synthesis, as with ^{14}C or ^{32}P, when the skeleton of the molecule is built up by successive reactions, the position is fixed unambiguously by the preparation; an example of such control is the synthesis of D-[1-^{14}C]glucose from potassium [^{14}C]cyanide and D-arabinose.

3. Control of Specific Activity

In the preparation of a radiochemical the specific activity (or isotopic abundance) obtained is usually maximized consistent with economy in the use of radioactive material and the scale of working; this is achieved by the prevention of unwanted dilution during the critical stages of the preparation. An example of the maintenance of specific activity during photosynthesis by the prevention of dilution with carrier has already been given for ^{14}C (see Section II A); the specific activity of the product is very dependent on the conditions of the reaction for many reductions or exchanges with tritium gas. Thus in the preparation of purine bases labeled in the 8-position by catalytic halogen-tritium replacement, prolongation of the reaction time leads to a lower specific activity product (contrary to the expectation that more exposure to radioactive gas would lead to a higher specific activity product) because under these conditions the catalyst causes exchange of the incorporated tritium with the solvent [9]. Conversely, when tritium atoms have been introduced by partial selective reduction of molecule with more than one double bond, an unexpectedly high specific activity in the product may indicate that it has been overreduced to (say) a dihydro derivative.

In the purification of radioactive products, it is sometimes necessary to add carrier (nonradioactive) product in order to "scavenge out" the radioactive product, for example, in the crystallization of ^{14}C-labeled sugars, and the amount added is limited to prevent undue dilution of the product. Similarly in the dispensing of high specific activity inorganic nuclides (so-called *carrier-free*) it is sometimes necessary to add a limited quantity of carrier to prevent adsorption of radiotracer on to active sites on the container.

4. Prevention of Contamination and Decomposition

Many external agents may affect the purity of a radiotracer compound during formulation and dispensing operations. The possibility of contamination and decomposition of the radiochemical, and impairment of its stability, have to be considered whenever the user subdivides or dilutes the radiotracer. The agents which may be responsible are:

1. Air, especially oxygen
2. Moisture
3. Temperature
4. Chemical, i.e., nonradioactive additives
5. Radioactive, i.e., radionuclidic or radiochemical impurities added by contamination
6. Particulate contaminants
7. Microorganisms

The importance attached to each factor depends on the susceptibility of the radiotracer compound, the methods used and the effect that the impurities introduced have on its use. Thus some compounds are sensitive to chemical degradation during dispensing; for example, [^{14}C]acetic anhydride is readily hydrolyzed by minute traces of water and [^{3}H]prostaglandins are thermally unstable. The purity of some materials is reduced by the process of thermal autoclaving necessary to sterilize them; indeed [^{203}Hg]chlormerodrin is not only slightly decomposed by autoclaving, but its stability on storage is also impaired. For this reason terminal sterilization by filtration is the preferred method for this material [10].

Radiochemical impurities introduced by contamination with other compounds are unlikely since it is possible to use different apparatus, solvents, etc. for each compound in dispensing. Radionuclidic impurities introduced into simple solutions of inorganic nuclides present more difficulty, since it is not uncommon to store and dispense different nuclides in the same shielded enclosure and cross-contamination may occur much more easily with the low concentrations involved. It is normal to arrange that the nuclides used together are easily identified and measured when there is cross-contamination.

The most difficult isotopes to detect as impurities are pure β emitters (^{14}C, ^{3}H, ^{35}S, for example); however, these are biologically the most important isotopes and justify separate enclosures in order to eliminate the possibility of accidental cross-contamination. Particulate contamination is normally only important for radiopharmaceuticals; however, it may be critical in the quality of, for example, [^{125}I]iodide solutions used in the iodination of proteins, when the iodinated products would include the adventitious particulate matter.

Cleanliness of working is also essential in the prevention of bacterial contamination. Although many radiotracers are present in very dilute concentration and are relatively poor substrates for microbial growth, it readily takes place in some of them, for example, in 5-[^{125}I]iodo-2'-deoxyuridine, and to prevent degradation they have to be sterilized. Special techniques are necessary to ensure the sterility of injectable radiopharmaceuticals and these are reviewed in Section IV D.

C. Quality of the Final Product

Before the research worker uses a radiochemical he needs to satisfy himself that its quality is suitable for his experiment. This is achieved both by verifying that it has passed all the appropriate controls and tests in the intermediate stages of the preparation and by analysis of the final product against a specification. In particular, for most applications the purity value or the content of specific impurities has to be ascertained. The term "purity" is, of course, relative and depends on the technique used to measure it and the impurities separated by the technique. It also depends on the use to which the product is to be applied; "pure for what?". This question is answered in detail in Section IV. Impurities in the final product arise, naturally, due to variations in the reagents used in the preparation, in the conditions of the preparation and, most importantly, in the purification procedures used to isolate the product, for example, crystallizations or column chromatography purifications.

It is almost impossible to carry out a broad screening for any possible impurity and hence the user must confine himself to a consideration of the most likely sources of impurity in his compound. The possible sources of *radiochemical* impurities are (cf. Ref. 11 and 12):

1. Those resulting from *impurities in the nonradioactive starting material* which become radioactive in the labeling process, e.g., homologs in an alkyl halide used to prepare fatty acids labeled in the carboxyl group by a Grignard reaction with $^{14}CO_2$ (see Chap. 5)
2. Those resulting from *radioactive by-products* of the synthesis, e.g., an [U-^{14}C]amino acid isolated by ion-exchange chromatography of algal protein hydrolysate may be contaminated with neighboring amino acids from the separation, D-[2-^{3}H]glucose prepared by reduction of the 2-keto compound may contain D-[2-^{3}H]mannose

3. Those resulting from *incomplete conversion of radioactive* intermediates or starting materials, e.g., [^{3}H]thymine in [^{3}H]thymidine prepared enzymatically, [1-^{14}C]arginine in [1-^{14}C]ornithine prepared by hydrolysis (these are the most frequent impurities in freshly prepared compounds and usually the most easily detected)
4. Those resulting from an *overrun of the reaction* producing a more highly reduced, oxidized, or substituted labeled compound, for example, [1-^{14}C]sorbitol in D-[1-^{14}C]glucose prepared by reduction of D-[1-^{14}C]gluconolactone
5. Those which *contain the label in a different position or with a different stereospecificity* than that expected from the method of synthesis
6. Those resulting from decomposition of the compound during the final purification and treatment of the product or in dispensing, e.g., [^{35}S]methionine sulfoxide in [^{35}S]methionine

Radionuclidic impurities may have one of the following causes:

1. *Daughter nuclides and decay products* produced from the major radionuclides (these will usually be radiochemical impurities as well because they will be present in a different chemical form)
2. Those produced by *nuclear side reactions* during irradiation, e.g., when a 2n,p reaction can occur simultaneously with an n,p reaction
3. Those produced from the *nuclear reactions of other isotopes* present in starting material, e.g., the presence of ^{60}Co in ^{58}Co produced by irradiation of ^{58}Ni
4. Those arising from *contamination* during processing or dispensing

Chemical (i.e., nonradioactive) impurities may arise from the following causes:

1. *Impurities in the starting material* carried through all the stages of the reaction and purification (these may appear in their original chemical form or modified by the reaction)
2. Those resulting from *reagents introduced during the preparation*, including their decomposition products and the nonradioactive by-products of the reaction
3. Those introduced by the *purification procedures* used in the preparation, for example, thin-layer chromatography adsorbents, solvent residues, desiccants
4. Those introduced from the *interaction of packaging material* with the contents of the vial, such as plasticizers and fillers in, for example, rubber inserts

Table 1 shows the important kinds of impurities present which have to be looked for in different classes of radiotracers. It is important to consider the implications of the fact that the preparation of a radiotracer usually results in a bulk stock of finished product. This is termed a "batch," a concept which is critical in the quality control of radiochemicals. A batch can be defined as a uniformly manufactured set of articles each having an identical risk with respect to the appropriate parameters of quality. Under this assumption a suitable random sample can be taken so that the quality of the batch can be assessed. If a uniform risk cannot be assumed then random sampling may not provide a representative sample of the batch. For example, in measuring radiochemical purity a small number of samples from a batch, probably only one or two vials, may be used. The answer may be assumed to give information on the batch as a whole only if all the vials have been treated similarly, if this is a truly random sample and if the assumption is valid that the vial to vial variation in purity is smaller than the variability in the method of measurement.

TABLE 1

Likely Impurities and Their Sources in Radiotracer Compounds

Radiotracer	Possible impurities	Source
Inorganic radionuclide, e.g., [^{35}S]sulfate	Radionuclidic	Target material, nuclear side reactions
	Chemical, e.g., metal ions	Contamination during work-up
Organic labeled compound, e.g., [^{14}C]leucine	Radiochemical	Chemical side reactions, decomposition
	Chemical, e.g., nonradioactive salts	Contamination during work up
e.g., [^{3}H]estradiol	Radiochemical	Starting material, decomposition
	Patterns of labeling other than specified chemical	Nonspecific reaction
	Chemical	Chromatography adsorbents, starting material
Radiopharmaceutical injectable, e.g., [^{75}Se]selenomethionine	Radionuclidic	Nuclear side reactions (removed in processing)
	Radiochemical	Chemical side reactions, decomposition
	Chemical	Processing contamination
	Bacterial contamination, pyrogens	Bacterial contamination during preparation
Saturation assay system, e.g., human placental lactogen radioimmunoassay	Not applicable in usual way; defects in kit parameters, e.g., background, slope, precision	Dispensing errors, decomposition of radiotracer, low activity of antisera, etc.

D. Stability Control

Except for those compounds whose use is limited solely by the half-life of the radionuclide, it is unusual for all of a radioactive compound to be used immediately on receipt and the user has to exercise control of its stability during storage. This subject is covered in some detail in Chapter 7; in general, the supplier's instructions for the best conditions of storage or the recommendations given in several reviews on the subject [13-15] should be followed. The material should be reanalyzed to measure the decrease in its radiochemical purity at intervals according to the user's experience of its stability and to the importance of the impurities formed in his experiments. The longer the material has been stored, the shorter the period between analyses should be, since the risk of catastrophic decomposition increases markedly with time [13]. The analytical systems should accordingly be selected so as to finely separate the likely decomposition products, if known, and to provide a direct comparison with the purity of the material when first prepared and used, so that a continual observation of purity can be maintained.

For some radiochemicals, especially radiopharmeceuticals, it is desirable to define an "expiry date." This is a time after preparation during which the radiochemical is expected to be suitable for a specified use. Such a period can only be based on past experience and hence involves a risk that the material may decompose faster due to the inherent variations in stability which occur with radiochemicals [13]. It is usual to ensure that there is a margin of safety in setting the expiry date although this may mean some wastage of acceptable material.

III. ANALYSIS OF RADIOCHEMICALS

A. Errors in Analysis

The statistical significance of the analysis of a radiotracer has always to be established before a conclusion may be made on its suitability for use; in particular any unusual result should be checked for reproducibility.

There are two types of error which may be encountered in analytical measurements: random error and bias. The former arises from the sampling techniques used, manipulations, chemical treatments and the counting and other measurements carried out. An appreciation of both the overall random error and the relative magnitude of its various components is advantageous. For example, there is little point in improving counting statistics to obtain a coefficient of variation of less than 1% if in an earlier stage of the analysis a chemical separation has a coefficient of variation of 5%. Except in very unusual circumstances, all analyses have to be carried out in replicate and these determinations should be as independent as time and manpower permit; thus in assaying the radioactive concentration of a solution which requires an intermediate dilution, two separate dilutions at least should be carried out and each one independently sampled rather than multiple sampling of one dilution.

Bias in a result means that an answer is precise but systematically wrong. This usually arises from lack of control in the measurement technique and part of the development of an assay procedure involves tests for its validity. Two general techniques are particularly useful: standard addition and dilutions. With the former, the difference in the measurement response with or without the addition of a known amount of standard is measured. In the case of dilution the unknown is measured at different dilutions or with accurately known amounts.

B. Measurement of Radioactivity

The final process in the analysis of a radiochemical is almost always a measurement of radioactivity. It is not intended in this chapter to cover the basic principles of counting techniques or the methods of absolute standardization; these are discussed in Chapters 4, 14, and 22. However, it is important to consider some of the advantages of and problems which may arise in the use of modern counting techniques.

For counting low radioactive concentrations, scintillation techniques are in general use, whereas ionization techniques are now used mainly for the measurement of high radioactive concentrations or large quantities of radioactivity, for scanning chromatograms, and a number of very specific purposes. The various applications of counting techniques are shown in Table 2. Bremsstrahlung counting for hard or moderate β emitters, such as ^{14}C, is a little-used technique. The conversion ratio to x-rays is low and this allows solutions at the mCi/ml levels to be measured without dilution by gamma counting techniques, thus allowing contents of finished packs of radiochemicals to be measured directly and providing a useful nondestructive quality control measurement. However as with many counting techniques the geometry and self-absorption losses are critical and have to be closely specified for reproducible and accurate results. High radionuclidic purity is also important since a small proportion of a γ emitter will have a disproportionate effect on the count rate observed.

Problems in the measurement of radioactivity during analysis usually arise from incorrect standardization and calibration

TABLE 2

Measurement Techniques

Technique	Level of radioactivity: Low (<1 μCi/ml)	Intermediate (1-100 μCi/ml)	High (>100 μCi/ml)
Liquid scintillation counting	α, excellent β, good γ, with moderation	α, by dilution β, by dilution γ, by dilution	α, by dilution β, by dilution γ, by dilution
Crystal scintillation counting	α, good (ZnS crystal) β, hard only γ, good	α, by dilution β, by bremsstrahlung γ, by dilution or low efficiency counters	α, by dilution β, by bremsstrahlung γ, by dilution or low efficiency counters
Ion chamber	Not applicable	α, possible β, for hard β only γ, excellent	α, possible β, good for hard β γ, excellent

or from difficulties in sample preparation, for example, in liquid scintillation counting. It is clear that no measurement of radioactivity can be nearer the "true value" than the standards used and in many cases may be in much greater error due to the statistical errors introduced in the comparison process (including the errors in the original calibration of the standards). In addition, systematic errors introduced by the method of sample preparation for the standard will lead to inaccuracies; thus the standard should have the same counting geometry, the same distribution in solution and the same attenuation by its container as the samples being standardized.

Liquid scintillation counting standards should be in a similar chemical form to the radioactive compound being measured, except in the case of ^{35}S where ^{14}C-labeled material may be used if necessary [16]. For example, significant errors are introduced when [^{3}H]hexadecane is used as a standard for tritium compounds soluble in the aqueous phase of an emulsion scintillant system. New standards are necessary for each new batch of scintillant and long-lived quench curves built into a counter or calculating machine show a tendency to drift and are checked frequently using standards at each end of the quenching range.

Crystal scintillation counters can be calibrated against the appropriate absolute standard. Drift in the gain of the counter, particularly when narrow windows are set on specific γ peaks, can lead to difficulties in the maintenance of these calibrations. The stability of a counter can be monitored by the use of low- and high-energy γ emitters. A suitable x-ray emitter for this is ^{129}I; a suitably long-lived γ emitter is ^{137}Cs.

It is good practice to design radiotracer experiments so that the results are expressed as relative ratios of count rates to compensate partially for calibration errors. In liquid scintillation counting, problems associated with the preparation of stable homogeneous samples are encountered. Counting solutions are chemically complex, especially in the case of gel or emulsion scintillant which consists of two phases [17,18]. Some causes of discrepancies in counting samples are tabulated in Table 3. It must be emphasized that many of the effects can be fairly small and difficult to detect; if they are critical to the experimental interpretation then a systematic validation of counting techniques is necessary.

One source of error which may occur with all counting techniques at high count rates is inadequate compensation for the dead time of the counting circuits. In modern counting equipment, live-timers are usually employed; however, the possibility of this effect should always be borne in mind. It is most easily diagnosed by the existence of a nonlinear relationship between count rate and quantity of sample; such a nonlinear relationship may also denote absorption or precipitation effects and is particularly useful as a criterion for judging the validity of the method of sample preparation and counting (see p. 103).

TABLE 3

Difficulties with Liquid Scintillation Counting

Assay value	CAUSE
Lower than expected	1. Plating out of compound due to either its insolubility in scintillant or chemical incompatibility 2. Active component in different phase from standard material 3. Deteriorated scintillant 4. Quenching agent inadequate to experimental material and standards 5. Loss of compound due to volatilization 6. Loss of counts due to inadequate dead-time compensation
Higher than expected	1. Chemiluminescence 2. Active component in different phase from standard material 3. Different quenching agent in experimental material and in standards 4. Contamination of the scintillation "cocktail"; check background counts

C. Determination of Radionuclidic Purity

There are no completely general techniques which permit the measurement of radionuclidic purity for all nuclides; the two major ones are gamma spectroscopy and chemical separation.

1. Spectroscopy (see also pp. 50 and 91)

Each radioactive isotope has a characteristic emission spectrum and this may be used to identify and measure it, either directly or after an initial chemical separation. α and γ spectra usually consist of discrete peaks which may be measured directly in a precalibrated spectrometer to estimate the amount of a specific isotope present. Several types of gamma spectrometer are available employing some form of multichannel analysis of the spectrum. For most simple separations of nuclide pairs, for example, ^{57}Co and ^{58}Co, low-resolution spectroscopy using a sodium iodide scintillation crystal is adequate. This does not allow the resolution of closely spaced peaks and in such cases a lithium-drifted germanium crystal is required. However, this technique is usually necessary only in the analysis of complex experimental mixtures containing numerous contaminants rather than in routine quality control of defined impurities. γ spectra of mixed β-γ emitters may be modified by use of absorbers to allow measurement of low levels of radioactive impurities which are otherwise difficult to quantitate. Thus, the ^{99}Mo content of ^{99m}Tc produced from a generator may be detected by measuring the γ-spectrum of the complete eluent in a lead pot which absorbs the ^{99m}Tc radiation but allows the harder ^{99}Mo γ-ray to pass through. The system must, of course, be calibrated in the same way.

For pure β and α emitters, liquid scintillation spectroscopy may be used directly to discriminate between different isotopes. For pure β emitters, the problems of efficiency variation arise, and it is necessary to use some method of quench correction so that the counting efficiency is known accurately for all the nuclides present. The channel windows on many commercial liquid scintillation counters are arranged to measure ^{14}C in the presence of tritium and vice-versa. It must be noted, however, that this technique is not suitable for the measurement of small amounts of tritium in ^{14}C because of the continuous nature of the ^{14}C spectrum; a better method in this case is chemical separation of the nuclides (see below).

2. Chemical Separation

Chemical techniques may be used to separate nuclides into different physical fractions so that they can then be more easily quantified. Separation requires a knowledge of the predictable radionuclidic impurities and their chemical forms. Different methods of chemical separation used include precipitation, solvent extraction, ion exchange, and other types of chromatography, distillation, and electrophoresis. The technique of choice depends on the level of impurity encountered. For example, the determination of the proportion of ^{14}C in tritium may be carried out by scintillation counting when there is a possibility of gross contamination; however, at low levels or when the proportion of tritium in ^{14}C is required, the sample should be combusted in oxygen and the two isotopes trapped as $^{3}H_2O$ and $^{14}CO_2$, respectively. These may be separated by distillation and counted separately. Such a determination has been carried out to refute results obtained from dual-isotope liquid scintillation counting which suggested the presence of tritium in a ^{14}C-labeled compound [19]. In this case the presence of chemiluminescence and highly quenched sample had given rise to an abnormally high proportion of counts in the "tritium window."

The general technique of combustion followed by distillation (from acid or alkaline solutions) is particularly useful for radionuclide separations since they are present in simple and defined chemical forms. In carrying out the separation, difficulties may arise in obtaining reproducible recoveries of trace amounts of radioactivity and "loading" the system with the appropriate carrier is necessary.

D. Determination of Radiochemical Purity

The measurement of the radiochemical purity of a labeled compound is usually the most important part of the analysis since it establishes the correct chemical form of the radionuclide. Two basic and complementary techniques are available: radiochromatography and reverse isotope dilution analysis.

1. Radiochromatography and Related Techniques

In general, radiochromatography is the most versatile and powerful technique. However, it has the fundamental limitation that *the purity value found refers only to the particular chromatography system used and its powers of separation.* What appears to be a pure compound in one system may in fact be several components and confirmatory measurements using other systems under preferably another technique are necessary. The technique involves separation of the radiotracer compound from its impurities by one of the established chromatographic techniques for the quantitation of the proportion of the radioactivity corresponding to the R_f, retention volume, retention time, mobility, etc. of the desired compound. The identity of the radioactive material can be confirmed by comparison of its movement with that of an internal standard of authentic carrier material chromatographed at the same time in the same system.

The general expression for the calculation of radiochemical purity is:

$$100 \times \frac{R_c - B_c}{R_t - B_t} \% \qquad (1)$$

where

R_c = the radioactivity corresponding to the desired compound
B_c = the background in same region of chromatogram
R_t = the total radioactivity on the chromatogram
B_t = the total background on the chromatogram

The reproducibility of the purity value is dependent on both the random errors inherent in the separation technique and those arising from counting errors in all the terms in Eq. (1). The accuracy of the purity value depends on all the systematic or bias errors introduced by the use of the separation and measurement techniques.

a. Paper and thin-layer chromatography. These, together with their associated electrophoretic techniques, are the most frequently used methods to determine the radiochemical purity of radiotracers because of their convenience and ready availability. They have the advantage that all the radioactive components in the original materials are exposed on the open chromatogram when it is quantitated, unless they are volatile. The measurement of the recovery of radioactivity on the chromatogram is of minor importance. Moreover, unlike closed column techniques, the chromatogram is retained for study at leisure, for example, remeasurement of the purity value or visualization of the inactive carriers by ultraviolet fluorescence or color reaction.

A wide variety of separation procedures using partition, adsorption, ion exchange, gel permeation, and electrophoretic mechanisms are possible, including combinations of these during the same separation, for which the reader is referred to standard texts [20-23]. They have been applied to all classes of organic compounds and many inorganic compounds, even for closely related compounds, such as those which only differ by the presence of a double bond (for example, cholesterol and cholestanol by argentation chromatography) or diastereoisomers, such as dipeptides [24].

Two-way elution techniques, that is first the elution in one direction with one eluent followed by the second in the perpendicular direction with another eluent, are possible with thin-layer and paper chromatography to carry out separations which are particularly difficult or complex [25]. For example, the accurate determination of cyclic AMP impurity levels in preparations of [^{32}P]ATP is best carried out by the technique of Bär and Hechter [26] which consists of two-way elution on a thin-layer plate with two eluents. Multiple elution, that is, elution in the same direction with the same eluent more than once, can also be used to improve separations.

Thin-layer and paper chromatograms can be quantitated by several techniques. First, the chromatogram can be autoradiographed to show the outline of the radioactive spots; a rough estimate of the radiochemical purity can be made directly from the film by comparison with standards or by densitometry. More often an accurate assessment of the purity is made by cutting up the chromatogram, or scraping off the thin-layer adsorbent, and counting the portions by crystal or liquid scintillation methods. This is a sensitive technique but can give rise to misleading results [27].

The chromatogram may be scanned directly by a gas-flow proportional counter or a collimated scintillation counter (for hard β and γ emitters) to give an analog trace (see Fig. 2).

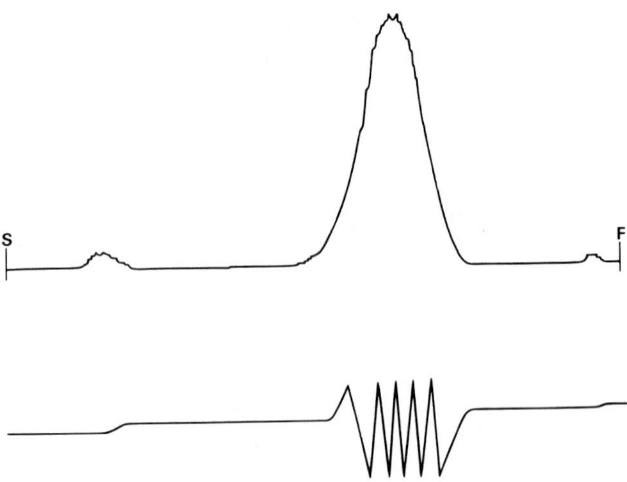

FIG. 2. Analog trace (and disc integration trace) obtained from a radiochromatogram by gas-flow proportional counter scanning.

This is the most popular technique at present and if used carefully is capable of good accuracy. More recently the spark-chamber technique has been refined and commercial instruments are available which give a qualitative picture of the chromatogram from which rough estimates of the proportions of radioactivity in the spots can be made.

The most sophisticated instrument available for the measurement of paper and thin-layer chromatograms is the 100-channel chromatogram analyzer which is based on an ordered array of 100 gas-flow proportional counters. Such an instrument may be used to analyze chromatograms statistically and to give highly accurate purity values (see Fig. 3). A critical review of the limitations and accuracy of the methods for the measurement of radiochromatograms has recently been published [27].

Because of the problems encountered in handling small amounts of sensitive compounds, radiochromatography may give rise to misleading results in the measurement of radiochemical purity and the sources of errors and problems in these measurements are

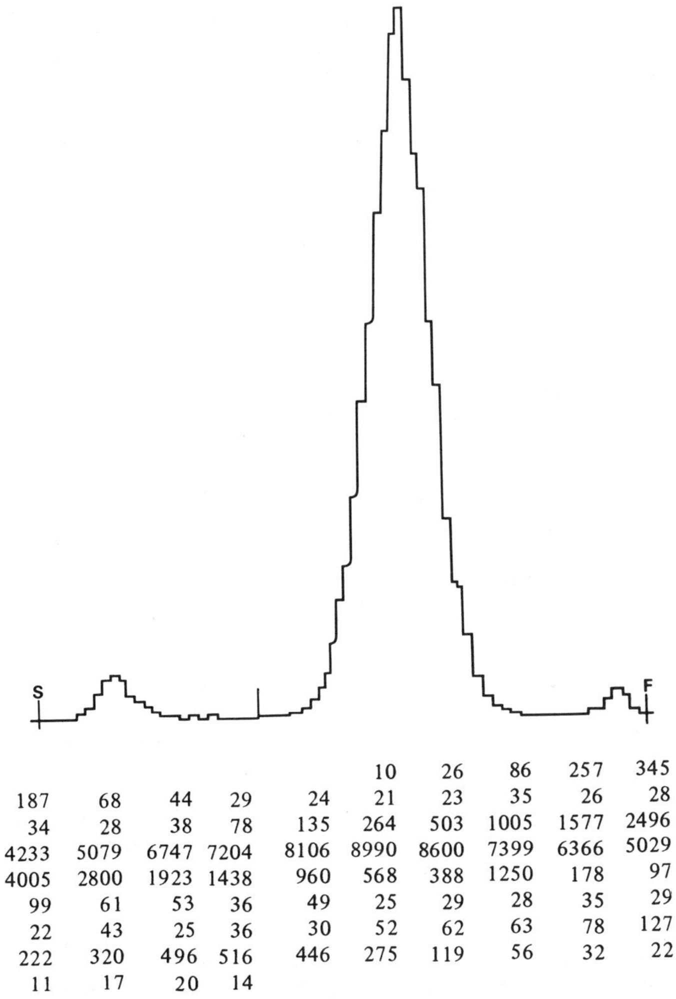

					10	26	86	257	345
187	68	44	29	24	21	23	35	26	28
34	28	38	78	135	264	503	1005	1577	2496
4233	5079	6747	7204	8106	8990	8600	7399	6366	5029
4005	2800	1923	1438	960	568	388	1250	178	97
99	61	53	36	49	25	29	28	35	29
22	43	25	36	30	52	62	63	78	127
222	320	496	516	446	275	119	56	32	22
11	17	20	14						

FIG. 3. Histogram of activity distribution and printout of counts obtained from a radiochromatogram by a 100-channel analyzer; cf. Fig. 2, which shows the same chromatogram scanned by gas-flow proportional counter.

not always easily recognized. Systematic errors are caused by:

1. Mistreatment and decomposition of the sample before chromatography

2. Decomposition during chromatography by attack from solvents, oxygen etc.
3. Peak-splitting due to an inadequate choice of the method of application or eluent, support, etc.
4. Use of an improper loading of carrier inactive material during chromatography
5. Inadequate resolution of peaks during chromatography or in the measurement process
6. Loss of the major components or impurities due to volatilization during chromatography or drying
7. Incorrect calibration of measuring instruments or inadequate allowance for such effects as static build-up during scanning
8. Misinterpretation of the scan, for example, inadequate compensation for background

Random errors may arise from variations in:

1. Method of application
2. Conditions of chromatography, e.g., the nature of the solvents, pH, support material, temperature, and time of elution
3. Measuring instruments, e.g., operating potentials, counting errors, background, integration errors, etc.
4. Interpretation of the scan or autoradiograph, for example, peak margins, the cutting-line between closely separated peaks, background level

A recent review had discussed in detail the type of artifacts and errors which can be introduced into radiochemical purity values [27], together with methods of diagnosing and overcoming them. Inadequate resolution leads to *falsely high* values of purity; artifacts and decomposition lead to *falsely low* values; errors in measurement can lead to both.

b. Radiogas-liquid chromatography. This technique is used for the determination of purity of volatile organic compounds such as those which boil or sublime below 150°C (and rarely for inorganic gaseous compounds). Typical examples are the lower fatty acids, formic to hexanoic, aromatic hydrocarbons and their derivatives, halogenated hydrocarbons and higher fatty acids as their methyl esters. Thus ^{14}C-labeled fatty acids are converted to their methyl esters for analysis by gas-liquid chromatography to determine the presence of homologs (see Fig. 4). The technique is also used to measure the purity of volatile intermediates in the preparation of nonvolatile compounds when the analysis of an

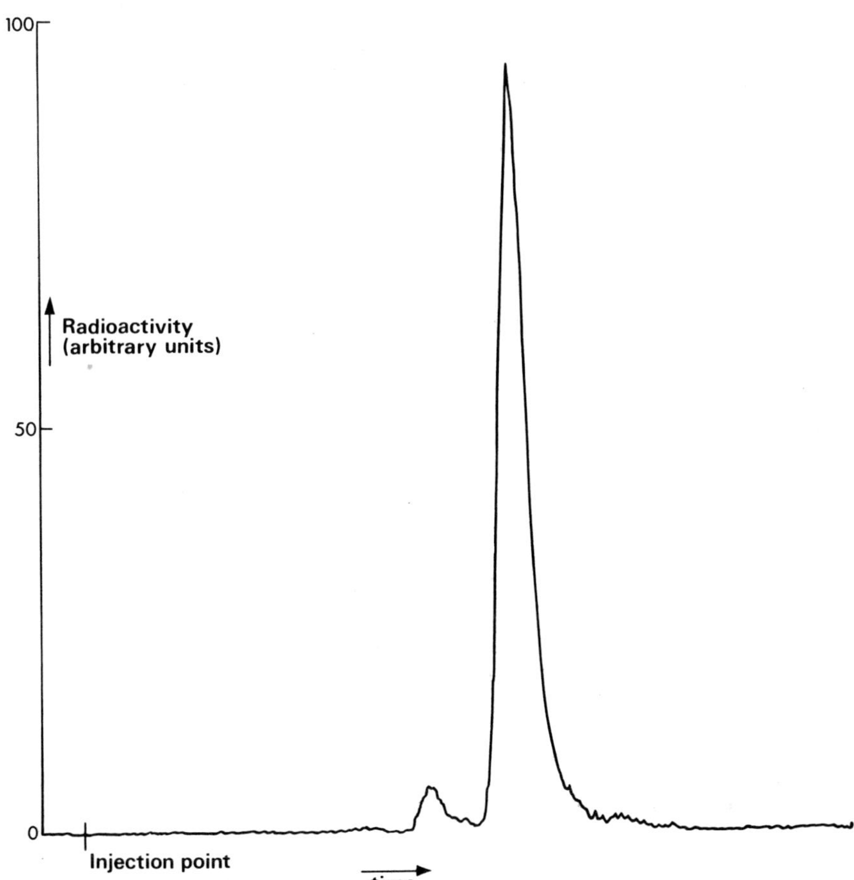

FIG. 4. Radiogas chromatograph trace of methyl[1-^{14}C]oleate; diethyleneglycol succinate packing in 6 in. × 1/4 in. glass column at 200°C carrier gas nitrogen 50 ml/min, injected in benzene solution.

intermediate is important to control the formation of unwanted impurities in the final product; thus in the preparation of [1-^{14}C]isopentenyl pyrophosphate the intermediate [1-^{14}C]-isopentenol is analyzed by gas-liquid chromatography to ensure that it is free of [1-^{14}C]dimethylallyl alcohol. It can be used for nonvolatile compounds if they are converted into volatile derivatives before injection; it is necessary to ensure that impurities are derivativized in the same way as the major component and are converted to volatile forms or alternatively to measure the recovery of the radioactive material from the derivativization process.

Detection systems for radiogas-liquid chromatography are of two types. In those based on trapping, the effluent from the column is condensed and either individual fractions sampled and counted, for example, by liquid scintillation counting, or more conveniently entrained in a continuous flow apparatus and conducted to a flow-type counting cell [28]. Otherwise the effluent is introduced into a gas-flow proportional counter and the activity is measured directly. Normally the effluent compounds are converted to a suitable gaseous form for counting by oxidation and/or reduction.

The major limitation of a "closed-column" chromatographic technique such as gas-liquid chromatography is that the recovery of the material from the column may be variable, and that important impurities which need to be measured may not be volatile. Thus, although gas-liquid chromatography is an excellent technique for separating the homologous series of lower fatty acids, it is not very useful for the quality control of [^{14}C]acetic acid which is unlikely to contain other fatty acids but forms nonvolatile condensation products on storage [13]. If a known amount of carrier corresponding to the radioactive compound is added before injection, so that the theoretical specific activity of the injected radioactive compound is known, then measurement of the recovery of the nonradioactive material can be used to calculate the recovery of radioactive material from the column, and thus a radiochemical purity value can be derived. The use of carrier to dilute the radioactive material also helps to prevent reaction with and absorption on to the column packing. A review of radio-gas chromatography has recently been published [29], and there are several standard texts [30,31] giving the conditions for separations.

c. Liquid column chromatography. This method is used to separate nonvolatile organic compounds in the same way that gas-liquid chromatography is used for volatile compounds. Until recently this technique has been poorly developed, but with the advent of high-pressure liquid chromatography, it is becoming more important. In general the technique has wider applicability than gas-liquid chromatography since many radiotracer compounds used in the biological sciences are not readily volatile, for example, carbohydrates, nucleic acids, and steroids. It can also be used for the separation of inorganic compounds and the existence of a wide variety of separations based on all four basic chromatographic mechanisms, together with variations such as affinity chromatography, means that the technique has great potential for the determination of purity of organic compounds. For example, ion exchange chromatography of individual amino acids labeled with ^{14}C has demonstrated the presence of small impurities which could not be detected by paper or thin-layer chromatography [32]; similarly, ion-exchange chromatography has been employed for confirmation of the distribution of ^{14}C-labeled amino acids in their mixtures for protein biosynthesis [19].

The radioactivity distribution in the column eluate may be measured by sampling the collected fractions and subsequent liquid scintillation counting. More conveniently the radioactivity can be measured using a flow-cell, either using a solid scintillator or mixing the flow with a scintillant mixture; the flow-cell system using a solid scintillator suffers from the build-up of contamination in the cell, and from limitations in the types of solvent which are compatible with it. Many researchers use column techniques to separate the products of enzyme assays involving labeled compounds, and an initial determination of the purity of the starting material by the same technique is necessary to measure the "blank" value for it. This frequently occurs in ion-exchange separations for studying enzyme reactions [33a].

Another advantage of liquid column chromatography is that easily oxidizable radioactive compounds, such as catecholamines, can be protected from attack by atmospheric oxygen during elution. It is possible, however, for decomposition of the radioactive material to take place by other mechanisms, for example hydrolysis on the column, and this possibility has to be considered when unusual or unexpected results are obtained.

Liquid column chromatography has the same fundamental disadvantage as gas-liquid chromatography, i.e., the recoveries of materials from the column are unknown, and hence impurities or the major component may be retained preferentially giving rise to incorrect purity values. It is usual to elute the radioactive material in the presence of carrier to prevent absorption effects, trailing, etc. in the chromatography and a measurement of this may be used to monitor the recovery of the radioactive material from the column. Such an experiment may be considered as a reverse isotope dilution analysis (see p. 206) by column chromatography and is used to measure the radiochemical purity of labeled cyanocobalamins [33b].

d. Chemical or enzymatic conversion reactions. These are often used in conjunction with chromatographic or electrophoretic separations to measure the purity of labeled compounds. Thus, when a radiotracer compound contains an impurity similar in structure to the desired one, but the chromatographic properties of both compounds are dominated by the presence of the same group and cannot be separated easily by chromatography, the compound and/or impurity may be selectively degraded by a chemical or enzymatic reaction, and the fragments separated by chromatography. Enzymatic techniques have the advantage that they are specific, though they may be unambiguous, especially where there is the possibility that the enzyme used may be contaminated with closely allied enzymes which react with either the major or minor component. It is usual to choose a technique which degrades the minor component which often can be done quantitatively, rather than the major one which may require repeated treatments for complete reaction.

In the enzymatic preparation of uridine-diphospho[U-^{14}C]-galactose (UDPGal) there is the possibility that uridine-diphospho[U-^{14}C]glucose (UDPG) is formed, which can only be separated with difficulty by chromatography [34]; the UDPG content of the sample is determined by two methods. Hydrolysis of the nucleotide sugars in dilute acid produces [U-^{14}C]glucose and [U-^{14}C]galactose, which can be separated and measured by paper chromatography. The other method involves enzymatic conversion of any UDPG present into uridine-diphospho[U-^{14}C]glucuronic acid (UDPGA) using UDPG dehydrogenase; the UDPGA can be separated easily from UDPGal by ion exchange or paper chromatography. This method is less satisfactory and tends to give higher values of the UDPG content compared with the hydrolytic one because the UDPG dehydrogenase is sometimes contaminated with UDPGal-4-epimerase, which will convert UDPGal to UDPG and causes an increased amount of UDPGA. This demonstrates the care necessary in using enzymatic conversion techniques, especially coupled ones, to determine purity values so that false answers are not obtained. The optical purity of enantiomers of amino acids can be determined by reverse isotope dilution analysis, but suitable carrier material is often unavailable and enzymatic techniques are preferable, especially for the accurate measurement of very small amounts (less than 1%) of optical impurities. D-amino acid oxidase specifically oxidizes many D-amino acids to keto acids which can be separated easily from unchanged L-amino acids and measurement of the keto acid indicates the content of D-amino acid in the original amino acid.

2. Reverse Isotope Dilution Analysis

Reverse isotope dilution analysis involves the dilution and homogenization of the radiochemical with pure carrier followed by a purification and separation procedure by which the radiotracer is removed from its radioactive impurities (this is commonly carried out chemically by the preparation of a specific derivative) and finally a comparison of the specific activity of the purified material with that of the crude diluted starting material. The technique can also be used to measure the purity of volatile or at least slightly volatile materials by formation of a nonvolatile derivative, for example, formic acid as benzimidazole or acetic anhydride as acetanilide.

Measurement of the overall radiochemical purity value by reverse isotope dilution analysis is often too inaccurate to be useful (the method includes two sets of random counting errors and two sets of mass measurement errors). However, due to the specificity of the method of preparation of the derivative it is used to confirm the identity of either the main component or impurities and particularly to measure the content of specific impurities such as optical enantiomers and isomers [35]. For example, reverse isotope dilution analysis showed the presence of approximately 10% methyl (β-D-[U-^{14}C]gluco)pyranoside in methyl-(α-D-[U-^{14}C]gluco)pyranoside, although chromatographic evidence had suggested that it was pure [36].

The principle of reverse isotope dilution analysis is such that when a radiotracer compound is diluted with a large excess of the carrier, the specific activity of the material is inversely proportional to the mass of the carrier and independent of the mass of the compound. The purity value is not dependent on the activity or mass recovered after the purification stage, but is dependent on knowing accurately the specific activity measured usually as a radioactive concentration (i.e., activity per unit mass of liquid or solid) for the crude and purified materials.

Hence it is necessary to know accurately both the chemical purity and the molecular weight of carrier and purified derivatives.

One disadvantage of reverse isotope dilution analysis is that it is difficult to assess the accuracy of the answer obtained and to diagnose the source of any errors because four different measurements are involved, each of which has its own source of error. Table 4 gives a guide to the diagnosis of the source of these errors. It is good practice to treat the crude and purified samples at the same time in the same apparatus at similar count rates with similar efficiencies to overcome errors in correcting for half-life, dead-time loss, and quenching.

3. Measurement of the Content of Labile Tritium

There are four kinds of tritium atoms which can be introduced into labeled compounds [36a].

1. Those freely exchangeable with hydroxylic solvents, such as those attached to hydroxyl, amino, and carboxyl functions; these are easily removed after preparation by successive dilutions with water and distillations, and are not usually a problem
2. Those bound to carbon atoms which have been introduced by such catalyzed reactions as reduction, halogen-tritium replacement exchange, and which can be removed only in the presence of a hydrogen-transfer catalyst
3. Those which are bound to carbon atoms and which are not normally considered to be labile, but whose rate of exchange with water may become significant under certain conditions of pH or temperature. These are the so-called "labile" tritium atoms
4. Those bound to carbon atoms which are stable under chemical conditions but may be exchanged under biological conditions; these are termed "biolabile"

The content of labile tritium in labeled compounds is usually measured by distillation of the tritiated water formed; measurements

TABLE 4

Errors in Reverse Isotope Dilution Analysis

Error	Reason
Random Poor agreement between replicates; poor reproducibility	Poor precision in counting, possibly low count rates Poor precision in measurement of mass; more sensitive technique or more mass required Poor sampling technique
Systematic Unexpectedly high answer, e.g., 100%, or much higher than chromatographic results	Carrier used was impure; needs purifying by same procedure as RIDA, if possible Too low molecular weight used for carrier; has correct chemical form been assumed? Radiochemically impure derivative; needs purifying to constant specific activity Too high molecular weight used for derivation; has the wrong or unexpected reaction taken place? Wrong stoichiometry assumed for reaction Counting errors; underestimated radioactivity in crude material (inadequate allowance for quenching, half-life, precipitation in vial or dead-time loss?). Overestimated in final sample (overcompensation for quenching, dead-time loss, half-life correction or chemiluminescence causing enhanced count rates) Error in mass measurement of derivative (e.g., too high extinction coefficient) or carrier (too low extinction coefficient) Incomplete homogenization of radioactive compound with carrier
Unexpectedly low answer, e.g., much lower than chromatographic results	Too high molecular weight used for carrier; has correct chemical form been assumed? Chemically impure derivative; is it present as a salt or a hydrate? Too low a molecular weight used for the derivative; has the wrong reaction taken place? Wrong stoichiometry assumed for reaction Counting errors; overestimated specific activity of starting material or underestimated specific activity of derivative (see above) Error in measurement of masses (see above) Incomplete homogenization of carrier and radioactive material Labilization and loss of tritium from derivatives of tritium-labeled compounds

of radiochemical purity by paper or thin-layer chromatography normally do not detect it, unless it is present in gross quantities. It is necessary to ensure that the compound is not being decomposed during measurement [37a]. The presence of labile tritium is usually not a problem in freshly prepared tritium compounds, since it is removed after preparation. It may accumulate on storage of certain tritium-labeled compounds, particularly nucleotides containing purine bases labeled in the 8-position [37b] and molecules containing tritium in "activated" positions in aromatic rings, e.g., 3,4-dihydroxy-L-[*ring*-2,5,6-^{3}H]phenylalanine [13].

4. Determination of Biological Activity

In many techniques using radiotracers it is necessary to establish the biological activity of the radiotracer before its use. Such a measurement determines a control value or "blank" for the assay procedure (this subject is covered more fully later; see Section IV) and often is the most important quality control measurement to be carried out on the compound. These measurements are frequently unique to the particular application of the radiotracer and depend intimately on the experimental procedures used, such as enzyme preparation and interferences, separation procedure, concentration, and counting technique.

In the quality control of [^{14}C]deoxyribonucleic acid (DNA), used for assays of systemic lupus erythematosus (SLE, the production of autoantibodies to DNA), it is necessary to measure the precipitatable activity of DNA in the presence of both normal and positive sera [38a], to give an assessment both of the blank and the discrimination possible.

^{125}I is frequently used to prepare iodinated protein hormones for immunoassay. Although estimation of its radiochemical purity and impurities, e.g., iodate ion, and chemical impurities, e.g., metal ion content, can be carried out by classical techniques, measurement of its ability to iodinate proteins (yield, substitution ratio, physical damage, etc.) is the most effective control of satisfactory quality. Test for biological activity are an indispensable part of the test of efficacy for radiopharmaceuticals, particularly in those cases where physicochemical properties are poorly defined (see Section VI).

5. Assessment of Radiochemical Purity

From the data obtained by different techniques the user of the radiochemical has to make an overall assessment of the radiochemical purity in order to determine its suitability for his experiment. In some cases, the different impurity values obtained by different methods are additive because each technique has separated a different impurity; for example, a paper chromatography system separating a by-product of the synthesis and a reverse isotope dilution analysis separating an optical isomer. In other cases they may duplicate each other and hence are not additive. If there is a single defined area of use for the main component and the impurities which interfere in this use can be identified, the purity determination has to take this into account and this particular impurity has to be measured. Several texts give examples of control measures used for radiopharmaceuticals [2,38b] and radiochemicals [11,12].

Analytical procedures used for some compounds are presented in Table 5, which shows the collation of measurements using different techniques to establish the purity and satisfactory nature of the product. In a few cases it may be possible to use evidence obtained from chemical purity measurements to supplement that obtained from purely radiochemical techniques. Only when the chemical purity has not been affected by the intentional or inadvertent introduction of carrier or nonradioactive impurities (which in turn

TABLE 5

Examples of Radiochemical Purity Determinations of Labeled Compounds*

Compound (use)	Methods of analysis	Labeled impurities separated	Source
L-[^{125}I]Thyroxine (thyroid function tests)	1. Electrophoresis at pH 8.6	Iodide ion	Starting material, decomposition
	2. TLC on silica gel G, eluent I	Liothyronine	Impurity in starting material labeled in reaction
	3. TLC on silica gel G, eluent II	Diiodotyrosine, etc.	Decomposition
[^{197}Hg]Chloromerodrin (brain and kidney scanning)	1. Filtration by membrane	Particulates, precipitate	Decomposition
	2. Paper chromatography on No. 1 paper, eluent III	Hydrolysis product	Decomposition
		Mercury ion	Starting material and decomposition
Sodium o-[^{131}I]-iodohippurate (kidney scanning and ERPF measurements)	1. Electrophoresis, pH 8.6	Iodide ion (>1%)	Starting material, decomposition
	2. Paper chromatography on No. 1 paper, eluent IV	o-Iodobenzoic acid	Impurities in starting material labeled in reaction
	3. Solvent extraction with chloroform	Iodide ion (<<1%)	--

TABLE 5 (continued)

L-[U-^{14}C]Leucine (protein biosynthesis)	1. TLC on cellulose eluents V, VI, VII	Other amino acids (acidic and basic); hydroxy acids	By-products of photosynthesis; decomposition
	2. Paper chromatography on No. 1 paper, eluent VIII - 3 days elution	Isoleucine	Contamination in same ion-exchange fraction
	3. Treatment with D-amino acid oxidase followed by TLC on PEI cellulose in eluent XII	D-leucine	Racemization during isolation
[G-^{3}H]Digitoxin	1. Reverse phase TLC on silica gel impregnated with propylene glycol in eluent IX	Dihydro compound	Overreaction
	2. Reverse phase PC on No. 1 paper impregnated with formamide in eluent X	Digoxin analogs	Impurities in starting material
	3. TLC silica gel in eluent XI	Digitoxigenin, mono- and bisglycosides	Impurities in starting material; decomposition

Methyl(α-D-[U-^{14}C]-gluco)pyranoside (transport studies, glycolysis)	1. PC in eluents XII, and VII, XIII	Free sugars	Decomposition of starting material [^{14}C]glucose
	2. PC in eluent XIV	Free sugars, glycosides of other sugars	Decomposition of starting material, by-products of preparation
	3. Reverse isotope dilution analysis with β-glucoside carrier	β-glycoside content	By-product of preparation
	4. Enzymatic hydrolysis with almond β-emulsin and PC in eluent XV	β-glycoside content	By-product of preparation

*Eluent Composition:

- I Chloroform:t-butanol:2 N ammonia (60:376:70)
- II Methyl acetate:isopropanol:25% ammonia (9:7:4)
- III n-Butanol:pyridine:water (10:3:3)
- IV n-Butanol:dioxane:ammonia:water(20:5:8:17)
- V n-Butanol:water:acetic acid (12:5:3)
- VI Ethanol:ammonia:water (20:1:4)
- VII n-Butanol:pyridine:water (1:1:1)
- VIII t-Amyl alcohol:acetic acid:water (20:1:20)
- IX Benzene:ethyl acetate (1:1) saturated with propylene glycol
- X Chloroform, saturated with formamide (lower phase)
- XI Dibutyl ether:methanol (9:1)
- XII 0.2 M ammonium bicarbonate solution
- XIII n-Butyl ethanol:water (52:33:15)
- XIV Ethyl acetate:acetic acid:water (containing 2% phenylboronic acid), (9:2:2)

PC = paper chromatography
TLC = Thin-layer chromatography

depends on rigorous process control), the radiochemical purity and the chemical purity will be the same and measurements of the latter may be helpful in estimating the former. For example, spectroscopic techniques are particularly useful in verifying the identity of compounds and of specific impurities; thus the trans content of [1-^{14}C]oleic acid is measured by infrared spectroscopy.

Classic chemical techniques, such as titrations, may also indicate the radiochemical purity of a compound. Thus the volume of gas evolved on acidification of sodium borotritide is a direct measure of the free hydrogen available for reduction.

E. Measurement of Distribution of Label

The presence in a radiochemical of the compound labeled in a different position from that specified must be regarded as a radiochemical impurity and thus the techniques used for measuring the distribution of the label consist essentially of specific degradations followed by separations. The degradation method used has to be controlled and have a well-established mechanistic pathway. A typical example is the oxidation of tritium atoms and isolation of the tritiated water formed to determine the positions labeled in [3',5',9-^{3}H]folic acid [39]. Similarly, periodate oxidation and other reactions have been used to degrade D-[U-^{14}C]-glucose to confirm that its distribution of labeling was truly uniform [40]. In this case microbiological degradation had suggested otherwise [41]; this shows the ambiguity which may be introduced by using degradations of uncertain pathway. Enzymatic and microbiological degradations may be particularly useful because of their specificity, but their pathway has to be established before use by unambiguous methods as described elsewhere [42].

Spectroscopic techniques have also been used to measure the distribution of labeling [43-45], a recent review [46] discusses the problems associated with the control and measurement of the position of labeling.

F. Measurement of Chemical Purity

Although, where possible, measurements of chemical purity are used to supplement radionuclidic and radiochemical purity measurements, especially for radiochemicals or radiopharmaceuticals to be used for in vivo studies, in many cases such measurements are associated with such a large mass of material that they would take up all the batch prepared or are prohibited by radiological considerations (e.g., 0.5 mg of a tritium compound containing 1 tritium atom per molecule, molecular weight 300, at 50% isotopic abundance contains 25 mCi). It is frequently necessary to determine the concentration of additives in any preparation used for in vivo use, e.g., the measurement of pH, the concentration of bacteriostats, stabilizers, buffers, etc. For example, the concentration of 2-phenylethanol used as a bacteriostat in some ^{133}Xe multidose syringes is measured by ultraviolet light spectroscopy.

Very few analytical techniques used for measuring chemical purity cover a wide range of possible impurities and it is almost impossible to use them to "screen" the radioactive compound for impurities. The best procedure for ensuring a high standard of chemical purity is the prevention of contamination during preparation followed by analysis for specific and unavoidable impurities; alternatively, but less satisfactorily, the chemical concentration of the radioactive product can be measured and related to the total mass of material. For many techniques it is necessary to measure the response of standards of the main component or of specific impurities in order to establish a calibration curve before their concentration can be estimated accurately. Measurements of chemical purity cannot usually be extrapolated to give an indication of radiochemical purity.

<u>Measurement of physical parameters</u>. The measurement of melting point and boiling point, for example, requires considerable quantities of compound and is more useful for identification than providing quantitative measurements of purity.

Infrared spectroscopy. IR spectroscopy can be used to make quantitative calculations of concentrations from the extinction coefficients of specific absorption bands. Impurities may be identified from the "fingerprinting" region of the spectrum, although conclusions are often complicated by the presence of isotope shifts [47]; also many important radiotracers used in biological chemistry are only soluble in aqueous solvents, and hence their infrared spectrum has to be determined in the crystalline form, where the lattice form has a considerable influence on the spectrum observed.

Ultraviolet light spectroscopy. UV light spectroscopy is restricted to those compounds with chromophoric groups and normally with extinction coefficients greater than 100; its sensitivity is usually sufficient for determining the chemical purity of organic compounds which conform to these criteria without requiring unduly large amounts of compound. However, it is not specific and interference arising from similar compounds may make interpretation difficult. This can be seen from Table 6, which lists the ultraviolet light absorption spectrum ratios for the common nucleic acid bases when incorporated into ribose phosphate derivatives. Although the ultraviolet light absorption spectrum of these compounds will give a useful indication of purity when peptides (absorption maximum 280 nm) may be present, it may be

TABLE 6

Ultraviolet Light Absorption Spectrum Absorbance Ratios for Nucleotides Derived from the Four Bases, Measured at pH 7 [48]

Base	Absorbance 250 nm / Absorbance 260 nm	Absorbance 280 nm / Absorbance 260 nm
Adenine	0.79	0.16
Cytosine	0.84	0.98
Uracil	0.75	0.39
Guanine	0.92	0.66

difficult to interpret the spectrum of one radioactive nucleotide derived from other bases. Colorimetric methods based on specific color reactions of the radioactive compound often have a greater specificity than ultraviolet light absorption measurements, but in many cases they are not completely quantitative. Polarimetry and spectrofluorimetry are also more specific than ultraviolet light spectroscopy, but these techniques do not seem to have been well developed for radiotracers yet.

Flame emission and atomic absorption spectrophotometry. Flame or spark emission spectrophotometry provides a broad spectrum screening technique for metal ions which is semiquantitative and is particularly useful for measuring the concentration of inorganic nuclides and their associated chemical impurities, which may have been introduced from the target material or in processing. Atomic absorption spectrophotometry can provide a specific and accurate measurement of the concentration of any particular element. Although these are destructive techniques their sensitivity does not involve the use of abnormally large amounts of sample but may introduce radiological hazards through the volatilization of the nuclide.

Titrations, pH measurements, polarographic measurements, etc. Unusual results from pH measurements may indicate the presence of ionizable impurities, the incorrect chemical form of the tracer (for example, an amino acid present as an anion rather than as a zwitterion) or a wrong strength of buffer etc. Titrations are used to measure the concentrations of ionizable labeled compounds or additives. An example is the measurement by iodimetry of the thiosulfate content of sodium [^{131}I]iodide-sodium thiosulfate solution. Such titrations may be carried out using indicators or more accurately potentiometrically. Estimations of inorganic ions can also be carried out by polarography when a reducible species can be produced.

Chromatographic and electrophoretic methods. These methods may be used to measure chemical purity. Nonvolatile impurities have to be measured by color reaction or ultraviolet light fluorescence, which may be too insensitive for impurities constituting only a small proportion of the mass of the compound. An example of a color reaction sensitive enough to detect impurities is the use of a light-green stain to measure the extent of protein denaturation in radioactive iodinated human serum albumin preparations, in which protein carrier has been added to dilute the labeled protein. Separation of denatured protein is carried out by electrophoresis [49].

Gas-liquid chromatography is valuable technique for determining the chemical purity of volatile compounds; thus the water content of ^{14}C-labeled alcohols can be determined with great sensitivity (less than 0.5%) using a katharometer detector and separating the components using a Porapak Q column. It is necessary to determine the proportion of water using standards, since the sensitivity of the detector is not uniform for all species.

Biological or enzymatic assays. These techniques may be used to measure the concentration of the radiotracer, or when the concentration is known to verify that its activity is unimpaired and that the level of chemical impurities is insignificant. Thus a very sensitive criterion of chemical purity for S-adenosyl-L-[^{14}C]methionine is its biological activity as a substrate for catecholamine-O-methyltransferase (see Section IV); similarly estimates of the biological activity of tritium-labeled prostaglandins have confirmed the value of specific activity obtained by ultraviolet light spectroscopy and have verified the absence of interfering nonradioactive compounds, such as other prostaglandins (see Chap. 18 for details).

G. Measurement of Specific Activity

Determination of an accurate value for the specific activity of a radiotracer is very important to the user who wishes to calculate its concentration in his experiments. The specific activity is normally expressed as the ratio of a measurement of radioactive concentration and a measurement of mass. An exception may occur when a direct estimate of isotopic abundance can be made by such techniques as mass spectrometry. The measurement of mass usually employs a technique such as those described in the previous section on chemical purity (Section III F). The specificity and accuracy of the mass measurement are all-important and hence a specific color reaction is often employed. Assays of the concentration based on enzymatic reactions or binding assays [50] may be particularly useful because of their specificity. Errors in the mass measurement usually arise from interference and a consequent underestimate of specific activity. Thus assays by simple weighings depend on chemical purity and ultraviolet absorbancies may be enhanced by the presence of highly absorbing impurities.

The specific activity value obtained should always be examined in the context of the most probable value for the product. For example, the specific activities per atom labeled of ^{14}C-labeled nucleotides isolated from algae which have photosynthesized in an atmosphere of $^{14}CO_2$ normally average out at a value corresponding to the isotopic abundance of the $^{14}CO_2$ used (allowing for dilution by the small inoculum); if the value for any nucleotide measured by ultraviolet light spectroscopy is found to be markedly below that average the presence of ultraviolet light absorbing chemical impurities is indicated. Because the value of a specific activity is a ratio of two measurements its inherent reproducibility will be poorer than that of the individual measurements. Thus a value for the specific activity of a tritium-labeled compound with a standard deviation of 5% would have an exceptionally high precision.

IV. APPLICATIONS OF RADIOTRACERS AND QUALITY CONTROL

The ultimate decision on the suitability of a radiochemical for use rests with the user, who must consider the information he has on its quality in the light of its intended application. Ideally the specification for the material should have been set with this in mind before an analysis is carried out. Retrospective analysis of a material after difficulties have been encountered in experimentation frequently means that effort has been wasted in producing ambiguous results and decomposition may have taken place so that the material has changed from its original state. In this section some of the effects of purity and impurities on the use of radiotracers are discussed first, followed by the specialized requirements of radiopharmaceuticals and saturation assay systems.

A. Purity and the Use of Radiochemicals

Purity is not an absolute term and always begs the question "pure for what?." The purity of a radiochemical can only be judged as satisfactory when compared with the requirements of the use to which it is to be put. Unless the purity and suitability for a specific use is stated in the information provided by the supplier the research worker has to determine whether the purity and evidence given are adequate for his purpose. After assessing the evidence presented, he may decide to accept the product and use it directly, to check it by more stringent criteria designed for his use and then accept it, to reject it, or to purity and recheck it. Decomposition of the labeled compound on storage will influence its suitability for use after a given time (see Chap. 7).

The purity of labeled compounds varies enormously from those compounds which are stable and can be produced in a high state of purity, for which the specifications may be quite exacting (say, 99% pure and specific impurities less than 0.5% each), to those unstable and sensitive compounds which are difficult to purify and for which a specification greater than 90% purity may be difficult to attain. This standard of purity is usually suitable provided that specific impurities are absent or their extent is known accurately.

1. Radiochemical Purity.

The overall radiochemical purity of the labeled compound gives an indication of its general level of purity. If the purity is markedly less than 100% calculations of reaction yields, uptakes, or conversions and their rates will be in error. When an impure radiochemical is used as a reagent, the yield of product is diminished and may be insufficient for further work. However, it is extremely rare for a radiochemical to be so extensively impure that the yield of product obtained is insignificant.

More serious errors occur when an assumption of 100% purity is necessary to calculate a specific activity value. In a double isotope dilution assay, such as the determination of steroids in physiological samples using ^{14}C-labeled steroids and tritium-labeled acetic anhydride, a decrease in the purity of the labeled steroid used results in an underestimate of the recovery, and hence an over-estimate of the amount of nonradioactive steroid present [51]. This kind of error applies to all systems where the dilution of labeled compounds or reagents is used to estimate recoveries. In this way, impure *p*-toluene-[^{35}S]sulfonyl chloride gave rise to difficulties in the estimation of steroid recovery [52] not only because of the low recovery obtained, which caused low count rates, but also errors in the estimated recovery.

Dilution with carrier of an impure labeled compound at high specific activity yields a product whose specific activity is lower than that calculated. This is more common with high specific activity tritium-labeled compounds on storage, when appreciable decomposition may have occurred. When such diluted material is used in enzyme assays or kinetic studies the rate of reaction observed will be lower than that calculated, giving rise to an error in the calculated enzyme concentration. Oldham [53] has shown that this explains an apparent isotope effect in the enzymatic phosphorylation conversion of [^{3}H]thymidine and [^{14}C]thymidine [54].

In enzymatic or binding assay systems using radiochemicals, where standards are used to plot a calibration curve from which the concentration of an unknown substrate can be interpolated, the effect of a reduction in the overall purity of a radioactive compound is less important, although specific impurities present may well interfere. The decrease in purity results in a decrease in the slope of the dose-response curve for saturation analysis, and hence a decrease in the precision of the assay [55]. This occurs in competitive protein binding assays (CPBA) of steroids or in the immunoassay of labeled hormones (see Chap. 21).

The content of a specific radiochemical impurity in a radiochemical is often more significant to the research worker for his use. The impurity level may be stated explicitly or its maximum value may be inferred from the overall purity specification of the product. The actual separation systems used should be considered carefully to decide whether the relevant impurities have been separated and whether they are meaningful for the use to which the compound is to be put. The uses of labeled compounds are so varied that any lengthy discussion of the effect of impurities on their use is beyond the scope of this chapter but it is useful to review the general principles of their influence on the use of labeled compounds.

Radiochemical impurities may interfere with the use of the radiochemical in two ways, either by behaving as a product of the reaction in which it is used, or as a substrate. In the former case they often appear as "blanks" in the subsequent separation and measurement process. Trace amounts of impurities such as those formed on radiation self-decomposition, which are well within the specification of material as prepared or supplied can cause difficulties; these are frequently not major impurities and may be difficult to detect by chromatography. The only satisfactory method of measuring them is by determining them directly in the particular reaction and separation under study, since in many cases the blank observed is a characteristic of the experimental protocol employed.

The blanks observed in the synthesis of proteins and measurement of amino acids using t-RNA may be due to small amounts of radioactive impurities [56]. Dihydroxyphenylalanine is a product of the radiation decomposition of [^{3}H]tyrosine in solution and interferes in the assay of tyrosine hydroxylase [57]. The presence of [^{14}C]methionine in S-adenosyl-L-[*methyl*-^{14}C]methionine interferes with its use in assaying methyltransferase activity when it has the same R_f value as the methylated product, for example normetanephrine, in the chromatographic system used to assay the reaction [58].

Free tritiated water which has been formed on storage of a tritium-labeled compound (labile tritium) constitutes a blank in tritium release assays, and the specific activity of the compound will be reduced so that calculations of the rate of reaction are in error [53].

In the enzymatic analysis of folic acid using folate reductase, radiation decomposition products of [^{3}H]folic acid are not precipitated with zinc sulfate and appear as tetrahydrofolic acid thus increasing the blank measurement [59]. Those impurities which act as substrates for the reaction of the labeled compound interfere by being incorporated in the same way, sometimes at a faster rate than the major component. Thus, crude ^{14}C-labeled algal protein hydrolysate is readily incorporated into macromolecular compounds [60]; but it has been shown that a significant proportion of the incorporated activity is present as nucleic acids from labeled nucleosides present in the crude mixture. Radiation self-decomposition products of [^{3}H]thymidine have been shown to be incorporated into macromolecules [61]. These effects are enhanced when an impure labeled compound at high specific activity is subsequently diluted with carrier to a lower specific activity, so that the impurities are preferentially incorporated (see also Chap. 18).

The presence of a small amount (less than 0.2% as measured by paper chromatography) of [^{14}C]glucose in 3-O-methyl-[U-^{14}C]glucose is not a significant impurity in its use to trace glucose transport. However, when this material is used to study starch biosynthesis, the incorporation of even this small amount of labeled glucose into macromolecular compound(s) was a severe embarrassment [62]. The use of 3-O-[^{14}C]methylglucose, which could not contain labeled glucose, obviated this difficulty.

In physiological systems specific radiochemical impurities may give rise to misleading results and conclusions and their level has to be determined before using the compound. Thus low molecular weight fragments in inulin (^{14}C- or ^{3}H-labeled) have given rise to the wrong results in measurements of extracellular space [63].

Finally, it should be pointed out that a labeled compound may be impure and still be quite acceptable for use. Previtamin D3 is present in vitamin D3 due to a thermal equilibrium [64]; even "pure" vitamin D_3 (labeled with ^{14}C or tritium) will exist in these two interequilibrating forms when used under physiological conditions. The presence of large amounts of nucleoside 5'-diphosphates is acceptable in nucleoside 5'-triphosphates when they are used in an enzyme system which regenerates triphosphates, especially in sequence studies. On the other hand, the presence of other bases (which are often present in nonradioactive nucleotides) would be very serious in such applications. In reactions which are specific quite large impurity levels may be unimportant. High specific activity [^{35}S]methionine used in studying t-RNA synthesis may contain up to 20-30% [^{35}S]sulfoxide, which is not taken up, and remains suitable for use. Insecticides labeled with ^{14}C, ^{32}P, or ^{36}Cl, may contain unnatural isomers (as do the corresponding nonradioactive compounds) which are difficult to remove and which do not interfere significantly in incorporation studies. It has been shown that oxidation products of [^{75}Se]-selenomethionine do not interfere with its distribution in the rat [65].

2. Chemical Purity

The main effects of chemical impurities on the use of radiotracers are concerned with interference in their biological activity or the toxicity of the impurities involved. Normally the specification required is not so critical as for radiochemical purity and, providing that no gross amounts of chemical contaminants are present, the radiochemical is often suitable for use. However, in some cases the limits of chemical content of impurities or additives have to be strictly specified in order that the efficacy of the radiochemical may be maintained. In sensitive assay techniques, the presence of chemical impurities, even in minor amounts, may be much more serious. Thus tritium-labeled steroids used in competitive protein binding assays may contain non-radioactive materials which result either from steroid-like materials not labeled in the preparation or from thin-layer absorbents used in their purification. These frequently interfere in the assay curves resulting in low efficiency of binding and a consequent reduction in the sensitivity for the assay [55]. In this case, the effect of the impurities is to decrease the "apparent specific activity" of the radiotracer and such an effect has to be considered whenever a binding assay is used to determine specific activity.

Particularly important are those chemical impurities which are closely related to the structure of substrates used in enzymatic assay techniques, such as products of the reaction or analogs of the substrates, which may act as inhibitors of the substrate and lead to false conclusions about enzymatic activity. For example, the presence in labeled S-adenosyl-L-methionine (SAM), of nonradioactive S-adenosylhomocysteine, a decomposition product of both labeled and unlabeled SAM, gives rise to strong inhibition of catecholamine-O-methyltransferase, (COMT), so much that many kinetic results using labeled SAM and COMT are in error by large amounts [66,67]. This material may be introduced by the use of impure SAM to dilute high specific activity labeled SAM and has to be removed by column chromatography after such a dilution during the preparation.

3. Radionuclidic Purity

In a similar manner to other types of purity, specifications for radionuclidic purity depend on the intended usage of the tracer.

The major difficulties with radionuclidic impurities usually occur when a minor impurity with a longer half-life increases in proportion to the major nuclide with a short half-life and can interfere in its use or be harmful. For example, in the measurement of iron metabolism, ^{59}Fe is injected into patients and its turnover is followed for 10 isotopic decay half-lives by whole-body counting to determine the biological half-life; minor long-lived impurities in the ^{59}Fe sample when injected become a significant proportion of the retained radioactivity and give rise to serious inaccuracies in the half-life measured [68]. In the use of [^{197}Hg]-chlormerodrin for kidney and brain scanning, the presence of [^{203}Hg]chlormerodrin which has a longer half-life and produces hard β emission and which therefore leads to a radiological hazard to the patient, has to be specified so that the radiological dose received is minimized [69].

The presence of radionuclidic impurities in labeled compounds may also lead to difficulties in measurement; in this case, of course, radionuclidic impurities are frequently radiochemical impurities as well. If there is the possibility that the same compound is labeled with different nuclides, for example, ^{14}C and tritium, the measurements of absolute amounts of radioactivity will be inaccurate when referred to standards of single nuclides; certainly the use of a compound containing significant amounts of another nuclide in a dual-isotope experiment would lead to serious errors.

4. Pattern of Labeling

The accurate establishment of the correct distribution of labeling of a radioactively labeled compound is often critical in its use. Thus metabolic pathways such as glycolysis are frequently elucidated from identification of the degradation products obtained from substrates specifically labeled with ^{14}C. With the use of tritium-labeled substrates for determining the kinetics of enzymatic reactions by, for example, tritium-release assays, apparently low rates of reactions will be observed if the position of labeling is not as specified [53]; moreover, the interpretation of such effects as the NIH shift [56] critically depends on the position of labeling. Similarly, the measurements of the uptake of [α-^{32}P]ATP into ribonucleic acid depend on a true α-labeling.

B. Use of Radiotracers as Pharmaceuticals

A number of isotopes processed into chemical forms can be administered to humans or animals for diagnostic or radiotherapeutic purposes. This administration brings the quality control requirements into the field of pharmaceuticals, where more stringent standards of control are normally required. In an ever-increasing number of countries, statutory pharmaceutical regulatory authorities are being set up which are required to impose strict controls on safety, efficacy and manufacture of materials intended for administration. Authorities in the United States of America and the United Kingdom have issued guides for pharmaceutical manufacturing practice, which should be followed. Special problems arise in the case of short-lived radiopharmaceuticals. Unlike other radiochemicals, radiopharmaceuticals are used for routine, well-determined purposes in most cases which permits the formulation of a fairly accurate specification.

The two most important factors in setting specifications are:

1. Safety, either toxicological or radiological
2. Diagnostic or therapeutic efficiency

For example, the content of ^{90}Sr impurity (a long-lived nuclide of high toxicity) has to be specified at very low levels in ^{90}Y (its short-lived daughter) when it is used in a colloidal state (see Chap. 8) for intracavity radiation therapy. A number of extra-quality attributes are required for radiopharmaceuticals which cannot be measured by the normal techniques of chemical analysis.

1. Sterility

To show that a batch of vials of a material is sterile would require that every vial is examined. To circumvent this problem a risk is introduced by testing a sample only, but this minimized by strict process control. It is vital that the batch is homogeneous and that a representative random sample is obtained; there have been occasions among inactive pharmaceuticals where loss of life has been traced to the nonobservance of these conditions [70]. For those radiopharmaceuticals containing a short-lived nuclide, for example, [^{197}Hg]chlormerodrin ($t_{1/2}$, 1.7 day), a full seven-day sterility test is not feasible and such material may be released for use if satisfactory indications are received before completion of the test. However, the strictest process controls are necessary to ensure sterility.

2. Apyrogenicity

Pyrogens are toxins which lead to febrile reactions on injection. They are produced by a wide range of bacteria and may be present although the bacteria themselves are destroyed. They are normally tested for by observing the temperature rise in rabbits after injection of the test substance. This is a time-consuming and material-consuming test, and recently an alternative test (Limulus test) using a lysate from the amebocytes of the blood of the horseshoe crab has been used [71], which offers promise for a simpler and faster test suitable for testing of short-lived radiopharmaceuticals.

The risk of pyrogen reactions can be much reduced by testing of raw materials for pyrogens, especially those which will support bacterial growth such as gelatin or ion-exchange resin and the maintenance of near-asepsis in processing and dispensing.

3. Animal Testing

For a number of pharmaceutical products it is necessary to use some type of animal test to confirm their physiological behavior; for example, particles of macroaggregated [^{131}I]iodinated human serum albumin which are used for indicating lung function by scintiscanning. Physical methods such as microscopic observation may be used to assess its particle size, but the best criterion is its physiological efficacy which is determined by injecting material into a rat, followed by scintiscanning or dissection to confirm that the material is localized in the lung. Similarly, the test may be used to confirm that no activity is located in other organs, which might retain smaller colloidal particles.

C. Quality Control of Saturation Assay Systems Using Radiotracers

One of the fastest expanding fields in the use of radiotracers is the determination of species of biological interest by substoichiometric (or saturation) assay systems, either employing a specific antibody (radioimmunoassay) or a specific serum or tissue protein (competitive protein binding assay) which are discussed in Chapter 21. The development of such an assay includes:

1. The establishment of the performance of the assay and from this the required specification of the labeled material
2. The maintenance of the performance of the assay to confirm that it is under control

The following criteria may be used to assess the quality of the radioactive compound.

1. Specific Activity

The specific activity of the radioactive tracer can determine the limit of detection of the assay. It should be sufficiently high that the amount of tracer used per tube is comparable with the limit of detection required. If the tracer has too high a specific activity an inordinately large quantity of radioactivity may be required to maintain saturation of the binding protein or antiserum [55].

It must be noted that the specific activity of the label is not the only factor affecting the limit of detection; for example the binding constant of the binder and the "blank" value affect it. In some assays indeed a low limit of detection is not required and the aim is to maximize precision over a certain range.

2. Purity of the Label

The purity of the label is specified by the requirements of the assay; for example, by the specific impurities which may interfere. Thus [^{125}I]thyroxine (T_4), and [^{125}I]triiodothyronine (T_3) can both be used as a tracer in the measurement of T_4 using a serum thyroxine-binding-globulin binder [72] and the presence of [^{125}I]T_3 in [^{125}I]T_4 used for this purpose is not critical. In other cases, however, the presence of a specific impurity may be important and the purity of the material should be checked before it is used.

3. Immunoreactivity and Biological Activity

In cases where a foreign label is introduced to a molecule it is important that the label material reacts with the same sites on the protein or antibody as the species being measured. This is especially important when a new batch of binding protein is being used [64].

4. Stability

Stability of the radiotracer compound is also important to ensure a valid assay; for example some knowledge of breakdown products is necessary to guarantee performance.

D. Control of Functioning Assays

There are three types of criteria which may be used to indicate that an assay is under control and functioning properly. They are those which show that:

1. The reagents are working properly (functional testing)
2. The assay has been carried out correctly (operational control)
3. There are indications that the expected result is being obtained on a typical sample over a range of assays (long term control).

Ideally controls 1 and 2 should make controls of type 3 redundant, if they are properly selected to monitor all the vital parameters of the assay. However because of the complexity of the situation this is rarely, if ever, possible and controls of type 3 are essential to show that the assay is producing satisfactory results.

Functional tests and controls have been proposed by several authors [73,74]; essentially the parameters which are specified are "initial values" and the slope of the dose-response line. Initial values include "blank value," total counts, nonspecific binding and fraction of total counts bound at zero dose. If the assay can be linearized, then the dose-response line over the whole operational range may be considered; if not, the changes in the response metameter over specified ranges of the dose metameter can be used to determine the slope and hence the discrimination of the assay.

Operational control calls for the monitoring of the performance of the test to ensure it has been correctly carried out. Typically, controls are applied to ensure the proper treatment of unknowns or to guard against errors in experimental technique, for example, the reproducibility of replicate determinations of both standards and unknowns [75]. A further indication of the quality of an assay comes from the results obtained from a sample which is treated as an unknown but whose performance has already been characterized previously. This technique using "quality control" samples is already well-established in clinical chemistry.

The validation of a single assay is the simplest case for the use of control samples. Such samples can also be used in the medium- and long-term to give an indication of any variations in the assay. The controls normally cover the typical range of samples to be examined and it is extremely important that any variations observed are attributable to the assay and not to variations in the control. For this reason, a set of control samples must be both homogeneous with respect to the species under assay, and stable with time. This may be achieved with a blood serum freeze-dried and stored at low temperature under nitrogen.

Provided these conditions are met, controls can be relied upon to provide vital feedback information on the performance of an assay over a long period of time.

REFERENCES

1. J. W. Cornforth, R. H. Cornforth, C. Donninger, and G. Popjak, Proc. Roy. Soc., 163B:492 (1966).
2. V. K. Tya, A. T. Ralaban, A. R. Palmer, and J. C. Maynard (eds.); Radioisotope Production and Quality Control, IAEA, Vienna, 1971.
3. R. J. Bayly and H. Weigel, Nature, 188:384 (1960).
4. L. E. Geller and N. Silberman, J. Label. Comp., 5:66 (1969). (1969).
5. F. C. Greenwood, W. M. Hunter, and J. S. Glover, Biochem. J., 89:114 (1963).
6. H. J. Brodie, K. Raab, G. Possanza, N. Seto, and M. Gut, J. Org. Chem., 34:2697 (1969).
7. Y. Osawa and D. G. Spaeth, Biochemistry, 10:66 (1971).
8. J. S. Glover, D. N. Slater, and B. P. Shepherd, Biochem. J., 103:120 (1967).
9. H. C. Sheppard, private communication (The Radiochemical Centre, Amersham).
10. British Pharmacopoeia 1973, p. 98.
11. P. Peyser, in Proc. 2nd Int. Conf. Methods of Preparing and Storing Labelled Compounds (J. Sirchis, ed.), Euratom 3746 d-f-e, Brussels, 1968, p. 1081.
12. J. R. Catch, Purity and Analysis of Labelled Compounds, Review Booklet 8, The Radiochemical Centre, Amersham, 1968.
13. G. Sheppard, Atomic Energy Rev., 10:3 (1972).
14. R. J. Bayly and E. A. Evans, J. Label. Comp., 2:1 (1966).
15. R. J. Bayly and E. A. Evans, J. Label. Comp., 3:349 (1967).
16. J. P. Buckley, Int. J. Appl. Radiat. Isotopes, 22:41 (1971).
17. M. S. Patterson and R. C. Greene, Anal. Chem., 37:854 (1965).
18. J. C. Turner, Int. J. Appl. Radiat. Isotopes, 20:499 (1969).

19. Unpublished observations at The Radiochemical Centre, Amersham, Bucks. England.

20. E. Stahl (ed.), _Thin-Layer Chromatography_, 2nd ed., Allen and Unwin, London, 1969.

21. E. Heftmann (ed.), _Chromatography_, 2nd ed., Reinhold, New York, 1967.

22. I. Smith (ed.), _Chromatographic and Electrophoretic Techniques_, Vols. 1, 2, 3rd ed., Heinemann, London, 1969.

23. I. M. Hais and K. Macek (eds.), _Paper Chromatography_, 3rd ed., Academic Press, New York and London, 1963.

24. A. V. Barooshian, M. J. Lautenschleger, J. M. Greenwood, and W. G. Harris, _Anal. Biochem._, _49_:602 (1972).

25. I. M. Hais, _J. Chromatog._, _48_:200 (1970).

26. H. P. Bär and O. Hechter, _Anal. Biochem._, _29_:476 (1969).

27. G. Sheppard, _Radiochromatography of Labelled Compounds_, Review Booklet 14, The Radiochemical Centre, Amersham, 1972.

28. L. Schutte and E. B. Koenders, _J. Chromatog._, _76_:13 (1973).

29. S. P. Cram, in _Adv. Chromatog._, _9_:377 (1970).

30. _Gas Chromatography Abstracts_, Institute of Petroleum, London and Applied Science Publishers, Barking, England, Quarterly.

31. _Compilation of Gas Chromatographic Data_, American Society for Testing and Materials, Philadelphia, 1967.

32. W-S. Chou, L. Kesner, and H. Ghadimi, _Anal. Biochem._, _37_:276 (1970).

33a. D. J. Reed, K. Goto, and C. H. Wang, _Anal. Biochem._, _16_:59 (1966).

33b. _British Pharmacopoeia_, 1973, p. 131.

34. H. Carminatti and S. Passeron, in _Methods Enzymol._, _8_:108 (1966).

35. R. J. Bayly, in _Radioisotopes in the Physical Sciences and Industry_, Vol. 2, IAEA, Vienna, 1962, p. 305.

36. G. M. Bartlett and G. Sheppard, _J. Label. Comp._, _5_:275 (1969).

36a. E. A. Evans, Chapter 6 in _Tritium and Its Compounds_, 2nd Ed., Butterworths, London, 1974.

37a. "_Carrier-free_", 1973 No. 3, The Radiochemical Centre, Amersham.

37b. E. A. Evans, H. C. Sheppard, and J. C. Turner, _J. Label. Comp._, _6_:76 (1970).

38a. T. Pincus, Arthritis Rheumatism, 14(5):623 (1971).

38b. Analytical Control of Radiopharmaceuticals, IAEA, Vienna, 1970, passim.

39. S. F. Zakrzewski, E. A. Evans, and R. F. Phillips, Anal. Biochem., 36:197 (1970).

40. J. C. Turner, J. Label. Comp., 3:217 (1967).

41. M. B. Kemp and J. R. Quayle, Biochem. J., 102:94 (1967).

42. H. Simon and H. G. Floss, Bestimmung der Isotopenverteilung in markierten Verbindunge, Springer-Verlag, Berlin, 1967.

43. J. Bloxsidge, J. A. Elvidge, J. R. Jones, and E. A. Evans, Org. Mag. Resonance, 3:127 (1971).

44. W. R. Waterfield, private communication (The Radiochemical Centre, Amersham).

45. S. Pinchas and I. Laulicht, Infra-red Spectra of Labelled Compounds, Academic Press, London, 1971.

46. J. R. Catch, Patterns of Labelling, Review Booklet 11, The Radiochemical Centre, Amersham, 1971.

47. Ref. 45, pp. 38-215.

48. Ultra-violet Absorption Spectra of 5'-Ribonucleotides, P. L. Biochemicals, Milwaukee, Wisc. (1956; reprinted 1969).

49. British Pharmacopoeia 1973, p. 62.

50. I. B. Rubin and G. Goldstein, Anal. Biochem., 33:244 (1970).

51. T. G. Brien and J. D. H. Slater, J. Endocrinol., 38:197 (1967).

52. B. A. Scoggins, J. P. Coghlan, C. J. Oddie, E. M. Wintour, M. D. Cain, and B. Hudson, in In Vitro Procedures with Radioisotopes in Medicine, IAEA, Vienna, 1969, p. 101.

53. K. G. Oldham, J. Label. Comp., 4:127 (1968).

54. L. Baugnet-Mahieu, R. Goutier, and M. Semal, J. Label. Comp., 2:77 (1966).

55. R. J. Ekins, in In Vitro Procedures with Radioisotopes in Medicine, IAEA, Vienna, 1969, p. 325.

56. K. G. Oldham, Radiochemical Methods of Enzyme Assay, Review Booklet 9, The Radiochemical Centre, Amersham, 1968.

57. B. Waldeck, J. Pharm. Pharmacol., 23:64 (1971).

58. G. H. Spiller and K. C. Tovey, private communication, (The Radiochemical Centre, Amersham).

59. S. P. Rothenberg, in In Vitro Procedures with Radioisotopes in Medicine, IAEA, Vienna, 1969, p. 150.

60. A. N. Sangster and C. Poort, Biochem. Biophys. Res. Commun., 20:218 (1965).

61. M. Wand, E. Zeuthen, and E. A. Evans, Science, 157:436 (1967).

62. P. Köhn, private communication.

63. G. Levi, Anal. Biochem., 32:348 (1969).

64. K. H. Hanewald, M. P. Rappoldt, and J. B. Roborgh, Recueil, 80:1003 (1961).

65. Y. Cohen, in Analytical Control of Radiopharmaceuticals, IAEA, Vienna, 1970, p. 9.

66. T. Deguchi and J. Barchas, J. Biol. Chem., 246:3175 (1971).

67. L. Flohé and K-P. Schwabe, Hoppe-Zeyler's Z. Physiol. Chem., 353:463 (1972).

68. H. C. Heinrich and E. E. Gabbe, Experientia, 26:467 (1970).

69. British Pharmacopoeia (1973), Appendix XXV, p. A140.

70. C. M. Clothier, Cmnd 5035, HMSO, London (1972).

71. J. F. Cooper and J. C. Harbert, J. Nucl. Med., 14:387 (1973).

72. R. P. Ekins, E. S. Williams, and S. Ellis, Clin. Biochem., 2: 253 (1969).

73. F. C. Greenwood, Principles of Competitive Protein Binding Assays, Odell and Daughaday, eds.), Lippincott, Philadelphia, Pa., 1971.

74. D. Rodbard, P. L. Rayford, and G. T. Ross, Statistics in Endocrinology, ed. J. W. McArthur and T. Colton, MIT Press, 1970.

75. G. S. Challand and T. Chard, Clin. Chim. Acta, 46:133 (1973).

Chapter 7

STORAGE AND STABILITY OF RADIOTRACER COMPOUNDS

E. Anthony Evans

Organic Department
The Radiochemical Centre, Limited
Amersham, Buckinghamshire
England

I. INTRODUCTION

Other chapters of this text describe the widespread use of radiotracer compounds and espcially in biological, medical and chemical research. In all such applications the purity and stability of tracer compounds are of utmost importance, and it is therefore necessary for investigators to understand the problems involved in the storage and stability of the labeled compounds. If the correct facts are to be obtained from uses of radiochemical tracers a knowledge of the radiochemical purity and radionuclidic purity, of the tracer, is often essential, unless it can be demonstrated that the presence of impurities is of no consequence in a particular study.

Chapter 6 emphasizes the need to control carefully the quality of radiochemicals for use in tracer investigations. It is imperative that information derived from the use of radiochemicals, which is often subsequently published, should be valid and not misleading due to the presence of impurities. As Kepler has observed "impurities are facts of life and the more one is aware of them the better able to deal with them." It is essential that the tracer compound has a purity value which is suitable for use in the experiments being undertaken and for which it has been prepared. It is, of course, not practical to guarantee absolutely either the purity of a compound for a particular use or the purity for several different uses. Absolute purity is an ideal which cannot be achieved in practice. The quality of a radiochemical is normally at its highest value immediately following its preparation and purification, but because of self-decomposition this quality is not maintained on storage of the compound. It is relatively easy for investigators to check the quality of radiochemical and its suitability for use immediately prior to the experiments by methods which are described in Chapter 6. There is no other safeguard for the investigator, the purity of the radiochemical must be checked immediately prior to use to be confident of meaningful results, or at least in order to eliminate one possible cause for obtaining erroneous results.

Unfortunately, most organic chemicals are unstable and decompose with varying degrees of sensitivity to external factors. The presence of radioactive atoms in a molecule makes the compound even more unstable and, in addition, the great sensitivity of detection of radioactivity makes the presence of impurities even more critical. Indeed, the presence of specific impurities in a tracer compound may have a more pronounced effect than an overall low or high radiochemical purity. The increased sensitivity of the methods now used for the analysis of radioactive products and for the measurement of radioactivity in general, are discussed in Chapters 6 and 4, respectively. This is making more investigators aware of the problems of self-decomposition and to be more critical in the interpretation of the data obtained from uses of radiochemical tracers. This awareness is very desirable, for not only are a very large number of labeled compounds, particularly organic compounds, used extensively as tracers, but many applications demand a very high purity of the tracer compound. In addition, many compounds are required for use at very high molar specific activity which also makes the control of self-decomposition more difficult to achieve.

Most users of compounds labeled with radionuclides now recognize that such compounds decompose on storage and that the decomposition rate is accelerated by self-irradiation. The degree of decomposition in relation to the structure of the compound, to the storage conditions used for the compounds, and the measures which can be taken to control and minimize the formation of impurities, are perhaps not always so well known. Although the interaction of radiation with a wide variety of compounds has been studied in great detail by many investigators, the subject of self-decomposition of radiochemicals is not one which has attracted wide scientific interest or study. For most investigators the decomposition of radiochemicals on storage is regarded as an unmitigated nuisance.

Much of the work on decomposition of radiochemicals by self-irradiation is empirical, in fact studied by manufacturers of such compounds as part of a routine quality control program. It is thus a record of observations and attempts to minimize decomposition without necessarily understanding all the processes occurring at the time. This has led to a general understanding of the basic principles of what is happening and the development of certain basic "do's" and "dont's" to be observed in the control of decomposition. The fact that complete comprehension of the phenomenon is lacking is, however, no reason for not considering what is known about it, so that the more general principles which are discussed may lead to the best use being made of the available protective measures. This chapter therefore discusses current knowledge of the self-decomposition of labeled compounds, methods for minimizing decomposition, and the effects of impurities on the interpretation of data from radiotracer experiments.

II. RADIONUCLIDES

Compounds labeled with the pure β-emitting radionuclides ^{14}C, ^{3}H, ^{35}S, ^{32}P, ^{33}P, and ^{36}Cl are most commonly used in tracer applications. Compounds labeled with the γ- or x-ray-emitting radionuclides such as ^{125}I, ^{131}I, ^{57}Co, ^{58}Co, ^{60}Co, ^{197}Hg, ^{203}Hg, and ^{75}Se usually have special in vitro and in vivo applications in medicine as illustrated in Chapters 20 and 21, respectively. Compounds labeled with ^{3}H, ^{125}I, and ^{75}Se are especially used in radioimmunoassay and competitive protein binding assays associated with routine in vitro testing in medical diagnosis. Some properties of these commonly used radionuclides are listed in Tables 1 and 2.

Decomposition depends in part on the amount of energy absorbed by the compound during its useful life, so that for a given amount of radioactivity the radiation energy emitted should be taken as a guide to the seriousness of the problems. If this were the only factor the extent of self-irradiation damage should increase as the list of pure β emitters in Table 1 descends. However, this is not so in practice, for several other factors are involved which include the following.

TABLE 1

Physical Properties of Some Beta-Emitting Radionuclides

Radionuclide		Beta energy (MeV)		Specific activity		Daughter nuclide (stable)
		Max.	Mean	Max. (mCi/mA)*	Common values for compounds (mCi/mmol)	
^{3}H	12.35 years	0.0186	0.0056	2.92×10^{4}	10^{2}–10^{5}	^{3}He
^{14}C	5730 years	0.156	0.049	62.4	1–10^{2}	^{14}N
^{35}S	87.4 days	0.167	0.049	1.50×10^{6}	1–10^{2}	^{35}Cl
^{33}P	25.5 days	0.248	0.077	5.32×10^{6}	10–10^{4}	^{33}S
^{36}Cl	3.01×10^{5} years	0.709	0.25	1.2	10^{-3}–10^{-1}	36
^{32}P	14.3 days	1.709	0.70	9.2×10^{6}	10–10^{5}	^{32}S

*A milliatom is the atomic weight of the element in milligrams.

TABLE 2

Physical Properties of Some γ- and X-ray-Emitting Radionuclides

Radionuclide	Half-life	Electromagnetic transitions* (MeV)	Specific activity: Max. (mCi/mA)	Specific activity: Common values for compounds (mCi/mmol)	Daughter nuclide (stable)
^{131}I	8.06 days	0.91(a) 0.36(b)	1.62×10^7	10^2–10^4	^{131}Xe
^{125}I	60 days	0.035(c)	2.18×10^6	10^2–10^4	^{125}Te
^{57}Co	270.5 days	0.13(c)	4.83×10^5	10^3–10^5	^{57}Fe
^{58}Co	70.8 days	0.475(a) 0.81(c)	1.84×10^6	10^3–10^5	^{58}Fe
^{75}Se	120 days	0.40(c)	1.08×10^6	10–10^3	^{75}As
^{197}Hg	64.4 hr	0.077(c)	4.8×10^7	10^2–10^4	^{197}Au
^{203}Hg	47 days	0.21(a) 0.279(b)	2.8×10	10 –10^3	^{203}Tl

*(a) Maximum β emission.
(b) γ photon emission.
(c) Electron capture photons

1. The fraction of the energy absorbed is much less than unity for the more energetic β emitters such as ^{32}P; on the other hand, almost total absorption of the β energy occurs with tritium compounds. It is not surprising therefore, that severe problems exist in the storage of tritium-labeled compounds. Gamma energy is, in general, little absorbed by the compound itself or its immediate environs. This is seen from the effects of γ-rays on freeze-dried thymidine shown in Table 3. The stability of thymidine and, rather surprisingly, other deoxynucleosides, bromo- and iododeoxyuridine, in the dry state to x-irradiation (10^8 rads) has been noted also by Tanovka [9]. In solution, of course, these compounds are all susceptible to attack by free radicals produced as a result of the interaction of the radiation with water (cf. Table 3).

2. The decomposition depends on the specific activity of the compounds. As can be seen from Table 1, the molar specific activity of tritium compounds is usually much higher than for compounds labeled with other pure β emitters. There are in fact no reports of significant decomposition by self-irradiation of compounds labeled with ^{36}Cl, which is perhaps not surprising because

TABLE 3

Effect of Gamma Radiation on Labeled Thymidine [34]

Compound	Weight (μg)	Activity (mCi)	Irradiation Dose (rads)	Irradiation Time (min)	Radiochemical purity (%)
[6-^{3}H]Thymidine (freeze-dried)	92	1	nil	--	100
	53	2	2.5×10^6	30	98
	53	2	2.5×10^6	30	92
[6-^{3}H]Thymidine (aqueous solution)	192	1	2.5×10^6	30	<5
[2-^{14}C]Thymidine (aqueous solution)	132	0.01	2.5×10^6	30	<5

of their very low molar specific activity.

3. The absorbed energy decreases exponentially with time, as the radionuclide decays. This is an important factor for compounds labeled with radionuclides of short half-life such as ^{131}I, ^{32}P, and ^{33}P.

III. MECHANISMS OF DECOMPOSITION

The observations of Tolbert et al. [1] published more than 20 years ago indicated that extensive self-decomposition can occur in compounds labeled even with the radionuclide of long half-life, ^{14}C, from which β particles of quite modest mean energy are emitted. These observations certainly shattered any illusions that radioisotopically labeled compounds were as stable as their nonradioactive counterparts.

The significance of these early observations was not fully grasped until commercial manufacturers began extensive investigations of the problems relating to the "off-the-shelf" supply of high quality radiochemicals of high radiochemical purity. As the demand for radiochemicals of high specific activity grew, especially for tritiated compounds, it was soon realized that the problems were severe and methods for the control of self-decomposition had to be found. Thus, besides the physical half-life of the radionuclide, the concept of the shelf-life of the radiochemical has emerged. This may be defined as "the time span during which a labeled compound may be used validly and safely."

The modes by which the decomposition of labeled compounds can arise were eminently classified by Bayly and Weigel [2] in 1960; this classification is summarized in Table 4. The radiation energy will commonly be absorbed by the compound itself and/or by its environs. Absorption of energy by the compound will excite the molecules which may break up in some manner or react with other molecules, and the excited decomposition fragments may also react with other molecules producing impurities. Absorption of

TABLE 4

Modes of Decomposition of Radiochemicals

Mode of decomposition	Cause	Method for control
Primary (internal)	Natural isotopic decay	None for a given specific activity
Primary (external)	Direct interaction of the radioactive emission (α, β, or γ) with molecules of the compound	Dispersal of the labeled molecules
Secondary	Interaction of excited products with molecules of the compound	Dispersal of labeled molecules; cooling to low temperatures; free radical scavenging
Chemical and microbiological	Thermodynamic instability of compound and poor choice of environment	Cooling to low temperatures; removal of harmful agents

the radiation energy by the environs can produce reactive species such as free radicals.

1. Primary (Internal) Decomposition

This arises as a result of the disintegration of the unstable nucleus of the radioactive atom. For compounds that are labeled with radionuclides which decay to stable isotopes, the decomposition fragment(s) so produced will be radioactive *only* if the molecule(s) decomposing contains two or more radioactive atoms. In many investigations labeled compounds are used which contain only a small fraction of multilabeled molecules [3] and for such compounds the radioactive impurities from primary (internal) decomposition can usually be neglected. In multilabeled molecules containing the radionuclides of long half-life, such as ^{14}C or ^{3}H, the contribution of radioactive impurities on storage is also very small from primary (internal) decay, except in the case of macromolecules, which is discussed later (see Section VI E). Thus, for example, Wolfgang et al. [4] have shown that only 0.001% per year of [^{14}C]methylamine, [Eq. (1)] is formed from [1,2-^{14}C]ethane. Similarly the yield of [1-^{14}C]glycine from [2,3-^{14}C]succinic acid is very small [5].

$$^{14}CH_3\cdot{}^{14}CH_3 \longrightarrow [^{14}CH_3\cdot{}^{14}NH_3] \longrightarrow {}^{14}CH_3\cdot NH_2 \qquad (1)$$

2. Primary (External) Decomposition

This involves the direct interaction of the particles (usually β particles) with the molecules surrounding the decaying atom. If these particles strike another labeled molecule and change it in some way, then a radioactive impurity results, whereas if the damaged molecule is unlabeled then only a trace of a chemical impurity is formed. As the molar specific activity of the compound is increased so also is the probability of the radiation (particles) interacting with other radioactive molecules. Thus primary (external) decomposition becomes increasingly significant as the molar specific activity of the compound is increased. From this it is also evident that by separating the radioactive molecules (atoms) with unlabeled molecules, for example, by the use of solvents, the primary (external) effects can be "diluted" out as illustrated diagrammatically in Fig. 1. However, because

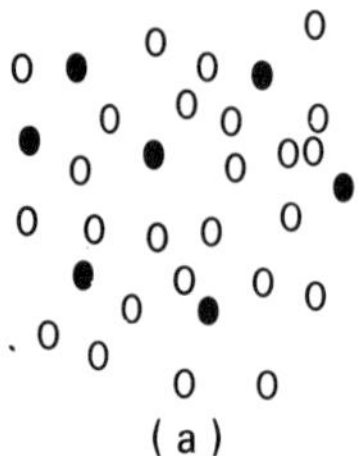

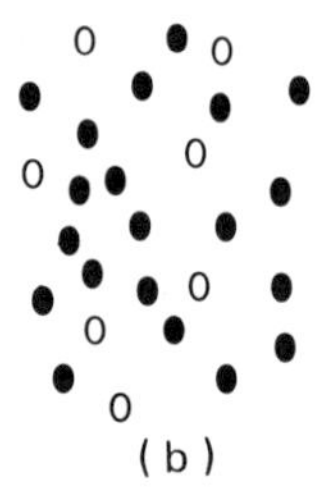

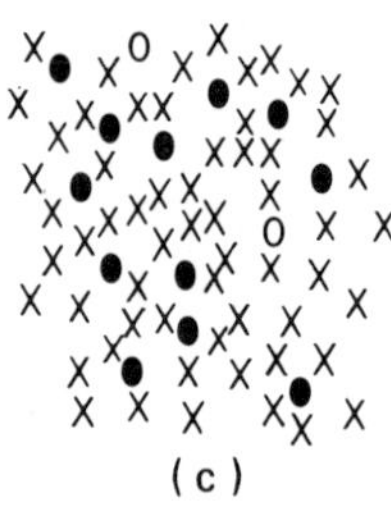

FIG. 1. Diagram representing molecules of radiochemicals at low (a) and high (b) specific activity, and at high specific activity in a diluent (c); 0 represents molecules of a compound; ●, labeled molecules of a compound; and X, diluent molecules.

of secondary decomposition the choice of the diluent is critical if the decomposition of the compound is to be minimal.

3. Secondary Decomposition

Secondary decomposition is commonly the most damaging and the most difficult to control. It is also the mode most susceptible to minor variations of the environmental conditions. The combination of a very low chemical mass of labeled compound and the great sensitivity of detection of the radioactivity, magnifies these problems which are discussed in detail later.

All the effects of radiation decomposition are related to the specific activity which is defined as the activity per unit mass of an element or a compound containing a radionuclide. In order to make comparisons between compounds it is the molar specific activity which should be considered, i.e., mCi/mmol rather than mCi/mg for example. This point is particularly important for macromolecules labeled, for example, with ^{14}C, where even a modest isotopic abundance may be associated with a very high molar specific activity, and disintegration causes extensive disruption of the chain. An example for this is in [^{14}C]DNA (deoxyribonucleic acid) [6]. Radiation decomposition is also related to radioactive concentration which is defined as the activity per unit quantity of any material in which the radionuclide is contained; this should not be confused with specific activity. Radioactive concentration often relates to the amount of active compound dissolved in a solvent and is usually expressed in terms mCi/ml or µCi/µl, for example.

4. Chemical and Microbiological Decomposition

In considering all the possible effects of radiation causing decomposition, ordinary chemical decomposition of a radiochemical is often overlooked. Smith [7] gives a brief but very illuminating account of factors, such as oxidation, hydrolysis, or biological reactions, which influence the decomposition of medical preparations. Chemical decomposition arising from such factors is even more likely to occur with radiochemicals because these are often stored and used in solution at very low chemical concentrations (μg/ml), as previously mentioned, and are likely also to be stored for a longer time. In addition, labeled compounds are usually prepared on such a small chemical scale (millimoles) and handled in such small amounts that it is difficult to ensure complete freedom from unlabeled (inactive) impurities which might be harmful by initiating or increasing the rate of decomposition of the compound. It is believed that this difficulty is at least part of the cause of batch-to-batch preparation variations in the decomposition rate of radiochemicals stored under the same environmental conditions and at the same specific activity and radioactive concentration.

Even surfaces of hard glass have been found to adversely affect the stability of radiochemicals, in particular carbohydrates, although the nonneutrality of the glass could only be demonstrated by extraction with boiling water [2]. Another example is the storage of acetic anhydride (with or without dry benzene) labeled with tritium or carbon-14 at high specific activity in all glass ampuls. Appreciable amounts (up to 15% or so) of acetic acid are observed to form within a few weeks of storage, which is believed to arise largely by interaction of the radiochemical with the surface of the glass. Thus even when careful cleaning and drying of all glass ampuls is carried out, it does not necessarily stop some chemical decomposition effects.

It is also necessary to take precautions against photochemical and microbiological decomposition of the radiochemical. Many chemicals are known to be susceptible to decomposition by light; folic acid, polycyclic aromatic hydrocarbons, aniline, and vitamin A are examples. Great care is needed in the handling and in the analysis of the radioactive forms of such compounds. Microbiological contaminants may also affect the rate of decomposition of a radiochemical in solution and therefore its solutions are normally sterilized to minimize this effect. At The Radiochemical Centre, sterilization of solutions of organic chemicals by bacterial filtration has been found, in general, to be superior and result in a longer shelf-life of the radiochemical, than by heat sterilization (autoclaving) even for compounds apparently stable to heating in solution [70]. Traces of chemical impurities arising from decomposition during heating of the solutions, is believed to account for the accelerated rate of decomposition, compared with bacterially filtered solutions, of the radiochemical on storage. The special problems of tritium-hydrogen exchange from tritiated compounds stored in solutions of hydroxylic solvents, are discussed later (see Section III B).

A. Quantitation of Decomposition

The phenomenon of self-decomposition of radiochemicals does not readily lend itself to accurate quantitative interpretations and predictions. Much of the data available show trends, but batch-to-batch variations and the large number of factors influencing decomposition make accurate quantitation impractical and of little direct or practical use. Early reviews on this subject [1,8] expressed the time dependence of radiation in terms of the total does received (rads or roentgens equivalent). Such a calculation does not take into account the specific activity or radioactive concentration of the radiochemical.

1. G(-M) Values

In radiation chemistry it is usual to express the yield or radiolysis products in terms of "G" values. The G value is defined as the yield in numbers of molecules changed (formed or destroyed) per 100 eV absorbed by the system. In studies of self-radiolysis it is sometimes useful to calculate (G-M) values, the "-M" referring to molecules of the compound irreversibly altered by the radiation process. The system may be the labeled compound itself when stored in the pure state, or a solution of it in a solvent (water, benzene, ethanol, etc.). G(-M) values are very dependent upon the storage conditions used for the compound.

2. Magnitude of Self-Decomposition

Theoretically, given a G(-*M*) value for a particular compound under specified conditions, it is possible to calculate the magnitude of the self-decomposition from Eq. (2)

$$P_d = f \cdot \overline{E} \cdot s \cdot 5.3 \times 10^{-9} \cdot G(-\underline{M}) \qquad (2)$$

where

P_d = the initial percentage decomposition per day

f = the fraction of the radiation energy absorbed by the system

$\overline{E}$ = the mean energy of the emission in electron volts

s = the initial specific activity of the compound in millicuries per millimole (mCi/mmol)

In order to calculate the degree of decomposition for a particular time, Eq. (2) can be used in its simple linear form providing the magnitude of decomposition is modest (say less than 10%) and the storage time is short compared to the half-life of the isotope. If these restrictions are not met, the exponential form (3) needs to be used [2].

$$P = 100[1 - \exp(-f \cdot \overline{E} \cdot G(-\underline{M}) \cdot s \cdot t \cdot 6.14 \times 10^{-16})] \qquad (3)$$

where P is the percentage decomposition

t is the time of storage in seconds

In the case of the shorter-lived radionuclides such as ^{35}S, for example, the "equivalent time of storage t_e" i.e., the time at which it can be regarded as being stored at its initial specific activity is given by (4)

$$t_e = 126[1 - \exp(\frac{-t_d}{126})] \text{ days} \qquad (4)$$

where t_d is the actual storage time in days. A similar adjustment must be made if Eq. (2) is used and the extent of decomposition exceeds about 10%.

For tritium compounds it is valid to take the value of f as unity because of the low penetrating power of the weak β energy resulting in virtually complete absorption of the radiation energy. However, for compounds labeled with other radionuclides the value of f will depend very much on how they are stored and could be considerably less than unity. Unfortunately it is very difficult to calculate or even to obtain a reasonable estimate for the value of f except in the simplest cases. It is therefore usual practice to assume a value of unity for f also for calculations involving compounds labeled with ^{14}C or ^{35}S, for example. This has been done in the figures quoted in this chapter, except where specifically indicated. It must be emphasized, however, that this basic assumption is often invalid in practice, expecially for ^{32}P-labeled compounds for example, and especially for compounds stored in their natural (solid, liquid, or freeze-dried) state.

If the appropriate $\bar{E}$ values are inserted in Eq. (2), the following equations are obtained for the commonly used β emitters

$$P_d = 2.9 \times 10^{-5} \cdot G(-\underline{M}) \cdot s \qquad (\text{for } ^3H) \tag{5}$$

$$P_d = 2.65 \times 10^{-4} \cdot G(-\underline{M}) \cdot s \cdot f \qquad (\text{for } ^{14}C) \tag{6}$$

$$P_d = 2.6 \times 10^{-4} \cdot G(-\underline{M}) \cdot s \cdot f \qquad (\text{for } ^{35}S) \tag{7}$$

$$P_d = 3.7 \times 10^{-3} \cdot G(-\underline{M}) \cdot s \cdot f \qquad (\text{for } ^{32}P) \tag{8}$$

3. Factors Affecting G(-M) Values

There are several important factors affecting the value of G(-M) which include:

1. In calculating the G(-M) value, it is implied that decomposition is entirely due to primary and/or secondary radiation effects. However, this is certainly not the case in practice as many compounds are chemically unstable in the nonradioactive form.
2. G(-M) value is affected by specific activity and by radioactive concentration. For a pure compound in the solid state the energy dose absorbed per unit mass of compound and hence

the rate of decomposition is proportional to the specific activity of the compound. However, in solution, although the rate of decomposition often increases with increasing specific activity, G(-M) values for compounds are quite different from those for the solid state and may not have a simple dependence on specific activity. In practice the rates of decomposition in solution and the corresponding G(-M) values depend on both the specific activity and the radioactive concentration of the solution. Investigations by Bayly and Shrimpton [10] on the self-decomposition of L-[U-^{14}C]phenylalanine which illustrates this point are shown in Table 5. At low radioactive concentrations the rate of decomposition is proportional to specific activity; at high radioactive concentrations it is independent of specific activity. At both high and low specific activities the rate of decomposition depends on the radioactive concentration but not linearly.

In general, at high specific activity the rate of decomposition is proportional to radioactive concentration and independent of specific activity, while at low specific activity it is proportional to specific activity and less dependent on radioactive concentration. These points have been discussed in detail by Sheppard [11] and Evans [12].

TABLE 5

Self-Decomposition Rates for L-[U-^{14}C]phenylalanine Stored in Aqueous Solution Containing 2% Ethanol at -20°C for 180 Days

Radioactive concentration (mCi/ml)	Specific activity (mCi/mmol)	Initial decomposition (% per day × 10^4)	G(-M)
0.5	495	236	0.2
0.5	248	236	0.4
0.0625	495	67	0.06
0.0625	248	35	0.06

It is an oversimplification to use G(-*M*) values to describe systems in which the labeled compound is diluted, either by another solid or more usually in a solvent.

3. As previously mentioned, the value of f is often falsely assumed to be unity without experimental verification or justification.
4. Much of the published data are taken from analyses of single batches of compounds. This means that in addition to the usual errors and inaccuracies in analysis (see Chapter 6), there may be batch-to-batch variation. This can be considerable, especially when the decomposition is largely secondary in nature, as such decomposition is notoriously dependent on even traces of impurities which, if they are not radioactive, are very difficult to detect.
5. G(-*M*) values assume the linearity of decomposition with time and radiation does. Evidence indicates that in general decomposition of a labeled compound does not necessarily increase linearly either with time or with radiation dose. This point was noted especially by Guarino and colleagues [13,14] in their studies on the decomposition of tritiated fatty acids and esters. Although G(-*M*) values have been studied for many systems by external radiation [15-17], they can rarely be correlated with G(-*M*) values obtained from self-radiolysis because of the factors previously mentioned (see above). As a result, external irradiation of compounds cannot be used and should not be used [18] a priori to predict the effects of self-decomposition in labeled compounds. At best the calculation of G(-*M*) value is useful in that it may be possible to predict the susceptibility of a compound to self-decomposition and published G(-*M*) values act only as a rough guide for the more sensitive compounds. External radiation studies have often indicated a maximum G(-*M*) value of 3 and it is likely that any compound with a G(-*M*) value above, say, 10 is decomposing by a chain reaction mechanism. An

example is [*methyl*-^{14}C]choline chloride which has a G(-$\underline{M}$) of ∿1000 when stored as a solid in vacuo [1].

B. Decomposition of Radiochemicals in Solutions

In general, two main parameters control the rate of decomposition of a radiochemical in solution. These are:

1. Reactive species produced per unit time per unit mass of *labeled compound* (this depends on the specific activity of the compound)
2. Reactive species produced per unit time per unit mass of *solution* (this depends on the radioactive concentration of the radiochemical in solution)

Although both these parameters are important, their effect overall depends very much upon the type of compound, the nature of the solvent used, and the storage conditions employed.

1. Decomposition in Organic Solvents

The detailed mechanism of self-decomposition of radiochemicals in organic solvents is unknown and is likely to be complex. The transfer and absorption of the radiation energy is undoubtedly quite different and produces different forms of "reactive species" from those produced in aqueous systems. The most widely used solvents include benzene, toluene, ethanol, methanol, ethyl acetate, and pentane. The chemical purity of the solvent used is critically important and only freshly purified solvents should be used. The presence of a trace of peroxide, for example, in the solvent may be sufficient to destroy completely the labeled compound. Other chemical impurities may cause an increase in the rate of self-decomposition of the radiochemical on storage.

Benzene is an especially good solvent for minimizing the decomposition of radiotracer compounds and many radiochemicals which are soluble in this solvent are stored in this way. Irradiation of benzene normally results in a small yield of polymers and no high yield of "reactive species." The actual species produced during the irradiation of benzene are not well understood but probably includes long-lived excited states and radicals stabilized by the π orbitals. Compared with water there are, of course, fewer ionic products because of the low dielectric constants. One of the main features of benzene is its ability to exercise a "protective effect" on other compounds [20,21]. Hence the solutions of many sensitive radiochemicals at high specific activity, e.g., steroids, fatty acids, and hydrocarbons, in benzene are remarkably stable toward self-radiolysis. The addition of a secondary solvent to the benzene solution of a radiochemical, normally added to increase the solubility of the compound in the solvent mixture, sometimes leads to an acceleration of self-decomposition. Thus the addition of methanol to benzene solutions of labeled steroids has been observed to increase the rate of decomposition of these compounds [22], possibly by increasing the lifetime of polar attacking species such as methoxyl radicals. Osinski [23] has also confirmed that polar solvents such as methanol or water greatly accelerate the rate of self-decomposition of labeled steroids. The presence of a secondary solvent may also alter the nature of the radiolysis products formed on storage of the compound. Thus, Evans and Phillips [24] have shown that the products of the self-decomposition of [5-*methyl*-^{3}H]tocopherol in benzene are altered by the presence of ethanol. Most radiochemicals which are stored in benzene are under an inert atmosphere (nitrogen, argon, or in vacuo). It should also be remembered that the reactive entities produced from benzene are also affected by the presence of air or oxygen [25].

Some radiochemicals stored in benzene solutions have shown an accelerated rate of decomposition after an initial period of apparent stability [22]. For example [11], in Fig. 2 the radiochemical purity of batches of [2,4,6,7-^{3}H]estradiol ($\sim$100 Ci/mmol) are shown. There is a marked deterioration in purity which begins after approximately 8 weeks storage. Time course studies [11] on the decomposition of other compounds labeled with

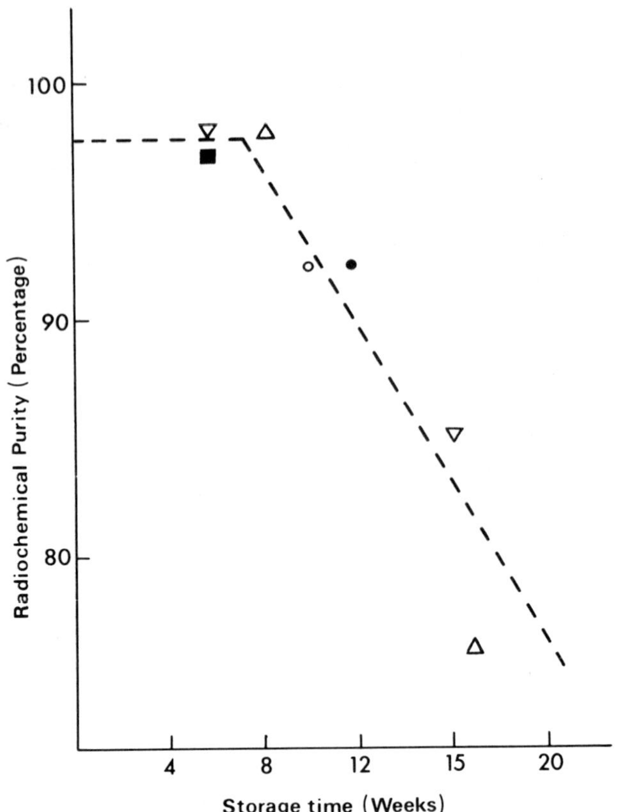

FIG. 2. The variation of radiochemical purity with time of storage for batches of [2,4,6,7-^{3}H]estradiol with 98% initial purity for all batches; batch 2 (△), batch 3 (○), batch 4 (▽), batch 5 (●), and batch 6 (■).

^{14}C, ^{3}H, and ^{125}I support the hypothesis that some radiochemicals deteriorate markedly after a period of apparent good stability. On the other hand, it has been suggested that the rates of self-decomposition of certain compounds decrease as the radioactive compound is used up and other competing species build up [13].

It is interesting to note that the self-radiolysis of [U-^{14}C]benzene itself at almost 100% isotopic purity (>360 mCi/mmol) results in polymerization within a few days to yield a yellowish polymer.

Many radiochemicals which are not soluble in benzene are stored in ethanol or methanol solution, and here the reactive species is likely to be ethoxyl or methoxyl radicals, respectively. It should be remembered that the radiolysis of alcohols produces aldehydes and other alcohols [15]. Hence radiochemicals which are likely to react with aldehydes should not be stored in alcoholic solutions. An interesting example is the comparison of the decomposition of [^{35}S]thiosemicarbazide stored in water and in methanolic solution shown in Table 6. The apparent rate of decomposition of [^{35}S]thiosemicarbazide in methanolic solution is about four times that observed in aqueous solution. Thiosemicarbazide is a reagent used in the analysis and identification (characterization) of aldehydes and ketones, and forms thiosemicarbazones with the aldehyde(s) produced during the radiolysis of the methanol by the β radiation, resulting in the effect observed. It should also be remembered that acidic compounds containing carboxyl groups, for example, may become esterified on prolonged storage in alcoholic solution. An

TABLE 6

Self-Decomposition of [^{35}S]Thiosemicarbazide [39]

Initial specific activity (mCi/mmol)	Storage conditions	Initial radio-active concentration (mCi/ml)	Temperature (°C)	Storage time (weeks)	Decomposition (%)	G(-M)
150	Solid*	--	0	9	1	0.5
150	Aq. soln	1.65	0	9	6	3.2
150	Methanol	1.65	0	9	12	6.6
196	Aq. soln	2	-30	9	7	2.8
196	Methanol	2.2	-30	9	20	8.5
196	Aq. soln	2	-30	16	8	2.2
196	Methanol	2.2	-30	16	30	9.4
196	Solid*	--	-30	16	8	2.2

*In vacuo over phosphorus pentoxide.

example is the formation of 24% [U-^{14}C]phenylalanine ethyl ester on storage of [U-^{14}C]phenylalanine in 95% ethanol at room (20°C) temperature for 22 months [10].

Thus organic solvent for the storage of a particular labeled compound must be carefully chosen in order to avoid problems such as those discussed. In addition, it is very important that such solvents are rigorously purified before use. Chemical impurities in the solvents have been demonstrated to be a cause of the variations in the stabilities of tritiated steroids for example [22,26,27].

2. Decomposition in Aqueous Solutions

Many radiochemicals are soluble only in hydroxylic solvents and water is necessarily used as the solvent. This fact applies to numerous tracer compounds of interest to biochemists and biologists, including amino acids, carbohydrates, and nucleic acids. The radiation chemistry of water is therefore an important consideration in the study of the stability of radiochemicals in aqueous solutions.

The action of ionizing radiation on water is well known [28]; the primary entities believed to result from the radiolysis of water are hydronium ions, hydrated electrons, hydrogen atoms, hydroxyl radicals, hydroperoxy radicals, molecular hydrogen, and hydrogen peroxide. It is these reactive species which cause self-decomposition of the radiochemical through secondary radiation effects. In solution ionization occurs along the paths of the (β) particles [29] in discrete pockets known as "spurs." The weaker the energy of the radiation, the closer together are the spurs, as is illustrated diagrammatically in Fig. 3. The very weak β radiation energy from tritium results in the spurs being very close together during the self-radiolysis of tritium compounds in solution. This effect has important consequences when considering frozen solutions of tritiated compounds and other compounds labeled with soft emitters (see Section IV).

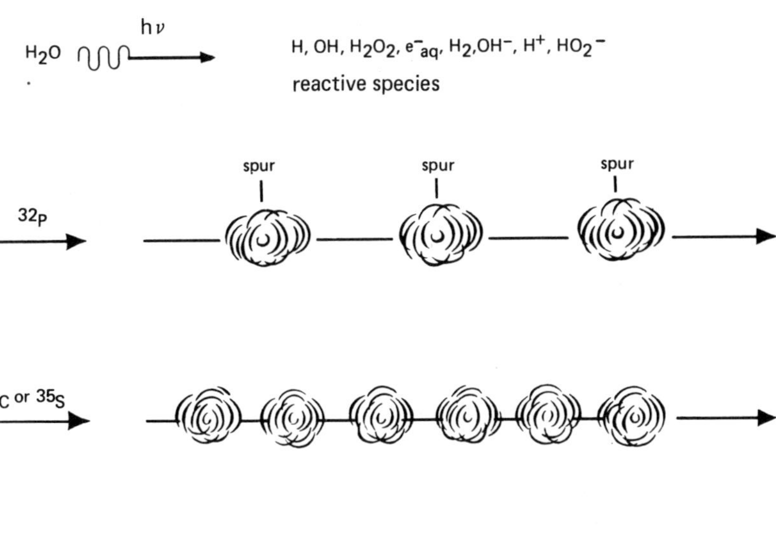

FIG. 3. Proximity of spurs or pockets of reactive species in solutions of radiochemicals labeled with beta emitting radionuclides.

In aqueous solutions of radiochemicals, it is believed from current evidence [12] that the hydroxyl radical is the most damaging species. Examples are the formation of tyrosine and dihydroxyphenylalanine on storage of solutions of phenylalanine labeled with ^{14}C or with tritium [30]; hydroxylated species, glycols, form on the storage of labeled pyrimidines in aqueous solutions [31]. Measures taken to minimize decomposition are therefore aimed at reducing the number of solute-radical interactions. This can normally be achieved by lowering the temperature of storage and by the use of radical scavengers. In order to be effective in minimizing self-decomposition, free radical scavengers must react preferentially and rapidly with the reactive species present in solution, and yet the products of the radical-scavenger reaction must not themselves react with the labeled compound. The interaction and rates of reaction of hydroxyl radicals, and other reactive species, produced in solution by x or β radiation, with many substances have been determined [15-17,28,32]. Such data can be used as a guide for the selection of suitable scavengers for reducing the self-decomposition of radiochemicals (see Section V 3).

C. Decomposition of Radiochemicals in Their Natural Physical State

The rate of self-decomposition of radiochemical in their natural physical state, i.e., solid, liquid, or gaseous, depends upon the radiation sensitivity of the compound, the energy of the emitted radiation and the storage conditions of the compound. In general, the weaker the energy of the radiation the more radiation absorbed by the compound and the greater the rate of self-decomposition. Thus for a labeled compound at a given specific activity under standard storage conditions in the natural physical state, the rate of self-decomposition increases in the order of $^{3}H > ^{14}C = ^{35}S > ^{33}P > ^{32}P$.

Perhaps one of the most interesting aspects of the storage of radiochemicals in their natural form, are the colors which are often observed. The visual appearance of a labeled compound can sometimes give a misleading indication of self-decomposition. While the presence of color in a normally colorless compound may indicate some impurity, it can often quantitatively represent a negligible amount and even this may not be radioactive. Some examples are the red-violet (iodine) coloration of tritiated or ^{14}C-labeled methyl iodide, and the straw color of labeled benzene or acetic anhydride. It is possible for a radiochemical, which is normally a white solid, to become highly colored, due to deformations in the crystal lattice, without signifying the presence of any detectable impurity. The dark green color produced on storage of solid [^{35}S]thiosemicarbazide, the purple color of [^{3}H]sodium borohydride and [^{3}H]dopamine hydrochloride, and the grey appearance of [^{14}C]guanine are examples of this phenomenon.

Radiochemicals stored as freeze-dried solids usually keep better than when stored as "bulk" solids because of the greater surface area in the freeze-dried state and the consequent reduction in the effective radiation energy absorbed by the compound.

IV. EFFECTS OF TEMPERATURE ON DECOMPOSITION

As a general rule, compounds, particularly organic compounds because of their thermodynamic instability, are best kept at low temperatures. Chemical reactions, such as radical-solute interactions, are normally decelerated by a decrease in temperature as the activation energy for such reactions is raised; hence the chemical stability of a radiochemical will be increased by lowering its temperature. It is normally beneficial to store radiochemicals in their natural physical state or in solution at the lowest practical temperature. An interesting exception to this general rule is the storage of certain radiochemicals which are normally gases or vapors, and which are prone to polymerization. Condensation of such compounds by lowering the temperature can often accelerate their rate of polymerization. This effect is most pronounced with compounds at a high molar specific activity. Some examples are [U-^{14}C]acethylene, [U-^{14}C]benzene, [^{3}H]methyl iodide, and [^{3}H]styrene.

Many investigators are uncertain as to whether solutions of labeled compounds should be kept frozen. The beneficial effect of lowering the temperature of storage is illustrated by a study [39] of the decomposition of [2-^{14}C]thymidine (^{14}C being used in this case as a tracer to aid in the analysis of the products) by the β radiation from tritiated water shown in the Table 7. At the lower temperature (-40°C) in the frozen state, radical-solute interaction is substantially reduced, only 10% of the [2-^{14}C]thymidine being decomposed compared with total

TABLE 7

Decomposition of [2-^{14}C]Thymidine with β Radiation from THO

Weight of [2-^{14}C]-thymidine* (μg)	Volume of solution (ml)	Tritium activity (mCi)	Storage temperature (°C)	Storage time (months)	Dose (rads)	Decomposition (%)
67.5	1	Nil (control)	2	16	1400	N.D.**
67.5	1	1	2	16	0.14×10^6	23
67.5	1	10	2	16	1.44×10^6	100
67.5	1	10	-40	16	1.44×10^6	10
67.5	1	10	-40	4	3.6×10^5	3

*Specific activity 35.9 mCi/mmol.
**N.D. = none detectable.

destruction of the compound at +2°C. However, another effect has to be considered when solutions, and especially aqueous solutions, of labeled compounds are frozen and then stored. This is the effect of molecular clustering of the solute as solutions are frozen.

Figure 4 shows a diagrammatic feature of freezing solutions of compounds. At +2°C in the unfrozen state, the solute molecules are free to move about the solution. In solutions that are frozen slowly, at say -20 to -40°C, molecular clustering of the solute occurs. This clustering of the solute molecules is much less marked in solutions that are frozen rapidly at -196°C (liquid nitrogen). Large analytical errors in measurements on samples taken for atomic absorption spectroscopy have been observed to be due to concentration gradients in frozen and thawed (without careful mixing) samples [33]. The phenomenon of *molecular clustering* is readily demonstrated by slowly freezing dilute solutions of a colored compound.

The close proximity of the spurs (see Section III B 2) in solutions of tritiated compounds, means that molecular clustering

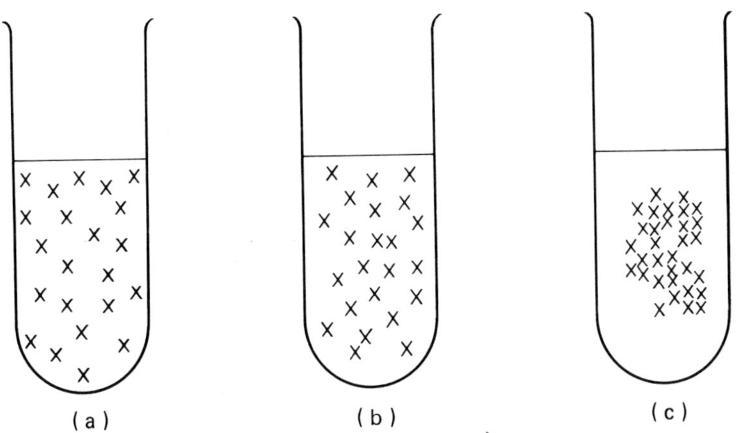

FIG. 4. Distribution of solute molecules (X) in aqueous solution and in frozen aqueous solution: molecular clustering effect on slow freezing; (a) +2°C, (b) frozen quickly at -196°C; (c) frozen slowly at -20°C to -40°C.

has a very pronounced effect in such solutions, compared with solutions of compounds labeled with higher radiation energy emitting radionuclides. In 1963 Evans and Stanford [34] observed an increased rate of decomposition in frozen solutions of tritiated thymidine when compared with the rate of decomposition in the unfrozen state at +2°C, which has recently been confirmed (see Table 21). This effect is seen from the results in Table 8 and it is now regarded as a fairly general phenomenon associated with molecular clustering and the consequential localization of reactive species. Apelgot and her colleagues [35] have carried out elegant studies to demonstrate the heterogeneity of solutions of labeled compounds that are stored frozen. Glycerol was effective in decreasing these heterogeneities but attempts to freeze tritiated compounds rapidly by immersion of their aqueous solutions in liquid nitrogen followed by storage at say -20 or -40°C have not been very successful; in many instances, decomposition is still observed to be faster than in the unfrozen state at +2°C, especially when a radical scavenger is present in the solution

TABLE 8

Decomposition of Tritiated Compounds in Frozen Solutions [39,55,70]

Compound	Specific activity (mCi/mmol)	Storage condi- tions*	Storage tempera- ture (°C)	Storage time (months)	Decompo- sition (%)
5-Bromo-2'-deoxy-[6-^{3}H]uridine	588	a	2**	7	1
	588	a	-40	4.5	5
[3',5',9-^{3}H]Folic acid	16,000	b	2**	3	16
	16,000	b	-20	3	30
L-[*methyl*-^{3}H]-Methionine	8,600	c	2**	1.5	27
	8,600	c	-40	1.5	47
2-Methyl-[6-^{3}H]-naphtho-1,4-quinol diphosphate	29,000	d	2**	3	25
	29,000	d	-80	3	40
2-Methyl-[5,6,7-^{3}H]-naphtho-1,4-quinol diphosphate	76,000	e	2**	3	20
	76,000	e	-80	3	30
[6-^{3}H]Thymidine	2,500	f	2**	8	10
	2,500	f	-40	12	28
[6-^{3}H]Thymidine	3,600	g	2**	16	18
	3,600	g	-40	10	20
[*methyl*-^{3}H]Thymidine	44,000	g	2**	1.2	4
	44,000	g	-20	1.2	17
[G-^{3}H]Uridine	3,000	h	2**	12	17
	3,000	h	-40	10	25
[5,6-^{3}H]Uridine	55,000	g	2**	2.5	6
	55,000	g	-20	2.5	25

*a: Aqueous solution at 0.4 mCi/ml.
b: Phosphate buffer pH 6.9 at 1.3 mCi/ml.
c: Aqueous solution at 5.5 mCi/ml.
d: Tetrasodium salt in isotonic saline under N_2 at 390 mCi/ml.
e: Tetrasodium salt in isotonic saline under N_2 at 360 mCi/ml.
f: Aqueous solution at 6.9 mCi/ml.
g: Aqueous solution at 1 mCi/ml.
h: Aqueous solution at 20 mCi/ml.
**2°C unfrozen.

[12]. Because of the low energy and penetrating power of the β particles from tritium, the effect of molecular clustering on the rate of decomposition of tritiated compounds is much more pronounced than with other labeled compounds [12].

Another effect to be avoided in frozen solutions is the separation of the solute from the solvent. This can be caused either by the compound crystallizing in the freezing process, or because a temperature gradient along the storage tube causes the solvent to sublime to the cooler parts of the tube. This effect is particularly noticeable in evacuated sealed ampuls and especially when compounds are stored in organic solvents such as benzene.

Apelgot et al. [36,37] first demonstrated the beneficial effects of very low storage temperatures (-196°C) on the stability of a radiochemical ([^{3}H]thymidine) in solution. It has been demonstrated subsequently that storage at -196°C can increase the useful shelf-life of many labeled compounds. This has been especially useful to commercial manufacturers of radiochemicals enabling them to provide an "off-the'shelf" service (supply) of sensitive radiochemicals at a high radiochemical purity. Some examples [11,12,39,55] are shown in Table 9. It is not always convenient to place ampuls in the liquid nitrogen and it has been found that storage of labeled compounds in the vapor above the liquid nitrogen (-176 to -140°C) is just as satisfactory in minimizing self-decomposition. This reduces the risk of ampuls breaking, or leakage of liquid nitrogen into the ampuls resulting in exploding on warming-up. It is always a wise precaution to allow the ampuls to warm up in a closed container in case of breakage.

TABLE 9

Self-Decomposition of Radiochemicals Stored at -196 to -140°C

Compound	Specific activity (mCi/mmol)	Storage conditions*	Activity conc. (mCi/ml)	Temp. (°C)	Time of storage (weeks)	Decomposition (%)
^{14}C Compounds						
[*methyl*-^{14}C]-Choline chloride	1.5	a	--	-196	6	N.D.
[4-^{14}C]Cortisol	57.9	b	0.05	-196	52	1
[4-^{14}C]Cortisone	57.1	b	0.05	-196	52	6
	57.0	b	0.05	-140	26	N.D.
L-[U-^{14}C]Phenylalanine	550	c	0.5	-196	26	N.D.
[4-^{14}C]Estradiol	51.4	b	0.05	-196	26	4
	58.0	b	0.05	-140	39	N.D.
[4-^{14}C]Estriol	53.4	b	0.05	-196	104	9
	53.0	b	0.05	-140	26	5
[4-^{14}C]Vitamin D_3	32.3	b	0.05	-140	50	N.D.
[*carbinol*-^{14}C]-Vitamin A	2.8	d	0.2	-196	82	18
^{3}H Compounds						
DL-[7-^{3}H]Epinephrine	6,100	e	1	-140	28	6
[G-^{3}H]Atropine	580	a	--	-196	6	N.D.
[3',5',9-^{3}H]Folic acid	16,000	f	--	-140	13	11
	1,500	f	--	-140	23	12
	27,000	f	--	-196	12	7
[G-^{3}H]Isonicotinic hydrazide	918	e	20	-196	40	5
[7-^{3}H]Isoprenaline HCl	3,640	e	1	-140	25	3
[3',5',9-^{3}H]-Methotrexate	6,150	f	--	-196	16	7
[G-^{3}H]Nicotine	310	g	--	-196	32	N.D.
L-[*methyl*-^{3}H]-Methionine	8,600	e	5.5	-196	7	6
	7,600	e	1	-140	17	8
	7,600	h	1	-140	17	1
1-[7-^{3}H]-Norepinephrine	5,800	e	1	-140	9	N.D.

TABLE 9 (continued)

Compound	Specific activity (mCi/mmol)	Storage condi- tions*	Activity conc. (mCi/ml)	Temp. (°C)	Time of storage (weeks)	Decompo- sition (%)
DL-[7-^{3}H]-Norepinephrine	10,200	e	1	-140	12	4
	3,800	e	1	-196	31	2
2-Methyl-[6-^{3}H]-naphtho-1,4-quinol diphos-phate	27,000	i	2	-196	24	7.5
	27,000	i	266	-196	26	12
L-[4,5-^{3}H]-Leucine	7,600	e	4.8	-196	64	25
L-3,4-Dihydroxy-[*ring*-2,5,6-^{3}H]-phenylalanine	23,500	e	1	-196	19	3
	1,000	e	1	-196	21	N.D.
25-Hydroxy[26(27)-*methyl*-^{3}H]-vitamin D_3	7,500	b	1	-196	16	5
[2,4,6,7-^{3}H]-Estradiol	100,000	b	1	-196	8	N.D.
[2,4,6,7-^{3}H]-Estrone	100,000	b	1	-196	26	6
[*methyl*-^{3}H]-Thymidine 5'-triphosphate	12,000	j	0.5	-196	8	N.D.
	12,000	j	0.25	-196	9	N.D.
[5-*methyl*-^{3}H]α-Tocopherol	870	k	0.6	-196	10	5
[7-^{3}H]Tetracycline HCl	1,000	a	--	-140	23	21
DL-[G-^{3}H]-Tryptophan	3,020	e	1	-196	38	N.D.
[5-^{3}H]Uridine	13,000	e	5	-196	24	5
S-Adenosyl-L-[*methyl*-^{3}H]-methionine	1,580	l	1	-196	52	10
[6-^{3}H]Thymidine	6,900	e	3.45	-196	25	3
	14,800	e	5.6	-196	92	22
[*methyl*-^{3}H]-Thymidine	44,000	e	1	-140	5	4
[5,6-^{3}H]Uridine	55,000	e	1	-140	10	N.D.

TABLE 9 (continued)

Compound	Specific activity (mCi/mmol)	Storage conditions*	Activity conc. (mCi/ml)	Temp. (°C)	Time of storage (weeks)	Decomposition (%)
[G-^{3}H]7,12-Dimethyl-1,2-benzanthracene	14,500	a	--	-196	10	40
[1α,2α-^{3}H]-Vitamin D_3	577	b	0.064	-140	56	N.D.
	595	b	0.125	-196	52	N.D.
	580	b	1	-196	60	N.D.
^{32}P Compounds						
Adenosine 5'-[α-^{32}P]-triphosphate	100,000	a	--	-140	1	1
	100,000	j	1	-140	1	3-4
	3,000	j	0.25	-140	5	1

*a. Freeze-dried solid.
b. Benzene/ethanol (9:1)
c. Aqueous solution containing 2% ethanol.
d. Benzene solution containing antioxidants (0.05% butylated hydroxyanisole + 0.05% butylated hydroxytoluene).
e. Sterilized aqueous solution.
f. Freeze-dried solid in phosphate buffer pH 10 or as potassium salt.
g. Liquid compound.
h. Aqueous solution containing 0.2% mercaptoethanol.
i. Tetrasodium salt in isotonic saline.
j. Ethanol/water (1:1).
k. Ethanol solution.
l. Aqueous sulfuric acid at pH 4.
**N.D. = none detectable.

V. CONTROL OF DECOMPOSITION OF RADIOCHEMICALS

The methods which can be used to minimize self-decomposition of radiochemicals are: reduce the molar specific activity of the compound; disperse the compound in a suitable medium, usually a solvent; add free radical scavengers to solutions where appropriate; store compounds at as low a temperature as possible taking care to avoid the effects of molecular clustering; keep solutions of compounds at their correct pH for maximum stability (chemical) of the compound. In practice combinations of these methods are used.

1. Reduction of Molar Specific Activity

For many experimental uses of radiochemicals, reduction of the molar specific activity of the compound is not practical. In cases where a low specific activity can be used, it is important either to prepare the compound at the required low specific activity (essential for macromolecules; see Section VI E) or to dilute a high specific activity product as soon as possible after preparation and isolation of the radiochemically pure product. If this is not done then dilution of stored material may merely enhance the effects of impurities because the radiochemical is reduced in specific activity, whereas the impurities are not.

Most commercial suppliers are pleased to dilute compounds for a small additional charge. However, it is often more convenient for the research investigator to undertake this operation himself, especially when the compound is required for use at both high and low specific activity. Further, unlabeled material of authenticated purity for the dilution may be more readily available to the investigator than to the commercial producer of radiochemicals.

The amount of carrier compound to be added can be calculated from Eq. (9):

$$W = Ma\left(\frac{1}{A'} - \frac{1}{A}\right) \tag{9}$$

where

W (mg) = weight of carrier compound to be added

M = molecular weight of compound

a (mCi) = total activity in sample being diluted

A (mCi/mmol) = molar specific activity of compound

A' (mCi/mmol) = molar specific activity of diluted compound

If the compound is nonvolatile, e.g., a freeze-dried solid, dissolve it in a suitable purified solvent and prepare a standard solution of carrier (unlabeled compound) in the same solvent. If the compound is already in solution, the standard carrier solution should be prepared using the same solvent. The calculated volume of standard carrier solution is then added to the radioactive solution and mixed thoroughly before using.

Example: To reduce 5 mCi of tritiated thymidine at 10 Ci/mmol to 500 mCi/mmol:

$$W = 242 \times 5\left(\frac{1}{500} - \frac{1}{10,000}\right) = 2.30 \text{ mg} \tag{10}$$

Suppose that 5 mCi of the above compound has been supplied in 5 ml of solution. Make up a standard solution of unlabeled thymidine by dissolving, for example, 25 mg (accurately weighed) in 10 ml of water (sterile, pyrogen-free if necessary). Therefore 0.92 ml of the standard solution (equivalent to 2.3 mg of carrier) should be added to the solution of the radioactive thymidine. Note, however, that the radioactive concentration is now changed from 1 mCi/ml to 0.84 mCi/ml. Radiochemicals which are gases or volatile liquids at room temperature are normally diluted with carrier using special equipment such as a calibrated vacuum manifold system [12,38]. Unless such equipment is readily available, it is advisable to obtain these radiochemicals already diluted.

2. Disperal in a Diluent

At present this technique of diluting the labeled molecules with unlabeled ones is the best. Numerous diluents have been investigated including paper, powdered cellulose, charcoal, sand, benzanthracene, clathrates, and various solvents [11,12,39]. While most of these diluents give some extra degree of protection from self-decomposition over storage in the natural form, they are all inconvenient, with the exception of the use of solvents. Radiochemicals are often supplied in solution because this is the most convenient form for dispensing and they are stabilized by lowering the radioactive concentration. As previously discussed, the chemistry of their self-decomposition results largely from the attack on the labeled compound by chemical reactive species produced by the effects of radiation on the solvent. The effect of the solvent on the stability of labeled compounds is therefore most marked, and the choice of solvent is very important. The solvent selected depends, of course, on the solubility and compatibility of the compound. Solvents that readily produce free radicals on irradiation with γ-rays or other forms of radiation are usually best avoided. Methylene dichloride, for example, would produce halogen radicals, but aromatic hydrocarbons, such as benzene, have a low radical yield, and are frequently used (see above). Solvents must be purified before use and it should always be remembered that most solutions of radiochemicals are very dilute. Any factors which contribute to the destruction of the compound are therefore magnified because the impurities produced are readily detected as a result of the compound being radioactive. For example, nucleoside triphosphates are very easily hydrolyzed, especially in acidic media; steroids may be oxidized by oxygen dissolved in benzene.

The most commonly used solvents are benzene, toluene, ethanol, and water. Although benzene is regarded as the best protective solvent, unfortunately many compounds, particularly those of interest to biochemists and biologists, such as carbohydrates, amino acids and nucleotides, are soluble only in hydroxylic solvents and one is obliged to use water, aqueous alcohol, or similar mixtures for their storage.

3. Free Radical Scavengers

The addition of free-radical scavengers to reduce radical/labeled molecule interactions in solution is an important advance in the steps taken to minimize self-decomposition of radiochemicals. In order to be effective in minimizing self-decomposition, scavengers must react preferentially and rapidly with the reactive species present in solution. Early experiments by Bayly and Shrimpton [10] and by Evans [12,40] showed that small amounts of ethanol (1 to 3%) can be used to reduce effectively the self-decomposition of radiochemicals in solution. Numerous free-radical scavengers have now been tested for minimizing the self-decomposition of radiochemicals in solutions. These include benzyl alcohol [11,12.39], sodium formate [12,40], glycerol [35-37], cysteamine [35-37], ascorbic acid [12], and mercaptoethanol [11,12]. Ethanol has the advantage that it can be removed rapidly and easily by freeze-drying (lyophilizing) or evaporating the solution. Benzyl alcohol is a very efficient inhibitor of self-decomposition, of [^{3}H]thymidine for example, presumably because of its high rate of reaction with hydroxyl radicals. However, its use is often less practical than the use of ethanol or even glycerol, because of its additional bacteriostatic action and the difficulty of its removal from solutions.

Three important criteria must be satisfied in choosing a free radical scavenger to minimize self-decomposition of radiochemicals in solution:

1. The scavenger must be compatible with the labeled compound, i.e., it must not react with the compound.
2. The scavenger must have no effect on the system in which the radiochemical is being used as a tracer.
3. The products of radiolysis of the scavenger must not react either with the labeled compound or have any effect on the system in which the radiochemical is being used.

It is obvious that the first of these criteria must be satisfied to prevent a series of radioactive impurities being formed. It is rare that the scavenger has any effect on the system because it is present usually in a small chemical amount. However, an example is the inhibition of the enzyme RNA polymerase from *Micrococcus lysodeikticus* by trace amounts of ethanol. Ethanol concentrations even as low as 0.5% in the reaction medium cause 25% inhibition, and 2% causes 80% inhibition [41]. It is possible that similar effects may occur in other enzyme systems [41a] and with the use of other scavengers. The cautious investigator will bear this possibility in mind. Waldeck [30] has pointed out that the effects of (β) radiation on ethanol solutions produces acetaldehyde (in low yield) which can react readily with substrates such as dihydroxyphenylalanine, to form tetrahydroisoquinolines and with 5-hydroxytryptamine, to form tetrahydro-β-carbolines.

Ethanol is now used routinely to minimize the self-decomposition of many labeled compounds and some examples [11,12, 55] are shown in Table 10. In general the concentration of ethanol (as scavenger) in the final solution actually used for tracer experiments will be very low (less than 0.1%); solutions of radiochemicals supplied by manufacturers are usually diluted before use. Note, however, that for effective protection of the labeled compounds against decomposition the concentration of ethanol (or other radical scavengers present) should not be reduced in stored solutions.

The reaction rate of various compounds, with hydroxyl radicals is shown in Table 11 [32]. Although ethanol reacts rapidly with hydroxyl radicals, the rate is considerably slower than that of methionine for example. This partly explains, for example, why ethanol does not afford any significant protection of [^{3}H]- or [^{14}C]methionine from self-decomposition in solution. Ethanol is not a universal panacea, but has proved to be extremely beneficial in the storage of many radiochemicals.

4. Low-Temperature Storage

The effects of temperature on the storage of radiochemicals have already been discussed. In general all radiochemicals should be kept at the lowest temperature but care must be taken with tritiated compounds and with benzene solutions of labeled compounds, to avoid the special effects previously discussed (see Section IV).

5. Inhibition of Chemical Decomposition

Apart from lowering the temperature of storage, it is possible to control chemical decomposition by adding inhibitors. Some examples are as follows:

TABLE 10

Self-Decomposition of Radiochemicals in Solutions Containing Ethanol as a Free Radical Scavenger

Compound	Specific activity (mCi/mmol)	Storage conditions*	Temperature (°C)	Time of storage (months)	Decomposition (%)**
^{14}C Compounds					
L-[U-^{14}C]Alanine	162	a	2	15	N.D.
4-Amino-n-[U-^{14}C]-butyric acid	204	a	-20	9	N.D.
L-[U-^{14}C]Arginine HCl	320	a	2	18	N.D.
L-[U-^{14}C]Asparagine	220	a	-20	20	2
L-[U-^{14}C]Aspartic acid	220	a	2	12	N.D.
L-[U-^{14}C]Cysteine	325	a	-20	21	N.D.
L-[U-^{14}C]Glutamic acid	250	a	2	15	N.D.
L-[U-^{14}C]Glutamine	40	a	-20	13	N.D.
[U-^{14}C]Glycine	110	a	2	15	N.D.
L-[U-^{14}C]Leucine	340	a	2	13	N.D.
L-[U-^{14}C]Isoleucine	310	a	2	18	2
L-[U-^{14}C]Lysine	310	a	2	21	N.D.
L-[U-^{14}C]Pheylalanine	475	a	2	18	N.D.
L-[U-^{14}C]Proline	265	a	2	17	2
L-[U-^{14}C]Serine	155	a	2	21	2
L-[U-^{14}C]Threonine	210	a	2	32	6
L-[U-^{14}C]Tyrosine	470	a	2	16	2
L-[U-^{14}C]Valine	260	a	2	19	N.D.
[U-^{14}C]Adenosine 5'-triphosphate	550	a	-20	5	3
[U-^{14}C]Deoxyguanosine 5'-triphosphate	510	a	-20	3	N.D.
[U-^{14}C]Guanosine 5'-triphosphate	575	a	-20	3	7
[U-^{14}C]Uridine 5'-triphosphate	420	a	-20	8	8

TABLE 10 (continued)

Compound	Specific activity (mCi/mmol)	Storage conditions*	Temperature (°C)	Time of storage (months)	Decomposition (%)**
L-[1-^{14}C]Arabinose	52	b	-20	21	N.D.
D-[U-^{14}C]Fructose 1,6-diphosphate	69	b	-20	36	2
D-[1-^{14}C]Galactosamine	53	b	-20	36	N.D.
D-[U-^{14}C]Glucosamine HCl	260	b	-20	18	N.D.
D-[1-^{14}C]Mannosamine HCl	55	b	-20	11	5
N-Acetyl-D-[1-^{14}C]-glucosamine	57	a	-20	24	2
D-[U-^{14}C]Galactose 1-phosphate	280	b	-20	12	2
D-[U-^{14}C]Glucose 1-phosphate	280	b	-20	22	N.D.
D-[U-^{14}C]Glucose	320	b	-20	15	N.D.
D-[1-^{14}C]Ribose	52.4	b	-20	4	N.D.
[U-^{14}C]Sucrose	600	b	-20	9	N.D.
α,α-[U-^{14}C]Trehalose	410	b	-20	20	N.D.
[*methyl*-^{14}C]Choline chloride	54	a	-20	6	N.D.
[U-^{14}C]Histamine	258	a	-20	12	2
^{3}H Compounds					
L-[3,3'-^{3}H]Cystine HCl	1,680	a	+2	20	11-18
L-3,4-Dihydroxy [*ring*-2,5,6-^{3}H] phenylalanine (L-DOPA)	28,700	c	+2	3.5	<5
L-[4,5-^{3}H]Isoleucine	2,100	a	+2	10	N.D.
DL-[2-^{3}H]Glutamic acid	1,100	a	+2	10	N.D.
L-[4,5-^{3}H]Leucine	23,000	a	+2	3	N.D.
L-[4-^{3}H]Phenylalanine	21,000	a	+2	6	N.D.

TABLE 10 (continued)

Compound	Specific activity (mCi/mmol)	Storage conditions*	Temperature (°C)	Time of storage (months)	Decomposition (%)**
DL-[G-^{3}H]Phenylalanine	1,530	a	+2	7	N.D.
	2,000	c	+2	30	3
DL-[*methyl*-^{3}H]-Selenomethionine	1,040	a	+2	12	N.D.
[G-^{3}H]Serotonin	5,650	a	+2	2.5	5
	8,500	a	-20	6	9
L-[3,5-^{3}H]Tyrosine	42,300	c	+2	3	N.D.
DL-[7-^{3}H]Epinephrine	1,400	c	+2	3.5	20
[8-^{3}H]Puromycin HCl	3,400	b	+2	10	N.D.
[5-^{3}H]Deoxycytidine	14,700	a	+2	5	N.D.
[5-^{3}H]Deoxyuridine 5'-monophosphate	7,600	a	+2	6	7
[6-^{3}H]Thymidine	20,000	d	+2	6	N.D.
[*methyl*-^{3}H]Thymidine	15,700	c	+2	0.75	N.D.
	20,100	d	+2	5	N.D.
	44,000	d	+2	8	N.D.
[5,6-^{3}H]Uracil	5,600	c	+2	3.5	<2
[5,6-^{3}H]Uridine	55,000	a	+2	4	3

*a. Sterilized aqueous solution containing 2% ethanol.
b. Sterilized aqueous solution containing 3% ethanol.
c. Sterilized aqueous solution containing 1% ethanol.
d. Sterilized aqueous solution containing 10% ethanol (^{14}C compounds approximately 50-100 µCi/ml and tritium compounds at 1 mCi/ml).

**N.D. = none detectable.

TABLE 11

Reaction Rates of Some Compounds with Hydroxyl Radicals

Compound	pH of solution	Reaction rate (k) (10^9 mol^{-1} sec^{-1})
Alcohols		
Ethanol	7	1.1
Methanol	7	0.48
sec-Butanol	7	1.86
Isopropanol	7	3.9
Tert-Butanol	7	0.25
Mercaptoethanol	6-7	5.1
Amino acids		
p-Aminobenzoic acid	6-7	9.9
Cysteine	1	7.9
Tryptophan	6.1-6.3	8.5
	2.0-2.2	6.5
Methionine	5.5-5.7	4.9
	6-7	5.1
Phenylalanine	5.5-6.0	3.5
	2.0-2.2	3.4
Leucine	5.5-6.0	0.98
Histidine	6-7	3.0
	2.0-2.2	1.15
Arginine	6.5-7.5	2.1
	2.0-2.2	0.4
Glycine	2.8-3.0	0.6
	5.8-6.0	0.01
Serine	5.5-6.0	0.19
Alanine	5.5-6.0	0.046
Aspartic acid	6.8-7.0	0.045
Nucleics		
Thymine	1	3.1
Thymidine	7.5	2.7
	2	2.8
Uracil	5.7	3.1

TABLE 11 (continued)

	pH of solution	Reaction rate (k) (10^9 mol^{-1} sec^{-1})
Miscellaneous		
Benzoic acid	6-7	3.0
Glucose	7	1.0
Hydroquinone	6-7	12.4
Phenol	6-7	10.6

1. Increased stability of nucleotide sugars in the presence of phosphate buffers [42]
2. Stabilization of nucleoside-5'-triphosphates by ammonium bicarbonate
3. Use of antioxidants, butylated hydroxyanisole, and butylated hydroxytoluene, to stabilize [*carbinol*-^{14}C]vitamin A and 5 mM mercaptoethanol to stabilize [^{35}S]methionine and [^{3}H]methionine

Some additional measures to be taken in the control of chemical decomposition, are mentioned on pages 248 and 249 (Section III 4).

VI. OBSERVATIONS OF SELF-DECOMPOSITION

Comprehensive reviews by Bayly and Evans [12,39,43-45] and by Sheppard [11] list the observed decomposition rates for many labeled compounds on storage, which includes their batch-to-batch variations. In this section are listed some examples which illustrate the protective measures used for the control of self-decomposition, discussed in Section V, and also some observations which may be of especial interest to readers. The methods of analysis and measurement of the self-decomposition of radiochemicals [12a] are described fully in Chapter 6.

A. Carbon-14 Labeled Compounds

Compounds labeled with ^{14}C are by far the most widely used of all radiochemicals. The molar specific activity of ^{14}C compounds is usually more modest than, for example, their tritiated equivalents, and this tends to make the problems of self-decomposition more manageable. However, the higher decay energy of the carbon isotope, its incorporation into numerous compounds at almost 100% isotopic abundance (62 mCi/mÅ), and the increasing sensitivity of tracer applications to small amounts of impurities, makes the problem severe enough.

All of the prophylactic measures to control self-decomposition mentioned in Section V have been usefully applied on occasions. In addition, the reader should recall the point made in Section III about the fraction of its own radiation energy which the labeled compound absorbs. In the examples given this is normally assumed to be unity except where additional data are available [2].

Observations of the self-decomposition of ^{14}C compounds have led to the development of optimum conditions for storage of the various classes of labeled organic compounds.

1. [^{14}C]Amino acids

The storage of [^{14}C]amino acids in dilute acid solutions, for example in 0.01-0.1 N hydrochloric acid, although affording some protection against the risk of bacterial growth in cases of accidental microbiological contamination, does little to protect the labeled compound from self-radiolysis. This is seen clearly from the results in Table 12.

Chow et al. [46] examined several commercial samples of [^{14}C]lysine which were found to contain as much as 18.5% of radiochemical impurities, with the exception of L-[U-^{14}C]lysine supplied in solution containing 2% ethanol, which contained only 0.6% impurity. The results in Table 12 clearly show the

TABLE 12

Self-Decomposition of [^{14}C]Amino Acids in Dilute Acid and in Aqueous Solutions Containing Ethanol

[^{14}C]Amino acid	Specific activity (mCi/mmol)	Storage conditions*	Temp. (°C)	Time of storage (months)	Decomposition (%)
L-[U-^{14}C]Glutamic acid	250	a	2	15	1
	205	b	5	13	1
L-[U-^{14}C]Isoleucine	310	a	2	18	2
	240	b	5	6	4
L-[U-^{14}C]Leucine	340	a	5	12	1
	275	b	5	5	5
L[U-^{14}C]Phenylalanine	282	c	20	4	0.5
	282	d	20	4	16
L-[U-^{14}C]Tyrosine	470	a	2	16	2
	334	b	5	9	8

*a, Aqueous solution containing 2% ethanol.
b, 0.1 N Hydrochloric acid.
c, Aqueous solution containing 3% alcohol.
d, 0.01 N Hydrochloric acid.

beneficial effects of ethanol as a radical scavenger in minimizing the self-decomposition of [^{14}C]amino acids.

At low molar specific activity (say, less than 30 mCi/mmol), ^{14}C-labeled amino acids are best stored as solids (freeze-dried) at 0°C or lower temperatures. It is preferable to replace the air in the ampuls by an inert gas such as nitrogen, especially when the compound is sensitive to air/oxygen as, for example, with [^{14}C]dihydroxyphenylalanine. At high molar specific activity amino acids are best stored in aqueous solutions containing 2% ethanol at a radioactive concentration of 50-100 µCi/ml and at +2°C or frozen at -20°C and lower temperatures. If these conditions of storage are followed, the observed decomposition rate for [^{14}C]amino acids does not normally exceed 2% per annum as shown in the review by Sheppard [11].

Of especial note in this class of labeled compound is S-adenosyl-L-[^{14}C]methionine, labeled either [^{14}C]_methyl_- or [*carboxyl*-^{14}C]. This compound is sensitive to chemical and radiochemical decomposition effects and must be stored in dilute acid (usually dilute sulfuric acid) at pH 2.5-3.5 at low temperatures. Thus S-adenosyl-L-[*carboxyl*-^{14}C]methionine at 50 mCi/mmol in dilute sulfuric acid at 10 μCi/ml decomposes at the rate of 3% per day at 25°C, 2% per week at +2°C, and less than 0.1% per week at -20°C. This example serves as a general warning not to leave solutions of sensitive labeled compounds at room temperature any longer than is absolutely necessary.

2. [^{14}C]Carbohydrates and Polyhydric Alcohols

When stored under similar conditions to those recommended for [^{14}C]amino acids (see above), [^{14}C]carbohydrates and polyhydric alcohols also normally show less than 2% decomposition per annum [11]. D-[U-^{14}C]glucose, for example, at 320 mCi/mmol in aqueous solution containing 3% ethanol, showed no detectable decomposition when stored at -20°C during 17 months. Similar results are reported for other [^{14}C]carbohydrates [11]. Storage of [^{14}C]dextran as the sulfate (solid) is an interesting example of secondary radiation effects; sulfuric acid is formed, which chars the sugar [47].

3. [^{14}C]Nucleics

The class of compounds called [^{14}C]nucleics include labeled purine and pyrimidine bases, nucleosides, nucleotides, and the nucleotide sugars.

[^{14}C]Purine and pyrimidine bases are among the most stable of labeled organic compounds showing no detectable self-decomposition when stored as (freeze-dried) solids at +2°C and lower temperatures during 12 months.

[^{14}C]Nucleosides are likewise reasonably stable but are usually best stored in aqueous solutions containing 2% ethanol at -20°C. For example, [U-^{14}C]uridine at 490 mCi/mmol in aqueous solution (no ethanol) at -20°C decomposes at the rate of about 1% per month, while [U-^{14}C]guanosine at 520 mCi/mmol in aqueous

solution containing 2% ethanol at -20°C shows less than 2% decomposition in 12 months [11]. Pritasil et al. [48] in their review of the stability of a number of labeled nucleics, confirm that [U-^{14}C]uridine is the least stable of the ^{14}C-labeled nucleosides even when stored in aqueous solution containing 2% ethanol at -10°C.

[^{14}C]Nucleotides, and especially the nucleoside 5'-triphosphates, are chemically unstable; considerable care is needed both in conditions used for their storage and for their analysis. The chemical decomposition rates per day of some [U-^{14}C]nucleoside triphosphates in sterile aqueous solution at 25°C are given in Table 13. The results in Table 13 give some indication of the decomposition that can be expected when working with these compounds at room temperature within a physiological pH range. In general, the presence of a dilute buffer can stabilize the triphosphates, the pyrimidine nucleotides hydrolyse at a slower rate than the purine compounds, and deoxyribonucleoside triphosphates hydrolyze at a faster rate than the ribonucleoside triphosphates. At pH values near neutrality (pH 7) the primary decomposition product of ribo- or deoxyribonucleoside triphosphates is usually the corresponding diphosphate. The

TABLE 13

Chemical Decomposition Rates Per Day of Labeled Nucleoside Triphosphates in Sterilized Aqueous Solution at 25°C

	Percent decomposition in:	
Compound*	Tris buffer 0.1 M pH 7.6	Distilled water pH 5.5
Uridine triphosphate	0.1	0.1
Adenosine triphosphate	0.3	0.6
Guanosine triphosphate	0.3	0.8
Deoxyguanosine triphosphate	1.0	2.0

*In order to facilitate the analyses the compounds used were labeled with ^{14}C at a specific activity of 50 mCi/mAt; chemical concentration of the compounds was 4.9×10^{-5} M.

hydrolysis of nucleotides in strong acids is well documented and the rate constants showing the relative stability of the N-glycosidic bond of ribonucleosides and deoxyribonucleosides, under such conditions, are given in a comprehensive review by Capon [49].

Tetas and Lowenstein [50] published data which demonstrates that the presence of small amounts of bivalent metal ions can considerably increase the rate of hydrolysis of nucleoside triphosphates. This catalytic effect passes through a peak at pH 5. Of 12 metal ions tested, only Ba^{2+} and Mg^{2+} are ineffective. A 60-fold increase in the rate of hydrolysis of ATP to ADP occurs in the presence of Cu^{2+} ions at pH 5. The apparent rapid self-decomposition of [U-^{14}C]nucleoside-5'-triphosphates (up to 10% in 2 weeks) stored in ethanol:1.5 N ammonia (1:1) at -10°C, observed by Nejedly et al. [51], is most likely due to chemical decomposition from the causes already mentioned (see above). In general, the rate of self-decomposition of [U-^{14}C]nucleotides at almost 100% isotopic abundance stored in sterilized aqueous solutions (50-100 μCi/ml; pH about 7) containing 2% ethanol, does not normally exceed 1-2% per month at -20°C. For specifically labeled [^{14}C]nucleotides stored under the same conditions, the rate is usually considerably less than 1% per month [11]. The corresponding [^{14}C]diphosphate is the major product formed on storage of solutions of [^{14}C]nucleoside triphosphates [11,42,51]. Self-decomposition of [^{14}C]nucleotides is also minimized by storage on filter paper at -15°C, but this is usually less convenient than storage in solution [48].

[^{14}C]Nucleotide sugars are even more sensitive to chemical decomposition than the [^{14}C]nucleoside triphosphates. Nucleoside diphosphate sugars are easily hydrolyzed by mild acid treatment. Uridine diphosphate glucose (UDPG), whose behavior is typical for this class of compound, is completely hydrolyzed to UDP and D-glucose in 0.01 M HCl at 100°C for 10 min [52]. Nucleoside diphosphate sugars which possess a free hydroxyl group at C_2 of the sugar group are also alkali labile [53]. For example, hydrolysis

of UDPG to glucose-1,2-cyclic phosphate and UMP is rapid in concentrated ammonia even at 0°C. Table 14 gives some indication of the stability of UDPG at various pH values and at 25°C. Figures in Table 14 refer to percentage radiochemical purity which at time zero was 98%. The UDP[U-^{14}C]glucose was dissolved in 0.05 M buffer at a radioactive concentration of 50 μCi/ml and the solution sterilized by filtration. At appropriate time intervals analyses were carried out by paper chromatography in ethanol: ammonium acetate pH 3.8 (5:2). Buffers employed were sodium phosphate (pH 4.6, 7.6, and 8), glycine-NaOH (pH 9.2) and sodium carbonate/bicarbonate (pH 10.7). At room temperature there is significant hydrolysis in a few hours even at pH values below 10; care must be taken during analysis of these compounds by paper chromatography in solvents containing ammonium acetate at pH 7.5 because they can be hydrolyzed by a trace amount of ammonia in the atmosphere of the chromatography tank [54]. Ammonia is formed by hydrolysis of the ammonium acetate in the chromatography solvent used to equilibrate the atmosphere in the tank. The problem is avoided by using only ethanol:water for the equilibration. When reactions involving nucleoside diphosphate sugars are followed by chromatographic analysis it is often convenient to spot samples at intervals on chromatography paper and subsequently develop the chromatogram overnight. Experiments [55] have shown that the

TABLE 14

Stability of Uridine Diphosphate [U-^{14}C]Glucose in Sterilized Aqueous Solution at 25°C at Various pH Values

	Time (hr)					
pH	1.5	4	24	48	72	96
4.6	--	--	97.5	98.0	97.5	97.7
7.6	--	--	97.0	97.4	97.1	96.6
8.0	--	--	97.6	97.0	96.5	95.7
9.2	97.2	96.7	90.4	83.8	--	--
10.7	93.0	80.9	29.9	14.5	--	--

decomposition of these compounds is accelerated on paper especially under mildly alkaline conditions. Table 15 shows the radiochemical purity of UDP [U-^{14}C]glucose measured after storage at various pH values in 0.2 M buffer solutions for 7 hr, and the radiochemical purity of samples (5 µl), from each of the various solutions, which had been applied to chromatography paper, dried with cold air, and stored for a similar period. UDP [U-^{14}C]-glucose at 28 mCi/mmol in solution at a radioactive concentration of 50 µCi/ml with an initial radiochemical purity at time zero of 99%, was used. Analyses were carried out in ethanol:ammonium acetate pH 3.8 (5:2) and the buffers employed were sodium acetate (pH 5.8), imidazole (pH 7.4 and 8.4) and diethanolamine (pH 9.5 and 10.2). From these results (Table 15) it is clearly advisable to store samples from reaction mixtures containing nucleoside diphosphate sugars in solution (after suitable steps have been taken to stop the reaction, such as low temperature storage) rather than on paper before chromatographic analysis.

Nucleoside diphosphate amino sugars are much more resistant to alkaline hydrolysis because there is no free hydroxyl group at C_2 in the amino sugar. For example, the initial rate of hydrolysis of UDP-N-acetyl glucosamine in 0.1 M NaOH at 60°C is about 0.7% per hour, whereas for UDPG in 0.1 M NaOH decomposition is virtually instantaneous at room temperature.

TABLE 15

Stability of Uridine Diphosphate [U-^{14}C]Glucose Stored in Solution and on Paper Under Various Conditions for 7 Hours

Storage conditions	pH 5.8	7.4	8.4	9.5	10.2
In solution at 20°C	--	99	99	95	94
On paper at 20°C	--	97	83	63	63
In solution at 5°C	99	99	99	97	96
On paper at 5°C	97	99	96	75	75

Table 16 shows the observed rates of decomposition of the commonly used ^{14}C-labeled nucleoside diphosphate sugars under what are believed to be optimal conditions of storage, viz., in sterilized aqueous solutions containing 2% ethanol at -20°C. The results are averaged over the first year but decomposition may accelerate after this period. Note also that bivalent metal ions such as Cu^{2+} cause significant increases in the rate of hydrolysis of nucleoside diphosphate sugars [55].

4. [^{14}C]Lipids, Fatty Acids and Other Benzene-Soluble Compounds

^{14}C-Labeled lipids, fatty acids, and other compounds which are soluble in purified benzene are best stored at room temperature in this solvent, and usually show less than 5% self-decomposition during 1 year. Thus, for example, [U-^{14}C]oleic acid at 1 Ci/mmol in benzene at 20°C showed 5% decomposition during 21 months, while [U-^{14}C]stearic acid at 92 mCi/mmol under the same conditions of storage showed no decomposition during 10 years [11]. In general, unsaturated lipids are more susceptible to self-decomposition than saturated compounds. It is especially important to store solutions of unsaturated labeled lipids in the absence of air/

TABLE 16

Self-Decomposition of [^{14}C]Nucleotide Sugars on Storage at -20°C

Compound*	Specific activity (mCi/mmol)	Decomposition rate per annum (%)
Adenosine diphospho[U-^{14}C]glucose	>200	6
Guanosine diphospho-L-[U-^{14}C]fucose	>75	5
Guanosine diphospho[U-^{14}C]mannose	>75	<2
Uridine diphospho-N-acetyl-[U-^{14}C]glucosamine	>200	<2
Uridine diphospho[U-^{14}C]galactose	>200	2
Uridine diphospho[U-^{14}C]glucose	>200	2
Uridine diphospho[U-^{14}C]glucuronic acid	>225	3

*In sterilized aqueous solution containing 2% ethanol.

oxygen under vacuum or under an inert gas such as nitrogen or argon.

^{14}C-Labeled phospholipids, phosphoryl choline, phosphoryl ethanolamine, phosphatides, and phosphatidyl cholines (lecithins), are not soluble in benzene and are best stored in aqueous solutions containing 2% (or more) of ethanol at -20°C and lower temperatures. Self decomposition is not normally more than 0.5% per month under these conditions.

5. [^{14}C]Steroids

[^{14}C]Steroids at both high and low molar specific activity are best stored in benzene or in benzene containing 5-10% ethanol at +2° to +20°C, not frozen. Note that the freezing point of pure benzene is 5.4°C. Under these conditions of storage the observed self-decomposition of [^{14}C]steroids is normally less than 2% per annum, with the exception of a few sensitive compounds [11], which include the [^{14}C]corticosteroids (cortisone and cortisol), estriol, and vitamin D_3. Thus [4-^{14}C]aldosterone at 57 mCi/mmol in benzene/ethanol (19:1) at 20°C showed 5% decomposition during 18 months, but [4-^{14}C]cortisol at 57 mCi/mmol showed 15% decomposition in 12 months under the same storage conditions. Note, however, that lower rates of decomposition of the sensitive [^{14}C]steroids are often obtained by storage at very low temperatures (-140°C and lower). For example, [4-^{14}C]vitamin D_3 at 32 mCi/mmol in benzene/ethanol (19:1) showed no detectable decomposition during 13 months at -140°C [11].

6. Miscellaneous ^{14}C Compounds

Among the miscellaneous [^{14}C] compounds are included labeled catecholamines, drugs, vitamins, and aliphatic and aromatic compounds. By following carefully the general methods recommended for the control of self-decomposition given in Section V, most ^{14}C compounds can be stored for long periods (12 months and more) with less than 2% decomposition [11]. ^{14}C compounds which are rather sensitive to decomposition in solution are often best stored as freeze-dried solids at -20°C and lower temperatures. For example, l-[*methylene*-^{14}C]norepinephrine stored as a freeze-dried bitartrate salt at 57 mCi/mmol showed no detectable decomposition during 20 months at -20°C. [^{14}C]Benzylpenicillin, an especially sensitive compound, showed 8% decomposition during 6 months as a freeze-dried solid at -20°C. Even lower storage temperatures (-140°C and below) are recommended in such cases.

A few other general observations with ^{14}C compounds of possible interest to readers are as follows:

1. In the solid or liquid form, it is frequently possible to obtain improved stability by storing a ^{14}C compound in a thin film rather than as bulk lump, i.e., dispersion increases stability and protects against self-decomposition.
2. Decomposition by free radical reactions can sometimes proceed at great speed and radiation catalyses the formation of free radicals. Thus [^{14}C]isobutene, even at 2.5 mCi/mmol, polymerizes within a few minutes on freezing in liquid nitrogen. Similarly, [1-^{14}C]acrylonitrile at 1.55 mCi/mmol was found to be converted into a solid polymer on standing at room temperature in the dark overnight. [U-^{14}C]Acetylene at high specific activity (over 100 mCi/mmol) tends to undergo spontaneous decomposition to a voluminous yellow powder "cuprene," on storage. N-[*methyl*-^{14}C]-N-nitroso-p-toluene sulfonic acid (11 mCi/mmol) and N-ethyl-[2,3-^{14}C]maleimide (2.6 mCi/mmol) are two other compounds which have been observed to undergo considerable decomposition during a few weeks storage at -40°C even at the low specific activities indicated. N-ethyl-[2,3-^{14}C]maleimide is best stored in pentane at room temperature (20°C); storage at lower temperatures appears to be less satisfactory in this case.
3. The rapid decomposition of [*methyl*-^{14}C]choline chloride [1] was originally believed to involve propagation of a chain reaction by free radicals. Later work [56] has shown that the radicals observed by electron spin resonance act as chain

initiators rather than propagators. Very low temperature storage (-196°C), storage on filter paper or dissolved in a suitable solvent successfully controls decomposition of this labeled compound [39]. It is also interesting to note that [N-*methyl*-^{14}C]choline iodide keeps better than the chloride [57].

4. The polymerization of labeled aliphatic iodides such as [^{14}C]methyl iodide at high specific activity (60 mCi/mmol), is not observed in contrast with the observations with tritiated methyl iodide which, at specific activities above 3 Ci/mmol may spontenaously form a yellow solid. A gradual darkening of the color due to the release of free iodine, is observed only together with the formation of [^{14}C]methane and [^{14}C]ethane as gaseous impurities.

5. Labeled aldehydes are usually sensitive to oxidation and are often better stored as derivatives. For example, [7-^{14}C]-phenylacetaldehyde, a compound very sensitive to oxidation and polymerization may be stored without significant decomposition for several months as the bisulfite adduct [58].

6. Barium [^{14}C]carbonate and sodium [^{14}C]carbonate are two raw materials from which are derived many organic compounds labeled with ^{14}C. Such compounds are not normally regarded as unstable in their radioactive form; however, decomposition can occur by loss of the isotope, usually as [^{14}C]carbon dioxide which is exchanged with the carbon dioxide in the atmosphere. Studies by Barakat and Farag [59] have shown that barium [^{14}C]carbonate is best stored in sealed cans, whereas sodium [^{14}C]carbonate, if stored as a solid, should be kept under vacuum in the presence of sodium hydroxide pellets. In contrast with what is normally recommended for the storage of labeled organic compounds in the solid state, care should always be taken to decrease the exposed surface area of the stored [^{14}C]carbonates. Sodium [^{14}C]carbonate is better stored as an aqueous solution in sealed ampuls and does not then show any loss of radioactivity, by exchange, on storage.

In summary, the observed self-decomposition of most organic compounds labeled with ^{14}C is usually quite small (less than 2-3% per annum) provided the recommended precautions are taken for the control of the decomposition rate.

B. Tritium-Labeled Compounds

The virtually complete absorption of the β radiation energy and the very high molar specific activities (1-200 Ci/mmol) which are attained in practice, have made the control of self-decomposition for tritium compounds much more difficult than for compounds labeled with other radionuclides. Because of the extensive use of tritium compounds in research [60], studies of the self-decomposition of tritiated compounds have been more intensive and conducted with perhaps a greater sense of urgency than for other labeled compounds. The quite long half-life of tritium means that investigators may tend to store compounds longer than the useful shelf-life, with the consequential problems discussed in Section VII.

Several special general features can be identified in the observed self-decomposition of tritiated compounds. These are:

1. All tritiated compounds at high specific activity (Ci/mmol), with a few specific exceptions, are required to be stored and manipulated in solution to minimize the effects of self-decomposition.
2. Tritiated compounds at high specific activity stored in solutions, are very sensitive to decomposition by chemical effects, because of their extremely small chemical concentration.
3. Tritiated water is formed by exchange on storage of many ^{3}H-labeled compounds in aqueous solutions.
4. A very pronounced acceleration of self-decomposition may be observed in stored frozen solutions of tritium-labeled compounds, because of molecular clustering (see p. 262).

5. The rate of self-decomposition of some tritiated compounds is adequately controlled only by storage at liquid nitrogen temperatures (-140 to -196°C).
6. A very marked influence of radical scavengers is observed on the decomposition rate of tritiated compounds stored in solutions.
7. The rate of self-decomposition of some tritiated compounds is increased suddenly after an unpredictable duration of induction period. The magnitude of this effect is quite unpredictable and the induction period of apparent good stability varies from weeks to months.

As with ^{14}C compounds, all of the prophylactic measures to control decomposition mentioned in Section V can be usefully applied to tritiated compounds.

Storage in their natural form. Although tritiated compounds are normally best manipulated in solutions, some are better stored in their natural form, viz., as solid, liquid, or gas. Examples are, [^3H]folic acid, [^3H]dimethylbenzanthracene, and sodium-, potassium-, and lithium [^3H]borohydrides. In such cases it is sometimes necessary and beneficial to dilute the compound with another solid (liquid or gas), usually (say) an inorganic compound such as a salt or a mixture of buffer salts. Thus, [^3H]folic acid is stored mixed with sodium (or potassium) bicarbonate and [^3H]cyclic AMP is stored as a freeze-dried solid with buffer salts [61]. Low temperatures are recommended for storage in all cases except for the [^3H]borohydrides where this has not been observed to be significantly beneficial and storage as solids at room temperature (+20°C) is usual. Observed decomposition rates with these compounds are shown in Table 17.

Storage in solutions. The beneficial effects of storing tritiated compounds in purified organic solvents are illustrated, for example, in the observed decomposition rates of tritiated steroids at very high specific activity shown in Table 18. Such

TABLE 17

Self-decomposition of Some Tritiated Compounds in Their Natural Form

Compound	Specific activity (mCi/mmol)	Storage conditions*	Storage temperature (°C)	Time of storage (weeks)	Decomposition (%)**
[^{3}H]Acetic anhydride	500	a	-20	8	3†
	5,600	a	-20	9	6†
	5,600	a	-20	9	25††
N-[Acetyl-^{3}H]-D-galactosamine	179	b	+2	100	5
N-[Acetyl-^{3}H]-D-glucosamine	243	b	+2	86	N.D.
[8-^{3}H]Adenine	500	b	-20	79	N.D.
[G-^{3}H]Adenosine	500	b	-20	47	3
L-[G-^{3}H]Asparagine	140	b	-20	55	N.D.
L-[G-^{3}H]Aspartic acid	250	b	+2	60	N.D.
[G-^{3}H]Atropine	580	b	-20	11	8
	580	b	-196	6	N.D.
[G-^{3}H]Benzoic acid	378	c	+2	30	N.D.
[G-^{3}H]Benzylpenicillin	218	b	-20	40	35
[2-^{3}H]Bromoacetic acid	139	c	-20	44	N.D.
[G-^{3}H]Bromobenzene	100	a	+20	40	N.D.
[G-^{3}H]Cyanocobalamin (vitamin B_{12})	1,650	b	-20	15	3
[7-^{3}H]Dehydroepiandrosterone sulfate, potassium salt	600	b	-20	49	8
[1,2-^{3}H]Decane	412	a	+20	7	N.D.
[^{3}H]Dihydrostreptomycin sesquisulfate	730	b	+2	48	N.D.
[G-^{3}H]7,12-Dimethyl-1,2-benzanthracene	500	c	-20	33	7
	5,700	c	-20	16	6
[^{3}H]Dimethyl sulfoxide	560	a	+2	22	8
L-[*ring*-2,5,6-^{3}H]DOPA	500	b	-20	33	N.D.
	500	b	-40	25	3
[*ring*-G-^{3}H]Dopamine	300	b	-20	23	N.D.

TABLE 17 (continued)

Compound	Specific activity (mCi/mmol)	Storage conditions*	Storage temperature (°C)	Time of storage (weeks)	Decomposition (%)**
[*side chain*-1,2-^{3}H]-Dopamine	370	b	-20	84	N.D.
[3',5',9-^{3}H]Folic acid potassium salt	250	b	-20	22	3
	16,000	b	-140	13	12
	12,000	b	-196	14	11
D-[6-^{3}H]Glucose 6-phosphate	244	b	+2	24	6
	244	b	-196	24	7
[8-^{3}H]Guanine sulfate	130	b	-20	23	N.D.
[8-^{3}H]Guanosine	500	b	+2	36	2
[2,5-^{3}H]Histamine dihydrochloride	500	b	-20	26	2
L-[2,5-^{3}H]Histidine	500	b	-20	32	12
[G-^{3}H]Hypoxanthine	435	b	-20	55	N.D.
[^{3}H]Inulin	200	b	+2	27	N.D.
L-[G-^{3}H]Leucine	250	b	-20	31	N.D.
Lithium [^{3}H]borohydride	280	c	+20	15	5§
	6,100	c	+20	28	12§
[3',5',9-^{3}H]Methotrexate potassium salt	250	b	-20	125	N.D.
[^{3}H]Methylamine hydrochloride	173	b	-20	52	N.D.
L-2-Methyl[3-^{3}H]DOPA	870	b	+2	11	10
[G-^{3}H]Nicotine	320	a	+2	19	9
	320	a	-196	32	4
[G-^{3}H]1-Naphthylacetic acid	307	b	-20	56	N.D.
[5-^{3}H]Orotic acid	1,000	b	-20	16	18
Potassium [^{3}H]borohydride	6,100	c	+20	28	12§
[G-^{3}H]Serotonin	500	b	-20	23	N.D.
Sodium [^{3}H]acetate	2,900	b	-20	20	N.D.

TABLE 17 (continued)

Compound	Specific activity (mCi/mmol)	Storage conditions*	Storage temperature (°C)	Time of storage (weeks)	Decomposition (%)**
Sodium [^{3}H]borohydride	600	c	+20	21	N.D.
	12,400	c	+20	31	10§
[7-^{3}H]Tetracycline	1,000	b	-140	22	21
[6-^{3}H]Thymidine	9,130	b	+20	12	8
	9,130	b	+2	16	6
	9,130	b	+2	26	35
	9,130	b	-40	26	25
	9,130	b	-80	26	30
[G-^{3}H]Tryptamine hydrochloride	250	b	-20	97	N.D.
L-[G- H]Tyrosine	434	b	-20	56	<2

*a, Liquid.
b, Freeze-dried solid.
c, Solid (or powder).
†As [^{3}H]acetic acid.
††Small package size, 25 mCi.
§Nonvolatile activity not available for reduction.
**N.D. = none detectable.

compounds would undergo almost total decomposition within a few days if stored as solids.

Self-decomposition of unsaturated compounds in benzene solutions have not yet been observed to result in any significant cis-trans isomerism. For example, less than 5% trans-isomer is observed during the storage of [9,10-^{3}H]oleic acid or of [2-^{3}H]-glyceryl trioleate, although about 10^5 rads of radiation energy were absorbed by the solutions [12,39]. In contrast, it is interesting to note that some unsaturated compounds in benzene solutions have been observed to undergo cis-trans isomerism when irradiated externally at comparable radiation doses [62,63], but at a higher dose rate, which may be a significant factor.

The effects of temperature and the rate of freezing of solutions of tritiated compounds have been discussed in Section

TABLE 18

Self-Decomposition of Some Tritiated Drugs, Prostaglandins, and Steroids [12,55]

Compound	Specific activity (Ci/mmol)	Storage conditions*	Storage temperature (°C)	Time of storage (weeks)	Decomposition (%)**
[^{3}H]Actinomycin D	6.9	a	-20	4	2
d-[*side chain*-^{3}H]-Amphetamine sulfate	10	b	+2	26	N.D.
[1-^{3}H]Codeine	22	c	-20	6	1
	22	c	-20	12	8
[1-^{3}H]Codeine hydrochloride	25	e	-20	8	4
Diacetyl [1-^{3}H]-morphine	25	c	-20	8	2
	25	c	-20	24	3
	25	c	-20	30	5
Diacetyl [1-^{3}H]-morphine hydrochloride	25	e	-20	8	2
[1,7,8-^{3}H]Dihydromorphine	76	c	-20	8	3
	76	c	-20	25	5
[1,7,8-^{3}H]Dihydromorphine hydrochloride	87	a	-20	8	9
[2-^{3}H]Lysergic acid diethylamide (LSD)	21	d	+2	14	1
	21	d	+2	22	3
	21	d	+2	40	4
[1-^{3}H]Morphine	28	c	-20	22	2
[1-^{3}H]Morphine hydrochloride	21	a	-20	12	4
Δ^1-[G-^{3}H]Tetrahydrocannabinol	14	c	-20	19	34
	14	f	-20	19	14
	14	g	-20	19	7
	14	g	-20	9	N.D.
[1,2-^{3}H]Aldosterone	15.7	h	+2	64	N.D.
[1,2,6,7-^{3}H]Corticosterone	106	c	-20	34	4
[1,2,6,7-^{3}H]Cortisol	110	d	+2	6	6
	97	d	+2	12	20
[2,4,6,7-^{3}H]Estradiol	100	i	+2	64	15

TABLE 18 (continued)

Compound	Specific activity (Ci/mmol)	Storage conditions*	Storage temperature (°C)	Time of storage (weeks)	Decomposition (%)**
[2,4,6,7-^{3}H]Estradiol	100	i	+2	16	8
	100	i	+2	20	2
[2,4,6,7-^{3}H]Estrone	100	i	+2	8	23
	100	i	-140	104	6
	100	i	-140	12	10
	100	i	-196	48	10
[6,7-^{3}H]-Ethinylestradiol [101]	1.82	j	+20	52	3-4
[1,2,6,7-^{3}H]Progesterone	110	j	+20	8	N.D.
	110	j	+20	22	9
[1α,2α-^{3}H]Progesterone	41	j	+20	12	N.D.
[2,4-^{3}H]Estriol	10	c	-20	26	N.D.
[1α,2α-^{3}H]Corticosterone	44	c	+2	17	<3
18-Hydroxy[1,2-^{3}H]-deoxycorticosterone	40	l	+2	21	7
[1α,2α-^{3}H]5α-Dihydrotestosterone	49	j	+20	17	N.D.
[1α,2α-^{3}H]Testosterone	56	j	+20	12	N.D.
[1,2,6,7-^{3}H]Testosterone	100	j	+20	16	N.D.
	100	j	+20	28	5
[1α,2α-^{3}H]Vitamin D_3	0.58	d	-140	60	N.D.
[1α,2α-^{3}H]Cholesterol	46	j	+20	52	<3
[5,6-^{3}H]PGA_1	40-60	k	-20	4	0.5-1
[5,6,8,11,12,14,15-^{3}H]PGA_2	100	k	-20	4	2
[5,6,11,14,15-^{3}H]PGB_2	80	k	-20	4	2
[5,6-^{3}H]PGE_1	53	k	-20	23	2

TABLE 18 (continued)

Compound	Specific activity (Ci/mmol)	Storage condi- tions*	Storage temper- ature (°C)	Time of storage (weeks)	Decompo- sition (%)**
[5,6,8,11,12,14,15-^{3}H]PGE_2	100	k	-20	5	5
[5,6-^{3}H]$PGF_{1\alpha}$	53	k	-20	22	2-3
[9-^{3}H]$PGF_{2\alpha}$	15	k	-20	23	1-2

*a, Ethanol/water (1:1).
b, Aqueous soln containing 2% ethanol.
c, Ethanol.
d, Benzene/ethanol (9:1).
e, Ethanol/water (9:1).
f, Toluene/ethanol (9:1).
g, Toluene/ethanol (1:1).
h, Benzene/ethanol (3:1).
i, Benzene/ethanol (95:5).
j, Benzene.
k, Ethanol/water (7:3).
l, Benzene/ethanol (9:1) containing 0.02% triethylamine.

**N.D. = none detectable.

IV; these probably represent additional sources of variability of decomposition rates between different batches (preparations) of the same tritiated compound. Observations of the decomposition rates of some tritiated compounds at liquid nitrogen temperatures (-140°C to -196°C) are shown in Table 19.

The ready ease of exchange of tritium atoms in certain molecules with hydroxylic solvents often results in the formation of tritiated water on storing compounds in aqueous solution [12, 64]. Even if the tritium atoms in the labeled compound are not directly exchanged, there is always a possibility of exchange occurring from the products of self-decomposition, as exemplified in Table 19. Some compounds, such as [5-^{3}H]orotic acid for example, can be stored in solution apparently without significant

TABLE 19

Isotopic Exchange During Storage of Some Tritiated Compounds in Neutral Aqueous Solution

Compound	Specific activity (Ci/mmol)	Storage conditions*	Time of storage (weeks)	Labile tritium (THO) (%)
Amines and amino acids				
5-Amino[G-^{3}H]levulinic acid	2.5	a	30	20
L-[G-^{3}H]Aspartic acid	0.22	a	40	2
L-3,4-Dihydroxy[*ring*-2,5,6-^{3}H]phenylalanine (L-DOPA)	34.7	a	11	15
	23.5	a	13	10
[*ring*-G-^{3}H]Dopamine	4.8	a	37	29
L-[2,5-^{3}H]Histidine	11.1	a	33	19**
L-[4,5-^{3}H]Leucine	1.0	a	25	1
DL-[4,5-^{3}H]Lysine	8.2	a	25	2
DL-[G-^{3}H]Methionine	0.25	a	22	1
L-[*ring*-4-^{3}H]Phenylaline	1.0	a	21	2
L-[*side chain*-2,3-^{3}H]-Tyrosine	3.6	a	30	<1
L-[3,5-^{3}H]Tyrosine	35.3	b	11	4
	28.2	b	30	8**
Nucleics				
[8-^{3}H]Adenine	15.0	a	56	20
[2,8-^{3}H]Adenine	3.6	a	30	5
[G-^{3}H]Adenosine	3.1	a	12	6
[8-^{3}H]Adenosine	0.64	a	47	18
[8-^{3}H]Adenosine 3',5'-cyclic phosphate	3.0	c	17	1
[8-^{3}H]Guanosine 3',5'-cyclic phosphate	0.88	c	32	6
[G-^{3}H]Deoxyadenosine	0.57	a	15	8
[8-^{3}H]Deoxyadenosine	13.0	a	10	3.5
[8-^{3}H]Guanosine	5.1	a	4	4
	4.9	a	16	13

TABLE 19 (continued)

Compound	Specific activity (Ci/mmol)	Storage condi- tions*	Time of storage (weeks)	Labile tritium (THO) (%)
[8-^{3}H]Guanosine 5'-monophosphate	2.04	c	76	6
[8-^{3}H]Inosine 5'-triphosphate	2.7	c	52	<2
5-[G-^{3}H]Methylcytosine	0.5	a	70	<1**
[5-^{3}H]Orotic acid	4.6	a	32	66**
[*methyl*-^{3}H]Thymidine	5.0	a	12	<2
[6-^{3}H]Thymidine	5.0	a	12	<2
[5-^{3}H]Uracil	1.0	a	43	6
[6-^{3}H]Uracil	6.2	a	65	<1
[5-^{3}H]Uridine	5.0	a	12	3
[6-^{3}H]Uridine	4.7	a	35	<1

*a, 1 mCi/ml at +2°C.
b, Solution containing 2% ethanol at 1 mCi/ml and +2°C.
c, Solution in ethanol/water (1:1), 1 mCi/ml at -20°C.
**Plus decomposition.

decomposition. However, a more detailed analysis indicates that the molecular structure is destroyed, giving rise to tritium atoms in a labile position, which then exchange to form tritiated water and unlabeled chemical impurities [12,39]. The presence of a few percent tritiated water is often unimportant but can, if necessary, easily be removed by freeze-drying (lyophilization) and redissolving the compound immediately before use.

The most sensitive test for the detection of labile tritium in nonvolatile compounds is to shake the compound with water or aqueous alcohol (if not already dissolved in such solvents), distil off the solvent (or a measured fraction of it) and measure the radioactivity in the distillate. However, great care is needed in determining and interpreting measurements of the labile tritium content of the compounds. Falsely high values can easily be obtained in the presence of dissolved oxygen with compounds which are sensitive to oxidation and decompose to products that readily exchange their tritium atoms, for example, [*ring*-2,5,6-^{3}H]-dopamine and L-[*ring*-2,5,6-^{3}H]DOPA. In such cases it is essential to use deoxygenated water (boiled and flushed with nitrogen) as diluent.

Observations of the self-decomposition of tritiated compounds have led to the development of optimum conditions for storage of the following classes of labeled organic compounds.

1. [^{3}H]Amino Acids

Tritiated amino acids at low specific activity (below 500 mCi/mmol) can often be stored as freeze-dried solids at +2°C and lower temperatures, with less than 5% decomposition per annum. At high specific activity stored in aqueous solutions at a maximum radioactive concentration of 1 mCi/ml, often containing a radical scavenger such as 2% ethanol, at +2°C, [^{3}H]amino acids have a decomposition rate of less than 1% per month [12]. Of course, for tritiated amino acids which are especially sensitive to self-decomposition effects, the rate can be much higher even under optimum conditions of storage [12].

An effect observed with tritiated optically active amino acids, which has not yet been observed with the corresponding ^{14}C-labeled compounds, is racemization on storage in solutions. Some examples are shown in Table 20. The cause of this effect is not known, but the addition of a few percent ethanol appears to decelerate and minimize racemization.

2. [^{3}H]Carbohydrates and Polyhydric Alcohols

The observed rates of decomposition of tritiated carbohydrates and polyhydric alcohols, stored under conditions similar to those used for the [^{3}H]amino acids, do not in general normally exceed 1-2% per month [12]. These rates of decomposition apply to compounds at low or high specific activity. It is also not always necessary to have ethanol present as a scavenger as many tritiated carbohydrates keep satisfactorily without alcohol. For example,

TABLE 20

Self-decomposition and Racemization of Optically Active Tritiated Amino Acids on Storage in Aqueous Solution [12]

Amino acid	Specific activity (Ci/mmol)	Activity (mCi/ml)*	Time of storage (months)	Decomposition (%)**	Racemization (%)**†
L-[2,5-^{3}H]Histidine	6.5	1.1(a)	3	--	26
L-[4,5-^{3}H]Leucine	7.0	1	10	30	36
	7.6	2.2	6	--	10
	0.96	1	7	--	34
D-[4,5-^{3}H]Leucine	16.0	1	9	--	60
L-[G-^{3}H]Methionine	0.16	1	13	20	98
L-[*methyl*-^{3}H]-Methionine	1.61	2.9	3.5	15	10
L-[G-^{3}H]Phenylalanine	0.825	1	9	9	10
L-[*ring*-4-^{3}H]-Phenylalanine	7.95	3.4	3	10	16
	2.0	1	7	15	40
L-[G-^{3}H]Proline	0.42	1	4	N.D.	N.D.
L-[G-^{3}H]Tryptophan	0.875	1	15	20	16
L-[3,5-^{3}H]Tyrosine	48.0	1	4.5	50	88
	48.0	1(b)	4.5	4	100
	42.3	1(c)	3	N.D.	N.D.

*(a) Solution containing 1% benzyl alcohol;
(b) solution containing 0.1% sodium formate;
(c) solution containing 1% ethanol.

**N.D. = none detectable.

†Percentage tritium in the DL form after repurification (usually by paper chromatography); rest of activity in original optically active form.

D-[6-^{3}H]glucose at 2.7 Ci/mmol in aqueous solution (1 mCi/ml) showed no detectable decomposition during 6 months at +2°C.

The 2-deoxysugars are more unstable than the sugars containing an hydroxyl group in the 2 position, which is consistent with the chemical stability of these compounds. Thus for example, while D-[1-^{3}H]glucose at 3.5 Ci/mmol in aqueous solution (1 mCi/ml) showed only 3% decomposition in 12 weeks at +2°C, 2-deoxy-D-[1-^{3}H]glucose, at a slightly higher specific activity of 5.3 Ci/mmol in aqueous solution (1 mCi/ml), showed 7% decomposition during 10 weeks at +2°C.

3. [^{3}H]Nucleics

The tritiated nucleics (purines, pyrimidines, nucleosides, nucleotides, and nucleotide sugars) constitute the most important class of labeled organic compounds and one which has received the most intensive study of self-decomposition.

[^{3}H]Purine and pyrimidine bases, although much more stable than the [^{3}H]nucleosides and [^{3}H]nucleotides, are not as stable as the corresponding ^{14}C-labeled compounds. Decomposition rates in aqueous solution (1 mCi/ml) at +2°C, without any scavenger present, are usually not more than 1% per month. The [^{3}H]purines exhibit a greater resistance to self-decomposition than the [^{3}H]pyrimidines. Thus [8-^{3}H]adenine at 15 Ci/mmol, for example, in aqueous solution showed no detectable decomposition during 1 year at +2°C, while [6-^{3}H]thymine, for example, at 24 Ci/mmol in solution at +2°C showed 4% decomposition in 15 weeks [12]. The site of attack of the hydroxyl radicals in the decomposition of labeled pyrimidines has been identified and confirmed as the 5,6-double bond, giving rise to the formation of glycols and hydroperoxy compounds [31,65,66].

[^{3}H]Nucleosides include the two most widely used [60] labeled compounds, [^{3}H]thymidine and [^{3}H]uridine; the former is incorporated efficiently into DNA and the latter into RNA. Radiation self-decomposition of [^{3}H]thymidine and [^{3}H]uridine has been observed to occur at a significant rate [11,34,35,37,39]

during storage in solutions and to result in the formation of glycols and hydrates [67-69].

Observations by Sheppard et al. [70], shown in Table 21 illustrate the stability of [methyl-^{3}H]thymidine and [5,6-^{3}H]-uridine, at high specific activity, under various conditions of temperature and solvent. The results show the great acceleration of decomposition observed in frozen solutions stored at -20°C compared with +2°C (by a factor of 3 or 4); this emphasizes the inadvisability of routinely storing solutions of tritiated nucleoside at -20°C in the deep freeze, as discussed in Section IV. The results also show the beneficial effects of a small amount of ethanol on the stability of the compounds, if the presence of an additive is acceptable, and also the beneficial effects of storage at very low temperatures (-140°C).

TABLE 21

Self-Decomposition of [*methyl*-^{3}H]Thymidine and [5,6-^{3}H]Uridine Under Various Conditions of Temperature and Solvent at 1 mCi/ml

Compound (specific activity)	Solvent	Temp. (°C)	Time of storage (weeks)	Decomposition (%)*
[*methyl*-^{3}H]Thymidine (44 Ci/mmol)	Water	-140	5	4
	Water	-20	5	17
	Water	+2	5	4
	2% Ethanol	+2	8	N.D.
	2% Ethanol	+2	23	6
	10% Ethanol	+2	8	N.D.
	10% Ethanol	+2	23	3
[5,6-^{3}H]Uridine (55 Ci/mmol)	Water	-140	10	N.D.
	Water	-20	10	25
	Water	+2	10	6
	2% Ethanol	+2	15	3
	50% Ethanol	-20	10	N.D.

*Mean of two determinations. N.D. = none detectable.

In general, tritiated nucleosides are normally stored in aqueous solutions at +2°C and 1 mCi/ml containing up to 10% ethanol. Under optimum conditions the rate of self-decomposition does not usually exceed 2% per month but varies, of course, with the compound and the actual conditions used for storage [12].

As with many other tritiated compounds, aqueous solutions of [^{3}H]nucleosides are filtered through a bacterial (membrane) filter to minimize decomposition due to any microbiological contamination. However, such methods may not always be effective in removing all such contaminants. Mycoplasmas, the smallest free living organisms known to science, which are virus-like or bacteria-like and characterized by the absence of a rigid cell wall, can pass through bacterial filters if present in solutions [71]. Frequently, mycoplasmas contaminate cultured tissue cells [71], and Lavelle [72] has suggested that mycoplasmas in [^{3}H]-uridine solutions are the cause of a decrease in cell growth and increased amounts of "debris" in the production of murine leukemic viruses. In view of the instability of mycoplasmas (see below) this suggestion is rather surprising. Although this case is an apparently isolated one, as there have been no other reported effects of suspected mycoplasmas contamination in the uses of labeled compounds, cautious investigators will bear this possibility in mind.

Mycoplasmas are not especially stable, they seldom cannot survive for long in distilled water, they normally prefer a pH slightly alkaline, they undergo lysis in solutions of primary alcohols and are normally inactivated by autoclaving solutions for a few minutes above 50°C (say at 100°C for 2 min) [71].

Autoclaving neutral solutions of tritiated nucleosides may produce a few percent (usually not more than 5%) of the corresponding purine or pyrimidine base, and tritiated water by exchange [64]. For example, autoclaving of [5-^{3}H]uridine (28 Ci/mmol) for 20 min at 121°C gave only 3% tritiated water, and only 1% tritiated water was formed from [6-^{3}H]thymidine during 120 min autoclaving. However, virtually all the tritium (98%) was exchanged from [8-^{3}H]guanosine (5.1 Ci/mmol) during 20 min autoclaving at 121°C [64], due to the ease of exchange of tritium from the 8-position of purines [68]. Therefore, caution is required before autoclaving tritiated compounds for sterilization, as it may be necessary to consider other methods. The radiochemical purity of the compound should always be checked after any such heat treatment.

Waterfield et al. [64] have investigated the exchange of tritium from a selection of tritiated compounds on autoclaving solutions at 121°C. The amounts of tritium exchanged depends upon the structure of the compound and the position of the tritium atom(s) in the labeled molecules. Some results are shown in Table 22.

[^{3}H]Nucleotides and *[^{3}H]nucleotide sugars* are normally observed to store satisfactorily in ethanol:water (1:1) at -20°C and 0.5-1 mCi/ml. The rate of self-decomposition under these conditions is normally less than 2% per month [12]. The problem of the chemical stability of nucleotides, and especially the nucleoside triphosphates, is discussed in Section VI A 3 and is even more critical with tritiated nucleotides because of the higher specific activities and the consequential lower chemical concentrations of these compounds. The presence of 1% ammonium bicarbonate has been observed to have beneficial effects of the stability of [^{3}H]nucleotides on storage [11,12], but the presence of bicarbonate may not be desirable for some uses of these tritiated compounds. For example, [5-^{3}H]uridine 5'-triphosphate (13 Ci/mmol) at 0.5 mCi/ml in 50% aqueous ethanol containing 1% ammonium bicarbonate showed only 3% decomposition during 6 months at -20°C. Uridine diphospho[6-^{3}H]glucose at 1000 mCi/mmol in 50% aqueous ethanol 1 mCi/ml showed only 3% decomposition during 1 year at -20°C [55].

TABLE 22

Tritium Exchange on Autoclaving
Solutions of Tritiated Compounds at 121°C [64]

Compound	Specific activity (Ci/mmol)	Autoclaving time (min)	Labile tritium (THO) (%)
[^{3}H]Amino acids			
DL-[G-^{3}H]Alanine	1.5	20	1
5-Amino [3,5-^{3}H]levulinic acid	2.5	20	95
L-[*ring*-2,5,6-^{3}H]DOPA	34.7	20	75*
L-[4,5-^{3}H]Leucine	24.3	20	<1
L-[4,5-^{3}H]Lysine	0.6	20	<1
L-[*methyl*-^{3}H]Methionine	9.3	20	<2
DL-[G-^{3}H]Ornithine	0.9	20	<1
L-[*ring*-4-^{3}H]Phenylalanine	5.0	20	2
L-[G-^{3}H]Serine	0.2	20	<2
L-[*side chain*-2,3-^{3}H]-Tyrosine	3.6	20	<1
L-[3,5-^{3}H]Tyrosine	28.2	20	18*
[^{3}H]Purines and purine nucleosides			
[2,8-^{3}H]Adenine	3.6	20	59
[G-^{3}H]Adenosine	2.7	20	54
[8-^{3}H]Guanosine	5.1	20	98
[G-^{3}H]Inosine	1.1	20	94
[^{3}H]Pyrimidines and pyrimidine nucleosides			
5-Bromo [6-^{3}H]Uracil	1.5	20	2
[G-^{3}H]Cytidine	1.4	20	<2
[6-^{3}H]Deoxyuridine	6.4	20	<1
[*methyl*-^{3}H]Thymine	1.0	120	1
[6-^{3}H]Thymine	19.3	20	<1
[*methyl*-^{3}H]Thymidine	25.2	120	1
[6-^{3}H]Thymidine	5.0	120	1
[5-^{3}H]Uracil	14.6	120	10

TABLE 22 (continued)

Compound	Specific activity (Ci/mmol)	Autoclaving time (min)	Labile tritium (THO) (%)
[6-^{3}H]Uracil	6.2	120	2
[5-^{3}H]Uridine	28.0	20	3
[5-^{3}H]Uridine	5.0	120	14
[6-^{3}H]Uridine	4.7	120	2

*Accompanied by decomposition of the compound.

4. [^{3}H]Lipids, Fatty Acids and Other Benzene-Soluble Compounds

These compounds are normally stored at room temperature (+20°C). Thus, for example, [^{3}H]mevalonic acid lactone at 250 mCi/mmol in benzene at +20°C showed no detectable decomposition on storage during 18 months; [9,10-^{3}H]oleic acid at 3500 mCi/mmol in benzene at +20°C showed only 8% decomposition during almost 1 year [12].

5. [^{3}H]Steroids

Tritiated steroids are normally stored in benzene solution or benzene/ethanol solution (up to 10% ethanol) at +2°C and 1 mCi/ml. Care is needed in selecting the temperature of storage to avoid crystallization of the compound or separation from the protecting solvent, as discussed in Section IV. In general, benzene solutions of [^{3}H]steroids should not be stored frozen; with unstable steroids it may be advantageous to store solutions at liquid nitrogen temperatures (-140 to -196°C). Because of the extensive use of tritiated steroids at very high specific activity in radioimmunoassays, there have been extensive studies of the self-decomposition of these compounds on storage [12]. Some observations of self-decomposition of tritiated steroids are given in Table 18.

6. Miscellaneous ^{3}H Compounds

In most cases the effects of self-decomposition on catecholamines, drugs, vitamins, etc. can be minimized by using the methods discussed in Section V. Thus, for example, the tritiated catecholamine 1-[7-^{3}H]norephinephrine, a compound very sensitive to decomposition and readily oxidized, at 5.8 Ci/mmol stored in aqueous solution (1 mCi/ml) at -140°C showed no detectable decomposition during 2 months [11,12].

For further information readers are referred to an extensive and detailed review of self-decomposition of tritium compounds by Evans [12].

C. Compounds Labeled with Phosphorus-32 or Sulfur-35

Compounds labeled with relatively short-lived radionuclides are less likely to be stored for long periods before use than compounds labeled with the longer lived radionuclides such as ^{14}C and ^{3}H. Impurities which may be present in compounds labeled with ^{32}P or ^{35}S are therefore more often likely to have arisen during the preparation of these compounds. Although the half-life of the two radionuclides is short, the primary (internal) effect of decay in producing impurities is minimal. This is because multiple labeling with ^{32}P or ^{35}S is exceptional and most compounds labeled with these radionuclides contain only one labeled atom per molecule.

1. Phosphorus-32 Compounds

^{32}P compounds are normally used within a few weeks of their preparation because of the 14.3 days half-life of the radionuclide. In general, compounds are best stored in their normal form as thin films (liquids or freeze-dried solids) at -20°C and lower temperatures. The high energy of the β particles results in only a small fraction of the radiation energy being absorbed by the compound.

The most widely used group of ^{32}P-labeled compounds are the α- and γ-[^{32}P]nucleotides and in many tracer applications in biochemistry, ^{32}P is the only useful label which can be employed [42]. [^{32}P]Nucleotides are prepared at very high specific activities (up to 100 Ci/mmol) and are usually handled in solutions for convenience and, in some cases, to decrease self-decomposition. The influence of various factors upon the decomposition and stability of labeled nucleotides has previously been described (see above). [^{32}P]Nucleotides in aqueous solutions are observed to undergo decomposition by both chemical and radiolytic processes; the higher the specific activity the more serious the radiolytic decomposition becomes; for example, after 3 days in aqueous solution at 20-25°C, [γ-^{32}P]ATP at 17 Ci/mmol undergoes more than 40% radiolytic decomposition, apart from chemical hydrolysis of the terminal phosphate group; the corresponding figure for [γ-^{32}P]ATP at 1.7 Ci/mmol being only 5% as seen from Table 23. The identity of the decomposition products has not yet been established although the triphosphate group is still intact [55]. Although [^{32}P]nucleotides at high specific activity store best as freeze-dried solids at low temperatures (see Table 25), these compounds are frequently handled and stored in solutions for convenience. In order to minimize self-decomposition, [^{32}P]nucleoside-5'-triphosphates in ethanol:water (1:1) solution should be stored at as low a temperature as possible. Under these conditions decomposition after storage for 1 month is usually observed to be less than 10% when the temperature is -30°C and less than 5% when the temperature is -140°C. In ethanol:water (1:1) solution at -30°C hydrolysis of the terminal phosphate group is less than 1% after 14 days whereas at 30°C it is about 7% after 3 days and 19% after 14 days, as seen from Table 24.

TABLE 23

Decomposition of [γ-^{32}P]ATP in Water at 20-25°C

Initial specific activity (mCi/mmol)	Percent radiolytic decomposition product on storage after: 3 days	7 days
170	5	9
1,700	5	16
17,000	43	66

TABLE 24

Formation of $[^{32}P]$Orthophosphate from $[\gamma\text{-}^{32}P]$ATP*

Time	Percent $[^{32}P]PO_4$ formed at a temperature of: +30°C	-30°C
After 1 day	∿5	<1
After 3 days	∿7	<1
After 7 days	∿10	<1
After 14 days	∿19	<1

*Sodium salt in ethanol/water (1:1 v/v) unbuffered.

$[\gamma\text{-}^{32}P]$ATP at high specific activity (over 10 Ci/mmol) stored as a freeze-dried solid at -30°C shows less than 5% decomposition after 2 weeks storage. $[\alpha\text{-}^{32}P]$ATP at 100 Ci/mmol stored as a freeze-dried solid showed less than 5% decomposition after 7 days at -140°C, less than 10% at -30°C, but about 30% decomposition at room temperature (20°C) even after 2 days storage [55].

$[^{32}P]$Diisopropylphosphorofluoridate (DFP) is an important labeled compound in research and in radiopharmacy, used for labeling red blood platelets. In propylene glycol (300 µCi/ml) at a specific activity of 80 mCi/mmol the rate of self-decomposition is about 1% per week at room temperature (20°C). Solutions of $[^{32}P]$DFP are sterilized by γ irradiation (2.5 Mrads) and this produces less than 0.2% impurity. Evidence [55] suggests that the presence of $[^{32}P]$diisopropylphosphorochloridate (the precursor in the preparation of DFP) accelerates the self-decomposition of $[^{32}P]$DFP. The radiochemical purity of labeled DRP is usually determined by gas-liquid chromatography and more recently by reaction with α-chymotrypsin [73]. The complex of DFP with α-chymotrypsin has zero R_f on thin layer chromatography (silica gel) in hexane:acetone (4:1) solvent while impurities have R_f values up to 0.2. The main impurity in labeled DFP is diisopropylphosphoric acid.

Sodium[^{32}P]pyrophosphate is stable in aqueous solution at pH 8.5 or above over a 4-week period; below this pH and in more acidic solutions the rate of decomposition increases sharply which may be aggravated by self-radiolysis.

Observations on the self-decomposition of some other [^{32}P] compounds are shown in Table 25. No information appears to have been published on the stability of compounds labeled with ^{33}P. Such compounds would be likely to have a stability which is very similar to that observed for the corresponding ^{32}P compounds. However, the lower beta energy of ^{33}P (see Table 1) is likely to produce quantitatively different results from those observed with ^{32}P compounds.

2. Sulfur-35 Compounds

The self-decomposition of ^{35}S-labeled compounds is similar to that observed for ^{14}C compounds and many [^{35}S] compounds are stored in their normal form at 20°C and lower temperatures with less than 1% decomposition during the half-life of the radionuclide (87 days).

[^{35}S]Methionine is the most widely used of the ^{35}S-labeled compounds. DL-[^{35}S]Methionine at 100 mCi/mmol is unstable at room temperature in the presence of moisture, but the anhydrous compound is quite stable as a crystalline solid stored in vacuo. However, L-[^{35}S]methionine undergoes up to 10% decomposition per month when stored as the anhydrous solid at a comparable specific activity; the decomposition rate is reduced to about 1% per week at -30°C. L-[^{35}S]Methionine therefore appears to be more sensitive to self-decomposition than DL-[^{35}S]methionine. This difference in observed behavior is probably due to the difference in the crystalline form of the two compounds; a similar phenomenon perhaps to the differing sensitivity of the two forms of choline chloride to radiation [74]. There is no significant difference between the rate of self-decomposition of DL- and L-[^{35}S]-methionine when stored in aqueous solution; storage at low

TABLE 25

Self-decomposition of Some [^{32}P]Compounds [11,55]

Compound	Initial specific activity (mCi/mmol)	Storage condi- tions*	Temp. (°C)	Time of storage (days)	Decompo- sition (%)**
2-Cyanoethyl [^{32}P]-phosphate barium salt	600	Solid	20	42	N.D.
Adenosine 5'-[γ-^{32}P]-triphosphate	800	a	-30	28	<2
	3,340	c	-20	18	N.D.
Adenosine 5'-[α-^{32}P]-triphosphate	255	a	20	20	7
	255	b	20	20	5
	255	c	20	20	4
	408	c	20	70	5
	3,000	c	20	35	21
	3,000	c	-30	35	6-9
	3,000	c	-140	35	N.D.
	100,000	c	20	2	∿30
	100,000	c	-30	2	3-7
	100,000	c	-140	2	1-2
	100,000	Solid	-140	7	1
	100,000	c	-30	7	2-4
	100,000	c	-140	7	2-3
Guanosine 5'-[γ-^{32}P]-triphosphate	2,960	c	-20	21	2
Thymidine 5'-[γ-^{32}P]-monophosphate	992	Solid	2	53	8
	992	a	20	11	2
	992	a	20	53	25
	992	a	2	53	20
Uridine 5'-[γ-^{32}P]-triphosphate	1,630	c	-20	21	N.D.

*a, Aqueous solution.
b, Aqueous solution containing 5% ethanol.
c, Ethanol/water (1:1) (at 3 Ci/mmol radioactive concentration is 0.25 mCi/ml and at 100 Ci/mmol is 1 mCi/ml).

**N.D. = none detectable.

temperature (-30°C) is satisfactory for both with less than 1% decomposition observed per month [39]. At -30°C in aqueous solution, less than 0.02% racemization of the L-isomer was observed during 152 days.

Although the yields of radiolysis products are different, there is a similarity between the compounds produced by self-radiolysis of labeled methionine and those produced by γ or x irradiation [75-77]. Products arise mainly through demethylation, deamination, and oxidation to the sulfoxide and sulfone.

The interest in mapping transfer RNA reactions has increased the use of L-[^{35}S]methionine at very high specific activity (over 100 Ci/mmol). Such labeled methionine is prepared biosynthetically from [^{35}S]sulfate and *E. coli* and after purification the product requires storage at low temperature in solutions containing mercaptoethanol to minimize self-decomposition. The addition of 5 mM mercaptoethanol inhibits the oxidation of the [^{35}S]methionine and also acts as a radical scavenger.

The effect of storage temperature on the stability of solutions of [^{35}S]methionine containing mercaptoethanol is shown in Table 26.

The sodium salts of long-chain aliphatic sulfates show a marked sensitivity to self-radiolysis which is similar to that observed for the decomposition of some long-chain aliphatic [^{14}C]carboxylic acids [39].

The stability of [^{35}S]thiosemicarbazide has been previously discussed (see Table 6). There appears to be little difference in decomposition rates between storage in aqueous solution at -30°C or at room temperature in vacuo over phosphorus pentoxide.

TABLE 26

Effect of Temperature on the Storage of L-[^{35}S]Methionine* [55]

Time of storage (days)	Percent decomposition at storage temperature		
	20°C	-20°C	-80°C
7	10	4	4
14	22	6.5	4
28	20	8	6
42	36	5	7
56	38	5	9
84	47	6	5
112	57	4.5	9
140	55	4	10
182	68	5	29

*Initial radiochemical purity 98% at a specific activity of 11.5 Ci/mmol stored in aqueous solution containing 5 mM concentration of mercaptoethanol. Ampuls sealed under nitrogen or in vacuo and radioactive concentration 4.2 mCi/ml.

Although often colored green when stored in the solid form, solutions of [^{35}S]thiosemicarbazide are colorless. The rapid decomposition of the compound when stored in methanol (see Section III B 1) does not necessarily preclude the use of [^{35}S]-thiosemicarbazide in this solvent for some applications.

Self-decomposition data for a number of ^{35}S compounds are given in Table 27. [^{35}S]Sulfate is frequently a product of the decomposition of ^{35}S compounds.

D. Compounds Labeled with γ- or X-Ray-Emitting Radionuclides

Compounds labeled with γ- or x-ray emitting radionuclides are not, in general, expected to undergo serious self-decomposition during the short period (usually days) over which such compounds are normally used. Many of the compounds are used as radiopharmaceuticals for in vivo and for in vitro diagnostic applications and the quality requirements of compounds used in this way are discussed in Chapter 6.

TABLE 27

Self-Decomposition of Miscellaneous ^{35}S Compounds on Storage [11,55]

[^{35}S]Compound	Initial specific activity (mCi/mmol)	Storage conditions*	Temp. (°C)	Time of storage (days)	Decomposition (%)**
Chlorpromazine	15	a	20	99	N.D.
	20	a	20	272	N.D.
	30	a	20	187	N.D.
Chlorpropamide	4	a	20	90	N.D.
Diaphenylsulfone (Dapsone)	18	a	20	102	N.D.
Dibenzyl disulfide	10	a	20	26	1-2
Dimethylthiourea	7.5	a	20	257	N.D.
L-Ethionine	5.7	a	20	92	2-8
S-Ethyl-L-cysteine	2	a	20	346	N.D.
_n_-Hexyl mercaptan	10	b	20	333	3
L-Homocystine	100	a	20	58	N.D.
L-Methionine	343	c	-20	55	4†
	11,920	d	-80	140	5†
	11,920	d	-20	140	N.D.
	11,920	d	+20	140	50†
S-Methyl-L-cysteine	4	a	20	58	N.D.
Phenol red	10	a	20	250	1-3
Sodium dodecyl sulfate	12.5	a	20	181	5
	17	a	20	70	1-2
Sulfaguanidine	4	a	20	209	4-5
Sulfamerazine	10	a	20	252	1-2
Sulfanilamide	8.5	a	20	154	N.D.
Sulfanilic acid	220	a	20	98	1-2
	5,000	e	20	5	17
	5,000	e	-20	5	6
	5,000	f	20	5	25
	5,000	f	-20	5	20
	5×10^5	g	20	14	45-50
Thiamine (vitamin B_1)	145	a	20	316	2-3
Thiophenol	10	b	20	290	N.D.
Thiourea	39	a	20	215	N.D.

TABLE 27 (continued)

$[^{32}S]$Compound	Initial specific activity (mCi/mmol)	Storage condi- tions*	Temp. (°C)	Time of storage (days)	Decompo- sition (%)**
p-Toluene sulfonic acid	190	a	20	143	1-2
Zinc N,N-dimethyldithio- carbamate (Ziram)	13	a	20	118	∿50

*a, Solid stored in vacuo over phosphoric oxide.
b, Liquid.
c, Sterilized aqueous solution at 2 mCi/ml.
d, Aqueous solution containing 5 mM mercaptoethanol, in sealed container.
e, Aqueous solution at 4.3 mCi/ml.
f, Aqueous solution at 4 mCi/ml.
g, Aqueous solution at 7 mCi/ml.
N.D. = none detectable.
†Sulfoxide the major impurity which increases if air admitted.

Most frequently used are compounds labeled with ^{75}Se, ^{57}Co, ^{58}Co, ^{125}I, ^{131}I, ^{197}Hg, ^{203}Hg, and ^{99m}Tc.

1. Compounds Labeled with Selenium-75

Compounds labeled with ^{75}Se have increased importance for use not only for in vivo studies but also for in vitro applications in saturation analysis (radioimmunoassay and competitive protein binding assays; see Chap. 21). Because of the fairly high energy of the electron capture x-ray emission from ^{75}Se, compounds labeled with this radionuclide would not be expected to undergo serious self-decomposition. This expectation is borne out in practice. L-[^{75}Se]Selenomethionine is currently the most important selenium-labeled compound used in pancreas-imaging studies, and is currently the only ^{75}Se-labeled compound to be studied in detail. Batches of this compound have an initial specific activity of 1 Ci/mmol stored in aqueous solution at 1 mCi/ml showed no detectable decomposition during 4 months at 20°C, or when stored for a similar time at -30°C even up to radioactive concentrations of 20 mCi/ml. Also, no racemization of the optically active amino acid was observed during the storage time. Over a longer storage time it is believed that the mode of decomposition is partly chemical, but the product is not only the selenoxide [55].

The nonisotopic labeling of several steroids with ^{75}Se has been achieved at about 20 Ci/mmol. These compounds are used in the RIA and CPB assay of steroid hormones [78] and the compounds appear to be satisfactory for such uses over a period of several weeks. Thus, 2-methyl-[^{75}Se]selenoprednisolone shows no serious self-decomposition during a period of several (6) weeks as supplied for use in a cortisol assay kit (Cortipac) [79].

[^{75}Se]Selenate is the main inorganic impurity formed on storing [^{75}Se] compounds.

2. Compounds Labeled with Cobalt-57 and Cobalt-58

The only compound labeled with cobalt radioisotopes which has been studied in detail is cyanocobalamin (vitamin B_{12}), used extensively in studies of pernicious anemia. The instability of cyanocabalamin was first reported by Smith [80], and a further series of observations on the decomposition of labeled cyanocobalamins in aqueous solution was later reported by Rosenblum [81]. The compound is readily converted into hydroxocobalamin and the so-called red acids by oxygen and light in aqueous solutions. Self-decomposition of the radioactive vitamin produces similar impurities. Cobaltous ions are the main inorganic impurity from compounds labeled with cobalt radioisotopes.

For cyanocobalamin labeled with ^{58}Co at a moderate specific activity (10 Ci/mmol), a good method of storage appears to be as a thin film of freeze-dried solid; no detectable decomposition being observed during 10 weeks at 2°C. Although freeze-dried [^{57}Co]cyanocobalamin has a satisfactory stability at low specific activity (1.4 Ci/mmol), at higher specific activities it is not as stable in this form as [^{58}Co]cyanocobalamin, nor is it stable in aqueous solution. However, the addition of 0.9% benzyl alcohol to aqueous solutions of either [^{57}Co]- or [^{58}Co]cyanocobalamin is beneficial both as a bacteriostat and as a radical scavenger, as shown in Table 28. Because of the sensitivity of cyanocobalamin to light and to oxygen, care must be taken to ensure ordinary chemical stability by storing the compound in a cool dark place free from contaminants which might be harmful.

3. Compounds Labeled with Iodine-125 and Iodine-131

Although the number of different types of investigations using iodine radioisotopes is probably fewer than for other

TABLE 28

Self-Decomposition of [^{57}Co]- and [^{58}Co]Cyanocobalamin [39]

Compound	Initial specific activity (Ci/mmol)	Storage conditions	Temp. (°C)	Time of storage (weeks)	Decomposition (%)*
[^{58}Co]Cyanocobalamin	10	Aqueous solution	+2	10	45
	10	Aqueous solution	-40	10	15
	10	Freeze-dried solid in vacuo	+2	10	N.D.
[^{57}Co]Cyanocobalamin	390	Freeze-dried solid in vacuo	+2	12	39
	390	Aqueous solution containing 0.9% benzyl alcohol	+2	12	N.D.
	400	Aqueous solution containing 0.9% benzyl alcohol	+2	11	N.D.

*N.D. = none detectable.

radionuclides such as ^{14}C or ^{3}H, the iodine radioisotopes are nevertheless the most widely used of the γ- or x-ray emitting radionuclides in medicine and in clinical research.

Stanford [82] has investigated the stability of 16 compounds labeled with either ^{125}I or with ^{131}I, but the number of investigations concerning the stability of compounds labeled with iodine isotopes is much fewer than for those labeled with other radionuclides. Many of these compounds are used in vivo, especially those labeled with ^{131}I and under these circumstances the important parameter to measure has usually been taken to be the liberation of free iodine rather than an overall drop in the radiochemical purity [82,83]. Clearly there is good reasoning behind this viewpoint, but it can be misleading; a faulty diagnosis due to the use of an impure radiochemical can be as damaging to the patient as an unwanted dose to the thyroid.

The carbon-iodine bond is relatively weak, compared with the carbon-hydrogen bond, for example, and it is therefore not surprising that one of the main impurities in the self-decomposition of iodine-labeled compounds is inorganic iodide. Chemical decomposition, self-radiolysis and radiolysis of solutions of iodine compounds [84] by external gamma irradiation, all yield iodide ions as impurity. Sorantin [85] has observed the chemical stabilization of organic radioiodine compounds by the addition of organic amines, but this method of stability is not acceptable for many uses of the compounds.

Some results obtained by observing the decomposition of a number of compounds labeled with radioactive iodine are given in Table 29 [39,82]. As with compounds labeled with other radioisotopes, the decomposition rates depend upon the exact storage conditions, but many compounds are observed to store satisfactorily in solutions at +2°C. In general, compounds labeled with ^{125}I are more stable than those correspondingly labeled with ^{131}I, as would be predicted from the lack of β radiation emission from ^{125}I.

TABLE 29

Self-Decomposition of ^{125}I- and ^{131}I-Labeled Compounds

Compound	Initial specific activity (mCi/mg)	Storage conditions*	Temperature of storage (°C)	Storage time (weeks)	Decomposition (%)	
					Iodide	Other**
3,5-Di[^{125}I]-iodo-L-tyrosine	10	a	20	2	2	N.D.
	10	a	20	4	1	N.D.
	10	a	20	8	1	4
5-[^{125}I]Iodo-2'-deoxyuridine	10	b	20	1	3	4-7†
	10	b	20	2	5	5-7†
	10	b	20	4	3	6-9†
5-[^{125}I]Iodo-2'-deoxyuridine	1000	b	+2	2	3-4	5-6†
5-[^{131}I]Iodo-2'-deoxyuridine	10	b	20	1	27	18-20†
	10	b	20	2	28	36†
	10	b	20	4	17	41-46†
	10	b	-80	1	2	N.D.
	10	b	-12 (or -25)	4	2	N.D.
3-[^{125}I]Iodo-L-tyrosine	10	a	20	6	1	2-4
L-Tri[3'-^{125}I]-iodothyronine (Liothyronine; T_3)	50	c	20	4	1	3††
	50	c	20	6	1	6††
L-Tri[3'-^{131}I]-iodothyronine	25	d	20	1	4	15††
	25	d	20	3	3	23††
	25	d	+2	1	3	13††
	25	d	-80	1	12	20††
[^{125}I]Polyvinylpyrrolidone	0.05	e	20	12	1-2	N.D.
[^{131}I]Polyvinylpyrrolidone	0.05	e	20	2	1	1
[^{125}I]Rose bengal	0.7	f	20	13	1-2	1-3
[^{131}I]Rose bengal	1.0	g	20	2	1-2	3-5
Sodium [^{125}I]-iodohippurate	0.03	h	20	13	<1	N.D.
Sodium [^{131}I]-iodohippurate	0.12	i	20	4	2-3	1-3
Sodium [^{125}I]-iothalamate	0.05	j	20	13	1	N.D.

TABLE 29 (continued)

Compound	Initial specific activity (mCi/mg)	Storage conditions*	Temperature of storage (°C)	Storage time (weeks)	Decomposition (%) Iodide	Other**
L-[3',5'-^{125}I]-Thyroxine	40-60	b	20	2	1-2	1-3[§]
	40-60	b	20	4	1-2	1-3
	40-60	b	20	8	2-3	5-7
L-[3',5'-^{131}I]-Thyroxine	40	k	20	3	2-3	3-7[§]
	33	a	20	0.5	2-3	3
	33	a	-80	0.5	16-20	5-9

*a, Solution in 50% aqueous propylene glycol at 0.2 mCi/ml.
b, Sterilized aqueous solution at 1 mCi/ml.
c, Solution in 50% aqueous propylene glycol at 1 mCi/ml.
d, Solution in 50% aqueous propylene glycol at 0.25 mCi/ml.
e, Sterilized aqueous solution containing 1% benzyl alcohol and succinate buffer at 1 mCi/ml.
f, Sterilized aqueous solution containing 1% benzyl alcohol at 0.4 mCi/ml.
g, Sterilized aqueous solution containing 1% benzyl alcohol at 1 mCi/ml.
h, Same as g at 0.5 mCi/ml.
i, Same as g at 3 mCi/ml.
j, Sterilized aqueous solution containing 1% benzyl alcohol and citrate-EDTA buffer at 1 mCi/ml.
k, Solution in 50% aqueous propylene glycol at 0.4 mCi/ml.

**N.D. = none detectable.

†Decomposition products are labeled iodide and 5-iodouracil.

††Decomposition products are iodide, 3,3'-diiodothyronine (reversed T_2) and unknowns; 3-8% of thyroxine forms on storage especially at low temperatures (-80°C).

§Decomposition products are iodide, reversed T_2 and unknowns.

For aqueous solutions of 3,5-di[^{131}I]iodo-L-tyrosine the presence of dissolved oxygen exerts a stabilizing effect [86], probably due to the scavenging properties of oxygen for hydrated electrons. A similar effect may be true of other related compounds labeled with ^{131}I.

Detailed studies have been carried out [55,82] on the stability of the radioiodine-labeled iodothyronines. Thus, tri-[^{125}I]iodothyronine (T_3) at 1500 mCi/mg stored in aqueous solution containing 75% ethanol at 300-400 mCi/ml showed about 2% decomposition during 1 week at 20 or -20°C. Only 1% iodide was observed to form during 4 weeks and a major impurity identified as 3,3'-di-[^{125}I]iodothyronine (reversed T_2). [^{125}I]Thyroxine (T_4) at 80 mCi/mg stored under similar conditions showed about 5% decomposition after 4 weeks at 20°C. Again only 1% iodide was observed together with T_3 and reversed T_2 as major identified products. These and other results are shown in Table 30 [55,82]. [^{131}I]Thyroxine in 50% aqueous propylene glycol stored at -80°C (solid carbon dioxide) was observed to decompose about 5-10 times faster than when solutions were stored at +2°C [55]. At -80°C the solvent becomes a two phase mixture of liquid and solid. It is

TABLE 30

Self-Decomposition of [^{125}I]Triiodothyronine (T_3) and [^{125}I]Thyroxine (T_4) in Ethanol/Water (3:1) at 20°C

Compound	Specific activity (mCi/mg)	Radioactive concentration (mCi/ml)	Time of storage (weeks)	Decomposition (%)	
				Iodide	Other
[^{125}I]Triiodothyronine (T_3)	1500	0.3-0.4	1	--	2
			2	--	5
			4	1	8-10
			8	2	12-15*
[^{125}I]Thyroxine (T_4)	80	0.3-0.4	1	--	1-2
			2	--	1-2
			4	1	3-5
			8	2	5-7**
	400	0.3	2	--	1-2
			4	1	7**

*Main decomposition products include [^{125}I]3,3'-diiodothyronine (reversed T_2) and two other equal percent [^{125}I] impurities of unidentified structure.

**Main impurities include [^{125}I]T_3, [^{125}I] reversed T_2 and one unknown.

likely the heterogeneous nature of the solution, resulting in locally increased concentrations of the solute, which gives the enhanced rate of decomposition. Similar effects are observed on freezing and storing solutions of L-tri-[3'-^{131}I]iodothyronine in 50% aqueous propylene glycol at -80°C [82].

Radioactive iodine isotopes are frequently used in nonisotopic labeling, especially ^{125}I, for example, to "tag" molecules such as proteins which cannot be conveniently labeled by the incorporation of an appropriate isotope, or for which very high molar specific activities may be required with a γ- or x-ray emitting radionuclide. In these cases it is not possible to describe radiochemical purity in the same sense as one can do for normal isotopically labeled compounds (see also Chap. 6). One can only speak of a certain fraction of the iodinated material behaving in an identical manner to the compound for which it is serving as a tracer under given test conditions. For example, one could measure the stability of iodinated insulin. One value would be obtained for its behavior in vivo and quite another might be obtained by measurement of the amount of activity bound to an antibody--a parameter of importance to those wishing to use the labeled hormone for radioimmunoassays. There are now many such examples with a wide variety of not only proteins, but also organic molecules such as steroids, labeled with iodine isotopes for RIA/CPB assays (see Chap. 21). Many of the general methods already described for the control of self-decomposition are effectively applied for minimizing the self-decomposition of such compounds; for example, storage (with care) at low temperatures and by the addition of buffers or proteins as protective agents.

4. Compounds Labeled with Mercury-197 and Mercury-203

The organomercurial 3-(chloromercurio)-2-methoxypropylurea (chlormerodrin, trade name Neohydrin), has a place in pharmacology as a diuretuc and in nuclear medicine as a kidney- and brain-scanning agent when labeled with ^{203}Hg or ^{197}Hg. The labeled compound is normally stored and supplied in isotonic saline as the presence of chloride ions have been observed to reduce the rate of self-decomposition [87]. Cree [87] and Burianek and Cifka [88] have shown that chlormerodrin labeled with ^{197}Hg or ^{203}Hg is hydrolysed in the presence of chloride ions to give a hydroxyl analog, 3-(chloromercurio)-2-hydroxypropyl urea; the molecule also undergoes self-radiolysis to yield free mercury as the inorganic impurity.

E. Decomposition of Labeled Macromolecules

Macromolecules include peptides, proteins, polysaccharides, polynucleotides, and other polymers. It is easy to appreciate that macromolecules will decompose at a much faster rate than simple molecules. Take D-[U-^{14}C]glucose, for example, and consider 500 labeled molecules, if two events occur and change irreversibly two of these labeled molecules then 0.4% radiochemical impurity is the result. If these 500 molecules were used to make 5 polysaccharide molecules, each therefore of 100 D-[U-^{14}C]-glucose units, two events occurring in different polymer units now cause a change to two out of five units or 40% radiochemical impurity.

In macromolecules of very large molecular weight and containing radionuclides at high isotopic abundance, even the primary (internal) effects of natural decay of long half-life isotopes such as ^{14}C, need to be considered. The primary (internal) effects of decay in simple molecules have been quantitatively studied in detail [12], but it is difficult to measure quantiatively the contribution of this natural decay of the radionuclide to the self-decomposition of macromolecules. In addition, the radiochemical purity of macromolecules (see Chap. 6) is not easily defined or measured, and must often be related to the change in biological activity of the molecules. For example, degradation of DNA by self-decomposition [89-91] is a likely explanation of the differences observed in the apparent reactivity of some samples of tritiated DNA to the enzyme DNase when compared with unlabeled DNA [92,93].

An interesting observation illustrating the effect of specific activity on the decomposition rate of (tritiated) polymers was made during the storage of labeled poly(methylmethacrylate) (Perspex), a polymer used in the preparation of tritiated reference sources for autoradiography [12]. A sheet of the polymer at 10 mCi/g stored for 15 months, during which time the radiation dose absorbed was about 1 Mrad, showed no coloration or evidence of decomposition. An identical piece of tritiated polymer at 316 mCi/g stored for the same period and absorbing about 45 Mrads of radiation, turned brown and readily crumbled to a fine powder when touched with a pair of forceps. The breaks that occurred along the polymer chains were clearly visible [12]. Another example is that of algal protein labeled with ^{14}C at over 50 mCi/mA, i.e., at over 80% isotopic abundance. The insoluble protein and cell walls become soluble after a few months storage [55].

Investigators should always bear in mind and be aware of the rapid decomposition that can occur in macromolecules, especially those labeled with radioisotopes at a high isotopic abundance. The consequential effects of the decomposition may be apparent in both in vitro and in vivo tracer experiments as discussed in Section VIII.

VII. EFFECTS OF IMPURITIES IN RADIOTRACER INVESTIGATIONS

Whether impurities in a labeled compound will affect the results of tracer experiments and the degree of possible interference, depend of course upon the nature of the impurities present (whether chemical or radioactive) and the type of tracer experiments being undertaken. However, it is clear that some impurities will be present when the investigator is ready to carry out his experiments - absolute purity is an unattainable ideal. Three sources of impurities must be considered:

1. Radiochemical (and chemical) impurities produced during the preparation of the labeled compound
2. Impurities, especially radioactive ones, produced during storage of the labeled compound and formed by self-decomposition
3. Impurities (chemical) introduced into the experimental system in solvents, chemicals, enzymes, etc.

Will any of these impurities really matter? This is a question that can normally be answered only by the investigator and often requires direct experimental verification. It is seldom possible to predict either qualitatively or quantitatively, the effects of impurities (chemical and radiochemical) on the results obtained from tracer experiments using impure radiochemicals. It is made even more difficult because in only a few cases (see Section VI) are the products of self-decomposition of compounds on storage known. Currently the only practical method is to isolate the impurities and to actually test them under the experimental conditions, as a type of control experiment. An example of this approach has already been mentioned briefly in Chapter 6.

Wand, Zeuthen, and Evans [94] isolated the self-decomposition products of [^{3}H]thymidine stored in aqueous solution, and studied their behavior with synchronized cells of *Tetrahymena pyriformis* (amicronucleate strain GL). These impurities were rapidly incorporated into the cells and associated with macromolecules other than DNA and RNA, probably protein. The impurities were not removed by treatment of the cells either with DNase or RNase, and autoradiography indicated extensive labeling of the cytoplasm of the cells, whereas only labeling of the nucleus (DNA) was expected from pure [^{3}H]thymidine. These observations were subsequently confirmed by Diab and Roth [95] during their studies of the autoradiographic differentiation of free, bound, pure, and impure [^{3}H]thymidine in mouse intestinal crypt cells after [^{3}H]thymidine injection. The authors emphasized the great care that is needed when interpreting the results of autoradiographic studies.

The observed incorporation of radioactivity into RNA and protein fractions during the in vitro labeling of rat liver with [^{3}H]thymidine has been ascribed to radiochemical impurities [96]. These and similar problems of radiochemical impurities in cell biology research, and especially when using autoradiographic techniques, were discussed by Evans [12,97,98].

Herrmann [99] drew attention to the nonenzymatic tight binding of radioactivity to macromolecular fractions of serum (horse or bovine) as a source of error in labeling experiments with chick embryo muscle cells. These effects were thought to be due to specific (unidentified) radioactive impurities, possibly from the self-decomposition of the labeled compounds being used ([1-^{3}H]-, [6-^{3}H]- and [U-^{14}C]glucosamine; [1-^{3}H]-, [2-^{3}H]- and [6-^{3}H]glucose; [1-,5,6-^{3}H]fucose). Data were reported which showed that a high proportion of radioactive material can be firmly bound to macromolecular components of serum in the absence of cells and under conditions which make enzymatic activity unlikely. Control procedures using cell free media by which this type of error can be identified and minimized, were suggested. Problems and methods for their identification and control in the radiometric assessment of protein and nucleic acid synthesis were described by Oldham [100] and by Monks, Oldham, and Tovey [42].

Tritiated tyrosine at high specific activity is often used as a precursor in studies of catecholamine synthesis. Labeled 3,4-dihydroxyphenylalanine (DOPA) is a radiolysis product of [^{3}H]tyrosine on storage in solution and is, in many tissues, incorporated much faster than is tyrosine. Waldeck [30] showed that even 1% of [^{3}H]DOPA present as an impurity in [^{3}H]tyrosine resulted in the production of three times as much [^{3}H]norepinephrine as was produced from pure [^{3}H]tyrosine. Thus the presence of certain specific impurities can be identified as being the most damaging in yielding erroneous results. Only 2.2% [^{14}C]glycine as impurity in [U-^{14}C]aspartic acid led to the erroneous conclusion that aspartic acid was an effective precursor of glycine in the dorsal root ganglion and spinal grey tissues [102]. In this case it is likely that the [^{14}C]glycine was an impurity not separated in the preparation of [U-^{14}C]aspartic acid, usually obtained from algal [^{14}C]protein hydrolysate, as glycine is not a major product in the γ radiolysis (or self-radiolysis) of aspartic acid [77].

Radiooptical purity, the fraction of the total radioactivity which is present as the stated optical form of the tracer compound, may be important where there are differences in the behavior between two or more optical isomers. For example, labeled amino acids are often used to assess amino acid turnover in vivo. Waterlow and Stephen [103,104] used [^{15}C]lysine to estimate lysine turnover in the rat and found that the specific activity of the lysine, measured by total radioactivity and by enzymatic decarboxylation, from the urine was apparently very high. This was attributed to the presence of 0.7% [D-^{14}C]lysine in the original [L-^{14}C]lysine [104]. The D-lysine was nonmetabolized and excreted in the urine where it contributed to the total radioactivity, but its weight was not included in the enzymatic lysine determination, hence giving a false high specific activity. In such experiments chromatographic analysis of the products following enzymatic decarboxylation would readily detect the presence of the D-isomer. The presence of labeled d(+)-norepinephrine in labeled l(-)-norepinephrine may result in some erroneous conclusions because of the difference in the metabolism of the d(+) and the l(-) forms [105]. The presence of labeled 17-isoaldosterone in [^{3}H]aldosterone can lead to false conclusions in the studies of the secretion rate of aldosterone [106]. The different handling of the two compounds by the body magnifies the effect of the impurity. However, [^{3}H]isoaldosterone is not normally observed to form on storage of [^{3}H]aldosterone in benzene or benzene/ethanol solutions [55]. The absence of the labeled 20,22-dihydro derivative was essential for certain investigations of metabolism using labeled cardiac glycosides [107,108].

Impurities, both radioactive and chemical, can seriously affect the sensitivity of enzyme assays. Radiochemical impurities in the substrate which may be isolated with the product of the enzymatic reaction, will contribute to the blank reading. Significant blanks are obtained even with less than 0.5% of a radiochemical impurity. For example, ethyl or methyl [9,10-^{3}H]-oleate as an impurity in glyceryl [9,10-^{3}H]trioleate can cause errors in the assay of low levels of triglyceride lipase activity, as extracts of human adipose tissue are found to hydrolyze the ethyl or methyl esters at a much faster rate than the glyceryl trioleate [109]. Tritiated water, as impurity, in solutions of labeled compounds used in the assay of enzymes by tritium release [60] will, of course, result in high blanks and lower the sentivity of the assay [12]. The presence of a small amount (few percent) of tritiated water usually presents no serious problems in many applications of tritiated compounds. These and other examples of problems in the radiometric assay of enzymes were discussed by Oldham [100,110] (see Chap. 18).

Although isotope effects may be apparent in the uses of labeled compounds, and especially with tritium compounds, these have not proved to give rise to any serious problems. However, in double isotope experiments, for example, when using a mixture of ^{3}H- and ^{14}C-labeled substrates, it is very important that the radiochemical purity of both labeled compounds should be identical or erroneous isotope effects may be apparent [12].

The presence of radioactive molecules labeled in position(s) not specified for a given tracer compound can present a different kind of impurity problem. This can also result in erroneous conclusions from observed experimental data. The importance of knowing the specificity of the labeling in radiochemicals, especially in those compounds labeled with tritium, were discussed by Evans [98,111]. In studies with tritiated compounds as tracers the problem has been largely overcome by the use of compounds with a known specificity and stereospecificity of labeling which can now be verified by tritium nuclear magnetic resonance spectroscopy [12,69,112,112a].

Nonuniformity of labeling in the use of compounds believed to be U (uniformly) labeled (see Chaps. 5 and 6) may give rise to erroneous results, especially in studies of catabolism and other experiments where splitting of the molecule carbon skeleton occurs [3]. The uniformity of labeling in a ^{14}C-labeled compound for example, depends upon the method used for its preparation (see Chap. 5) and has been firmly established for some U ^{14}C-labeled compounds. Some examples are, [U-^{14}C]glucose obtained from *Canna indica* leaves by photosynthesis with $^{14}CO_2$ [118], and L-[U-^{14}C]-lysine [119] produced by hydrolysis of algal [^{14}C]protein obtained by growing algae on $^{14}CO_2$ (or [^{14}C]bicarbonate) as the only source of carbon [120].

Other examples and a discussion of the effects of impurities in tracer investigations are presented in Section IV A of Chapter 6.

VIII. STABILITY OF LABELED COMPOUNDS DURING RADIOTRACER STUDIES

Because of endogenous dilution factors and the fact that many tracer experiments are conducted over a short time span (hours or days), there are few examples where the decomposition of a tracer compound in the experimental system, due to the radiation effects, has resulted in erroneous data. In general, errors from this source can usually be neglected; however, one should remember that the introduction of radioactivity into biological material both in vivo and in vitro, for example into cells, may affect the viability of the cellular structures. Special problems may arise in the labeling of macromolecules (polynucleotides, proteins, etc.) with two different radioisotopes or with the same isotope but at different times and at different specific activity levels. Each isotope will contribute differently to the self-decomposition of the macromolecules. For example, in the use of [^{14}C]- and [^{3}H]-thymidine for double labeling of the DNA in cells, radiation effects may cause the surviving cell population to contain more ^{14}C because of the greater rate of self-decomposition of the [^{3}H]-DNA [91]. The proceedings of a conference on the biological effects of transmutation and decay of incorporated radioisotopes has been published [113].

Many labeled compounds, especially those labeled with ^{125}I, ^{75}Se, and ^{3}H, are used in competitive protein binding assays and in radioimmunoassays. Labeling is often nonisotopic (see Section VI D 3) and it is difficult to define purity in the strict sense. Stability of such compounds can only be tested in the experimental system (see Chapter 6).

Another type of stability is the unexpected loss of the labeled atoms from molecules. This is more likely to occur with tritium compounds than with those compounds labeled with ^{14}C for example. Tritiated compounds are used for tracing carbon structures where the carbon-tritium bond may unexpectedly be broken and the tritium atoms replaced (exchanged) for hydrogen atoms. Some examples are, the nonenzymatic exchange of tritium from [5-^{3}H]-deoxyuridylate in the thymidylate synthetase assay caused by the presence of 2-mercaptoethanol [114], and the unexpected biolability [12] of the tritium atoms in some tritiated bile acids used in vivo in studies of pool sizes and turnover [115]. The latter example also illustrates the need to know the specificity of labeling in tritiated compounds and especially those used in metabolism (catabolism) studies. The stability of tritium atoms in molecules under various experimental conditions was discussed by Evans [12].

In concluding this chapter, readers should remember that the study of self-decomposition of radiochemicals, the control and effects of this problem, are continually under investigation by commercial manufacturers, as well as by users. As each new radiochemical is prepared so its stability has to be studied under various conditions of storage which must obviously include the control of decomposition under transit conditions to the user.

Most commercial suppliers now issue technical specifications or data sheets with each batch of their labeled compounds, which normally lists their recommendations for controlling the self-decomposition of the compound on storage. Sometimes it is even possible to advise the investigator of the nature and the structure of impurities likely to be present or known (quantified) to be present. The methods which have been described in this chapter for controlling the instability of radiochemicals, and the data for any particular radiochemical, may therefore change as new evidence becomes available.

During the past few years several reports have described various technical problems encountered by research investigators with purchased radiochemicals from commercial suppliers. Many of these problems have generally related to the purity, analysis and instability of the compounds reflecting technical problems in the user's laboratory as well as in the manufacturer's laboratory, and were reviewed by Merrill and Lewis [116]. The problems were remarkably few relative to the large number of compounds used, and the wide and diversified uses of radiochemicals generally. No supplier can claim to be omniscient and honesty does not imply infallibility, so both the supplier and the user of radiotracer compounds have a common goal and sense of responsibility in communicating effectively with one another to achieve scientific integrity, as indeed observed by Catch [117].

REFERENCES

1. B. M. Tolbert, P. T. Adams, E. L. Bennett, A. M. Hughes, M. R. Kirk, R. M. Lemmon, R. M. Noller, R. Ostwald, and M. Calvin, J. Am. Chem. Soc., 75:1867 (1953).
2. R. J. Bayly and H. Weigel, Nature, 188:384 (1960).
3. J. R. Catch, Patterns of Labelling, Review No. 11. The Radiochemical Centre, Amersham, Bucks., England, 1971.

4. R. L. Wolfgang, R. C. Anderson, and R. W. Dodson, J. Chem. Phys., 24:16 (1956).

5. V. D. Nefedoc, K. N. Skorobogatov, K. Novak, G. Pluchennik and Yu. K. Gusev, Zh. Obsch. Khim., 33:339 (1963).

6. R. E. Krisch and M. R. Zelle, Adv. Radiat. Biol., 3:177 (1969).

7. G. Smith, Pharm. J., 194:219 (1965).

8. P. Rochlin, Chem. Rev., 65:685 (1965).

9. H. Tanovka, Radiat. Res., 21:26 (1964).

10. R. J. Bayly and S. Shrimpton, Unpublished experiments at The Radiochemical Centre, Amersham, England, 1968.

11. G. Sheppard, Atomic Energy Rev., 10:3 (1972).

12. E. A. Evans, in Tritium and Its Compounds, 2nd ed., Butterworths, London, 1974, Chap. 6.

12a. G. Sheppard, The Radiochromatography of Labelled Compounds, Review No. 14, The Radiochemical Centre Ltd., Amersham, Bucks., England, 1972.

13. A. Guarino, R. Pizzella, and E. Possagno, J. Label. Comp., 4:147 (1968).

14. M. Colosimo and A. Guarino, J. Label. Comp., 8:257 (1972).

15. A. J. Swallow, Radiation Chemistry of Organic Compounds, Pergamon Press, Oxford, 1960.

16. R. O. Bolt and J. G. Carroll (eds.), Radiation Effects on Organic Materials, Academic Press, New York, 1963.

17. M. Ebert, J. P. Keene, A. J. Swallow, and J. H. Baxendale (eds.), Pulse Radiolysis, Academic Press, London, 1965, 319 pp.

18. C. A. Wachtmeister, B. Pring, S. Osterman, and L. Ehrenberg, Acta Chem. Scand., 20:908 (1966).

19. A. J. Swallow, in Radiation Chemistry of Organic Compounds, Pergamon Press, Oxford, 1960, p. 124.

20. A. Ekstrom and J. L. Garnett, J. Label. Comp., 3:167 (1967).

21. J. Kroh and J. Mayer, Bull. Acad. Pol. Sci. Ser. Sci. Chim., 16:377 (1968).

22. L. E. Geller and N. Silberman, J. Label. Comp., 5:66 (1969).

23. P. Osinski, Influence of Molecular Structure on the Decomposition and Self-radiolysis of Steroids in Solution, Euratom Report EUR 2435 f, 1965.

24. E. A. Evans and R. F. Phillips, J. Label. Comp., 5:12 (1969).

25. A. J. Swallow, in Radiation Chemistry of Organic Compounds, Pergamon Press, Oxford, 1960, p. 141.

26. A. I. Frankel and A. V. Nalbandov, Steroids, 8:749 (1966).

27. L. E. Geller and N. Silberman, Steroids, 9:157 (1967).

28. J. K. Thomas, Adv. Radiat. Chem., 1:103 (1969).

29. E. Collinson and A. J. Swallow, Quart. Rev. Chem. Soc. (Lond.), 9:311 (1955).

30. B. Waldeck, J. Pharm. Pharmacol., 23:64 (1971).

31. A. Apelgot, B. Ekert, and A. Bouyat, J. Chim. Phys., 60:505 (1963); cf. G. A. Infante, E. J. Fendler, and J. H. Fendler, Radiat. Res. Rev., 4:301 (1973).

32. M. Anbar and P. Neta, Int. J. Appl. Radiat. Isotopes, 18:493 (1967).

33. S. H. Omang and O. D. Vellar, Clin. Chim. Acta, 49:125 (1973).

34. E. A. Evans and F. G. Stanford, Nature, 199:762 (1963).

35. S. Apelgot and M. Frilley, J. Chim. Phys., 62:838 (1965).

36. S. Apelgot, B. Ekert, and M. R. Tisne, Proc. Conf. Methods Preparing Storing Marked Molecules, Brussels 1963. European Atomic Energy Community, Euratom Report EUR 1625 e, 1964, p. 939.

37. S. Apelgot and B. Ekert, J. Chim. Phys., 62:845 (1965).

38. Guide for Users of Labelled Compounds, The Radiochemical Centre, Amersham, Bucks., England, 1972.

39. R. J. Bayly and E. A. Evans, Storage and Stability of Compounds Labelled with Radioisotopes Review No. 7, The Radiochemical Centre Ltd., Amersham, Bucks., England, 1968.

40. E. A. Evans, Nature, 209:169 (1966).

41. P. A. Straat, P. O. P. Ts'o, and F. J. Bollum, J. Biol. Chem., 243:5000 (1968).

41a. A. Perin, G. Scalabrino, A. Sessa, and A. Arnaboldi, Biochim. Biophys. Acta, 366:101 (1974).

42. R. Monks, K. G. Oldham, and K. C. Tovey, Labelled Nucleotides in Biochemistry Review No. 12, The Radiochemical Centre, Ltd., Amersham, Bucks., England, 1971.

43. R. J. Bayly and E. A. Evans, J. Label. Compounds, 2:1 (1966).

44. R. J. Bayly and E. A. Evans, J. Label. Compounds, 3(suppl.): 349 (1967).

45. R. J. Bayly and E. A. Evans, Radioisotopes (Tokyo), 16:293 (1967).

46. W. S. Chou, L. Kesner, and H. Ghadimi, Analyt. Biochem., 37: 276 (1970).

47. E. J. Bourne, D. H. Hutson, and H. Weigel, J. Chem. Soc., 5153 (1960).

48. L. Pritasil, J. Filip, J. Ekl, and Z. Nejedly, Radioisotopy (Czech.), 10:525 (1969).

49. B. Capon, Chem. Rev., 69:449 (1969).

50. M. Tetas and J. M. Lowenstein, Biochemistry, 2:350 (1963).

51. Z. Nejedly, J. Ekl, K. Hybs, and J. Felip, J. Label. Compounds, 5:320 (1969)

52. L. F. LeLoir and A. C. Paladini, Methods Enzymol., 3:968 (1967).

53. A. C. Paladini and L. F. LeLoir, Biochem. J., 51:426 (1952).

54. K. C. Tovey and R. M. Roberts, J. Chromatog., 47:287 (1970).

55. The Radiochemical Centre Ltd., Amersham, England, Unpublished observations.

56. R. O. Lindblom, R. M. Lemmon and M. Calvin, J. Am. Chem. Soc., 83:2484 (1961)

57. R. M. Lemmon, M. A. Parsons, and D. M. Chin, J. Am. Chem. Soc., 77:4139 (1955).

58. W. Y. Cobb, J. Label. Compounds, 5:378 (1969).

59. M. F. Barakat and A. N. Farag, Arab Republic of Egypt, Atomic Energy Establishment Report AREAEE/Rep. 156, Scientific Information Division, Cairo, 1972.

60. E. A. Evans, in Tritium and Its Compounds, 2nd ed., Butterworths, London, 1974, Chap. 2.

61. K. C. Tovey, K. G. Oldham, and J. A. M. Whelan, Clin. Chim. Acta, 56:221 (1974).

62. R. B. Cundall and P. A. Griffiths, J. Am. Chem. Soc., 85:1211 (1963).

63. J. M. Nosworthy, Trans. Faraday Soc., 61:1138 (1965).

64. W. R. Waterfield, J. A. Spanner, and F. G. Stanford, Nature, 218:472 (1968).

65. C. L. Greenstock, J. W. Hunt, and M. Ng, Trans. Faraday Soc., 65:3279 (1969).

66. B. S. Hahn and S. Y. Wang, Biochem. Biophys. Res. Commun., 54:1224 (1973).

67. J. Cadet and R. Teoule, J. Chromatog., 76:407 (1973).

68. E. A. Evans, H. C. Sheppard, and J. C. Turner, J. Label. Comp., 6:76 (1970).

69. J. Bloxsidge, J. A. Elvidge, J. R. Jones and E. A. Evans, Org. Magnetic Res., 3:127 (1971).

70. G. Sheppard, H. C. Sheppard, and J. F. Stivala, Carrier Free (1) 1974, The Radiochemical Centre Ltd., Amersham, Bucks., England.

71. P. F. Smith, The Biology of Mycoplasmas, Academic Press, London and New York, 1971.

72. G. C. Lavelle, Science, 186:870 (1974).

73. R. E. Christopher, The Radiochemical Centre Ltd., Amersham, England. Unpublished results.

74. P. Shanley and R. L. Collin, Radiat. Res., 16:674 (1962).

75. J. Kolousek, J. Liebster, and A. Babicky, Nature, 179:521 (1957).

76. J. Kopoldova, J. Kolousek, A. Babicky, and J. Liebster, Nature, 182:1074 (1958).

77. A. Ohara, J. Radiat. Res., 7:18 (1966).

78. V. E. M. Chambers, R. Tudor, and A. L. M. Riley, J. Steroid Biochem., 5:298 (1974). Abstract of paper presented at the 4th Int. Congr. on Hormonal Steroids, Mexico City, 2-7 Sept., 1974.

79. 'Cortipac' - Cortisol Assay Kit, A product of The Radiochemical Centre, Ltd., Amersham, Bucks., England.

80. E. L. Smith, Lancet, 1:387 (1959).

81. C. Rosenblum, in Vitamin B_{12} and intrinsic Factor (H. L. Heinrich, ed.), 2nd European Symp., Hamburg, 1961. Enke, Stuttgart, 1962, p. 294 et seq.

82. F. G. Stanford, The Radiochemical Centre Ltd., Amersham, Bucks., England, Report No. 354, Feb. 1974, 18 pp.

83. S. Kato, K. Kurata, and Y. Sugisawa, Yakugaku Zasshi., 85:935 (1965).

84. Z. M. Potapova, V. T. Kharlamov, A. K. Pikaev, and A. F. Gus'kov, Khim. Vysokikh. Energii, 7:509 (1973).

85. A. Sorantin, Radiochem. Radioanalyt. Lett., 17:21 (1974).

86. A. Appleby, The Radiochemical Centre Ltd., Amersham, Bucks., England, Report No. RCC-R209, 1967, 25 pp.

87. G. M. Cree, J. Organometal. Chem., 27:1 (1971).

88. J. Cifka and J. Burianek, J. Label. Compounds, 4:107 (1968).

89. S. Person and M. H. Sclair, Radiat. Res., 33:66 (1968).

90. L. Ledoux, C. Dvaila, R. Huart and P. Charles, Proc. Conf. on Methods of Preparing and Storing Marked Molecules, Brussels, November 1963. European Atomic Energy Community, Euratom Rep. EUR 1625e, 1964, p. 1123.

91. C. Paoletti, European Atomic Energy Community, Euratom Rep. EUR 3628f, 1967, 14 pp.

92. L. Dimitrijevie and P. Launay, J. Label. Compounds, 2:411 (1966).

93. A. B. Robins, Nature, Lond., 215:1291 (1967).

94. M. Wand, E. Zeuthen, and E. A. Evans, Science, 157:436 (1967).

95. I. M. Diab and L. J. Roth, Stain Technol., 45:285 (1970).

96. M. Castagna, Biochim. Biophys. Acta, 138:598 (1967).

97. E. A. Evans, J. Microsc., 96:165 (1972).

98. E. A. Evans, Methods Cell Biol., 10:291 (1975).

99. H. Herrmann, Anal. Biochem., 59:293 (1974).

100. K. G. Oldham, Anal. Biochem., 44:143 (1971).

101. E. J. Merrill and G. C. Vernice, J. Label. Compounds, 9:769 (1973).

102. J. L. Johnson, Brain Res., 77:513 (1974).

103. J. C. Waterlow and J. M. L. Stephen, Clin. Sci., 33:489 (1967).

104. J. C. Waterlow and J. M. L. Stephen, Clin. Sci., 35:287 (1968).

105. L. L. Iversen, B. Jarrott, and M. A. Simmonds, Brit. J. Pharmacol., 43:845 (1971).

106. C. Flood, G. Pincus, J. F. Tait, S. A. S. Tait, and S. Willoughby, J. Clin. Invest., 46:717 (1967).

107. S. Dutta, S. Goswami, D. K. Datta, J. O. Lindower, and B. H. Marks, J. Pharm. Exp. Therap., 164:10 (1968).

108. S. Dutta, B. H. Marks and C. R. Smith, J. Pharm. Exp. Therap., 142:223 (1963).

109. J. Boyer, J. Arnaud-Le Petit, and M. Charbonnier, Biochim. Biophys. Acta, 239:353 (1971).

110. K. G. Oldham, Int. J. Appl. Radiat. Isotopes, 21:421 (1970).

111. E. A. Evans, in Tritium and Its Compounds, 2nd ed., Butterworths, London, 1974, Chap. 5.

112. J. M. A. Al-Rawi, J. P. Bloxsidge, C. O'Brien, D. E. Caddy, J. A. Elvidge, J. R. Jones, and E. A. Evans, J. Chem. Soc. Perkin II, 1635 (1974).

112a. J. M. A. Al-Rawi, J. P. Bloxsidge, J. A. Elvidge, J. R. Jones, V. E. M. Chambers, Y. M. A. Chamber, and E. A. Evans, Steroids, 28:359 (1976).

113. Proc. Symp. Biological Effects of Transmutation and Decay of Incorporated Radioisotopes, IAEA, Vienna, 1968.

114. M. Kawai and B. L. Hillcoat, Anal. Biochem., 58:404 (1974).

115. D. K. Panveliwalla, D. Pertsemlidis, and E. H. Ahrens, J. Lipid Res., 15:530 (1974).

116. E. J. Merrill and A. D. Lewis, Anal. Chem., 46:1114 (1974).

117. J. R. Catch, J. Label. Compounds, 9:737 (1973).

118. J. C. Turner, J. Label. Compounds, 3:217 (1967).

119. I. J. Christensen, P. Olsen Larsen, and B. L. Møller, Anal. Biochem., 60:531 (1974).

120. K. C. Tovey, G. H. Spiller, K. G. Oldham, N. Lucas, and N. Carr, Biochem. J., 142:47 (1974).

Chapter 8

BEHAVIOR OF CARRIER-FREE RADIONUCLIDES

František Kepák

Department of Radioactive Wastes
Nuclear Research Institute
Řež, Czechoslovakia

I. INTRODUCTION

In conventional chemistry we usually deal with solutions in concentrations exceeding 10^{-5} M. In radiochemical and radiotracer experiments, we work with radionuclide concentrations several orders of 10 lower, e.g., 1 mCi/liter of $^{199}Au^{3+}$ corresponds to 10^{-11} M solution. A radionuclide without artificial isotopic dilution can be considered to be in the carrier-free state. The radionuclide usually accompanies a microamount of carrier which may be isotopic, chemically very similar, or isomorphous. The exact concentration of the radionuclide and carrier is not always known. The extremely low concentration of the radionuclide may affect reaction rates, equilibrium, and the reproducibility of the process studied. The behavior of carrier-free radionuclides or of those accompanied by minute amounts of isotopic carrier is often not a simple extrapolation from that of ordinary concentration. The presence of nonradioactive impurities in small amounts can usually be disregarded at ordinary concentrations but not at low concentrations of carrier-free radionuclides, where they may be coprecipitated with those impurities or be adsorbed on them or on the vessel and apparatus walls [1-7]. Besides these findings, colloidal behavior of carrier-free radionuclides was observed [1-7]. It is necessary to say that adsorption on walls and impurities is not negligible even when isotopic carrier is present at higher concentration, e.g., 10^{-7}-10^{-6} mol/liter and not only when radionuclide is carrier-free or accompanied by a minute amount of carrier [1,3,6,7]. Studies on all these phenomena are still in a stage of accumulation of experimental results, qualitative analysis, and interpretation. Theoretical and experimental results appearing in the literature often are largely scattered and contradictory, which will be revealed in this chapter. It is rather easy to assume a plausible hypothesis for a peculiar phenomenon. However, it is extremely difficult to establish the significant condition to support adequately the postulation that excludes other possibilities. Negation of a plausible positive explanation is much harder.

II. SOME THERMODYNAMICAL ASPECTS

As mentioned previously the concentration of carrier-free radionuclide may be extremely low, and we must see if it fulfills the statistical conditions for equilibrium based on the Maxwell-Boltzmann distribution law [8]. Statistical conditions are satisfied and fluctuations are very small when at least 10^6 molecules are present in 1 liter of a system, the value corresponding to 10^{-16} g of the substance if its molecular weight is 100 [9,10]. This condition is possibly acceptable in the majority of systems containing carrier-free radionuclides, since their concentration is not usually lower than 10^{-14} mol/liter [3]. Establishment of equilibrium distribution in these systems can also be exemplified by negligible fluctuations in a 10^{12} molecules/cm^3-dense gas having a volume of 2 cm^3, for which the probability that the number of molecules in one-half of the whole volume is higher by 0.001 than in the other half, is only -10^6 [11]. Calculations presented in Ref. 10 prove that fluctuation from equilibrium distribution is small even when the number of molecules is by many orders of magnitude lower. In spite of the fact that these concentrations are much lower than the solubility of sparingly soluble electrolytes, there are certain indications [4,5,12-14] that radionuclides such as ThB (^{212}Pb), ^{131}I, and ^{90}Y at very low concentrations behave sometimes as colloids without exceeding the solubility of their sparingly soluble compound. In these cases, however, the concentrations were not low enough to destroy the equilibrium distribution. Under such conditions, colloidal behavior of carrier-free radionuclides may be explained by their adsorption on impurity particles (Fig. 1). In this case the

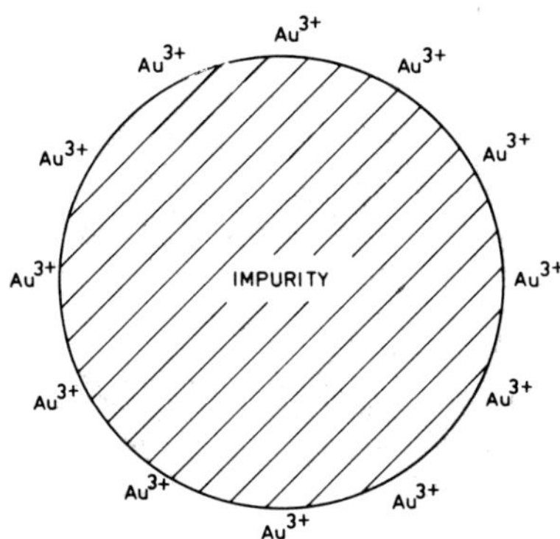

FIG. 1. An impurity carrying radionuclide at its surface. Behavior of such a radiocolloid is governed by the nature of impurity.

observed behavior represents the properties of impurity but not of radionuclide. Besides, there are some phenomena explicable by formation of unstable aggregates composed, e.g., of some hydrolysis products. The possibility of forming such aggregates in unsaturated solutions is reported by Dunning [15] and Walton [16]. Such a possibility increases with concentration and may be large near the saturation [15,16], as revealed in the dependence of self-diffusion coefficient of $^{131}I^-$ on the concentration of coexisting Ag^+ ions (see Fig. 2), one experimental point (open circle) may suggest the formation of $Ag^{131}I^-$ containing unstable aggregates in unsaturated solution below solubility product of AgI. On the other hand, remainders (filled circles) indicate formation of $Ag^{131}I$ particles suspended in saturated solution, i.e., ordinary stable colloidal dispersion.

Fluctuations from equilibrium distribution need not be negligible for an extremely low concentration such as less than

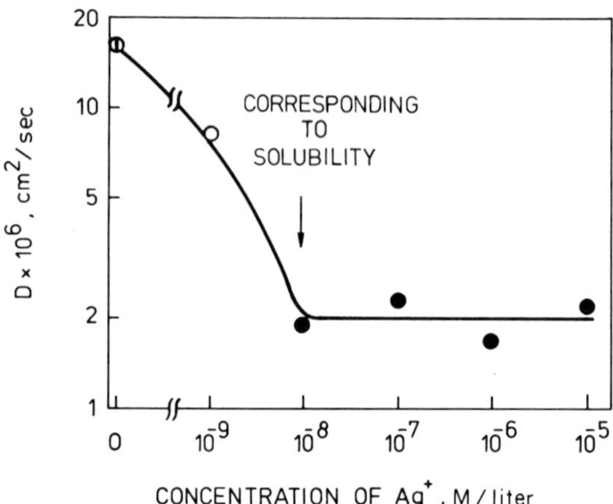

FIG. 2. Dependence of self-diffusion coefficient D of $^{131}I^-$ on the concentration of coexisting Ag^+ at $[I^-]_{const} = 10^{-8}$ M. (◐) ionic dispersion of $^{131}I^-$ and Ag^+; (○) transient aggregates of $Ag^{131}I$; (●) colloidal dispersion of $Ag^{131}I$. (Reproduced from Ref. 13 by permission.)

10^2 molecules/liter [9-11], at which the number of molecules is small enough to disregard the second law of thermodynamics; spontaneous decrease in entropy and fluctuations can occur [11, 17-19]. The lower is the number of molecules; the higher are the fluctuations [11]. In these extremely dilute solutions, fluctuations may lead to the formation of transient aggregates which may exhibit apparently anomalous behavior, such as extraordinarily strong adsorption responsible for an unexpected decrease of radionuclide concentration in the solution. From the thermodynaics aspect, formation of accidental aggregates at such dilute solutions is only possible when entropy accidentally decreases within the limit of reasonable fluctuations of the number of complexions without dissipation of enthalpy. Such transient aggregates formed at one position disintegrate and at the next moment either they will or will not appear at another position as illustrated in Fig. 3. Taking an average, they must be considered to form ionic solution, except the case that there is some perturbation invalidating the averaging procedure. This can be a side reaction giving rise to the formation of other species such as metal hydrolysates or a strong potential field at solid/solution interfaces provided by vessel walls and/or microdust (Fig. 1).

III. THE STATE OF CARRIER-FREE RADIONUCLIDES IN AQUEOUS SOLUTIONS

A. Formation of Radiocolloidal Dispersions

All substances in trace concentrations can exist in solutions as ions, molecules, or colloidal particles [3,6,7]. The last are formed only when the solubility is negligibly small so that the isothermal distillation (growth of larger particles at dissolution of smaller ones; see Refs. 20 and 173 for details) is sufficiently suppressed. In solutions without colloidal particles (called analytical dispersions) partition of ionic and molecular forms of

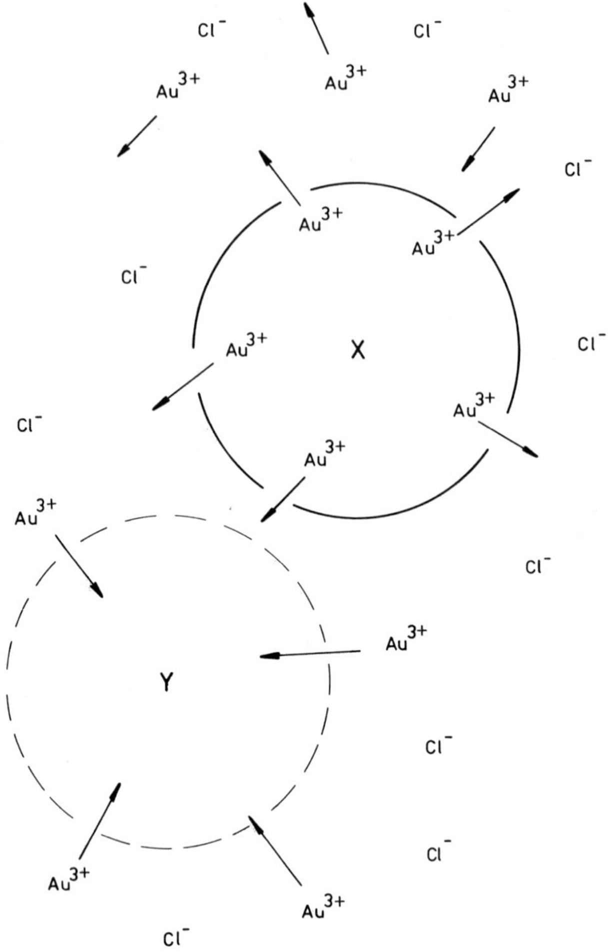

FIG. 3. Radiocolloid of Au^{3+} temporarily formed at position X. At the next moment, it will disappear and either will or will not appear at position Y.

radionuclides depends on pH, redox potential, concentration of the trace element and of complexing substances, ionic strength, as well as on the presence of impurities [3,7]. The oxidation state of carrier-free radionuclides, which can exist in aqueous solution in several stable oxidation states, may be uncertain due to the effect of oxidizing or reducing impurities [1]. At an elevated pH some radionuclides may be hydrolyzed to form colloidal hydroxides [3,6,7]. Very low concentration of radionuclides may cause decreases in rates of the hydrolytic and complexing reactions [3].

Studies on properties of radionuclides have shown that they behave under certain conditions as if they were in a colloidal state [1-7]. To explain these phenomena two assumptions have been proposed [1-7]. The first is based on the postulate that the particles are composed only of insoluble hydrolysate of metal ions in microamounts ("true colloid"). In the second they are formed by adsorption of ionic and molecular radionuclide at the surface of colloidal particles present by chance in microconcentrations. Radiocolloids formed in this manner are called "adsorption colloids" or "pseudocolloids." In the former case the radionuclide concentration must be high enough to surpass the solubility product of the hydroxide [3] whose values are exemplified in Table 1. In such dilute solutions statistical conditions of equilibrium distribution are fulfilled and formation of a solid phase in the solutions is explained by nucleation theory [15,16, 24-26] based on the hypothesis that not only single molecules, but also molecular aggregates forming an embryo which acquires or

TABLE 1

Solubility Products of Some Hydroxides

Hydroxide	$Ru(OH)_3$	$Au(OH)_3$	$Po(OH)_4$	$Pu(OH)_4$	$Zr(OH)_4$	$Ru(OH)_4$	$Nb(OH)_5$
Solubility product	10^{-36}*	10^{-45}**	10^{-37} - 10^{-38}*	7×10^{-56}*	10^{-48} - 10^{-52}*	2×10^{-44}*	10^{-70}*
Reference	21	21	22	23	21	22	22

*Temperature not stated.
**At 22°C.

loses a single molecule are present in supersaturated solutions. Attraction of single molecules by the embryo prevails, and a critical nucleus is formed in transient equilibrium with the solution. When solution is sufficiently supersaturated the critical nuclei are likely to grow to macroscopic dimensions [15,16,24-29]. At an optimum value of low concentration nucleation does not necessarily lead to the formation of a macroscopic solid phase, but probably of small colloidal aggregates [16,27,28], as it is probably in the case of radiocolloid dispersions [3,6,7]. The presence of various ions as impurities is always unavoidable [3, 26,30] and it is likely that they also initiate the formation of radiocolloidal dispersion through heterogeneous nucleation [15, 16,25,26].

Critical nuclei of sparingly soluble electrolytes, with a probability of growing to macroscopic dimensions, can be comprised of several tens or more than a hundred, of countercharged ions [25,31,32]; e.g., critical nucleus of barium sulfate probably contains 33 molecules of $BaSO_4$, that of strontium sulfate 63 molecules, and that of thallium bromide 165 molecules [32]. The radius of such a critical nucleus is believed to be about 1-5 nm [26,32] (see p. 463).

Radiocolloidal dispersion is said to contain about 10-100 molecules of yttrium hydroxide per aggregate in an yttrium solution [33] and less than 100 molecules of polonium hydroxide in polonium solution [34]. Comparison of these data with the number of molecules in a critical nucleus suggests that the size of radiocolloidal particles may be small enough to compare with that of critical nuclei. These small aggregates can be assumed to be "true colloids" [3,7,34,35]. Similar kinds of "true colloids" have been presumed for polonium [3,36,37], thorium [3,36], americium [3,37], plutonium [3], europium [35], promethium [35], cerium [35], bismuth [3], niobium [3], zirconium [3,38], and some others in carrier-free state [3,5,7].

The smallest size of thermodynamically distinguishable colloidal particles would be 1 nm, which corresponds to the dimension of the smallest possible aggregate with one inner atom surrounded by 12 outer, with assumption of spherical atoms in close contact with each other [28]. Some radiocolloids may approach this size, as has been determined by self-diffusion [33, 35,39,40] and centrifugation [41] of radiocolloidal dispersions. These, however, are shown to be polydispersed [3,34,36,42,43], containing larger particles of radius of about 10 nm [3,36,42,43].

Starik [37,44,45] insists that the radiocolloid is "true" when conditions for maximum adsorption of a radionuclide on glass surface differ from those for maximum formation of radiocolloid determined by another method. This was the case of polonium, for which the pH for maximum adsorption on glass did not agree with the pH for maximum formation of colloidal particles [37,44,45]. On the other hand, another report [46] claims that silicic acid from vessel walls is in solution in monomeric, not colloidal, form. It is thus possible that the silicate anion may in some cases form an insoluble compound with the hydrolysed radionuclide [47]. Besides hydrated silica the solution may contain also other microprecipitates which could adsorb the radionuclide or affect its adsorption at glass surface [5]. It is reported to be 10^6-10^8 impurity particles/1 ml solution [16].

In addition to dust particles and accidental impurities of colloidal nature, foreign metal ions from added chemical reagent may be present in the solution in low concentration [30]. Such ions at increased pH may yield insoluble colloidal hydroxides on which radionuclide can sorb to form an "adsorption colloid" [1-7]. Colloidal insoluble hydroxide, present even in microamounts may be capable of adsorbing the radionuclide. The adsorption capacity of such a hydroxide for bi- or trivalent cations was found to be about 0.1-1 mmol/g [48-51]. Properties of such "an adsorption colloid" can to some extent or wholly represent the properties of the core substance [52]. Such "adsorption colloids" have been reported for americium [3,37], bismuth [3], protactinium [53], uranium [54,55], ruthenium [56], yttrium [33], polonium [3,57], strontium [33], phosphorus [58], mercury [59], zirconium [60], and others [3,5,7]. Experimental results reveals that the formation of "true" and "adsorption colloids" depends on concentration of radionuclide, pH and properties of impurities present [3,5,7,61]. Unlike "true colloids," "adsorption colloids" can be formed at a concentration much lower than the solubility product of the hydroxides [3,7]. It is said that the particles of such an "adsorption colloid" are generally much larger than those of "true colloid," giving in some cases several tenths of nanometers for particle radius [3,7,36]. The effect of a foreign element added for simulating an impurity on the self-diffusion of a radionuclide reflects mutual interaction resulting in coprecipitation [33] or adsorption [40] of the radionuclide forming "adsorption colloid." For example, the self-diffusion of $^{85}Sr(II)$ was slowed down, probably due to adsorption, in the presence of microamounts of Mn(IV)- or Fe(III)-hydroxides [40]. A similar effect was observed with $^{91}Y(III)$ in the presence of microamounts of Fe(III)-hydroxide, presumably because of coprecipitation of the former hydroxide with the latter [33]. Formation of a trace amount of an $Ag^{131}I$ radiocolloid was deduced from the course of $^{131}I^-$ self-diffusion in the presence of an increasing concentration of Ag^+ ions [13].

Haissinsky [62] and Davydov [63] pointed out that some necessary conditions such as quick establishment of equilibrium between the solid and liquid phases for determining the solubility product of hydroxides are not fulfilled in the solutions of radiocolloids. Their solubility product varies with pH of the solution and period of aging [62-64]. The situation thus becomes too complex for a proper explanation in a simple form unless one considers the radiocolloidal state.

Table 2 gives the pH values for precipitation of hydroxides, pH ranges for the formation of radiocolloidal states and isoelectric points of possible radiocolloidal particles. The formation of a colloidal state at "low pH" may be indication of

TABLE 2

pH Range for Observation of Radiocolloidal States and Isoelectric Points of Possible Radiocolloidal Particles

Radioactive element	pH for onset of hydroxide precipitation	Ref.	pH range of formation of radiocolloidal dispersion	Ref.	pH of isoelectric point	Ref.
Y(III)	6.8	65	5-13	3,33,66	8.4-9.6	66
Ce(III)	7.1	65	5-10	35,47	--	--
Pm(III)	--	--	5-10	3,35	--	--
Eu(III)	6.7	65	5-10	35,67	5.8-8.0	67
Am(III)	--	--	6-11	3	5.5-8.5	3
Ru(IV)	3.3	68	3.5-11	69	--	--
Po(IV)	7	3	2-12	3,70,71	4.6-12.0	3,57
Zr(IV)	2	72	2-12	3,73	2.1-3.2	3,73
Pu(IV)	2	23	2.8-12	3,74,75	3.4;7.5	3,74,75
Nb(IV)	2	3	2-6	3,76	2	76
Pa(V)	--	--	5-14	3,77	--	--
U(VI)	4 (UO_2^{2+})	23,55	3-8	3	7.7	55

the formation of "pseudocolloids." Such a "pH shift" could depend on the nature of impurities.

B. Properties of Radiocolloidal Dispersions

Although radiocolloidal state is mostly attributable to insoluble hydroxides of radioactive cations, there are reports dealing with different kinds of radiocolloids such as $Ag^{131}I$ [13], ^{95}Zr with phenylarsenic acid [78], and ^{137}Cs with tetraphenylborate [79], as well as radiocolloids in nonaqueous media [1,80].

Properties of radiocolloids are often much the same as those of ordinary colloids [5,12,43,81,82]. For example, colloidal properties of UX_1 (^{234}Th) at 10^{-13} mol/liter were analogous in their dependences on pH and various electrolytes to those of Th at 10^{-4} mol/liter concentration [5]. Distribution of centrifugible fraction of radiocolloidal ThC (^{212}Bi) was almost the same as that of colloidal Bi at ordinary concentration [5]. Conditions for coagulation of radiocolloidal zirconium at 10^{-11} mol/liter by sodium chloride were not much different from those of zirconium at 1.5×10^{-6} mol/liter [81]. Ichikawa and Sato [83] reported that there are no differences between ^{155}Eu and nonradioactive europium in the colloidal properties, giving the conclusion that a low level of radioactivity does not affect the physical and chemical properties of radiocolloids. Kepák and Kaňka [56] found that the separable fraction of ^{106}Ru was lower in the presence of bivalent cations than in the presence of univalent ones and that the ruthenium colloid is formed by adsorption on impurities. Similar to ordinary colloids [84], redispersion following secondary recrystalization has also been reported in some radiocolloidal systems [43,45]. In general, however, radiocolloidal particles spontaneously grow to larger units in amount unsufficient to produce sedimentary sludge after the coagulation [56,58,78,85]. In the case of polonium a centrifugation-separable fraction is increased by aging [85]. The results of centrifugation and ultrafiltration of polonium proved that its colloid was comprised by two fractions; coarse particles of an "adsorption colloid" and fine particles of a "true colloid" [36]. Polonium dispersity depended also on pH and the fraction of highest dispersity was mostly in the pH range 8.0-9.3 [71]. Coagulation during aging seems to be superimposed with gradual adsorption of the radiocolloid on vessel walls or on accidental impurities [3,7,34].

With pH increasing beyond the isoelectric point, charge reversal of radiocolloidal hydroxide takes place [50,57,66,67,74, 86]. The change in pH may cause peptization of the radiocolloid, as was observed for ^{155}Eu radiocolloid which peptized to finer particles above pH 7 [83]. Acidification of solution or complexing of radionuclide usually causes disappearance of the radiocolloidal state [3,4,52]. The addition of simple electrolyte depresses the thickness of electric double layer, which affects the electrophoretic mobility, ζ potential, and stability of the colloid [27]. On the other hand, addition of a polyvalent cation to pseudocolloids apparently results in their decreased occurrence as if they were destabilized [56,73]. This, however, has been attributed to exchange reaction between the cation and the pre-adsorbed radionuclide at the solution/impurity interface [56,73].

It is assumed that specific activity does not affect the chemical and physical properties of radiocolloids [1,3,83,87], unless it is so high as to cause the secondary self-radiolysis (Chap. 7). The same is true for external irradiation that causes an increase of self-diffusion coefficient with increasing dose [88,89]. Outer irradiation of a solution containing radionuclides in colloidal state resulted in an increase in the self-diffusion rate (see Table 3), suggesting fragmentation of colloidal particles to smaller ones [88,89]. Such a radiation effect increases with increasing radiation dose [89]. The radiation-induced chemical reactions between ions in diffused layer that stabilizes the radiocolloid and the water radiolysis products may lead to perturbation of existing equilibria between the solution and the colloidal aggregates and induce partial disintegration [88,89]. In the case of ordinary colloids also, internal and external irradiation causes an increase in dispersity [90].

IV. METHODOLOGY OF STUDYING CARRIER-FREE STATE

The study of radionuclides in carrier-free state concerns to a great extent radiocolloid formation and adsorption. The colloidal state of radionuclides has been studied for 10^{-6}-10^{-14} M solutions [3] by such methods as self-diffusion, centrifugation, electrophoresis, electrochemical deposition, dialysis, ultrafiltration, etc. [1,3,7]. Needless to say, adsorption of the radionuclides on the walls of vessels and laboratory equipment must be calibrated. To suppress such an adsorption effect addition of a complexing agent or acidification of the radionuclide solution are often more effective than the presaturation of vessel walls with the radionuclide or with a corresponding nonradioactive element [4,52]. Another unfavorable factor may be the presence of microamounts of impurities which can adsorb radionuclides and affect their behavior. Such impurities are likely to be responsible for the low reproducibility and discrepancies among different reports. In order to reduce these unfavorable effects on phenomena thereby studied, redistilled water and refined chemicals should be used and all experimental procedures should be made in a dust-free atmosphere.

A. Self-diffusion

Differing from diffusion under concentration gradient, self-diffusion provides direct information about mobility of kinetic units without perturbation. It is suitable for the estimation of an average size of ion, molecule, polymer, or solvated product of these species, the state of which depends on the nature of surrounding phase [91,92]. It is obvious that the radiocolloidal dispersion can be studied by means of self-diffusion techniques which are treated mainly in Chapter 12. In addition, many articles [3-5,13,33,35,38-40,42,88,89,91-96] describe detailed techniques for studying diffusion in tube or capillary, through diaphragm and across solution/gel interface. The present subsection deals with extra techniques for radiocolloid studies and the experimental results thereby obtained.

The most serious problem arises from the fact that the concentration of carrier-free radionuclide is so low as to be strikingly decreased by a small amount of adsorption at glass/solution interface when we use both the capillary and diaphragm method. In the case of gel method, the radionuclide is likely bound to the matrix of gel-like material which is usually agar or gelatin. All these effects can be calibrated to some extent by the aid of cell constant obtained with solutions of known diffusion coefficients [1,3]. This calibration, however, is rationalized only when the cell constant is independent of the concentration in the range from carrier-free to adsorption-unaffectable concentration. Influence of adsorption in the capillary method may be eliminated by experimental conditions [13,33, 35,39,40,88,89,95].

Table 3 lists the values of self-diffusion coefficient D of some radionuclides in carrier-free state. The equivalent radius r of radiocolloidal kinetic unit obtained by [1,27,42]

$$r = \left(\frac{D_i}{D}\right)^{2/3} r_i \qquad (1)$$

where D_i and D are the diffusion coefficients of the radionuclide in ionic and radiocolloidal forms, respectively, and r_i is the radius of the unit in ionic state. The r values are tabulated in the fifth column of Table 3.

TABLE 3

Self-Diffusion Coefficients of Radionuclides

Radionuclide	Concentration (C) (mol/liter)	pH	Self-diffusion coefficient $D \times 10^6$ cm^2 sec^{-1}	Radius (r,Å)	Ref.	pH*	$D \times 10^6$ cm sec^{-1}*	Ref.
$^{144}Ce(III)$	10^{-12}	3.0	4.53		39	3.1	7.82	88
$^{144}Ce(III)$	10^{-12}	9.6	0.65	16.5	39	7.2	6.86	88
$^{147}Pm(III)$	10^{-12}	3.0	4.87		39	3.1	6.50	88
$^{147}Pm(III)$	10^{-12}	9.6	1.07	12.5	39	6.6	5.26	88
$^{155}Eu(III)$	10^{-12}	3.0	4.70		39			
$^{155}Eu(III)$	10^{-12}	10.0	1.47	9.6	39			
$^{91}Y(III)$	10^{-8}	3.0	5.61		33			
$^{91}Y(III)$	10^{-8}	8.5	1.42	11.3	33			
$^{95}Zr(III)$	10^{-7}	--	4.84		38			

*Under a γ irradiation of 2.8×10^6 rad absorption by solution [88].

B. Centrifugation

The colloidal state of radionuclides has been studied by centrifugation of angular velocity (radians/sec) up to 65,000 g either at the sedimentation equilibrium or, more conveniently, in a stationary state between the settling particles and the fluid medium [3]. The radius r(cm) of the separating particles is given [41,43] by

$$r^2 = \frac{9\eta \log x_2/x_1}{2\omega^2 t(\rho_p - \rho)} \tag{2}$$

where x_1 and x_2 are the distances of settling particles (cm) from the rotation center before and after, respectively, the centrifugation, ρ_p and ρ are the densities of settling particles and fluid medium (g/cm^3), respectively, t is period of centrifugation (sec) and η is viscosity of fluid medium (poise). For a polydisperse system, the average radius of the actual particles will equal or exceed the value calculated from Eq. (2). Some results obtained by this method are summarized in Table 4. King [100] found that colloidal Pu was formed in a weakly acidic medium, whereas Np did not form a colloid at all. Beneš [59] investigated colloidal behavior of $^{203}Hg(II)$ by centrifugation and came to the conclusion that $^{203}Hg(II)$ formed only an "adsorption colloid." In addition there are some reports dealing with the centrifugation techniques for studying the colloidal behavior of Nb [37], P [58], Zr [60,101], Ag [97,102], and Au [98]. The pH dependences of the separable fraction for ^{155}Eu and ^{203}Hg are presented in Fig. 4.

C. Electrophoresis

The electrophoretic mobility of colloidal particles in aqueous solutions under an electric field can provide information about the particle charge and the isoelectric point [27]. The techniques for the determination of the mobility of colloidal

TABLE 4

Results of Ultracentrifugation and Centrifugation of Radionuclides

Radionuclide	Concentration (C) (mol/liter)	pH	Fraction of centrifugated radionuclide (%)	Conditions of centrifugation*	Ref.
^{111}Ag	10^{-8}	8.0	100	25,000 *g*	97
^{199}Au	$<10^{-8}$	5.0	96	25,000 *g*	98
^{7}Be	10^{-9}	11.8	90	25,000 *g*	99
^{155}Eu	10^{-9}	7.4	98	65,000 *g*	83
^{203}Hg	3×10^{-8}	3.8	58.7	1,300 *g*	59
^{95}Nb	10^{-11}**	2.0	10	2,000 *g*	37,76
Po	Trace	3.0	56	2,500-3,000 rpm	36
^{95}Zr	10^{-10}-10^{-11}	4.0-10.0	95	3,000 rpm	37
^{90}Y	10^{-8}	7.0	100	25,000 *g*	97
Th	Trace	10.0	17.6	2,500-3,000 rpm	36

*g, gravity acceleration.

**Concentration in g-atom/liter.

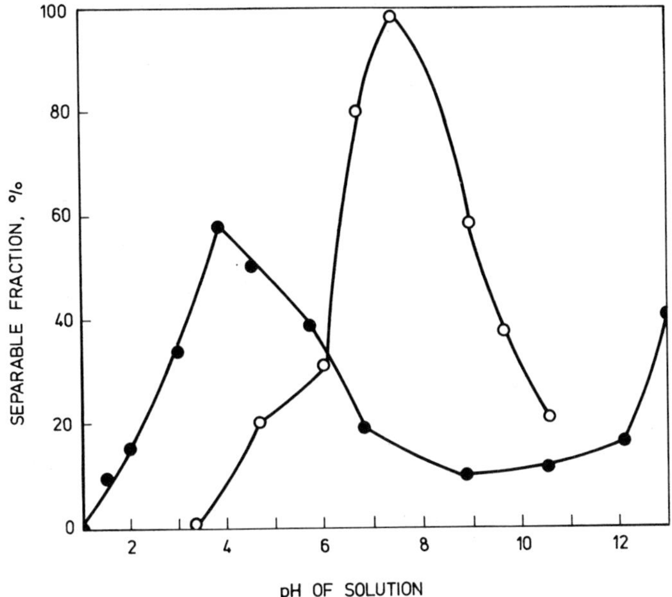

FIG. 4. pH dependences of the fraction of carrier-free $^{155}Eu(III)$ [83] (o) and 3×10^{-8} M $^{203}Hg(II)$ [59] (●) separable by centrifugation. (Reproduced by permission from Refs. 83 and 59.)

particles [3,103,104] are extended to the determination of the isoelectric point and dependence of electrophoretic mobility on pH for aqueous solutions of Y(III) [66], Ru(IV) [105], Po(VI) [22, 57], Eu(III) [67], Am(III) [106], Zr(IV) [73], Mn(II,III) [107], Pa(V) [53,77], U(VI) [54,55], and Pu(IV) [74,75]. In Table 2 the last two columns list the isoelectric points of some radiocolloids obtained from the pH dependence of their electrophoretic mobility. The value of the isoelectric point of hydroxides may be affected besides other factors by the presence of impurities [108] and the differences observed in experimental values for identical radionuclides may be partly explained in this way. The electrophoresis was used for investigation of the interaction between radionuclides and inactive substances [66,109], as well as for separation of radiocolloids from radionuclides in ionic state [110].

^{137}Cs, ^{86}Rb, ^{24}Na, ^{42}K, and ^{131}I were separated by paper electrophoresis from ^{137}Ba, ^{90}Y, and ^{144}Ce. The latter radioelements were transformed to colloids and did not show electrophoretic movement [110].

D. Electrochemical Deposition

The amount of cationic radionuclide deposited on a less noble metal spontaneously or on a cathode electrolytically should be proportional to its concentration in the solution when present in perfectly ionic form. A negative deviation from the proportionality can sometimes be explained by the formation of radiocolloid [3]. A great deal of attention has been paid to electrochemical deposition of polonium on copper electrode [3,57]. In the pH range 8-11, i.e., approximately in the pH range for appearance of colloidal polonium(IV)-hydroxide, the relative amount of spontaneously deposited polonium on a copper electrode decreased. The deposited fraction of polonium under these conditions did not depend at all, or only very slightly, on its concentration in the solution. Similar results were obtained in electrolysis cell with protactinium on gold plated copper electrode [111] and curium on platinum electrode [112].

E. Dialysis

Radiocolloidal behavior of many elements in trace concentrations, such as Fe [113], Mn [107], Au [107], P [58], Pu [4], La [4], and Nb [114] has been studied by equilibrium dialysis. Electrodialysis was used for investigation of the colloidal properties of Po [3]. The decrease of the dialyzable fraction of Mn(II,III) at pH 8.5 proved transition of Mn(II,III) to a colloidal state [107]. The course of dialysis of radioyttrium in the presence of microamounts of iron was similar to that of iron, the result leading to the conclusion that iron affected the state of yttrium [66] (Fig. 5). Some other results are summarized in Table 5 and Fig. 5.

TABLE 5

Fraction of Nondialyzable Radionuclide

Radionuclide	Concentration (C) (mol/liter)	pH	Fraction of nondialyzable radionuclide (%)	Ref.
$^{52,54}Mn$	5×10^{-8}	10-12	90-100	107
^{198}Au	10^{-8}	2	86	107
^{95}Nb	10^{-16}	6.9	100	114

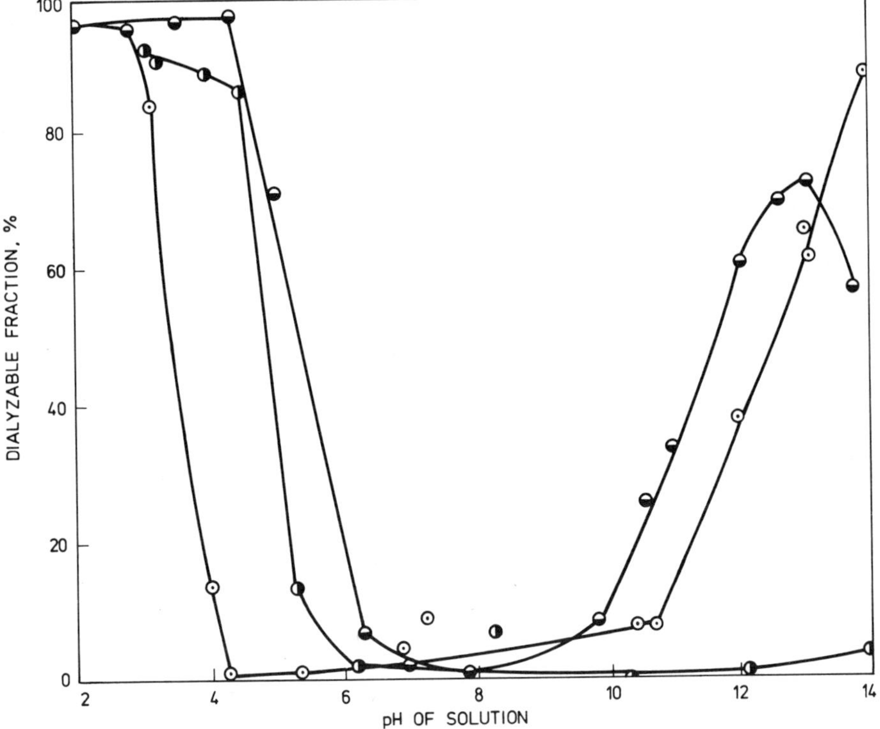

FIG. 5. pH dependence of radionuclide separable by dialysis of 3×10^{-7} M Fe(III) [113] (◒), a mixture of 10^{-9} M ^{88}Y and 3×10^{-7} M Fe(III) [66] (◑), and carrier-free ^{32}P (phosphate) [58] (⊙). (Reproduced by permission from Refs. 66 and 58.)

F. Ultrafiltration

Particle size may be roughly estimated by ultrafiltration through membranes with various pore sizes. The permeation, however, is determined not only by spatial sieve effect, but also by electrostatic interaction between the particle and the membrane [27]. Extension of this technique to the system of radiocolloidal ruthenium led to the conclusion that the separable fraction of the kinetic unit of ruthenium in aqueous solution sharply increased at pH 5 and decreased in alkaline medium [105]. For aqueous polonium solution, a pH range of 5-9 was reported to be the optimum for separation of colloidal polonium [34]. Such examples are summarized in Table 6.

TABLE 6

Fraction of Colloidal Radionuclide Separable by Ultrafiltration

Radionuclide	Concentration (C) (mol/liter)	pH	Separation (%)	Fil. Mat.*	Ref.
^{241}Am	10^{-9}	7-11	98	a	106
^{95}Nb	10^{-10}-10^{-11}	4-6	100	a	37
^{233}Pa	10^{-10}-10^{-11}	5-12	96	a	77
^{210}Po	10^{-10}	8-12	90	a	115
^{210}Po	8.3×10^{-11}	7	100	c	34
^{147}Pm	10^{-10}	9	90	a	116
UO_2^{2+}	10^{-8}	4.5-6.5	98	a	54
Th	Trace	6.9	57	b	36
^{95}Zr	10^{-11}**	2.8-11	90-96	a	73
Po	$\sim 10^{-12}$	7-13	95	a	71

*Filtration material: a, cellophane diaphragm; b, membrane ultrafilter; and c, molecular sieve.

**Concentration in g-atom/liter.

G. Autoradiography

When a radioactive solution is in close contact with photographic emulsion its darkening is, in principle, homogeneous for ionic radionuclides and heterogeneous for colloidal ones. The principle, however, will be invalid if the darkening is caused by chemical or mechanical fogging [3]. Colloidal polonium particles were estimated by this method to consist predominantly of less than 100 atoms of polonium and only a small portion contained 700-2500 atoms [34]. This method was also applied to studying the colloidal state of ^{239}Pu [42], $^{231,233}Pa$ [111], and of ^{155}Eu [117], as well as the nature of the radionuclide adsorbed on solid surfaces [48,118].

H. Isotopic Exchange

The presence of radionuclide in colloidal state can be assumed from its unusual behavior in isotopic exchange (see Chap. 10) in comparison with ionic and molecular radionuclides [1,3,52,119]. For example, isotopic exchange between the radionuclide in the radiocolloid and ionic inactive isotopic carrier is incomplete [1,3,52,119].

I. Adsorption

The adsorption of radionuclides in ionic and colloidal states on ion exchangers [3,4,6,49,120] and on various materials such as glass [37,54,67,69,73,74,101,106,115,117,121-129], silicon [37, 76,130], metals [129,131-136], polyethylene [48,114,117,127,128, 137-139], polystyrene [48,117], filter paper [97-99,102,126,139-144], and animal skin [145-147] has been studied. The study of adsorption mechanism may provide information about the state of carrier-free radionuclides in solution [3,4].

The adsorption of radionuclides on ion exchangers takes place under the conditions of low sorbent saturation [120,148-150]. Radionuclides in the form of ions and hydrolytic products are adsorbed by ion exchange and by chemisorption [3,4,49,120], while radiocolloids by physical adsorption or by retention in sorbent pores [3,4,114,120,151,152]. Relations between the distribution coefficient and such variable parameters as the amount of ion exchanger [148-150,153], pH [150,153,154], the effect of inert [150, 153,155], and complexing [149,153,156] electrolytes and the degree of hydrolysis [153,157-160], were derived for the adsorption of radionuclides present in solution in hydrolytic and colloidal forms. Contrary to the ordinary colloids, radiocolloids may be efficiently concentrated from aqueous phase on ion exchangers [4, 120,161-165]. Colloidal forms of ^{106}Ru(IV), ^{106}RuNO(III) [56,154, 162], ^{144}Ce(III), and ^{147}Pm(III) [50] were efficiently sorbed on Fe(II)-, Fe(III)-, and Mn(IV)-hydroxides, and on dried ferric oxides. An increased adsorption of radiocolloid with increasing

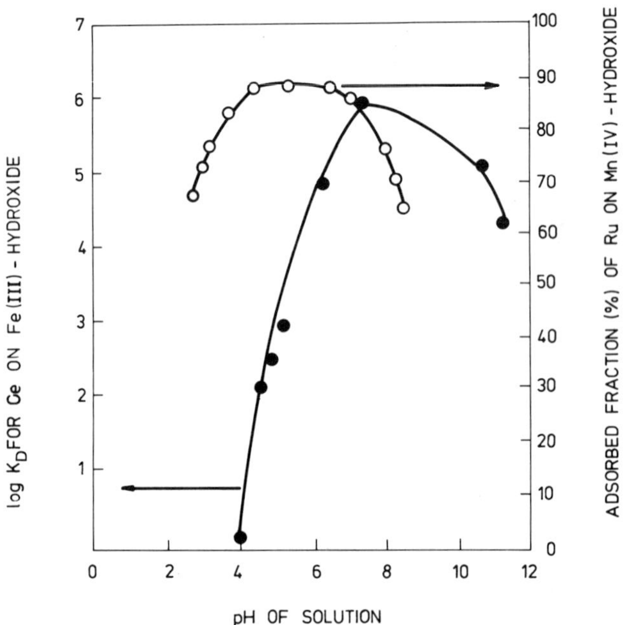

FIG. 6. pH dependences of the distribution coefficient K_d (ml × g^{-1}) for 10^{-8} M ^{144}Ce adsorbed on hydrated ferric oxide [50] (●) and of the fraction of radioruthenium adsorbed on Mn(IV)-hydroxide [153] (○). (Reproduced by permission from Refs. 50 and 154.)

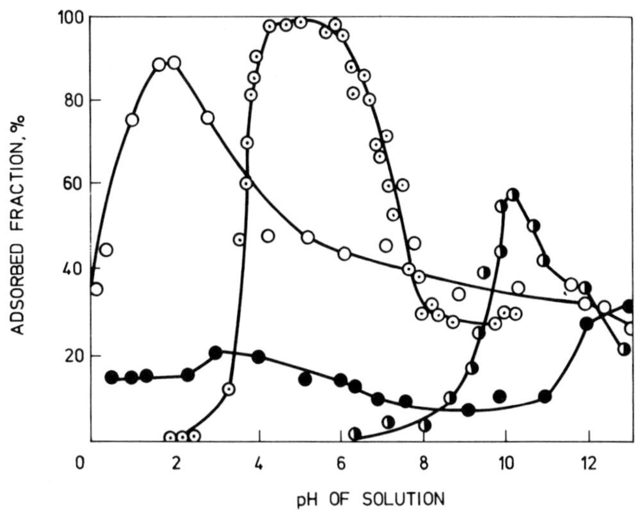

FIG. 7. pH dependence of the fraction of radionuclide adsorbed on glass, polyethylene, and platinum. (○) 10^{-8} M ^{198}Au(III) on polyethylene [137]; (◑) 5×10^{-8} M $^{52(54)}$Mn(II,III) on glass [168]; (⊙) 1.5×10^{-12} M ^{242}Cm on platinum [131]; (●) 3×10^{-8} M ^{203}Hg(II) on glass [125]. (Reproduced by permission from Refs. 137, 168, 131, and 125.)

concentration of inert electrolyte was observed [4,120,155,166, 167]. This was attributed to a coagulation action of the inert electrolyte [120,166,167]. The pH dependence of sorption of radionuclides of cerium and ruthenium on hydrated oxides is shown in Fig. 6. Figure 7 illustrates the pH dependence of radionuclide adsorption at various surfaces.

The adsorption of radionuclides in ionic and molecular forms at various surfaces is explained by ion exchange [3,117,122, 126,127,130,139,140], chemisorption [139], electrostatic attraction, and physical adsorption [127,137,168]. Some examples are summarized in Table 7. The radiocolloidal adsorbate is often firmly bound to the surface and can be desorbed only with concentrated acids [3,52,117,122,137,168]. The extent of the surface contamination may depend on whether the contaminant is in the radiocolloidal state or ionic dispersion [145,147,169]. The contamination of solid surface by radiocolloid is sometimes re-

TABLE 7

Radionuclide Adsorption on Various Solid Surfaces

Radionuclide	Concentration (C) (mol/liter)	pH	Material	Fraction of adsorbed radionuclide (%)	Geometrical surface area (cm^2)	Volume of solution (ml)	Ref.
^{198}Au	10^{-8}	2.4	Glass	46	66	10	122
^{242}Cm	1.5×10^{-12}	5.3	Glass	84	100	100	131
^{60}Co	10^{-8}	9.0	Glass	95	1186	8	123
^{59}Fe	3×10^{-7}	11	Glass	55	66	10	121
$^{52,54}Mn$	$\leq 5 \times 10^{-8}$	10.0	Glass	64	66	10	168
^{241}Am	10^{-9}	6.8	Quartz glass	60*			106
^{95}Nb	10^{-10}-10^{-11}	2.0	Quartz glass	14*			42,101

^{233}Pa	Trace	5.0	Quartz glass	60*			42
Po	5×10^{-13}	5.4	Quartz glass	14*			144
^{95}Zr	10^{-10}-10^{-11}	3.1	Quartz glass	74*			101
^{198}Au	10^{-8}	1.8	Polyethylene	88	74	10	137
^{59}Fe	3×10^{-7}	6.5-9.2	Polyethylene	98-99	74	10	113
^{203}Hg	3×10^{-8}	12.0	Polyethylene	94.8	74	10	138
^{95}Zr	10^{-11}	3.3	Filter paper	80	28	20	140
^{242}Cm	1.5×10^{-12}	4.0-6.0	Platinum	100	100	100	141

*Adsorption per 1 cm^2 of surface from 1 ml of surface from 1 ml of solution.

moved with a solution of such a strong electrolyte as NaCl or KCl rather than detergent or EDTA (see p. 55).

Radionuclides in colloidal state may be sufficiently separated from those in ionic state [14,81,87,130,141,143,152,167]. For example, stronger adsorption of hydrolytic products of radioyttrium on Teflon, glass and metals as compared to that of strontium was used for the separation of radioyttrium from radiostrontium; after the adsorption, the latter was washed out from the surface of these materials, while radioyttrium remained adsorbed [87].

J. Flotation

Flotation, which has been used for separation of pulverized or colloidal material [170], can sometimes provide a means to see whether solute is in ionic (or molecular) or in colloidal state [153,171,172]. A great fraction of ^{106}Ru was thus separated from its solution of pH 5.6 [171], where radiocolloidal state had been presumed [39]. The extent of separation depends on the concentration and pH of the radionuclide solution, as well as on the choice of surface-active substance and collector [153]. By an optimum combination of gelatin for the surface-active substance and ferric hydroxide as a collector, a value of 97-99% separation was achieved for ^{144}Ce, ^{95}Nb, ^{95}Zr, $^{90,91}Y$, and ^{106}Ru [153].

REFERENCES

1. A. C. Wahl and N. A. Bonner, Radioactivity Applied to Chemistry, Wiley, New York, Chapman and Hall, London, 1951, pp. 2, 62, 75, 102-178.
2. M. Haissinski, La Chimie Nucléaire et ses Applications, Masson, Paris, 1957, p. 524.
3. I. E. Starik, Osnovy Radiokhimii, 2nd ed., Izd. Nauka, Leningrad, 1969, pp. 38-293.
4. J. Schubert and E. E. Conn, Nucleonics, 4:2 (1949).
5. G. K. Schweitzer and J. Jackson, J. Chem. Educ., 29:513 (1952).
6. F. Kepák, Chem. Rev., 71:357 (1971).
7. P. Beneš, Chem. Listy, 66:561 (1972).
8. S. Glasstone, Theoretical Chemistry, Van Nostrand, New York, 1960, p. 288.
9. A. A. Benedetti-Pichler and J. R. Rachel, Ind. Eng. Chem., Anal. Ed., 12:233 (1940).
10. J. Mayer and M. Goeppert Mayer, Statisticheskaya Mekhanika, Izd. Innostr. Lit., Moscow, 1952, pp. 79, 94, 99.
11. J. Stauff, Kolloidchemie, Springer-Verlag, Berlin, 1960, p. 116.
12. J. D. Kurbatov and M. H. Kurbatov, J. Phys. Chem., 46:441 (1942).
13. F. Kepák and J. Křivá, J. Inorg. Nucl. Chem., 34:2543 (1972).
14. T. Sasaki, M. Muramatsu, H. Hotta, and Y. Wadachi, Radioisotopes (Tokyo), 7:47 (1958).
15. W. J. Dunning, in Nucleation, (A. G. Zettlemoyer, ed.), Dekker, New York, 1969, pp. 1, 51.
16. A. G. Walton, in Nucleation (A. G. Zettlemoyer, ed.), Dekker, New York, 1969, p. 225.
17. A. Sheludko, Colloid Chemistry, Elsevier, Amsterdam, 1966, pp. 19, 52, 93, 100.
18. H. B. Callen, Thermodynamics, Wiley, New York, 1960, pp. 280, 315.
19. J. C. Slater, Introduction to Chemical Physics, 1st ed., McGraw-Hill, New York, 1939, pp. 32, 101.
20. S. Glasstone and D. Lewis, Elements of Physical Chemistry, Van Nostrand, 1960, p. 366.

21. J. Bjerrum, G. Schwarzenbach, and L. G. Sillén, Stability Constants Part III: Inorganic Ligands, The Chemical Society, London, 1958, p. 1.

22. N. I. Ampelogova, Tr. Tashkentsk. Konf. po Mirnomu Ispol'z. At. Energii, 2:353 (1960).

23. V. M. Vdovenko, Sovremennaya Radiokhimiya, Atomizdat, Moscow, 1964, pp. 296, 336.

24. G. H. Nancollas and N. Purdie, Quart. Rev., 18:1 (1964).

25. A. E. Nielsen, Kinetics of Precipitation, Pergamon Press, Oxford, 1964, pp. 1, 11, 29, 40.

26. A. G. Walton, The Formation and Properties of Precipitates, Wiley-Interscience, New York, 1957, pp. 1, 44, 183, 184.

27. J. Th. G. Overbeek, in Colloid Science, Vol. 1 (H. K. Kruyt, ed.), Elsevier, Amsterdam, 1952, pp. 17, 63, 76, 78, 79, 80, 83, 86, 264, 271, 301, 302.

28. V. Kellö and A. Tkáč, Fyzikálna Chémia (in Slovak), ALFA, Bratislava, 1969, p. 572, 575, 591.

29. V. K. LaMer, Ind. Eng. Chem., 44:1270 (1952).

30. R. E. Thiers, in Trace Analysis (S. J. Hoe and H. J. Koch, eds.), Wiley, New York, 1957, p. 637.

31. H. A. Laitinen, Chemical Analysis, New York, 1960, p. 131.

32. V. A. Garten and R. B. Head, J. C. S. Faraday Trans. I, 69: 514 (1973).

33. F. Kepák, J. Inorg. Nucl. Chem., 35:145 (1973).

34. P. E. Morrow, R. J. Della Rosa, L. J. Casarett, and G. J. Miller, Radiat. Res. Suppl., 5:1 (1964).

35. F. Kepák and J. Křivá, J. Inorg. Nucl. Chem., 33:1741 (1971).

36. A. P. Ratner, N. G. Rozovskaya, and V. Gokhman, Tr. Radievogo Inst. Akad. Nauk SSSR, 5:148 (1957).

37. I. E. Starik, N. I. Ampelogova, Yu. A. Barbanel, F. L. Ginzburg, L. I. Ilmenkova, N. G. Rozovskaya, I. A. Skulskii, and L. D. Sheidina, Radiokhimiya, 9:105 (1967).

38. I. E. Starik, F. L. Ginzburg, and B. N. Raevskii, Radiokhimiya, 6:474 (1964).

39. F. Kepák and J. Křivá, J. Inorg. Nucl. Chem., 32:719 (1970).

40. F. Kepák and J. Křivá, J. Inorg. Nucl. Chem., 34:185 (1972).

41. J. Takagi and H. Shimojima, J. Inorg. Nucl. Chem., 27:405 (1965).

42. D. W. Ockenden and G. H. Welch, J. Chem. Soc., 3358 (1956).

43. F. Kepák and J. Křivá, Radiochim. Acta, 22:60 (1975).

44. I. E. Starik, Tr. Radievogo Inst. Akad. Nauk SSSR, 1:29 (1930).

45. I. E. Starik and L. V. Komlev, Tr. Radievogo Inst. Akad. Nauk SSSR, 2:91 (1933).

46. G. S. Sinitsyna, L. I. Ilmenkova, B. N. Radievskii, and Yu. P. Tarlakov, Radiokhimiya, 9:397 (1967).

47. V. I. Spitsyn, R. N. Barnovskaya, and N. I. Popov, Dokl. Akad. Nauk SSSR, 182:855 (1968).

48. G. E. Melish, J. A. Payne, and G. Worral, Radiochim. Acta, 2:204 (1964).

49. M. J. Fuller, Chromatog. Rev., 14:45 (1971).

50. F. Kepák, M. Nuderová, and J. Kaňka, J. Radioanal. Chem., 14:325 (1973).

51. F. Kepák. R. Caletka, and I. Nová, J. Radioanal. Chem., 25: 247 (1975).

52. A. K. Lavrukhina, T. B. Malysheva, and F. I. Pavlotskaya, Radiokhimicheskii Analiz, Izd. Akad. Nauk SSSR, Moscow, 1963, p. 5.

53. I. E. Starik, L. D. Sheidina, and L. I. Ilmenkova, Radiokhimiya, 1:391 (1959).

54. I. E. Starik and L. B. Kolyagin, Zh. Neorg. Khim., 2:1432 (1957).

55. I. E. Starik, F. E. Starik, and A. N. Apollonova, Zh. Neorg. Khim., 3:121 (1958).

56. F. Kepák and J. Kaňka, Int. J. Appl. Radiation Isotopes, 18: 673 (1967).

57. I. E. Starik and N. I. Ampelogova, Radiokhimiya, 1:414 (1959).

58. P. Beneš, V. Urbanová and E. Vidová, Radiochim. Acta, 17:209 (1972).

59. P. Beneš, J. Inorg. Nucl. Chem., 31:1923 (1969).

60. I. E. Starik and I. A. Skulskii, Radiokhimiya, 1:379 (1959).

61. Yu. P. Davydov, Radiokhimiya, 9:94 (1967).

62. M. Haissinski, Radiokhimiya, 11:479 (1969).

63. Yu. P. Davydov, Radiokhimiya, 9:52 (1967).

64. M. Haissinski, Acta Physicochim. URSS, 3:517 (1935).

65. T. Moeller and H. E. Kremers, Chem. Rev., 37:97 (1965).

66. P. Beneš and J. Kučera, J. Inorg. Nucl. Chem., 33:103 (1971).

67. Yu. P. Davydov, Radiokhimiya, 9:89 (1967).

68. O. E. Zvyagincev, H. I. Kolbin, A. N. Ryabov, T. D. Avtokratova, and A. A. Goryunov, Khimiya Ruteniya, Izd. Nauka, Moscow, 1965, pp. 63, 153.

69. V. P. Khvostova and V. K. Shlenskaya, Izv. Sibir. Otd. Akad. Nauk SSSR, Ser. Khim. Nauk, 4:116 (1970).

70. I. E. Starik and N. I. Ampelogova, Radiokhimiya, 1:419 (1959).

71. I. E. Starik and N. I. Ampelogova, Radiokhimiya, 1:425 (1959).

72. I. M. Kolthoff, P. J. Elving, and E. B. Sandell, Treatise on Analytical Chemistry Part 1: Theory and Practice, Vol. 1, Interscience, New York, 1959, p. 806.

73. I. E. Starik, A. A. Skulskii, and A. I. Yurtov, Radiokhimiya, 1:66 (1959).

74. L. D. Sheidina and E. N. Kovarskaya, Radiokhimiya, 12:253 (1970).

75. V. I. Grebenshchikova and Yu. P. Davydov, Radiokhimiya, 3:155 (1961).

76. I. E. Starik and I. A. Skulskii, Radiokhimiya, 1:77 (1959).

77. I. E. Starik, L. D. Sheidina, and L. I. Ilmenkova, Radiokhimiya, 3:690 (1961).

78. M. Kyrš, P. Selucký, and P. Pištek, Radiochim. Acta, 6:72 (1966).

79. I. A. Skulskii and V. V. Glazunov, Radiokhimiya, 9:602 (1967).

80. L. M. Yurchenko, E. S. Filatov, and A. N. Nesmeyanov, Radiochem. Radioanal. Lett., 11:129 (1972).

81. M. H. Kurbatov and J. D. Kurbatov, J. Chem. Phys., 13:208 (1945).

82. J. Bresler, Radioaktivni Prvky (Czech Transl.), Nakl. ČSAV, Prague, 1959, p. 71.

83. F. Ichikawa and T. Sato, Radiochim. Acta, 6:128 (1966).

84. Z. Ya. Berestneva and V. A. Kargin, Usp. Khim., 24:249 (1955).

85. C. Chamié and M. M. Haissinski, Compt. Rend., 198:1229 (1934).

86. N. D. Betenekov, Yu. V. Egorov, and V. D. Puzako, Radiokhimiya, 16:20 (1974).

87. N. W. Kirby, J. Inorg. Nucl. Chem., 25:483 (1963).

88. F. Kepák and J. Křivá, J. Inorg. Nucl. Chem., 36:220 (1974).

89. F. Kepák, J. Teplý, and J. Křivá, Radiochem. Radioanal. Lett., 21:365 (1975).

90. M. Haissinski, Action Chim. Biol. Radiat., 1971, p. 185.

91. H. J. Tyrrel, Diffusion and Heat Flow in Liquids, Butterworths, London, 1961, p. 118.

92. R. A. Robinson and R. H. Stokes, Electrolyte Solutions, 2nd ed. (R), London, 1959, pp. 12, 314.

93. H. Reinhardt, J. O. Liljenzin, and R. Lindner, Radiochim. Acta, 1:199 (1963).

94. F. Paneth, Kolloid-Z., 13:297 (1913).

95. H. Reinhard, J. O. Liljenzin, and R. Lindner, Radiochim. Acta, 3:215 (1964).

96. R. P. Gupta and G. Prasad, Z. Phys. Chem., N. F., 72:255 (1970).

97. G. K. Schweitzer and W. M. Jackson, J. Am. Chem. Soc., 76:3348 (1954).

98. G. K. Schweitzer and W. N. Bishop, J. Am. Chem. Soc., 75:6330 (1953).

99. G. K. Schweitzer and J. W. Nehl, J. Am. Chem. Soc., 75:4354 (1953).

100. E. L. King, The Transuranium Elements Research Papers, Part 1, 1st ed. (G. T. Seaborg, J. J. Katz, and W. M. Manning eds.), McGraw-Hill, New York, 1949, p. 434.

101. Yu. P. Davydov, Radiokhimiya, 9:84 (1967).

102. G. K. Schweitzer and J. W. Nehl, J. Am. Chem. Soc., 74:6186 (1952).

103. H. W. Hoyer, K. J. Mysels, and D. Stigter, J. Phys. Chem., 58:385 (1954).

104. J. Kučera and P. Beneš, Chem. Listy, 65:644 (1971).

105. I. E. Starik and A. V. Kositsyn, Zh. Neorg. Khim., 2:444 (1957).

106. I. E. Starik and F. L. Ginzburg, Radiokhimiya, 3:685 (1961).

107. P. Beneš, J. Inorg. Nucl. Chem., 29:2889 (1967).

108. G. A. Parks, Chem. Rev., 65:177 (1965).

109. P. Beneš and J. Kučera, J. Inorg. Nucl. Chem., 33:4181 (1971).

110. L. Lafuma, Ch. Sachs, and J. Funck-Brentano, Rev. Fr. Etud. Clin. Biol., 8:700 (1963).

111. M. Sakanoue, T. Takagi, and M. Maeda, Radiochim. Acta, 5:79 (1966).

112. A. G. Samartseva, Radiokhimiya, 4:696 (1962).

113. P. Beneš and J. Smetana, Coll. Czech. Chem. Commun., 34:1360 (1969).

114. V. I. Paramonova and V. B. Kolychev, Zh. Neorg. Khim., 1:1896 (1956).

115. I. E. Starik, N. K. Aleksenko, and N. G. Rozovskaya, Izv. Akad. Nauk SSSR, Otd. Khim. Nauk, 755 (1956).

116. I. E. Starik and M. S. Lambet, Zh. Neorg. Khim., 3:136 (1958).

117. F. Ichikawa and T. Sato, Radiochim. Acta, 12:89 (1969).

118. G. E. Melish and J. A. Paine, Radiochim. Acta, 7:153 (1967).

119. C. E. Crouthamel and R. R. Heinrich, in Treatise on Analytical Chemistry, Part I, Vol. 9 (I. M. Kolthoff, P. J. Elving, and E. B. Sandell, eds.), Wiley-Interscience, New York, 1971, pp. 5474, 5493.

120. F. Kepák, J. Radioanal. Chem., 20:159 (1974).

121. P. Beneš, J. Smetana, and V. Majer, Coll. Czech. Chem. Commun., 33:3410 (1968).

122. P. Beneš, Radiochim. Acta, 3:159 (1964).

123. T. Seimiya, H. Kozai, and T. Sasaki, Bull. Chem. Soc. Japan, 42:2797 (1966).

124. I. E. Starik and F. L. Ginzburg, Radiokhimiya, 1:435 (1959).

125. P. Beneš, Coll. Czech. Chem. Commun., 35:1349 (1970).

126. I. E. Starik, B. S. Kuznetsov, and N. I. Ampelogova, Radiokhimiya, 5:304 (1963).

127. P. Beneš and J. Kučera, Coll. Czech. Chem. Commun., 37:523 (1972).

128. P. Benes and E. Vidová, Coll. Czech. Chem. Commun., 37:2864 (1972).

129. J. Belloni, M. Haissinski, and N. L. Salama, J. Phys. Chem., 63:881 (1959).

130. P. Beneš and A. Riedel, Coll. Czech. Chem. Commun., 32:2547 (1967).

131. A. G. Samartseva, Radiokhimiya, 8:269 (1966).

132. I. E. Starik, F. G. Ginzburg, and L. Sheidina, Radiokhimiya, 8:19 (1964).

133. E. Hercynska, and I. G. Cambell, Z. Phys. Chem., 215:248 (1962).

134. I. E. Starik, L. D. Sheidina, and L. I. Ilmenkova, Radiokhimiya, 4:44 (1962).

135. A. G. Samartseva, Radiokhimiya, 6:230 (1964).

136. A. G. Samartseva, Radiokhimiya, 5:28 (1963).

137. P. Beneš and J. Smetana, Radiochim. Acta, 6:196 (1966).

138. P. Beneš and I. Rajman, Coll. Czech. Chem. Commun., 34:1375 (1969).

139. J. Burclová, J. Prášilová, and P. Beneš, J. Inorg. Nucl. Chem., 35:909 (1973).

140. I. E. Starik, A. P. Ratner, I. A. Skulskii, and K. A. Gavrilov, Zh. Neorg. Khim., 2:1175 (1957).

141. G. K. Schweitzer and W. M. Jackson, J. Am. Chem. Soc., 74:4178 (1952).

142. G. K. Schweitzer, B. R. Stein, and W. M. Jackson, J. Am. Chem. Soc., 75:793 (1953).

143. J. E. Duval and M. H. Kurbatov, J. Am. Chem. Soc., 75:2246 (1953).

144. I. E. Starik, B. S. Kuznetsov, and N. I. Ampelogova, Radiokhimiya, 5:304 (1963).

145. S. Tashiro, Y. Wadachi, and M. Muramatsu, J. At. Energy Soc. Japan, 8:642 (1966).

146. S. Tashiro, Y. Wadachi, and M. Muramatsu, Radioisotopes, 15:224 (1966).

147. S. Tashiro, Y. Wadachi, and M. Muramatsu, J. Nucl. Sci. Technol., 5:160 (1968).

148. Yu. V. Egorov, in Soosazhdenie i Adsorptsiya Radioaktivnykh Elementov (V. M. Vdovenko, ed.), Izd. Nauka, Moscow-Leningrad, 1965, pp. 111, 117.

149. Yu. V. Egorov and V. M. Nikolaev, Radiokhimiya, 7:273 (1965).

150. Yu. V. Egorov, A. S. Lyubimov, V. M. Nikolaev, and B. N. Khrustalev, Izv. Sibir. Otd. Akad. Nauk SSSR, Ser. Khim. Nauk, 3:33 (1965).

151. Yu. P. Davydov, Radiokhimiya, 14:210 (1972).

152. H. Cheng, J. Chim. Soc. (Taipai), 7:154 (1960).

153. V. V. Pushkarev, Yu. V. Egorov, and B. N. Khrustalev, Osvetlenie i Desaktivatsiya Stochnykh vod Pennoi Flotatsiei, Atomizdat, Moscow, 1969, pp. 9, 13, 21, 64, 78, 79.

154. Yu. V. Egorov, A. S. Lyubimov, and B. N. Khrustalev, Radiokhimiya, 7:386 (1965).

155. Yu. V. Egorov, V. N. Nikolaev, and A. S. Lyubimov, Radiokhimiya, 8:8 (1966).

156. Yu. V. Egorov, A. S. Lyubimov, V. M. Nikolaev, and B. N. Khrustalev, Radiokhimiya, 8:397 (1966).

157. Yu. V. Egorov, Radiokhimiya, 13:357 (1971).
158. Yu. V. Egorov, Radiokhimiya, 13:364 (1971).
159. N. D. Benetekov, V. D. Puzako, and Yu. V. Egorov, Radiokhimiya, 13:751 (1971).
160. N. D. Betenekov, V. D. Puzako, Yu. V. Egorov, and A. G. Lisienko, Radiokhimiya, 13:821 (1971).
161. J. A. Ayres, J. Am. Chem. Soc., 69:2879 (1947).
162. F. Kepák and J. Kaňka, Int. J. Appl. Radiation Isotopes, 19:485 (1968).
163. I. E. Starik and F. L. Ginzburg, Radiokhimiya, 3:45 (1961).
164. V. I. Paramonova and S. A. Bartenev, Zh. Neorg. Khim., 3:74 (1958).
165. V. M. Nikolaev, E. I. Krylov V. F. Bagretsov, and Yu. V. Egorov, Radiokhimiya, 5:622 (1963).
166. I. E. Starik and I. A. Skulskii, Izv. Akad. Nauk SSSR, Otdel. Khim. Nauk, 1278 (1958).
167. J. Schubert and J. W. Richter, J. Colloid. Sci., 5:376 (1950).
168. P. Beneš and A. Garba, Radiochim. Acta, 5:99 (1966).
169. Y. Wadachi, S. Tashiro, Y. Inoue, and M. Muramatsu, J. At. Energy Soc. Japan, 7:492 (1965).
170. P. Somasundaran, Separation Purification Meth., 1:117 (1972).
171. F. Kepák and J. Křivá, Separation Sci., 5:385 (1970).
172. F. Kepák and J. Křivá, Separation Sci., 7:433 (1972).
173. D. A. F. Fridrichsberg, Kurs Kolloidnoi Khimii, Izd. Khimia, Leningrad, 1974, pp. 71, 73, 239.

Chapter 9

REACTION KINETICS--MECHANISMS AND ISOTOPE EFFECTS

J. R. Jones

Chemistry Department
University of Surrey
Guildford, England

I. INTRODUCTION

In recent years the use of kinetic isotope effects has become a widely employed and powerful tool in furthering our understanding of reaction kinetics both in the gas phase and in solution. The stage when work was strictly undertaken just to demonstrate the existence of an isotope effect has passed--the emphasis now is on explaining what the magnitude of the effect signifies. This state of affairs represents an intermediate stage, as it is not yet possible to predict with any certainty what the magnitude of, for example, the primary kinetic hydrogen isotope effect k^H/k^D, for a particular reaction under certain conditions of temperature, solvent, base, etc. will be. The same, but only more so, applies to the calculation of reaction rates. For how long this state of affairs will exist is a matter for conjecture.

Most of the early work on isotope effects was undertaken by physical chemists; nowadays chemists interested in bioinorganic and physical-organic chemistry are prepared to undertake isotopic studies in order to supplement information obtained from other experiments. With this increase in interest (it has been estimated [1] that well over 300 papers are published annually on the subject of isotope effects), it becomes impossible within the confines of a single chapter to deal satisfactorily with all the aspects involved. Fortunately there is, apart from Melander's [2] pioneering monograph, another [3] that gives an excellent account of isotope effects in chemical reactions, as well as the proceedings of a symposium [4]. In addition, chapters on primary kinetic hydrogen isotope effects [5,6], solvent isotope effects [7-9], and secondary hydrogen isotope effects [10,11] are available.

In this chapter emphasis is placed on first, how the isotope effects are obtained, and second, the factors that govern their magnitude. Most of the discussion is concerned with hydrogen isotope effects partly because the much bigger mass differences between hydrogen, deuterium, and tritium leads to larger effects than are observed for other isotopes and partly because of the wide occurrence and importance of proton transfer reactions in chemistry and biochemistry. Of the other isotopes much work has been done on carbon, as many organic reactions involve the formation or rupture of carbon-carbon bonds. Less extensive studies have been reported using the isotopes of chlorine (^{35}Cl, ^{37}Cl), sulfur (^{32}S, ^{34}S), nitrogen (^{14}N, ^{15}N), and oxygen (^{16}O, ^{17}O, ^{18}O). Some of the properties of these isotopes are given in Table 1. Many are stable (the radioactive isotopes of nitrogen ^{12}N, ^{13}N, ^{16}N, and ^{17}N all have half-lives less than 10 sec, and the same applies to the oxygen isotopes ^{13}O, ^{14}O, ^{15}O, ^{19}O, ^{20}O), so that an account of isotope effects must concern itself with both varieties.

TABLE 1

Properties of Some Important Isotopes

Element	Isotope	Properties	Natural abundance (%)	Nuclear spin
Hydrogen	Deuterium (^{2}H)	Stable	1.1	1
	Tritium (^{3}H)	Weak β emitter	$<10^{-16}$	$\frac{1}{2}$
		E_{max} = 18.6 keV		
		Half-life = 12.35 years		
Carbon	^{13}C	Stable	1.1	$\frac{1}{2}$
	^{14}C	β emitter		
		E_{max} = 0.156 MeV	0 to 10^{-10}	0
		Half-life = 5730 years		
Nitrogen	^{14}N	Stable	99.6	1
	^{15}N	Stable	0.4	$\frac{1}{2}$
Oxygen	^{16}O	Stable	99.8	0
	^{17}O	Stable	0.037	$\frac{5}{2}$
	^{18}O	Stable	0.2	0

II. EXPERIMENTAL METHODS

For hydrogen the isotope effects fall into three categories. First, there is the primary isotope effect which results from the cleavage of the X-H(D) bond in the rate-determining step of the reaction. Most of the work done refers to carbon-hydrogen bonds and these compounds are usually referred to as carbon acids as distinct from nitrogen acids (e.g., aniline) or oxygen acids (e.g., phenol) where the rates of bond cleavage are usually much faster. In addition there is the secondary isotope effect which arises when the position of isotopic substitution is not directly involved in the bond-making or -breaking part of the reaction. Finally, there is the solvent isotope effect which is brought about by the replacement of H_2O by D_2O or, more rarely, MeOH by MeOD.

Because primary hydrogen isotope effects are usually much higher than those obtained for other isotopes, the methods used to determine k^H/k^D or k^H/k^T are experimentally less demanding. One method, extensively used by Bell and coworkers [12] entails measuring the rates of halogenation of a β-diketone or keto ester (RH):

$$RH + B^- \rightarrow R^- + BH \qquad (1)$$

$$R^- + Br_2 \rightarrow RBr + Br^- \qquad (2)$$

Step (1) is usually rate-determining so that the rate of bromination (or iodination) represents the rate of ionization of the $\geq$C-H bond. The corresponding rate for the deuterium-labeled compound can be followed by the same method. In some cases more than 1 mol of bromine is consumed as, for example, in the case of malononitrile [13] and nitroacetone [14]; usually, however, the first step remains the rate-determining one. In days gone by it was customary to follow the rates of bromination by withdrawing reaction samples at appropriate time intervals, adding to allyl alcohol to remove any excess bromine and then taking advantage of the fact that the bromo compound liberated iodine from solutions of potassium iodide. Nowadays it is far more convenient to follow changes in the characteristic absorption spectrum of bromine, e.g., at 390 nm [15] or 500 nm [12]. Similarly, for studies of iodination changes in the triiodide absorption at 352 nm are most frequently used [16].

The bromination method also makes possible the measurement of both very slow ($k_2 \sim 10^{-8}$ M^{-1} sec^{-1}) and very fast ($k_2 \sim 10^6$ M^{-1} sec^{-1}) reactions. In the first case a large excess of ketone is employed and low concentrations ($\sim 10^{-7}$ M) of bromine measured using the Br_2-Br_3^- redox potential at a platinum electrode [14]. In the second case reactions can be studied by following changes in bromine concentrations through several powers of 10. In circumstances where the method cannot be applied (e.g., when the basic anions react with bromine) a simple alternative may be used [17]--many β-diketones exist as mixtures of keto and enol forms, the exact proportion depending on the solvent. Addition of a solution of the β-diketone to an aqueous buffer system is accompanied by spectral changes that represent the change keto $\rightleftarrows$ enol so that if the equilibrium constant is known the rate of enolization (or ionization) can be determined.

Aliphatic nitro compounds tend to ionize much more slowly than one would expect from their acidities, and in some circumstances the process can be followed by monitoring the appearance of the anion [18]. If the compound is optically active, then the racemization [19] of the hydrogen form and its deuteriated analog provide a most direct way of obtaining k^H/k^D.

The most widely used and generally applicable of methods is that of isotopic exchange--either hydrogen-tritium or hydrogen-deuterium. Since the development of the technique of liquid scintillation counting for detecting weak β emitters, an increasing number of detritiation studies have been reported [20]. Because the tritium is present in trace concentrations, it becomes possible to measure rates of very slow reactions--this is done by the initial rate method [21] in which the rate of increase of radioactivity in the solvent is followed over the first 1-2% of the reaction. In this way the familiar equation for the pseudo-first-order rate constant for detritiation

$$k^T = \frac{1}{t} \log \frac{1}{1 - x} \tag{3}$$

reduces to

$$k^T = \frac{x}{t} \tag{4}$$

The increase in radioactivity with time is therefore linear and the zero-order rate constant so obtained can then be converted to a first-order rate constant, provided the total radioactivity of the substrate in solution is known.

An interesting method introduced by Margolin and Long [22] involves following the uptake of tritium by the protio form and the deuterio form of the substrate from an aqueous solution during the initial zero-order phase of the reaction. The relevant equations in the case of a hydroxide-catalyzed reaction are

$$RH + OH^- \underset{k_{-1}^H}{\overset{k_1^H}{\rightleftharpoons}} R^- + H_2O \tag{5}$$

$$RD + OH^- \underset{k_{-1}^D}{\overset{k_1^D}{\rightleftharpoons}} R^- + HDO \tag{6}$$

$$R^- + HTO \underset{k_{-2}}{\overset{k_2}{\rightleftharpoons}} RT + OH^- \tag{7}$$

It is easily shown that, if the initial concentrations are adjusted such that $[RH]_o = [RD]_o$ and $[OH^-]^H = [OH^-]^D$, then the initial rates of tritium uptake by the two reactants gives directly the primary isotope effect k^H/k^D. The method can also be used for a reaction which is subject to general catalysis.

Providing the tritium concentration is high enough, rates of detritiation can be followed by nuclear magnetic resonance (NMR) spectroscopy [23], although this method is usually employed for studying hydrogen-deuterium exchange reactions. Changes in spectra are measured as a function of time, with a nonexchanging standard being used as a reference. The method, although somewhat lacking in accuracy, is not confined to the hydrogen nuclei. Thus rates of hydrogen and deuterium exchange have been determined [24,25] by ^{19}F NMR, e.g., the fluorinated carbon acid, $C_6H_5CH(CF_3)_2$, exhibits a doublet resonance, and as the deuterium becomes incorporated, a broad singlet resonance appears upfield of the right-hand member of the doublet. Area measurements of both resonances enable the calculation of mole fractions of protio- and deuterio-compound at any time during a kinetic run [25]. Similarly, for molecules having several different acidic hydrogens of similar kinetic properties, and where many of the usual techniques can no longer be applied, the use of ^{13}C NMR to monitor deuterium incorporation provides a satisfactory alternative [26] provided the individual carbon signals are separate and assignable. The method has been employed to follow deuterium incorporation into acenaphthalene, 2,3-dihydrophenalene, and 1,8-dimethylnaphthalene.

Competitive methods of determining isotope effects are usually applied to the heavier elements, but in at least one case [27], namely, proton transfer from carbon acids to organometallic compounds, the method has been successfully used. A large excess of an equimolar mixture of AH and AD was allowed to react with the organometallic compound RM:

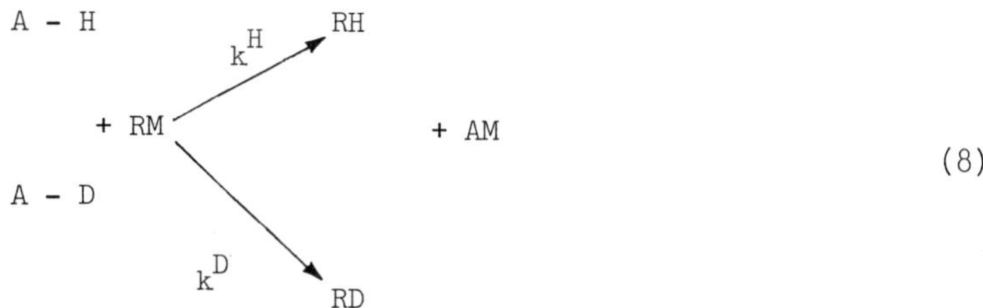

(8)

TABLE 2

Methods Employed for Determining Kinetic Isotope Effects

Element	Method	Comments	Refs.
Hydrogen	Halogenation	Can be used for H and D compounds; very versatile as both slow and fast reactions can be followed	12,13,14, 15,16,17,30
	Tracer	Tritiation into H and D compounds, otherwise detritiation rates must be combined with some other method; very precise but confined to relatively slow reactions	22,20
	NMR	Can be employed in a variety of ways, e.g., to study dedeuteriation, deuteriation, detritiation; line broadening, ^{19}F and ^{13}C methods also; relatively low precision	31,32,33, 20 24,25,26
	Racemization	Confined to optically active H and D compounds	19,34
	Conductance	Suitable for secondary isotope effect determinations in solvolytic and other reactions	10,11
	Differential	Well suited for accurate studies of solvent isotope effects in mixtures of H_2O and D_2O	35,36
Heavy atoms (C, O, N)	Competitive	Extensively used for heavy-atom isotope effects--analysis of product ratios by isotope-ratio mass spectrometer	37-44

The composition of the products in terms of the ratio [RH]/[RD] was then determined and as long as the reaction is irreversible this is equivalent to the isotope effect k^H/k^D. Detailed accounts of the use of competitive methods in carbon and other heavy atom isotope effects have been given [28,29]. The various methods are summarized in Table 2.

III. THEORETICAL ASPECTS

A satisfactory theory of kinetic isotope effects, based on the transition state theory of reaction kinetics, has been available for many years [45]. Although it is usually employed to illustrate various aspects of hydrogen isotope effects, it is equally valid for other isotopes. Several accounts are available [3,5,6] and only the essential features will be given here.

The energy of a molecule A-H is made up of electronic, translational, rotational, and vibrational contributions. Within the limits of the Born-Oppenheimer approximation, the electronic energy is independent of isotopic substitution so that the motion of A-D will take place on the same potential energy surface as that of A-H. In addition, the vibrational contribution is much higher than the sum of the rotational and translational contributions so that it is this first term that is the major factor in the emergence of an isotope effect.

If both AH and AD are made to react with a common base B according to the equation

$$\begin{matrix} \text{A-H} & & & \text{A---H---B} & & \text{A + HB} \\ & + \text{B} & \rightarrow & & \rightarrow & \\ \text{A-D} & & & \text{A---D---B} & & \text{A + DB} \end{matrix} \qquad (9)$$

then, by assuming that the preexponential term in the Arrhenius equation has the same value for both reactions, the primary isotope effect is given by

$$\frac{k^H}{k^D} = \exp \frac{\Delta E_{vib}}{RT} \tag{10}$$

In other words, the magnitude of the isotope effect is governed by what fraction of the vibrational energy is used up in getting over the energy barrier. If we assume that only stretching frequencies are involved, ΔE_{vib} is approximately 1100 cal/mol and k^H/k^D at 25°C would be close to 7. Similarly, if only bending frequencies (which are much lower) have to be taken into account, k^H/k^D under the same circumstances is only 2.4. In the unlikely event of both the stretching and bending vibrational energies being required in order to get over the barrier, k^H/k^D would now be in the region of 18, a value that can be compared [46] to that for $^{12}O/^{14}O$ (1.50), $^{14}N/^{16}N$ (1.25), and $^{16}O/^{18}O$ (1.19). Because the bending and stretching frequencies depend on the atom or group to which hydrogen (or its isotope) is attached, the magnitude of the effect will also vary with the nature of this atom or group. Primary hydrogen isotope effects can therefore vary widely in magnitude, as in many cases considerable energy is retained in the transition state.

This very simplified picture of primary isotope effects is inadequate in many respects. Thus it assumes a 3-center model, makes no reference to the solvent, and does not take into account the need (because the mass of the hydrogen is low and its wavelength comparable to the width of the energy barrier) to discuss its movement in terms of quantum mechanics.

In order to obtain further information from isotope effects, there are two possible approaches. The experimental approach is based on measuring the value of k^H/k^D and the vibrational frequencies of the reactants, so that it is possible to infer something about the vibrational frequencies and configuration of the transition state, as well as the reaction mechanism. The theoretical approach is based on some kind of model and by means of computational analysis these frequencies can be calculated.

The results of theoretical calculations [47-49] show that the primary hydrogen isotope effect k^H/k^D should go through a maximum as the transition state varies from being reactant-like (A-H---B) to product-like (A---H-B). If, as in the symmetric situation (A---H---B) the hydrogen is equally firmly bound to A and to B it will remain motionless and the "symmetrical" stretching frequency will become independent of the isotope. Under these circumstances there will be no vibrational energy in the transition state to offset that in the reactants, and the isotope effect will be at its maximum. This concept is a particularly useful one as it provides a qualitative correlation between the magnitude of the primary hydrogen isotope effect and transition state structure. Furthermore it is no longer confined to hydrogen isotope effects as Fry et al. [50] have recently carried out calculations (Table 3) which support an earlier suggestion [51] that carbon isotope effects in S_N2 displacement reactions should exhibit similar behavior, whereas the isotope effect for the labeled atom X should increase monotonically as the stretching force constant for bond 1, decreases:

$$Y + R\overset{*}{C}H_2\ X \rightarrow [Y\overset{2}{---}R\overset{*}{C}H_2\overset{1}{---}X]^{\ddagger} \rightarrow R\overset{*}{C}H_2Y + X \qquad (11)$$

TABLE 3

Calculated Carbon and Chlorine Isotope Effects for S_N2 Reactions [50]

n_2 (bond order)	0.1	0.3	0.5	0.7	0.9
$^{12}C/^{14}C$	1.04531	1.06907	1.06788	1.04939	1.01689
$^{35}Cl/^{37}Cl$	1.00048	1.00607	1.01265	1.01784	1.02493

Model calculations [52] of another kind show that under certain circumstances the temperature dependence of primary hydrogen isotope effects may be more complicated than usual and result in maxima, minima, or points of inflection. This is expected to occur, however, only when the values are low and when, e.g., the force constant and geometry changes between reactants and transition state at the isotopic position are very small, or when quantum-mechanical tunneling is operative in an inverse isotope effect. In circumstances where the primary isotope effect is large, its dependence on temperature regular, and no tunneling, the Swain-Schaad equation [53] relating k^H/k^D and k^H/k^T,

$$\frac{\log k^H/k^T}{\log k^H/k^D} = 1.442 = r \tag{12}$$

should be well obeyed with the value of r restricted to the range $1.33 \leq r \leq 1.58$, as previously specified by Bigeleisen [54]. Further calculations [55] indicate that the effective lower limit to the preexponential ratio A_H/A_D should be 0.7-0.5.

The same type of temperature-dependent irregularities that have been predicted for low primary kinetic hydrogen isotope effects may also be found for secondary hydrogen isotope effects and primary heavy-atom effects [56,57]. It has also been shown [58] that relative ^{14}C-^{13}C kinetic isotope effects, defined as $r = \log (k_{12}/k_{14})/\log (k_{12}/k_{13})$ should be restricted to the range $1.8 \leq r \leq 2.0$ except where individual ^{14}C and ^{13}C kinetic isotope effects are unusually small and/or associated with temperature-dependent anomalies.

Although the theories of primary kinetic isotope effects apply to all isotopes [59,60], there is one factor, namely, quantum-mechanical tunneling, which is far more important for hydrogen than it is for any other isotope. Although its potential importance was predicted in the late 1930s, differences of opinion still remain as to how to calculate its contribution to the observed rate of a chemical reaction. Wigner [61], Eckart [62],

Pitzer [63], Johnson [64], and Bell [65,66] have all made important contributions to the problem and it seems agreed that:

1. Its importance decreases in the sequence H >> D >> T.
2. The effect will become increasingly important at lower temperatures, and in the case of the hydrogen results, it could be sufficient to cause curvature in the plot of $\log k^H$ vs. $1/T$.
3. Activation energy differences $(E^D_{obs} - E^H_{obs})$ in excess of 1.1 kcal/mol and $A_H/A_D < 0.5$ are indicative of tunneling.
4. The differences $(E^T_{obs} - E^D_{obs})$ and the ratio A_D/A_T should not deviate as seriously from the theoretical limits.

Solvent isotope effects which result from first of all carrying out a reaction in, say H_2O, and then repeating the measurements in D_2O, provide a very sensitive method of ascertaining details of transition state structure and hence reaction mechanism [7-9,67,68]. The effect comes about as a result of differences in solute-solvent interactions and in particular changes in the frequencies and zero-point energies of hydrogen (deuterium)-bonded solute and solvent molecules. It can be considered in terms of an exchange effect which represents the fact that hydrogen and deuterium are not randomly distributed between the various species in solution, and a medium (or transfer) effect that arises from changes in solvation in going from one solvent to another.

The theory of solvent isotope effects is basically the same for an equilibrium process as it is for a kinetic process. If one considers the dissociation of an acid HA in H_2O and then in a H_2O-D_2O mixture it is readily shown that

$$\frac{K_H}{K_n} = \frac{\prod^{\text{Reactants}} (1 - n + n\phi)^r}{\prod^{\text{Products}} (1 - n + n\phi)^p} \tag{13}$$

n refers to the deuterium atom fraction, and ϕ is the fractionation factor for the aqueous hydrogen ion and is in fact the equilibrium constant for the reaction

$$\frac{1}{3} H_3O^+ + \frac{1}{2} D_2O \rightleftharpoons \frac{1}{3} D_3O^+ + \frac{1}{2} H_2O \qquad (14)$$

It has a value of 0.69, whereas that for the aqueous hydroxide ion is 0.42. Each of the terms $(1 - n + n\phi)$ is raised to a power (r for reactants, p for products) corresponding to the number of equivalent hydrogens or deuteriums in the relevant species. For a rate process the ratio of the rate constants (k_n/k_H) is equal to the ratio of the equilibrium constants (K_n/K_H).

In the one case [69] where model calculations of solvent isotope effects have been made they refer to a reaction of the kind

$$H_3O^+ + \gt C = C\lt \rightarrow [H_2O\text{---}H\text{---}C \overset{\text{---}}{=} C] \rightarrow H_2O + H\text{-}\overset{|}{\underset{|}{C}}\text{-}\overset{|}{\underset{|}{C}}^+ \qquad (15)$$

and the results show that the solvent isotope effect decreases monotonically with the degree of proton transfer in the transition state and correlates very well with the calculated primary isotope effect.

IV. RESULTS

The customary way of altering the symmetry of the transition state is to vary the strength of the base B, and there are now several thoroughly investigated studies [70-72] which show the primary kinetic hydrogen isotope effect k^H/k^D passing through a clearly defined maximum in the region where A^- and B are approximately of equal base strength. A recent example [72] of such studies is the general base-catalyzed ionization of phenylnitromethane (Table 4), there being a factor of 2 between the maximum and minimum isotope effects. An alternative, and in many ways preferable, way of

TABLE 4

Primary Kinetic Hydrogen Isotope Effects for the Ionization of Phenylnitromethane [72] (pK ∿ 6.8)

Base	pK(BH)	ΔpK	k^H/k^D
Hydroxide	15.7	-8.9	7.4
Guanidine	13.6	-6.8	6.7
Piperidine	11.1	-4.3	8.5
Carbonate	9.8	-2.5	9.4
Imidazole	6.9	-0.4	11.5
Acetate	4.7	+2.3	8.3
Chloroacetate	2.8	+4.2	7.4
Water	-1.7	+8.5	5.5

varying transition state symmetry is by employing highly basic media. Very large increases in the basicity of an aqueous hydroxide ion solution can be brought about by adding variable amounts of dipolar aprotic solvents such as dimethyl sulfoxide [73]. In this way the nature of the base is kept the same only its solvation sheath varying. The racemization of (-)-menthone [19], the ionization of (+)-methyldeoxybenzoin [74] and also nitroethane [75] have been studied in this way. In the first two cases increasing medium basicity brought about a more reactant-like transition state and the isotope effects passed through a well-defined maximum.

The ionization of (+)-methyldeoxybenzoin has recently been studied in ethanol-ethoxide solutions containing dimethyl sulfoxide [76] and the results (see Fig. 1) show that the variation in isotope effect k^H/k^T, as well as its magnitude, is very much the same as in the corresponding water-hydroxide solutions. However, a markedly different situation exists in the base-catalyzed proton transfer reaction of 4-nitrophenylnitromethane [77]. Here the isotope effect k^H/k^D is markedly dependent on the solvent (Table 5) being particularly large for tetramethylguanidine in

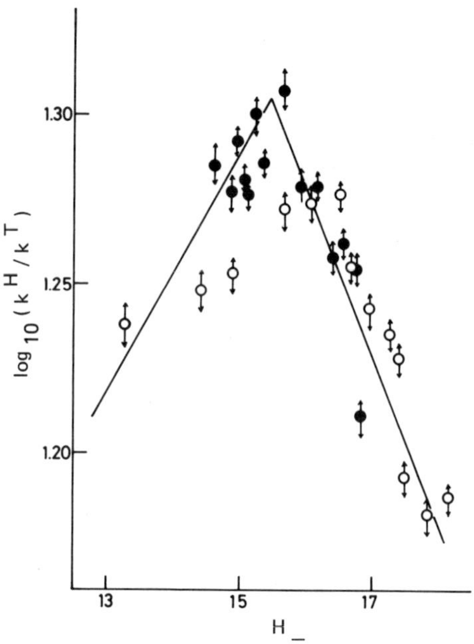

FIG. 1. Plot of log k^H/k^T against H_- for the ionization of (+)-methyldeoxybenzoin in OH^- — H_2O — DMSO (o) and in OEt^- — EtOH — DMSO (●).

toluene. The authors suggest that solvation changes are mainly responsible for this behavior being more important for the dipolar molecules CH_2Cl_2 and CH_3CN. If the transfer of the proton is accompanied by the motion of heavier atoms then tunneling will become less important. Both the $E^D_{obs} - E^H_{obs}$ and A^D_{obs}/A^H_{obs} values for the tetramethylguanidine-toluene system suggest the tunneling is a very important factor for this reaction, whereas for both triethylamine and tri-n-butylamine in acetonitrile this is not so.

Very large primary kinetic hydrogen isotope effects have also been observed for other nitrocompounds. Thus in the ionization of 2-nitropropane by 2,4,6-trimethylpyridine [78] the k^H/k^D values are 24 and 19.5, respectively, considerably higher than for catalysis by unsubstituted pyridines. A k^H/k^D value in excess

TABLE 5

Dependence of Primary Kinetic Hydrogen Isotope Effects on Solvent Composition in the Base-Catalyzed Proton Transfer Reaction of 4-Nitrophenylnitromethane [77]

	Tetramethylguanidine		Tri-n-butylamine		Triethylamine	
Solvent	Toluene	Dichloromethane	Toluene	Acetonitrile	Toluene	Acetonitrile
k^H/k^D at 25°C	45	11.7	14	2.2	11.0	3.1
$E^D_{obs}-E^H_{obs}$ (kcal/mol)	4.3	1.9	2.0	0.8	2.0	1.0
A^D_{obs}/A^H_{obs}	31	2.2	2.2	1.7	2.4	1.8

of 20 at 30°C for proton transfer from methyl 4-nitrovalerate to trimethylpyridine is consistent with these findings [79]. It is thought that for these sterically hindered reactions the energy barriers are both high and narrow, conditions which favor tunneling. For a number of nitroalkanes of the general form $ArCHMeNO_2$, $ArCH_2NO_2$, and $CH_2{=}CHCH_2NO_2$ there is no great change in the k^H/k^D values as the bases varied in strength from pyridine to the hydroxide ion [80]. Either the isotope effect is relatively insensitive to the symmetry of the transition state or the symmetry does not change over a wide range of ΔpK. Similar conclusions have been drawn from a study of primary kinetic hydrogen isotope effects for a number of hydrocarbons [81].

Some hydrogen atom reactions which may also be characterized by small solvation changes exhibit extremely large k^H/k^D values. Thus in the initiation step in the autoxidation of 4a,4b-dihydrophenanthrene in octane [82-84] the k^H/k^D values lie in the range 64 and 95 over the temperature range -10 to -31°C, reaching a value close to 250 at -52°C. At low temperatures (77 K) hydrogen atom abstraction by methyl radicals in solid γ-irradiated acetonitrile proceeds nearly entirely by quantum-mechanical tunneling [85].

$$CH_3CN + \cdot CH_3 \rightarrow CH_4 + \cdot CH_2CN \qquad (16)$$

The activation energy of 1.4 kcal/mol is much smaller than the value of 10 kcal/mol which has been reported for the same reaction in the gas phase at a much higher temperature (373-573 K). In another example of hydrogen atom transfer, Lewis and Butler [86] have reported a k^H/k^T value of 14.9 at 25°C for the reaction between tritium radicals and thiophenol, its temperature dependence being given by $k^H/k^T = 0.19 \exp(2590/RT)$, again signifying an appreciable tunneling contribution. For reactions of the kind

$$R\dot{C}HCH_2Br + {}^{*}HBr \rightarrow RCH{}^{*}HCH_2Br + Br\cdot \qquad (17)$$

it has been shown [87] that the magnitude of the isotope effect is related to the strength of the forming C-H bond but unfortunately some of the earlier findings are now considered suspect [88]. Isotope effects from a series of reactions involving the abstraction of a hydrogen atom from the S-H position of isotopically labeled thiols pass through a maximum when the heat of reaction is zero [89]. This observation is in line with Hammond's postulate [90] that the most symmetric transition state in a series of reactions should occur when ΔH is zero.

Large primary kinetic hydrogen isotope effects have been observed for several hydride transfer reactions. Thus for the oxidation of 1-phenyl-2,2,2-trifluoroethanol by alkaline permanganate [91,92] k^H/k^D and k^H/k^T values of 16 and 57, respectively, have been reported; k^H/k^D is also in excess of 10 in the quinone oxidation of leucotriphenylmethane dyes using acetonitrile or methanol as the solvent. Because these reactions involve the transfer of an electron-deficient atom, it is thought that the energy barrier is narrower than those encountered in proton transfer reactions.

At the other extreme, there are many examples of low primary kinetic hydrogen isotope effects being observed. Thus in the methoxide-catalyzed racemization of 2-methyl-3-phenylpropionitrile [34] k^H/k^D only changes from 1.15 to 1.67 as the basicity of the medium is increased by adding dimethyl sulfoxide. The reason for this low value lies in the fact that the transition state resembles closely the products of the reaction--the reverse reaction is very fast and probably diffusion controlled so that the small changes observed in the k^H/k^D values probably reflect changes in the equilibrium values. Recent calculations [93] for this reaction show that very weak kinetic isotope effects can be obtained from highly unsymmetric transition states. There are several other instances of reactions with unsymmetric transition states which exhibit very low kinetic hydrogen isotope effects [22,94,95].

Another possible reason for low k^H/k^D values stem from a consideration of the mechanism of proton transfer. Eigen [96] has chosen to consider it in terms of three steps:

1. The diffusion of the reactants toward one another to a distance close enough to facilitate hydrogen bridge formation between the donor and the acceptor
2. The formation of the hydrogen bridge and the transfer of the proton; the tendency to form hydrogen bonds is in the order

   OH---O > OH---N, NH---O > NH---N >> CH---O, OH---C

3. The separation of the products

For many oxygen and nitrogen acids the rates are usually fast and close to diffusion control and the isotope effects expected to be small. For carbon acids which do not form hydrogen bonds very readily, the rates are much slower and much more sensitive to changes in basicity. The isotope effects are often large and remain so over a range of basic strengths. However, this will not be so if internal return [97] is important:

$$\gt C\text{-H} + B^- \underset{k_{-1}}{\overset{k_1}{\rightleftharpoons}} \gt C^- \text{---HB} \overset{k_2}{\rightarrow} \text{exchange} \qquad (18)$$

A steady-state treatment gives

$$k_{obs} = \frac{k_1 k_2}{k_{-1} + k_2} \qquad (19)$$

For carbon acids such as ketones and nitro compounds where the process of ionization involves considerable structural and solvent reorganization, exchange is expected to be much faster than the protonation of the resonance-delocalized anion (i.e., $k_2 \gg k_{-1}$) so that $k_{obs} = k_1$ and the isotope effect k^H/k^D will be fairly large (3-8 at 25°C). However, in certain circumstances, e.g., when the energy barrier for reprotonation is low, $k_2 \ll k_{-1}$ so that $k_{obs} = Kk_2$, the product of an equilibrium for which k^H/k^D would be small and a rate-determining step that involves the separation of a hydrogen-bonded complex for which k^H/k^D

should be negligible. This could well be the reason why isotope effects reported for heterocyclic carbon acids are so low [98] (Table 6).

Phenylacetylene is a moderately weak carbon acid ($pK_a \sim 20$) which if it behaved like other carbon acids (e.g., ketones, nitrocompounds) would be expected to give a k^H/k^D value at 25°C of between 3 and 5 for the hydroxide-catalyzed reaction. The recently reported value [100] of 0.95, however, shows that the proton transfer does not occur during the rate-determining stage of the reaction, but probably before it. The slow step in the reaction is either a diffusive separation from the hydrogen-bonded complex or rotation of another hydrogen atom into a hydrogen-bonding position,

$$\begin{array}{l} ArC{\equiv}CH + B \rightleftarrows ArC{\equiv}CH{\cdot}B \rightleftarrows ArC{\equiv}C^{-}{\cdot}HB^{+} \\ \qquad\qquad\qquad\qquad\qquad\qquad\quad \upharpoonleft\downharpoonright \quad \text{rate determining} \\ ArC{\equiv}CD + B \rightleftarrows ArC{\equiv}CD{\cdot}B \rightleftarrows ArC{\equiv}C^{-}{\cdot}DB^{+} \end{array} \qquad (20)$$

Although the ionization of toluene in cyclohexylamine [101] (lithium cyclohexylamide being the base) gives a k^H/k^D value in excess of 10, the same reaction when studied in dimethyl sulfoxide [102] with t-butoxide as base gave k^H/k^D of unity and this was also taken to mean that ionization was not the rate-determining step in the latter solvent.

In some cases the isotope effect k^H/k^D may depend on the base concentration [103]. Thus in the kinetics of iodination of 4-nitrophenol and 4-nitrophenol-2,6-^{2}H k^H/k^D changes from 5.4 at an iodide concentration of 22×10^{-4} M to 2.3 when $[I^-] = 0.15 \times 10^{-4}$ M. The mechanism of the reaction is thought to involve an equilibrium followed by a generally rate-determining step:

$$E + ArH \underset{k_{-1}}{\overset{k_1}{\rightleftarrows}} E\text{-}Ar\text{-}H \qquad (21)$$

$$E\text{-}Ar\text{-}H + B \overset{k_2}{\rightarrow} Ar\text{-}E + HB \qquad (22)$$

In these circumstances $k_{-1} \gg k_2[B]$, whereas if $k_2[B] \gg k_{-1}$ no isotope effect should be observed.

TABLE 6

Primary Kinetic Isotope Effects in the Base-Catalyzed Exchange of Various Heterocyclic Compounds [99]

Compound	Position of exchange	k^H/k^T	k^D/k^T	β	Temp. (°C)	Base
Benzthiazole	H-2		1.1		0	OEt^-
Purine	H-8	3.8		∼1	85	OH^-
N-Methylpyridinium iodide	H-2(6)		∼1.0		55	$(C_2H_5)_3N$
3-Benzylbenzthiazolium bromide	H-2	4.8		>0.9	30	OH^-
3-Benzyl 4,5-dimethylthiazolium bromide	H-2	2.7		>0.9	30	OH^-

Secondary deuterium isotope effects were first observed as early as 1957 and were readily recognized as an additional test for the unimolecular and bimolecular mechanisms in nucleophilic substitutions [104-6]. Although hyperconjunction [107], inductive effects [108], and nonbonded interactions [109] are all possible reasons for the observation of a secondary isotope effect, in model calculations they are first treated as differences between the force field of reactants and transition states. Secondary isotope effects are therefore like primary isotope effects, a means of determining force constant changes at the isotopic position occurring along the reaction coordinate.

Recently the main application of secondary hydrogen isotope effects has been in the study of the ability of neighboring groups to alter the energetics of organic reactions. Scheppele [110], for example, has considered the situation that arises when a hypothetical reactant undergoes reaction via a unimolecular, or a bimolecular mechanism, or a unimolecular reaction which involves anchimeric assistance by a neighboring group. Sunko and Borcic [111] have also given an excellent account of secondary deuterium isotope effects as applied to the question of neighboring-group participation.

Solvent isotope effects for reactions which exhibit catalysis by hydronium ions [67] such as those of the A-1 type of mechanism:

$$S + H_3O^+ \rightleftarrows SH^+ + H_2O \tag{23}$$

$$SH^+ \rightarrow \text{products} \tag{24}$$

give k^{H_2O}/k^{D_2O} values in the range 0.25-0.5. The main reason for this is the fact that acids are stronger in H_2O than in D_2O, so that the concentration of SH^+ in H_2O is less than SD^+ in D_2O. For reactions like the solvent (base)-catalyzed ionization of carbon acids, e.g., t-butylmalononitrile [94] k^{H_2O}/k^{D_2O} are in the range 1-4. In other words, the reaction is slower in D_2O than in H_2O, in agreement with other findings that D_2O is less basic than H_2O. The importance of the medium effect as distinct from the exchange effect can be estimated by studying, e.g., the acetate-catalyzed reaction in H_2O and D_2O. In all the cases studied [112] so far the reactions are between 10 and 25% faster in H_2O than D_2O. When measurements of both primary and solvent effects have been made for reactions of this type it has been shown that the solvent isotope effect is large when the primary isotope effect is low because of a very product-like transition state. Conversely, with a primary isotope effect that is low because of a reactant-like transition state the solvent isotope effect is also low [94].

Solvent isotope effects may also be used to distinguish between different mechanisms [36], e.g., the A-1 and A-2 type:

$$S + H_3O^+ \rightleftarrows SH^+ + H_2O \quad \text{fast} \tag{25}$$

$$SH^+ + H_2O \rightarrow \text{products} \quad \text{slow} \tag{26}$$

In this mechanism water molecules are involved in the rate-determining step and by studying the reaction in mixtures of H_2O and D_2O the variation in rate (expressed as k_n/k_H) can be compared with that predicted by various models. Similarly for an A-S_E2 type of reaction the simplest formulation

$$L_3O^+ + S \qquad \begin{matrix} L^{\phi_2} \diagdown & & & \\ & O{-}{-}{-}\vec{L}{-}{-}{-}\overset{+}{S} \\ L_{\phi_2} \diagup & \quad \phi_1 & \end{matrix} \tag{27}$$

leads to

$$\frac{k_n}{k_H} = \frac{(1 - n + n\phi_1)(1 - n + n\phi_2)^2}{(1 - n + n\ell)^3} \tag{28}$$

As ℓ is known (0.69) analysis of the variation of k_n/k_H as a function of n allows the possibility of determining both ϕ_1

and $\emptyset_2$. The reaction is eminently well suited for study using the differential method as very accurate results are necessary [113].

For hydroxide-catalyzed reactions the combination of OD^- in D_2O is a more basic system than OH^- in H_2O and several values of k^{H_2O}/k^{D_2O} in the range 0.6-0.7 have been reported [112]. In mixtures of H_2O and D_2O the dependence of rate is consistent with a trihydrated hydroxide ion $OH^-(H_2O)_3$ and similar studies in methanol-methoxide solutions lead to a similar structure for the methoxide ion. However, the solvent isotope effects in this medium are lower (k^{MeOH}/k^{MeOD} = 0.4-0.5) than for water and this has led to speculation that in the latter a chain conduction mechanism may be in operation [114].

REFERENCES

1. M. Wolfsberg, Ann. Rev. Phys. Chem., 20:449 (1969).
2. L. Melander, Isotope Effects on Reaction Rates, Ronald Press, New York, 1960.
3. C. J. Collins and N. S. Bowman (eds.), Isotope Effects in Chemical Reactions, Van Nostrand, New York, 1971.
4. W. Spindel (ed.), Isotope Effects in Chemical Processes, American Chemical Society, Washington, D. C., 1969.
5. R. P. Bell, The Proton in Chemistry, 2nd ed., Chapman and Hall, London, 1973, Chaps. 11 and 12.
6. J. R. Jones, The Ionisation of Carbon Acids, Academic Press, London, 1973, Chap. 9.
7. V. Gold, Adv. Phys. Org. Chem., 7:259 (1969).
8. R. L. Schowen, Progr. Phys. Org. Chem., 9:275 (1972).
9. Chap. 10 of Ref. 6.
10. E. A. Halevi, Progr. Phys. Org. Chem., 1:109 (1963).
11. Chaps. 2, 3, and 4 of Ref. 3.
12. R. P. Bell and J. E. Crooks, Proc. Roy. Soc., A286:285 (1965).
13. R. G. Pearson and R. L. Dillon, J. Am. Chem. Soc., 75:2439 (1953).

14. R. P. Bell and R. R. Robinson, Proc. Roy. Soc., A270:411 (1962).

15. T. Riley and F. A. Long, J. Am. Chem. Soc., 84:522 (1962).

16. M. L. Bender and A. Williams, J. Am. Chem. Soc., 88:2502 (1966).

17. A. J. Kirby and G. Meyer, J.C.S. Perkin II, 1446 (1972).

18. F. G. Bordwell, W. J. Boyle Jr., and K. C. Yee, J. Am. Chem. Soc., 92:5926 (1970).

19. R. P. Bell and B. G. Cox, J. Chem. Soc. (B), 194 (1970).

20. Chap. 1 in Ref. 6.

21. A. J. Kresge and Y. Chiang, J. Am. Chem. Soc., 83:2877 (1961).

22. Z. Margolin and F. A. Long, J. Am. Chem. Soc., 94:5108 (1972).

23. J. Bloxsidge, J. A. Elvidge, J. R. Jones, and E. A. Evans, Org. Magnet. Resonance, 3:127 (1971).

24. S. Andreades, J. Am. Chem. Soc., 86:2003 (1964).

25. K. J. Klabunde and D. J. Burton, J. Am. Chem. Soc., 94:820 (1972).

26. D. H. Hunter and J. B. Stothers, Can. J. Chem., 51:2884 (1973).

27. Y. Pocker and J. H. Exner, J. Am. Chem. Soc., 90:6764 (1968).

28. C. J. Collins, Adv. Phys. Org. Chem., 2:3 (1964).

29. V. F. Raaen, G. A. Ropp, and H. P. Raaen, Carbon-14, McGraw-Hill, New York, 1968, Chap. 15.

30. E. S. Lewis, J. D. Allen, and E. T. Wallick, J. Org. Chem., 34:255 (1969).

31. J. Warkentin and C. Barnett, J. Am. Chem. Soc., 90:4629 (1968).

32. W. H. Sachs, Acta Chem. Scand., 25:2643 (1971).

33. J. Hine, K. G. Hampton, and B. C. Menon, J. Am. Chem. Soc., 89:2664 (1967).

34. L. Melander and N. A. Bergman, Acta Chem. Scand., 25:2264 (1971).

35. W. J. Albery and B. H. Robinson, Trans. Faraday Soc., 65:980 (1969).

36. W. J. Albery and M. H. Davies, Trans. Faraday Soc., 65:1067 (1969).

37. A. Fry, Chap. 6 of Ref. 3.

38. A. G. Loudon, A. Maccoll, and D. Smith, J.C.S. Faraday 1, 69:894 (1973).

39. A. G. Loudon, A. Maccoll and D. Smith, J.C.S. Faraday 1, 69:899 (1973).

40. J. W. Hill and A. Fry, J. Am. Chem. Soc., 84:2763 (1962).

41. G. A. Ropp and V. F. Raaen, J. Chem. Phys., 22:1223 (1954).

42. C. G. Swain and E. R. Thornton, J. Org. Chem., 26:4808 (1961).

43. A. F. Cockerill and W. H. Saunders, Jr., J. Am. Chem. Soc., 89:4985 (1967).

44. L. L. Brown and J. S. Drury, J. Chem. Phys., 43:1688 (1965).

45. J. Bigeleisen, J. Chem. Phys., 17:675 (1949).

46. J. Bigeleisen, Science, 110:14 (1949).

47. F. H. Westheimer, Chem. Rev., 61:265 (1961).

48. R. A. More O'Ferrall and J. Kouba, J. Chem. Soc. (B), 985 (1967).

49. R. P. Bell, W. H. Sachs, and R. L. Tranter, Trans. Faraday Soc., 67:1995 (1971).

50. L. B. Sims, A. Fry, L. T. Netherton, J. C. Wilson, K. D. Reppond, and S. W. Crook, J. Am. Chem. Soc., 94:1364 (1972).

51. A. Fry, Pure Appl. Chem., 8:409 (1964).

52. P. C. Vogel and M. J. Stern, J. Chem. Phys., 54:779 (1971).

53. C. G. Swain, E. C. Stivers, J. T. Reuwer, Jr., and L. J. Schaad, J. Am. Chem. Soc., 80:5885 (1958).

54. J. Bigeleisen, Tritium in the Physical and Biological Sciences, Vol. 1, IAEA, Vienna, 1962, p. 161.

55. M. E. Schneider and M. J. Stern, J. Am. Chem. Soc., 94:1517 (1972).

56. M. Wolfsberg and M. J. Stern, Pure Appl. Chem., 8:325 (1964).

57. M. J. Stern, M. E. Schneider, and P. C. Vogel, J. Chem. Phys., 55:4286 (1971).

58. M. J. Stern and P. C. Vogel, J. Chem. Phys., 55:2007 (1971).

59. E. F. Caldin, Chem. Rev., 69:135 (1969).

60. E. F. Caldin, Reaction Transition States (J. E. Dubois, ed.), Gordon and Barach, 1972, p. 247.

61. E. P. Wigner, Z. Physik Chem., 19B:903 (1932).

62. C. Eckart, Phys. Rev., 35:1303 (1930).

63. E. M. Mortensen and K. S. Pitzer, The Transition State,

Special publication No. 16, The Chemical Society, London, 1962, p. 57.

64. H. S. Johnston and D. Rapp, J. Am. Chem. Soc., 83:1 (1961).

65. R. P. Bell, Proc. Roy. Soc., 148A:241 (1935).

66. R. P. Bell, Trans. Faraday Soc., 54:1 (1959).

67. P. M. Laughton and R. E. Robertson, Solute Solvent Interactions, (J. F. Coetzee and C. D. Ritchie, eds.), Dekker, New York, 1969, Chap. 7.

68. A. J. Kresge, Pure Appl. Chem., 8:243 (1964).

69. R. A. More O'Ferrall, G. W. Koeppl, and A. J. Kresge, J. Am. Chem. Soc., 93:9 (1971).

70. R. P. Bell and D. M. Goodall, Proc. Roy. Soc., 294A:273 (1966).

71. D. J. Barnes and R. P. Bell, Proc. Roy. Soc., 318A:421 (1970).

72. J. R. Keefe and N. H. Munderloch, J.C.S. Chem. Commun., 17 (1974).

73. D. Dolman and R. Stewart, Can. J. Chem., 45:911 (1967).

74. D. W. Earls, J. R. Jones, and T. G. Rumney, J.C.S. Faraday 1, 68:925 (1972).

75. R. P. Bell and B. G. Cox, J. Chem. Soc. (B), 783 (1971).

76. D. W. Earls, J. R. Jones, and T. G. Rumney, unpublished results.

77. E. F. Caldin and S. Mateo, J.C.S. Chem. Commun., 854 (1973).

78. E. S. Lewis and L. H. Funderburk, J. Am. Chem. Soc., 89:2322 (1967).

79. H. Wilson, J. D. Caldwell, and E. S. Lewis, J. Org. Chem., 38:564 (1973).

80. F. G. Bordwell and W. J. Boyle, Jr., J. Am. Chem. Soc., 93:512 (1971).

81. A. Streitwieser Jr., W. B. Hollyhead, G. Sonnichsen, A. H. Pudjaatmaka, C. J. Chang, and T. L. Kruger, J. Am. Chem. Soc., 93:5096 (1971).

82. A. Bromberg, K. A. Muszkat, and E. Fischer, Chem. Commun., 1352 (1968).

83. A. Bromberg and K. A. Muszkat, J. Am. Chem. Soc., 91:2860 (1969).

84. A. Bromberg, K. A. Muszkat, E. Fischer, and F. S. Klein, J.C.S. Perkin 11, 588 (1972).

85. R. J. Le Roy, E. D. Sprague, and F. Williams, J. Phys. Chem., 76:546 (1972).

86. E. S. Lewis and M. M. Butler, J. Org. Chem., 36:2582 (1971).

87. E. S. Lewis and S. Kozuka, J. Am. Chem. Soc., 95:282 (1973).

88. E. S. Lewis and M. M. Butler, J.C.S. Chem. Commun., 292 (1973).

89. W. A. Pryor and K. G. Kneipp, J. Am. Chem. Soc., 93:5584 (1971).

90. G. S. Hammond, J. Am. Chem. Soc., 77:334 (1955).

91. R. Stewart and R. van der Linden, Disc. Faraday Soc., 29:211 (1960).

92. E. S. Lewis and J. K. Robinson, J. Am. Chem. Soc., 90:4337 (1968).

93. N. A. Bergman, W. H. Saunders, Jr., and L. Melander, Acta Chem. Scand., 26:1130 (1972).

94. F. Hibbert and F. A. Long, J. Am. Chem. Soc., 93:2836 (1971).

95. Z. Margolin and F. A. Long, J. Am. Chem. Soc., 95:2757 (1973).

96. M. Eigen, Angew. Chem., Int. edn., 3:1 (1964).

97. D. J. Cram, C. A. Kingsbury, and B. Rickborn, J. Am. Chem. Soc., 83:3688 (1961).

98. J. A. Zoltewicz and L. S. Helmick, J. Am. Chem. Soc., 92:7547 (1970).

99. J. A. Elvidge, J. R. Jones, C. O'Brien, E. A. Evans, and H. C. Sheppard, Adv. Heterocyclic Chem., 16:1 (1974).

100. A. J. Kresge and A. C. Lin, J.C.S. Chem. Commun., 761 (1973).

101. A. Streitwieser, Jr., W. C. Langworthy, and D. E. Van Sickle, J. Am. Chem. Soc., 84:251 (1962).

102. J. E. Hofmann, A. Schriesheim, and R. E. Nickols, Tetrahedron Lett., 22:1745 (1965).

103. E. Grovenstein and N. S. Aprahamian, J. Am. Chem. Soc., 84:212 (1962).

104. W. H. Saunders, Jr., S. Asperger, and D. H. Edison, Chem. Ind., 1417 (1957).

105. A. Streitwieser, Jr. and R. C. Fahey, Chem. Ind., 1417 (1957).

106. K. Mislow, S. Borcic, and V. Prelog, Helv. Chim. Acta, 40:2477 (1957).

107. E. S. Lewis, Tetrahedron., 5:143 (1959).

108. E. A. Halevi, Tetrahedron, 1:174 (1957).

109. L. S. Bartell, J. Am. Chem. Soc., 83:3567 (1961).

110. S. E. Scheppele, _Chem. Rev._, _72_:511 (1972).

111. Chap. 3 of Ref. 3.

112. Chap. 10 of Ref. 6.

113. W. J. Albery and A. N. Campbell-Crawford, _J.C.S. Perkin II_, 2190 (1972).

114. V. Gold and S. Grist, _J. Chem. Soc._ (B), 2282 (1971).

Chapter 10

ISOTOPIC EXCHANGE PROCESSES

Francis J. Johnston

Department of Chemistry
University of Georgia
Athens, Georgia

I. INTRODUCTION

The detection of isotopic exchange and the measurement of isotopic exchange rates in chemical and biological systems have proven a valuable probe into the elementary processes occurring in such systems. Through the use of radioactive or stable isotopes it is possible to follow an interchange of molecules, groups, or atoms between different physical states or between chemical species having common constituents. Such interchange occurs whether or not labeled material is present. It can be detected, however, only when one of the reactants differs initially in isotopic composition from the other. The rate of such an exchange process may be determined by measurement of the change in isotopic composition in the system as a function of time. In order that the measured exchange rate represent the behavior of the normal molecules, it is necessary that the behavior of the labeled and unlabeled species should not differ significantly. In many experimental situations this is true. In some cases, however, large errors may be introduced by ignoring isotopic fractionation effects.

When the isotopic label is radioactive the exchange process may be monitored by conventional counting techniques following a physical or chemical separation of the exchanging species. For stable isotopic species, mass spectrometry must be used.

Examples of molecular interchange processes that can be measured by isotopic exchange include the following:

1. An interchange of water molecules between the liquid and vapor states using a gas phase labeled with ^{3}H, ^{2}H, or ^{18}O
2. An exchange of SO_4^{2-} ions between aqueous Na_2SO_4 and solid $PbSO_4$ in which either the aqueous or solid phase initially contains $^{35}SO_4^{2-}$
3. An exchange of iodine atoms between I_2 and HI with one of the molecular species initially labeled with ^{131}I

Isotopic exchange rate measurements can provide information of a basic nature in a variety of chemical systems. The following are several examples.

1. The occurrence of an exchange of X between a species AX and, for example, X^-, is evidence for the existence of a possible dissociation equilibrium such as

$$AX \rightleftarrows A^+ + X^- \tag{1}$$

When no other mode for exchange exists, the exchange rate may be equated to the dissociation (or recombination) rate.

2. The physical displacement of one group in a molecule by another chemically identical to it may occur as a distinct process in the system,

$$AX + {}^*X^- \rightleftarrows A{}^*X + X^- \tag{2}$$

Isotopic exchange may be used to study the mechanisms and energetics of such reactions. These are goals of the usual investigations in chemical kinetics.

3. In a chemical system at equilibrium,

$$AX + BY \rightleftarrows AY + BX \tag{3}$$

the reaction process is accompanied by an exchange of X species between AX and BX and by an exchange of Y between BY and AY. If an independent mode for exchange does not exist, then the exchange rate may be equated to the reaction rate under equilibrium conditions. If AX and BX do undergo an independent exchange reaction, then the observed rate of exchange in the equilibrium system will be the sum of the direct exchange and the equilibrium reaction rate.

4. Specific information, such as surface area measurements and diffusion rates in solids, can be obtained from exchange measurements carried out in heterogeneous systems.

It should be emphasized at this point that the reaction process being monitored is an exchange of X between the AX and BX states. The isotopic label is only a tool for following this process. For the reaction of interest, ΔH and ΔS are zero. For the redistribution of isotopes, ΔH and ΔS are not zero but cannot be associated with a "driving force" for the reaction.

Myers and Prestwood [1] have outlined the mathematical treatment of isotopic exchange rates and have summarized experimental results to 1950. Hundreds of exchange studies on a wide variety of systems have been carried out since then. Kahn [2] has compiled a bibliography that includes work up to 1962. Stranks and Wilkins [3] have summarized applications of exchange studies in investigations of reaction mechanisms in inorganic chemistry. Boyer [4] has discussed the application of isotopic exchange processes to studies of enzymatic reactions. No attempt will be made here to give a comprehensive review of work done in these areas. The purpose of this chapter will be that of mathematically describing exchange behavior under a variety of conditions, outlining the evaluation of exchange rates from experimental data, and illustrating the basic information that can be obtained from such studies.

II. MATHEMATICAL DESCRIPTION

An isotopic exchange process may be followed experimentally by physically or chemically separating the exchanging species and measuring the extent of isotopic redistribution as a function of time. When radioactive tracers are used, the experimental data will express the specific activity of either or both of the exchanging species as a function of time. The time dependence was mathematically first described by McKay [5]. In order to define and clarify some of the terms used in this chapter, this derivation will be outlined in some detail.

A. The First-order Exchange Law

Equation (4) represents the result of some process or series of processes which involve an exchange of X between the AX and BX forms. The asterisk represents a labeled atom or group.

$$\overset{*}{A}X + BX \rightleftarrows AX + \overset{*}{B}X \tag{4}$$

Let a and b represent the total molar concentrations (labeled and unlabeled) of AX and BX in the system. In a given experiment, these will be constant. Let x and y represent the molar concentrations of $\overset{*}{A}X$ and $\overset{*}{B}X$, respectively. In the usual case, a >> x and b >> y. Assume that AX is the initially labeled species with the initial concentration of $\overset{*}{A}X$ designated by x_0. If isotopic fractionation is negligible, x and y at exchange equilibrium will be given by

$$\frac{x_\infty}{y_\infty} = \frac{a}{b} \tag{5}$$

and from a material balance (after correction for decay),

$$x + y = x_\infty + y_\infty = x_0 \tag{6}$$

The rate of formation of $\overset{*}{B}X$ is then given by

$$\frac{dy}{dt} = R\left(\frac{x}{a} - \frac{y}{b}\right) \tag{7}$$

where R represents the rate, in mole liter^{-1} sec^{-1}, at which X is exchanged between the AX and BX states. In a system of fixed chemical composition and temperature, R will be constant. The significance of the terms in the parentheses is as follows: x/a is the probability that the X transferred from AX to BX is labeled. y/b represents the fraction of X's transferred from BX to AX that is labeled. Introduction of Eqs. (5) and (6) gives

$$\frac{dy}{dt} = \frac{R(a + b)}{ab}(y_\infty - y) \tag{8}$$

or

$$-\log(y_\infty - y) = \frac{R(a + b)}{ab} t - \log y_\infty \tag{9}$$

By use of the fractional exchange, defined as

$$F = \frac{y}{y_\infty} \tag{10}$$

Eq. (9) becomes

$$\log(1 - F) = -R \frac{(a + b)}{ab} t \tag{11}$$

In the above derivation, it was assumed that the experimentally measured quantity was the appearance of the label or radioactivity in the initially unlabeled species BX. If the decrease of radioactivity in the AX form is measured, the fractional exchange becomes

$$F = \frac{x_0 - x}{x_0 - x_\infty} \tag{12}$$

Equation (11) is similar in form to that describing a first-order rate process and is frequently referred to as the first-order exchange law.

In theory, a single measurement of F would allow the evaluation of the exchange rate. In practice, R is usually determined graphically from the slope of a plot of $\log(1 - F)$ vs. time or from a semilogarithmic plot of $1 - F$ vs. time. In the latter case, the half-time for the exchange is determined from the plot and the value of R is given by

$$R = \frac{0.693ab}{(a + b)t_{1/2}} \tag{13}$$

When one of the molecular species participating in the exchange reaction contains more than one exchangeable group, the treatment can become complex. In the simplest case, that of a molecule containing two chemically and structurally similar X groups, e.g., BX_2, the groups will exchange at the same rate in the process

$$\overset{*}{A}X + BXX \rightleftarrows AX + B\overset{*}{X}X \quad (\text{or} \quad BX\overset{*}{X}) \tag{14}$$

so that a treatment similar to that used in the derivation of Eq. (11) leads to

$$\log(1 - F) = -R\,\frac{(a + 2b)}{2ab}\,t \tag{15}$$

R is the rate of exchange of X between the molecular species AX and BX_2. The exchange rate per X group in BX_2 is R/2.

Equation (15) may readily be extended to the general case of exchange of X between AX_m and BX_n providing that the X groups within each molecule are of equal exchangeability. The case in which the groups differ in exchangeability will be discussed in Section V A.

B. Evaluation of Exchange Rate from Experimental Data

As an example of an exchange in a chemical system, we shall consider an aqueous solution containing bromoacetic acid and bromide ions. Over prolonged periods of time the bromoacetic acid will hydrolyze forming glycollic acid and HBr. The hydrolysis reaction is slow at temperatures below 60°C, and for time periods as long as 2 hr, decomposition is insignificant at 50°. Under such conditions, the chemical composition may be considered as static. An exchange of bromine between bromide and bromoacetic acid,

$$BrCH_2COOH + \overset{*}{B}r^- \rightleftarrows \overset{*}{B}rCH_2COOH + Br^- \tag{16}$$

does occur, however, and has been studied using ^{82}Br as a tracer. The data shown in Table 1 were obtained [6] in one series of experiments at 50.4°C. The second column represents the total radioactivity in a 1-ml aliquot of the reaction mixture, and the third column gives the radioactivity in a corresponding volume after removal of the bromide by precipitation and filtering. The radioactivity present in the bromoacetic acid form after exchange equilibrium is attained would be

TABLE 1

The Exchange of Bromine Between Bromide and Bromoacetic Acid*
$[BrCH_2COOH] = 0.2100$ M, $[Br^-] = 0.0197$ M

Time (sec)	Radioactivity (cpm/10 ml)		F
	Total	$CH_2BrCOOH$	
0	48,900	--	0
900	48,455	9,934	0.224
1800	48,127	17,393	0.396
2700	48,048	23,540	0.535
3600	47,800	27,702	0.634
4500	47,716	31,371	0.718

*From Ref. 6.

$$\frac{0.2100}{0.2100 + 0.0197} \times 48{,}048 = 44{,}000 \text{ cpm}$$

The value of F at 2700 sec is

$$F = \frac{23{,}540}{44{,}000} = 0.535$$

Figure 1 shows a semilogarithmic plot of $1 - F$ vs. t. The half-time for the exchange is 2440 sec and the corresponding exchange rate is

$$R = 0.693 \frac{0.2100 \times 0.0197}{0.2297} \frac{1}{2440}$$

$$= 5.12 \times 10^{-2} \text{ mol liter}^{-1} \text{ sec}^{-1}$$

This result means that under these reaction conditions the bromine atom in a bromoacetic acid molecule is exchanged, on the average, every 4 sec with a bromide ion.

In order to obtain information concerning the mechanism by which the exchange occurs, R must be studied as a function of the concentrations of bromide and bromoacetic acid. In this case, the experimental rate law was found to be $R = k[CH_2BrCOOH][Br^-]$, with k, the second-order rate constant, determined to be $2.1 \times 10^{10} \exp(-19{,}600/\underline{R}T)$ l mol^{-1} sec^{-1}.

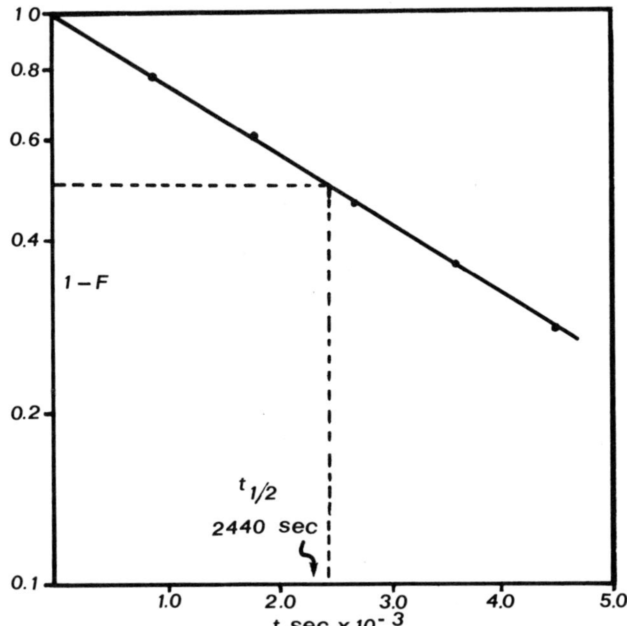

FIG. 1. A semilogarithmic plot of 1-F versus t for bromine-exchange between 0.210 M $BrCH_2COOH$ and 0.0197 M Br^- at 50.4°C. (Adapted from Ref. 6.)

There are two most probable mechanisms for this exchange. One of these involves a dissociation of the acid into bromide and CH_2COOH^+ ions,

$$BrCH_2COOH \rightleftarrows CH_2COOH^+ + Br^- \tag{17}$$

The second involves a direct bimolecular displacement process,

$$^*Br^- + H_2C(COOH){-}Br \rightleftarrows {}^*Br{-}CH_2(COOH) + Br^- \tag{18}$$

The first mechanism would result in a rate of exchange that is first-order in the bromoacetic acid concentration and independent of that of bromide. The second mechanism corresponds to second-order kinetics and is in agreement with the experimental rate equation. The experimental evidence suggests that this is the mechanism by which the exchange occurs.

C. Limitations of the First-order Exchange Law

In the derivation of Eq. (11), it was assumed that R, the rate of the exchange of X between AX and BX, could always be used to describe the distribution rate of $\overset{*}{X}$ between AX and BX. This, of course, is the actual process being experimentally followed. There are two situations for which this assumption is not correct. In the first of these, the masses of X and $\overset{*}{X}$ are sufficiently different that the corresponding exchange rates differ by more than experimental variations. The second situation may result when the exchange occurs through one or more intermediate processes in which chemical species of a transient existence are involved. If a significant time is required for isotopic equilibration between the initially labeled reactant and any such intermediate, Eq. (11) will not represent the isotopic redistribution process.

1. Isotope Effects in Exchange Reactions

In general, for isotopic species of mass numbers of approximately 30 or greater, differences in their physical and chemical behavior are within experimental variations. For lighter isotopes and for reactions in which the bond to the tagged atom or group is directly involved in the exchange process, this may no longer be true. In such cases rates of the elementary reactions involved in the exchange can differ significantly for the isotopic species. (Isotope effects on reaction rates are described in detail in Chap. 9.) For an exchange reaction proceeding through a complex mechanism, the experimental rate expression will have the form, $R = ka^m b^n$, or possibly a sum of such terms. The rate constant, k, will be comprised of rate constants characteristic of the elementary reaction steps. Isotope effects, then, will depend upon the reaction mechanism. The corresponding rate constant differences will also be reflected in the equilibrium constant or fractionation factor for the exchange reaction.

Two recent results give an indication of the magnitudes of observed isotopic fractionation factors in exchange reactions. For the reaction

$$HCOOH(\ell) + HD(g) \rightleftarrows HCOOD(\ell) + H_2(g) \tag{19}$$

the equilibrium constant,

$$K = \frac{[HCOOD]_\ell [H_2]_g}{[HCOOH]_\ell [HD]_g} \tag{20}$$

was found to be 3.86 at 25°C [7]. Similarly, for the exchange reaction

$$^{34}SO_2(g) + H^{32}SO_3^-(aq) \rightleftarrows {}^{32}SO_2(g) + H^{34}SO_3^-(aq) \tag{21}$$

the equilibrium constant,

$$K = \frac{[H^{34}SO_3^-]_{aq}[^{32}SO_2]_g}{[H^{32}SO_3^-]_{aq}[^{34}SO_2]_g} \tag{22}$$

has been reported to be 1.01-1.02 at 25-45°C [8].

Harris [9] has discussed such situations and has derived equations describing isotopic exchange rates in systems where isotope effects cannot be ignored. As an example, let us consider an exchange process that occurs through a mechanism resulting in second-order kinetics,

$$\overset{*}{A}X + BX \underset{k_2}{\overset{k_1}{\rightleftarrows}} AX + \overset{*}{B}X \tag{23}$$

The exchange rate for unlabeled molecules is given by

$$R = k_0[AX][BX] \tag{24}$$

where $k_0 = k_1 = k_2$. In the absence of an isotope effect, the distribution of the labeled species at exchange equilibrium will be described by

$$\frac{[AX][B^*X]_\infty}{[BX][A^*X]_\infty} = \frac{y_\infty/b}{x_\infty/a} = \frac{k_1}{k_2} = 1 \qquad (25)$$

If, on the other hand, $k_1 \neq k_2$, then

$$\frac{y_\infty/b}{x_\infty/a} = \frac{k_1}{k_2} = K \neq 1 \qquad (26)$$

Using the same notation that was used in the derivation of Eq. (11), the rate of appearance of the label in the BX form will be:

$$\frac{dy}{dt} = k_1 ab \left[\frac{x}{a} - \frac{y}{Kb}\right] \qquad (27)$$

Equation (27) assumes a negligibly small tracer concentration and that only AX is initially labeled. Substitution for x in terms of y_∞ gives

$$\frac{dy}{dt} = \frac{k_1}{K} (a + bK)(y_\infty - y) \qquad (28)$$

which yields upon integration,

$$\log (1 - F) = - \frac{k_1}{K} (a + bK)t \qquad (29)$$

$F = y/y_\infty$, $y_\infty = bKc/(bK + a)$ and c is the total concentration of the tracer in the system. Equation (29) also predicts a linear dependence of $\log (1 - F)$ upon t. In order to obtain the exchange rate constants, however, K must be known.

When the concentration of the tracer is not negligible with regard to that of the unlabeled material, Eq. (27) must be modified to

$$\frac{dy}{dt} = k_1 [x(b-y) - y\frac{(a-x)}{K}] \qquad (30)$$

Equation (30) does not integrate to a simple form like Eqs. (11) and (29). Thus, if isotopic effects are significant, the first-order rate law will not be applicable in this case.

Harris [7] has also discussed the more general case for which the exchange rate for unlabeled molecules is of the form

$R = ka^m b^n$. He shows that as long as the concentration of the tracer is small,

$$\log(1 - F) = -\frac{k'a^{m-1}b^{n-1}}{K'}(a + bK')t \tag{31}$$

where K' represents an isotope effect factor, not necessarily equal to the equilibrium constant for the isotopic exchange reaction. The effective rate constant for the labeled species is k'.

2. A "Bottleneck" Equilibration Step

If isotopic equilibration between a labeled reactant and a reaction intermediate is not fast compared to other steps involved in the reaction, there will be an apparent induction period in the appearance of tracer in the initially unlabeled reactant. In this case, also, the rate of isotopic redistribution will not be represented by Eq. (11).

As an illustration of this situation, consider an exchange of X between AX and BX occurring through a reaction process in the equilibrium system represented by Eq. (3) and involving a catalyst, C. Assume that the mechanism by which this reaction proceeds is that corresponding to Eqs. (32) - (34).

$$C + AX \rightleftarrows A + CX \tag{32}$$

$$CX + BY \rightleftarrows BX + CY \tag{33}$$

$$A + CY \rightleftarrows AY + C \tag{34}$$

The rate of exchange of X between AX and BX will be the rate of reaction (33), R_2. Assume that to such a system containing equilibrium quantities of C, AX, BY, AY, A, CX and CY, a very small quantity of labeled AX (A^*X) is added. Let d represent the total concentration of CX, and z, that of C^*X. As in the development of Eq. (11), a and b represent the total concentrations of AX and BX and x and y, the corresponding concentrations of the labeled species. The rate of formation of B^*X will be

$$\frac{dy}{dt} = R_2 \left[\frac{z}{d} - \frac{y}{b}\right] \tag{35}$$

If isotopic equilibrium is at all times maintained between AX and X, then

$$\frac{z}{d} = \frac{x}{a} \tag{36}$$

and Eq. (35) is the same as Eq. (7), and Eq. (11) is the appropriate solution. If, on the other hand, the exchange of X between AX and CX is relatively slow, Eq. (7) will not be applicable and the first-order exchange law will not describe the isotopic redistribution. Fleck [10] has discussed this limitation of the first-order rate law in more detail.

III. SOME EXPERIMENTAL CONSIDERATIONS

In the usual experiments, conventional techniques are used to separate the participants in an exchange reaction. These include solvent extraction, precipitation, ion-exchange, and evaporation. In some systems, the reaction may be "quenched" prior to separation by conversion of one or other of the reactants to a nonexchangeable form. In order for the measurements to accurately reflect the rate, the total time for separation must be short to the half-time for the exchange. Exchange reactions that can be quantitatively studied by use of separation procedures such as those mentioned above, are limited to those having half-times greater than ∿100 sec. The counting procedure (Chap. 4) used will depend upon the tracer and its chemical form. Several problems and techniques of specific applicability in exchange experiments will be briefly discussed in the following sections.

A. Background Exchange

If the process used to separate the reactants is not an efficient one, there will be an apparent exchange at zero reaction time. In certain systems some exchange may occur as a result of the separation process itself. If, for example, the process involves a precipitation of one of the exchanging species, a surface-catalyzed exchange process may contribute significantly to the homogeneous exchange being measured. Myers and Prestwood [1] have quantitatively discussed the apparent exchange that may result from incomplete reactant separation or from a separation-induced exchange process. They point out that if such effects are reproducible, then

$$\log(1 - \overline{F}) = \frac{R(a + b)}{ab} t + \log(1 - \overline{F}_0) \tag{37}$$

where $\overline{F}$ represents the total observed exchange at time t and $\overline{F}_0$, the apparent fractional exchange observed following the separation process at zero reaction time. This useful result shows that in such systems, the variation of $\ell n\ (1 - \overline{F})$ with time may still be used to evaluate the rate of the exchange process. A plot of $\log\ (1 - F)$ vs. time, however, will have a nonzero intercept.

In the usual exchange experiment, the radioactivity of the tracer is at a sufficiently low level that there will be an insignificant contribution from a radiation-induced exchange. Typically, microcurie quantities of radioactivity are used in reaction volumes of approximately 100 ml. If, for example, the tracer is ^{35}S with an average β energy of 0.049 MeV [11] the rate of energy absorption in a reaction system containing 0.01 $\mu Ci\ ml^{-1}$ is approximately 1.8×10^7 eV ml^{-1} sec^{-1}. Typical yields for radiation-induced reactions are of the order of one molecule reacting per 100 eV of absorbed energy. A radiation-induced exchange rate in this system would, therefore, be of the order of 3×10^{-16} mol $liter^{-1}$ sec^{-1}. This is negligible when compared to any process occurring at a measurable rate in the system.

If high levels of activity are required for an experiment, the possibility of a radiation-induced exchange cannot be discounted. Gromov and Spitsyn [12], for example, report marked effects of auto- and preirradiation of the solid phase on heterogeneous exchange reactions.

B. Electrochemical Separation Methods

Beronius [13] has developed electrochemical procedures for measuring exchange rates in systems where one of the participants can be electrodeposited for direct counting. In one application of this technique, one of the exchanging species is sampled at appropriate time intervals by deposition of a very small, known quantity at an electrode. The amount of material sampled, which must be a very small fraction of the total, is known from the current and its time of flow. The specific activity is obtained from a direct count of the material on the electrode. To use this technique, it is necessary for the exchange to be sufficiently slow that a negligible change in specific activity occurs during the electrodecomposition. By a modification of this procedure, termed the integral method by Beronius, it is possible to measure exchange rates much faster than can be studied by usual techniques. In this procedure, electrolysis is initiated at the beginning of the reaction and continued through 1-3 half-times. Again, the electrodeposited sample must be a small fraction of the total reactant. The measured activity on the electrode is then compared with that obtained in an identical electrolysis after exchange equilibrium is attained. To illustrate the application of the technique, consider the exchange reaction corresponding to Eq. (4). Assume that BX (or X from BX) can be deposited by reaction at an electrode. Using the same terminology as was used previously (a and b representing total concentrations of AX and BX, respectively, and y the concentration of $\overset{*}{B}X$ at time t), the rate at which $\overset{*}{B}X$ is deposited at the electrode is given by

$$\frac{dw}{dt} = \frac{iy}{Fb} \tag{38}$$

F is the Faraday constant, i the current and w the number of moles (an identity of moles with equivalents is assumed here for simplicity) of $\overset{*}{X}$ deposited. The total amount, A, of $\overset{*}{X}$ deposited during an electrolysis time t is

$$A = \int_0^t dw = \frac{i}{bF} \int_0^t y_\infty \{1 - \exp [- \frac{R(a + b)}{ab} t]\} dt$$

$$= \frac{iy_\infty}{bF} \{t + \frac{ab}{R(a + b)} (\exp [- \frac{R(a + b)}{ab} t] - 1)\} \qquad (39)$$

The amount of $\overset{*}{X}$ deposited during the same electrolysis time after exchange equilibrium is attained is

$$A_\infty = \frac{iy_\infty}{bF} t \qquad (40)$$

The ratio of these quantities gives

$$\frac{A}{A_\infty} = 1 - \frac{1 - \exp [-R \frac{(a + b)}{ab} t]}{R(a + b)t} ab \qquad (41)$$

from which the exchange rate can be recovered.

When such an electrolytic separation of reactants is feasible, it should be possible to measure exchange reaction rates with half-times of 10 sec or less.

C. Other Methods

In theory, an experimental sampling procedure which does not completely separate the exchanging species may be used for the determination of exchange rates. If, in the sampling, a fixed ratio of AX to BX is obtained that is different from that in the reaction system, the total specific activity of the sample will reflect both the separation ratio and the extent of exchange. Gosman [14] has developed such procedures using diffusion of the exchanging species through a capillary or through glass or cellophane membranes as a means of fractionating the reactants. The most promising technique resulting from this work involves an electrodeposition of both reactants at an electrode in a ratio that is reproducible and significantly different from that in the reacting mixture. This method has been applied to the Tl(I)-Tl(III) exchange.

In other recent experiments, homogeneous gas phase exchanges have been studied in shock tubes. The application of a pressure pulse in a gaseous system can result in localized temperatures as high as 3800 K for periods corresponding to several milliseconds. This technique has been used to study the exchange of N atoms between N_2 molecules [15],

$$^{14}N - {}^{14}N + {}^{15}N - {}^{15}N \quad 2\,{}^{14}N - {}^{15}N \tag{42}$$

and the exchange of iodine between molecular iodine and methyl iodide [16],

$$CH_3I + {}^{131}I_2 \quad CH_3{}^{131}I + I_2 \tag{43}$$

Besides permitting the study of reactions under conditions of extremely high temperature, the procedure has the advantage of minimizing heterogeneous contributions to a gas phase reaction.

IV. SOME HOMOGENEOUS EXCHANGE REACTIONS

The rate of an exchange process, R, is a function of the concentrations of the exchanging species. In general, it will be of the form

$$R = k[AX]^m[BX]^n \tag{44}$$

or, less frequently of a form comprised of a sum of such terms. Concentrations of other species effecting the reaction rate will also be appropriately incorporated into Eq. (44). The mathematical form of the rate equation may be determined by the usual techniques of experimental chemical kinetics. If a single term expression such as Eq. (44) is applicable, the exponents m and n may be determined as the slopes of logarithmic plots of the measured exchange rates vs. reactant concentrations for a series of experiments in which only one of the reactant concentrations is varied. If, for example, R is measured for a series of AX concentrations at a fixed BX concentration, $[BX]_0$,

$$\log R = m \log [AX] + \log k[BX]_0^n, \qquad (45)$$

and m may be determined from the slope of a plot of $\log R$ vs. $\log [AX]$.

The mathematical form of the rate equation reflects the mechanism by which the exchange occurs and its determination is essential to the definition of the elementary steps involved in the exchange (or any other reaction) process.

The following paragraphs summarize the results of investigations of several homogeneous exchange reactions and illustrate the chemical information that can be obtained through exchange studies.

A. Exchange Mechanisms

1. The U(IV)-U(VI) Exchange Reaction

The exchange of uranium between the IV and VI oxidation states,

$$U^{4+} + {}^{*}UO_2^{2+} \rightleftarrows {}^{*}U^{4+} + UO_2^{2+} \qquad (46)$$

has been the subject of early experiments by Rona [17], and more recently by Masters and Schwartz [18]. ^{233}U was used as a tracer for these investigations. The latter workers studied the exchange at lower uranium concentrations and at temperatures from 25 to 47°C. Some of their data, plotted according to Eq. (45), are shown in Fig. 2. The results show that the reaction is first-order each in U^{4+} and UO_2^{2+} and minus third-order in H^+. The corresponding rate equation is:

$$R = k \frac{[U^{4+}][UO_2^{2+}]}{[H^+]^3} \qquad (47)$$

The reaction mechanism suggested by the experimental rate equation is:

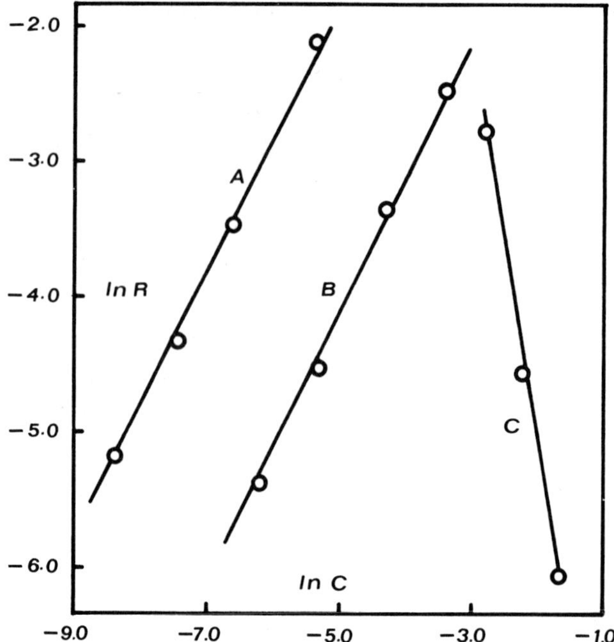

FIG. 2. Rates of the U(IV)-U(VI) exchange reaction at 39.4°C with (A)[U(IV)] varied at [U(VI)] = 0.0132 M and $[H^+]$ = 0.122 M; (B)[U(VI)] varied at [U(IV)] = 0.00132 M and $[H^+]$ = 0.112 M; (C)$[H^+]$ varied at [U(IV)] = 0.00132 and [U(VI)] = 0.00484 M. (Adapted from Ref. 18.)

$$U^{4+} + UO_2^{2+} + 2H_2O \rightleftarrows (U_2O_3{\cdot}OH)^{3+} + 3H^+ \qquad (48)$$

$$(U_2O_3{\cdot}OH)^{3+} \rightleftarrows 2UO_2^+ + H^+ \qquad (49)$$

$(U_2O_3{\cdot}OH)^{3+}$ represents, stoichiometrically, an intermediate in which the identities of the uranium atoms are lost with the result that uranium exchange may occur upon dissociation. As indicated by Eq. (48), the intermediate is that involved in the disproportionation reaction,

$$U(IV) + U(VI) \rightleftarrows 2U(V) \qquad (50)$$

The exchange rate is proportional to the concentration of the intermediate complex,

$$R = k_1[U_2O_3\cdot OH^{3+}]$$

$$= \frac{k_1K[U^{4+}][UO_2^{2+}]}{[H^+]^3} \tag{51}$$

where K is the effective equilibrium constant for reaction (48). The exchange rate equation, then, has the form of Eq. (47) with the experimental rate constant corresponding to k_1K. The value of k at 25°C was 2.13×10^{-7} mol^2 $liter^{-2}$ sec^{-1}. The activation enthalpy was 37.5 kcal mol^{-1} and the corresponding entropy of activation was 36 cal mol^{-1} deg^{-1}.

The agreement of the experimental rate equation with that based upon a postulated mechanism is an indication, though not a proof, of the correctness of the mechanism. In this study, the results permitted an evaluation of the equilibrium constant for reaction (50). This was found to be in excellent agreement with the value experimentally obtained by independent means and thus lends confidence in the correctness of the suggested mechanism.

2. Isotopic Exchange and Dissociation Equilibria

Extensive work has been carried out in studies of halogen exchange between halide ions and inorganic complexes containing halogen ligands. Stranks and Wilkins [3] have summarized earlier experimental results in this area. Recent studies [19-20] have involved the halopentamine complexes of rhodium, iridium, and ruthenium. With the halide ion initially labeled, the exchange may be represented by

$$M(NH_3)_5X^{2+} + {}^*X^- \rightleftarrows M(NH_3)_5{}^*X^{2+} + X^- \tag{52}$$

For chloride, bromide, and iodide and for all three metals, the experimental rate law was found to have the form of

$$R = k[M(NH_3)_5X^{2+}] \tag{53}$$

independent of the halide ion concentration. The results are consistent with a dissociative mechanism involving the replacement of a halide ion by a molecule of water,

$$M(NH_3)_5X^{2+} + H_2O \underset{k_2}{\overset{k_1}{\rightleftarrows}} M(NH_3)_5H_2O^{3+} + X^- \tag{54}$$

The exchange rate is equal to the equilibrium rate of (54) and may be associated with either the forward or reverse direction. In terms of the aquo complex, the exchange rate will be

$$R = k_2[M(NH_3)_5H_2O^{3+}][X^-] \tag{55}$$

Since

$$[M(NH_3)_5H_2O^{3+}] = \frac{k_1}{k_2}\,\frac{[M(NH_3)_5X^{2+}]}{[X^-]} \tag{56}$$

Eq. (55) becomes

$$R = k_1[M(NH_3)_5X^{2+}] \tag{57}$$

Thus, the experimental rate constant in Eq. (53) is, according to this mechanism, equal to the pseudo-first-order rate constant for the aquation process.

The system just described is an example of the determination of a dissociation rate from isotopic exchange rate measurements. In the general case,

$$AB \rightleftarrows A + B \tag{58}$$

the equilibrium dissociation (or association) rate is equal to the rate of an isotopic exchange between AB and A or B provided, of course, that no independent mode for exchange exists. The measured exchange rate, when combined with a known equilibrium constant for the reaction, permits the evaluation of the rate constants for both the dissociation and association processes. Exchange measurements may be used to characterize a dissociation process such as (58) only if there is convincing evidence that exchange does not occur by any other path. A rapid exchange between [1-^{14}C]acetate and acetic acid [21],

$$CH_3COOH + CH_3{}^{14}COO^- \rightleftarrows CH_3{}^{14}COOH + CH_3COO^- \qquad (59)$$

for example, has been attributed to a direct proton transfer rather than to an autoionization of the acid,

$$CH_3COOH + CH_3COOH \rightleftarrows CH_3COOH_2^+ + CH_3COO^- \qquad (60)$$

The lack of a measurable exchange between added $\overset{*}{A}$ and AB can, however, serve as one criterion for an absence of a direct dissociation process such as reaction (58).

3. The Fluoride-Catalyzed Fe(II)-Fe(III) Exchange

In aqueous solutions of iron(III) containing fluoride ions, the following equilibria occur:

$$Fe^{3+} + F^- \rightleftarrows FeF^{2+} \qquad K_1 \qquad (61)$$

$$FeF^{2+} + F^- \rightleftarrows FeF_2^+ \qquad K_2 \qquad (62)$$

$$FeF_2^+ + F^- \rightleftarrows FeF_3 \qquad K_3 \qquad (63)$$

$$Fe^{3+} + H_2O \rightleftarrows FeOH^{2+} + H^+ \qquad K_h \qquad (64)$$

The K's represent the equilibrium constants for the respective reactions. In the absence of fluoride or other complexing agents, the Fe(III)-Fe(II) exchange in acid solutions is relatively slow. It occurs rapidly in the presence of fluoride, presumably due to an enhanced exchangeability of Fe(II) with certain or all of the above fluoride complexes. Hudis and Wahl [22] have studied the Fe(III)-Fe(II) exchange in fluoride-containing solutions at 0 and 10°C and at an ionic strength of 0.5. At a given temperature and fluoride concentration, the observed exchange rates were first-order each in total Fe(II) and Fe(III), i.e.,

$$R = k'[Fe(II)][Fe(III)] \qquad (65)$$

A complex dependence of the rate upon the fluoride concentration was observed, however, and is shown graphically in Fig. 3.

The results were interpreted on the basis of the following simultaneous paths for the exchange:

$$^{*}Fe^{3+} + Fe^{2+} \xrightarrow{k_0} Fe^{3+} + {}^{*}Fe^{2+} \tag{66}$$

$$^{*}FeF^{2+} + Fe^{2+} \xrightarrow{k_1} FeF^{2+} + {}^{*}Fe^{2+} \tag{67}$$

$$^{*}FeF_2^{+} + Fe^{2+} \xrightarrow{k_2} FeF_2^{+} + {}^{*}Fe^{2+} \tag{68}$$

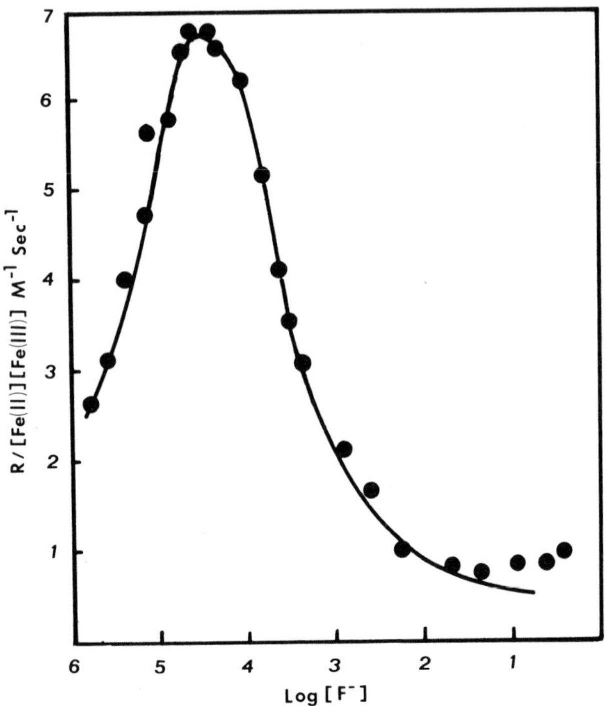

FIG. 3. The Fe(II)-Fe(III) exchange rate as a function of $[F^-]$ at 0°C. The circles denote experimental points and the smooth curve represents calculated rates based upon Eq. (74) and using the values listed in the text for k_0, k_1, k_2, k_3, K_1, K_2 and K_3. (Adapted from Ref. 22.)

$$^{*}FeF_3 + Fe^{2+} \xrightarrow{k_3} FeF_3 + {}^{*}Fe^{2+} \tag{69}$$

and

$$^{*}FeOH^{2+} + Fe^{2+} \xrightarrow{k_h} FeOH^{2+} + {}^{*}Fe^{2+} \tag{70}$$

Based upon this mechanism, the exchange rate expression is:

$$R = (k_0[Fe^{3+}] + k_1[FeF^{2+}] + k_2[FeF_2{}^{+}] + k_3[FeF_3])[Fe^{2+}] \tag{71}$$

where

$$k_0 = (k + k_hK)/[H^+] \tag{72}$$

Upon substitution for the concentrations of the fluoride complexes in terms of the equilibrium constants and $[Fe^{3+}]$, and upon introducing the total Fe(III) concentration as

$$[Fe(III)] = [Fe^{3+}] + [FeF^{2+}] + [FeF_2{}^{+}] + [FeF_3] \tag{73}$$

Eq. (71) becomes

$$R = [Fe(II)][Fe(III)] \frac{k_0 + k_1K_1[F^-] + k_2K_1K_2[F^-]^2 + k_3K_1K_2K_3[F^-]^3}{1 + K_1[F^-] + K_1K_2[F^-]^2 + K_1K_2K_3[F^-]^3} \tag{74}$$

From experiments in the absence of fluoride, the value of k_0 was obtained. At low fluoride concentrations, Eq. (74) simplifies to

$$R = [Fe(II)][Fe(III)] \frac{k_0 + k_1K_1[F^-]}{1 + K_1[F^-]} \tag{75}$$

and experiments under these conditions allow the determination of k_1 and K_1. Values of k_2, K_2, k_3, and K_3 were obtained from experiments at higher fluoride concentrations. The smooth curve of Fig. (3) was obtained by using values of 1.50, 9.7, 2.5, and

TABLE 2

Examples of Rate Laws and Mechanisms for Exchange Reactions

Exchange reaction	Experimental rate law	Proposed Mechanism	Ref.
$CH_2XCOOH + {}^{*}X^{-} \rightleftarrows CH_2{}^{*}XCOOH + X^{-}$ X = Cl, Br, I.	$R = k[CH_2XCOOH][X^{-}]$	$CH_2XCOOH + {}^{*}X^{-} \rightleftarrows CH_2{}^{*}XCOOH + X^{-}$	23
$PCl_5 + {}^{*}PCl_3 \rightleftarrows {}^{*}PCl_5 + PCl_3$	$R = k[PCl_5]$	$PCl_5 \rightleftarrows PCl_3 + Cl_2$	24
		${}^{*}PCl_3 + Cl_2 \rightleftarrows {}^{*}PCl_5$	
${}^{*}I_2 + CH_3I \rightleftarrows I_2 + CH_3{}^{*}I$	$R = k[I_2]^{1/2}[CH_3I]$	$I_2 \rightleftarrows 2I$	
		$I + CH_3I \rightleftarrows CH_3 + I_2$	16,25
		$CH_3 + {}^{*}I_2 \rightleftarrows CH_3{}^{*}I + I$	
${}^{*}SbCl_3 + SbCl_5 \rightleftarrows SbCl_3 + {}^{*}SbCl_5$	$R = k_1[SbCl_5]$	(1) $SbCl_5 \rightleftarrows SbCl_3 + Cl_2$	26
	$+ k_2[SbCl_3][SbCl_5]^2$	$Cl_2 + {}^{*}SbCl_3 \rightleftarrows {}^{*}SbCl_5$	
		(2) $2SbCl_5 \rightleftarrows Sb_2Cl_{10}$	
		$Sb_2Cl_{10} + {}^{*}SbCl_3 \rightleftarrows {}^{*}SbSb_2Cl_{13}$	
		${}^{*}SbSb_2Cl_{13} \rightleftarrows SbCl_3 + {}^{*}Sb_2Cl_{10}$	

$Br^{*}O_3^- + H_2O \rightleftarrows BrO_3^- + H_2{}^{*}O$	$R = (k_1 + k_2[OH^-] + k_3[H^+]^2)[BrO_3^-]$	(1) $Br^{*}O_3^- + H_2O \rightleftarrows BrO_3^- + H_2{}^{*}O$ (2) $OH^- + Br^{*}O_3^- \rightleftarrows BrO_3^- + {}^{*}OH^-$ (3) $Br^{*}O_3^- + 2H^+ \rightleftarrows H_2Br^{*}O_3^+$ $H_2Br^{*}O_3^+ + H_2O \rightleftarrows H_2BrO_3^+ + H_2{}^{*}O$	27
$S_2O_3^{2-} + {}^{*}S^{2-} \rightleftarrows {}^{*}S_2O_3^{2-} + S^{2-}$	$R = k[S^{2-}]^m[S_2O_3^{2-}]^n$ $m < 1, n > 1$	${}^{*}S^{2-} + S_2O_3^{2-} \rightleftarrows {}^{*}S_2^{2-} + SO_3^{2-}$ ${}^{*}S_2^{2-} + S_2O_3^{2-} \rightleftarrows {}^{*}S_2O_3^{2-} + S_2^{2-}$ $S^{2-} + {}^{*}S^{2-} \rightleftarrows {}^{*}S^{2-} + S^{2-}$	28

0.5 l mol^{-1} sec^{-1} for k_0, k_1, k_2, and k_3, respectively, and values of 9.7×10^4, 7.8×10^3 and 1×10^3 for K_1, K_2, and K_3, respectively. The deviation of the experimental points in Fig. 3 from the calculated curve at high fluoride concentrations is presumably due to the participation in the exchange of higher fluoride complexes of Fe(III).

4. Other Exchange Examples

In Table 2 are summarized the results of several other exchange reaction systems.

B. Exchange in Biochemical Systems

Measurements of isotopic exchange processes accompanying enzymatic reactions have proven to be a useful probe for studying the mechanisms of such processes. Boyer [4] has summarized the applicability and limitations of exchange measurements in obtaining such information.

While enzyme-catalyzed reactions encompass a wide range of complexity, many can be described in terms of a series of consecutive reactions. In order to illustrate this application of isotopic exchange, let us assume a reaction proceeding by way of a mechanism through which an atom or group, X, is transferred from one chemical form, A_1X, to a final form, A_nX. Let us express the reaction steps (not necessarily limited to enzymatic reactions) by

$$A_1X + B_1 \rightleftarrows A_2X + B_2 \tag{76}$$

$$A_2X + B_3 \rightleftarrows A_3X + B_4 \tag{77}$$

$$\cdots\cdots\cdots\cdots\cdots\cdots$$

and

$$A_mX + B_m \rightleftarrows A_nX + B_n \tag{78}$$

Imagine the system to be in equilibrium with rate of the ith step represented by v_i. Assume the exchange of X proceeds no further along the series than A_3X. At equilibrium, X is being transferred from A_1X to A_2X at a rate of v_1 mol liter^{-1} sec^{-1}. Of the X's transferred from A_1X to A_2X, $v_2/(v_1 + v_2)$ will be transferred to A_3X. The rate of exchange, R, of X between A_1X and A_3X will be

$$R = \frac{v_1 v_2}{v_1 + v_2} \tag{79}$$

or

$$\frac{1}{R} = \frac{1}{v_1} + \frac{1}{v_2} \tag{80}$$

Yagil and Hoberman [29] have proven the generality of this result for n such consecutive exchange steps. The rate of transfer of X from A_1X to A_nX is

$$\frac{1}{R} = \frac{1}{v_1} + \frac{1}{v_2} + \ldots + \frac{1}{v_n} \tag{81}$$

Flossdorf and Kula [30] have treated isotopic exchange rates in such systems by analogy with electrical circuits containing capacitors in series or resistors in parallel. For a series of reactions such as (76)-(78), for example, the exchange rate between A_1X and A_nX corresponds to the reciprocal of the resultant resistance of n resistors in parallel and having individual resistances $v_1, v_2, \ldots, v_n$.

In an interesting application of isotopic exchange in enzymatic (E) reactions, Dinovo and Boyer [31] have studied the ^2H, ^3H, ^{14}C, and ^{18}O exchange rates in the 2-phosphoglycerate (PGA) - 2-phosphoenolpyruvate (PEP)-enolase system. The overall reaction in this system is the dehydration of the glycerate to form the pyruvate,

$$\begin{array}{c} COOH \\ | \\ H\text{-}C\text{-}OPO_3H \\ | \\ CH_2OH \end{array} + E \rightleftharpoons \begin{array}{c} COOH \\ | \\ C\text{-}OP_3H \\ \| \\ CH_2 \end{array} + H_2O + E \tag{82}$$

The following exchange rates were measured under equilibrium conditions in the system:

$$\begin{matrix} \overset{*}{C}OOH \\ | \\ H\text{-}C\text{-}OPO_3H \\ | \\ CH_2OH \end{matrix} + \begin{matrix} COOH \\ | \\ C\text{-}OPO_3H \\ \| \\ CH_2 \end{matrix} \overset{R_C}{\rightleftarrows} \begin{matrix} COOH \\ | \\ HC\text{-}OPO_3H \\ | \\ CH_2OH \end{matrix} + \begin{matrix} \overset{*}{C}OOH \\ | \\ C\text{-}OPO_3H \\ \| \\ CH_2 \end{matrix} \qquad (83)$$

$$\begin{matrix} COOH \\ | \\ {}^{*}H\text{-}C\text{-}OPO_3H \\ | \\ CH_2OH \end{matrix} + HOH \overset{R_H}{\rightleftarrows} \begin{matrix} COOH \\ | \\ H\text{-}C\text{-}OPO_3H \\ | \\ CH_2OH \end{matrix} + {}^{*}HOH \qquad (84)$$

and

$$\begin{matrix} COOH \\ | \\ H\text{-}C\text{-}OPO_3H \\ | \\ CH_2\overset{*}{O}H \end{matrix} + HOH \overset{R_O}{\rightleftarrows} \begin{matrix} COOH \\ | \\ H\text{-}C\text{-}OPO_3H \\ | \\ CH_2OH \end{matrix} + H\overset{*}{O}H \qquad (85)$$

R_H, the hydrogen exchange rate between water and PGA was estimated from the measured rates for ^2H and ^3H. The results were consistent with the reaction mechanism schematically indicated below.

$$PGA + E \xrightarrow{v_1} E \cdot PGA \qquad (86)$$

$$E \cdot PGA \xrightarrow{v_2} E \cdot X^- + H^+ \qquad (87)$$

$$E \cdot X^- \xrightarrow{v_3} E \cdot PEP + OH^- \qquad (88)$$

$$E \cdot PEP \xrightarrow{v_4} E + PEP \qquad (89)$$

In Ref. 32, reaction (88) is represented as a two-step process. It is not done here for clarity. Based upon the indicated mechanism, the exchange rate, R_C, between PGA and PEP will be

$$\frac{1}{R_C} = \frac{1}{v_1} + \frac{1}{v_2} + \frac{1}{v_3} + \frac{1}{v_4} \qquad (89a)$$

The rate of oxygen exchange between water and PGA will be

$$\frac{1}{R_0} = \frac{1}{v_1} + \frac{1}{v_2} + \frac{1}{v_3} \tag{90}$$

and for the corresponding hydrogen exchange rate,

$$\frac{1}{R_H} = \frac{1}{v_1} + \frac{1}{v_2} \tag{91}$$

Dinovo and Boyer also used the observed hydrogen isotope effects as an indicator for the postulated mechanism. Their results indicated that reaction (88) was the major rate-limiting step in the overall reaction, occurring (at pH 6.5) approximately 0.1 as fast as (89) and about 0.2 as fast as (86) and (87).

Reaction parameters for enzymatic reactions measured by isotopic exchange in the equilibrium systems are usually in reassuringly good agreement with those obtained from initial rate studies under conditions far removed from equilibrium.

C. Exchange in the Solid State

There have been relatively few reports of isotopic exchange processes in the solid state. A number of studies have been carried out in which solid samples of a compound have been "doped" with a carrier-free radioisotope of an element of the compound but in a different oxidation state. The sample is then heated or exposed to visible or ultraviolet radiation. The appearance of the tracer in the parent compound may result from a true exchange process, or from a reduction or oxidation of the tracer atom that occurs during a dissolution process. Since the chemical separation process involves dissolving the sample, true exchange cannot always be established. There are, however, at least two systems in which exchange in the solid state has been observed.

1. tris-Dipyridyl Co(III) perchlorate-Co^{2+} Exchange

Nath et al. [32] have observed the incorporation of ^{57}Co into tris-dipyridyl Co(III) perchlorate in solid samples that had been doped with carrier-free $^{57}Co^{2+}$. They suggested that this process involved a true exchange occurring through the formation of $^{57}Co^{+}$

as a result of a reaction with thermally detrapped electrons,

$$^{57}Co^{2+} + e \rightarrow {}^{57}Co^{+} \tag{92}$$

$$^{57}Co^{+} + Co(III)(dipyridyl)_3(ClO_4)_3^- \rightleftharpoons {}^{57}Co(III)(dipyridyl)_3(ClO_4)_3^- + Co^{2+} \tag{93}$$

The existence of exchange was established by measuring [33] the Mossbauer spectrum of the sample before and after an exchange experiment. The spectrum characterizes the chemical state of the excited ^{57}Fe, daughter of the ^{57}Co decay. In the freshly doped sample, the Fe was present as Fe^{2+}. After 4 days at 25°C, the Fe was almost entirely in the chemical form of the trisdipyridyl complex, establishing the exchange nature of the process.

2. Sulfur Exchange in Anhydrous Thiosulfate

The thiosulfate ion, both in aqueous solution and in the anhydrous sodium or potassium salts, has the following tetrahedral structure,

$$\left[S - S \begin{matrix} \diagup O \\ - O \\ \diagdown O \end{matrix} \right]^{2-}$$

The sulfur atoms maintain their identity within the molecule under normal conditions and throughout chemical changes. For example, "outer labeled" thiosulfate may be prepared by the reaction of aqueous sulfite with elemental sulfur labeled with ^{35}S,

$$SO_3^{2-} + {}^*S \rightarrow {}^*SSO_3^{2-} \tag{94}$$

The radioactive sulfur may be recovered without change in specific activity by acidification of the thiosulfate

$$^*SSO_3^{2-} + H^+ \rightarrow {}^*S + HSO_3^- \tag{95}$$

The thiosulfate ion may be heated in aqueous solution for

prolonged periods to temperatures at least as high as 100°C with no detectable intramolecular exchange. However, when solid anhydrous sodium thiosulfate is heated to 300°C or above, the sulfur exchange,

$$Na_2{}^{*}SSO_3 \overset{k_x}{\rightleftarrows} Na_2S{}^{*}SO_3 \qquad (96)$$

occurs at an easily measurable rate. This exchange was first studied by Buntrock and Neumann [34], using both inner and outer labeled thiosulfate. The first-order rate expression for this reaction is given by

$$\log (1 - F) = k_x t \qquad (97)$$

Exchange occurs much faster than thermal decomposition of the solid and it was possible to study the exchange rates at temperatures as high as 350°C with negligible chemical change. These workers found that

$$k_x = 10^{20\pm2} \exp\left[\frac{-(69.5 \pm 5) \times 10^3}{RT}\right] \sec^{-1} \qquad (98)$$

For the same reaction, but using improved counting methods it was found that [35]

$$k_x = 10^{17.4\pm0.6} \exp\left[\frac{-(59.7 \pm 1.9) \times 10^3}{RT}\right] \sec^{-1} \qquad (99)$$

It is unlikely that the exchange occurs by a single concerted process within the solid, and McAmish and Johnston [36] have suggested the following mechanism:

$${}^{*}SSO_3^{2-} \rightleftarrows {}^{*}S + SO_3^{2-} \qquad (100)$$

$${}^{*}S + SO_3^{2-} \rightleftarrows {}^{*}SO^- + SO_2^- \qquad (101)$$

$${}^{*}SO + SO_2^- \rightleftarrows {}^{*}SO_2^- + SO^- \qquad (102)$$

V. SOME COMPLEX EXCHANGE SITUATIONS

A variety of exchange processes cannot be treated by the equations and procedures described to this point. Examples include reactions involving molecules with more than one exchangeable group and in which these groups differ in reactivity, and reactions in which chemical change is occurring simultaneously with exchange. Two such cases will be described in the following paragraphs.

A. Chemically Dissimilar Groups

When a molecular participant in an exchange reaction contains two groups, which for structural reasons differ in their exchangeability, $\log(1 - F)$ will not longer be a linear function of time, and except for special cases, it may be quite difficult to assign distinct values to the individual rates. For such a dual exchange as

$$\overset{*}{A}X + BX_1X_2 \rightleftarrows AX + \overset{*}{B}X_1X_2 \qquad (103)$$

and, simultaneously,

$$\overset{*}{A}X + BX_1X_2 \rightleftarrows AX + BX_1\overset{*}{X}_2 \qquad (104)$$

the dependence of the fractional change upon the time can be derived in the following manner [37].

Let a and b represent the total concentrations of AX and BX_1X_2. Using $[\overset{*}{A}X] = x$, $[\overset{*}{B}X,X_2] = y$, $[BX,\overset{*}{X}_2] = z$, $x + y + z = c$, and R_1 and R_2 for the rates of reactions (103) and (104), respectively, the rates of appearance of the label in the X_1 and X_2 positions are given by

$$\frac{dy}{dt} = R_1\left(\frac{x}{a} - \frac{y}{b}\right) \qquad (105)$$

and

$$\frac{dz}{dt} = R_2\left(\frac{x}{a} - \frac{z}{b}\right) \qquad (106)$$

Elimination of x from these equations gives

$$\frac{dy}{dt} + \frac{R_1(a + b)}{ab} y + \frac{R_1}{a} z = \frac{R_1}{a} c \tag{107}$$

and

$$\frac{dz}{dt} + \frac{R_2(a + b)}{ab} z + \frac{R_2}{a} y = \frac{R_2}{a} c \tag{108}$$

These simultaneous linear differential equations lead to

$$y = C_0 + C_1 e^{-\gamma_1 t} + C_2 e^{-\gamma_2 t} \tag{109}$$

and

$$z = C_3 + C_4 e^{-\gamma_1 t} + C_5 e^{-\gamma_2 t} \tag{110}$$

with

$$C_0 = C_3 = y_\infty = z_\infty = \frac{b}{a + 2b} c \tag{111}$$

$$\gamma_1 = \frac{E_1 + (E_1^2 - 4E_2)^{1/2}}{2} \tag{112}$$

$$\gamma_2 = \frac{E_1 - (E_1^2 - 4E_2)^{1/2}}{2} \tag{113}$$

$$E_1 = \frac{(R_1 + R_2)(a + b)}{ab} \tag{114}$$

$$E_2 = \frac{R_1 R_2 (a + b)}{ab^2} \tag{115}$$

$$C_1 = \left(\frac{R_1 c}{a} - \frac{\gamma_2 bc}{a + 2b}\right) \frac{1}{\gamma_2 - \gamma_1} \tag{116}$$

$$C_2 = -\frac{bc}{a + 2b} - C_1 \tag{117}$$

$$C_4 = \left(\frac{R_2 c}{a} - \frac{\gamma_2 bc}{a + 2b}\right) \frac{1}{\gamma_2 - \gamma_1} \tag{118}$$

and

$$C_5 = -\frac{bc}{a + 2b} - C_4 \tag{119}$$

The quantity evaluated from experimental data is

$$F = \frac{y + z}{y_\infty + z_\infty} \tag{120}$$

Insertion of Eqs. (109) and (110) in Eq. (120) results in,

$$F = 1 - \frac{C_1 + C_4}{y_\infty + z_\infty} e^{-\gamma_1 t} - \frac{C_2 + C_5}{y_\infty + z_\infty} e^{-\gamma_2 t} \tag{121}$$

To illustrate behavior in systems of this type, consider three situations corresponding to simplified concentration of $a = b = 1M$. In the first case,

$$1 - F = e^{-3Rt} \tag{122}$$

for $R_1 = R_2 = R$; in the second,

$$1 - F = 0.93e^{-4.73Rt} + 0.07e^{-1.27Rt} \tag{123}$$

for $R_2 = 2R_1 = 2R$ and in the third,

$$1 - F = 0.79e^{-20.5Rt} + 0.21e^{-1.46Rt} \tag{124}$$

for $R_2 = 10\ R_1 = 10\ R$. Semilogarithmic plots of $1 - F$ vs. t for these three cases are shown in Fig. 4 with an assumed value for $R = 1$ mol liter^{-1} sec^{-1}. R, as used in the above treatment, is an exchange rate per X group in the BX_1X_2 molecule. As used in Eq. (15), it represents the exchange rate per BX_2 molecule. For $R_2 \gg R_1$, the semilogarithmic plots of $1 - F$ vs. t can be resolved into two linear components with half times characteristic of the two exchange rates. The half-time for the faster exchange process will be $0.693/\gamma_1$, for the slower it is $0.693/\gamma_2$.

It should be noted that even for a twofold difference in exchange rates, deviations from linearity of the $\log(1 - F)$ vs. t plot are not evident until considerably beyond 50% exchange.

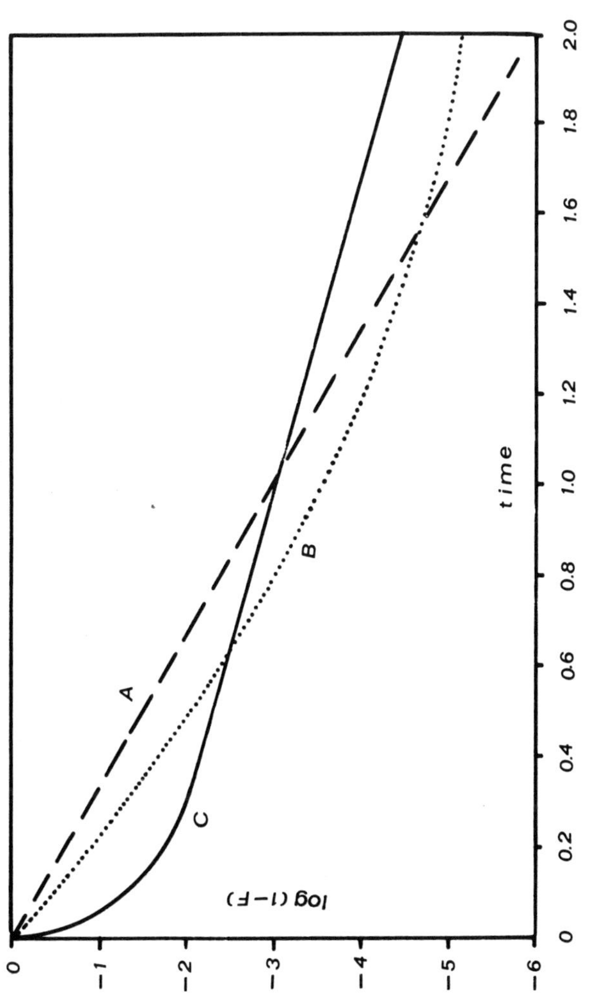

FIG. 4. Plots of $\log(1 - F)$ versus t, based upon Eq. (121), for an exchange reaction in which one of the molecules contains two exchangeable groups. In A, both groups at the same rate, R. In B and C, the rates differ by factors of 2 (R and 2R) and 10 (R and 10R), respectively. For convenience in plotting, $a = b = 1$ M and $R = 1$ M sec^{-1} are assumed.

Conclusions concerning similarities in exchangeability of two groups based upon the apparent linearity of such a plot should be made with caution.

B. Isotopic Exchange in Nonstatic Systems

In the situations that have been described so far, isotopic exchange was assumed to occur in the absence of any net chemical change. An exchange process may well occur simultaneously with chemical reaction in a system. In at least one important reaction situation, the exchange rate may be determined despite changing reactant concentrations. In this case, one of the participants in the exchange reaction is produced by a chemical transformation of the other. Examples include an alkyl halide-halide ion exchange during hydrolysis of the alkyl halide,

$$RX + {}^{*}X^{-} \rightleftarrows R^{*}X + X^{-} \tag{125}$$

$$RX + H_2O \rightarrow ROH + H^{+} + X^{-} \tag{126}$$

and a Fe^{3+}-Fe^{2+} exchange during an oxidation of Fe^{2+} to Fe^{3+},

$$Fe^{2+} + {}^{*}Fe^{3+} \rightleftarrows {}^{*}Fe^{2+} + Fe^{3+} \tag{127}$$

$$Fe^{2+} + Ox \rightleftarrows Fe^{3+} + Red \tag{128}$$

The general case has been treated and applications made to several systems [38-40]. Reactions of this type may be represented by

$$A^{*}X + BX \rightleftarrows AX + B^{*}X \tag{129}$$

and

$$BX + M \rightarrow AX + N \tag{130}$$

The rates of reactions (129) and (130) are given by $R(t)$ and $\alpha(t)$, respectively, with both now functions of time. With the initial concentrations of AX and BX given by a and b, the concentrations of AX and BX at time t are given by

$$[AX] = a + \rho(t) \tag{131}$$

and

$$[BX] = b - \rho(t) \tag{132}$$

respectively, where

$$\rho(t) = \int_0^t \alpha(t)\, dt \tag{133}$$

With $[A^*X] = x$ and $[B^*X] = z$ and $x + z = c$, the rate of formation of B^*X is given by

$$\frac{dz}{dt} = -\left\{\frac{R(t)(a + b)}{[a + \rho(t)][b - \rho(t)]} + \frac{\alpha(t)}{[b - \rho(t)]}\right\} z + \frac{cR(t)}{a + \rho(t)} \tag{134}$$

which leads to

$$\frac{z}{[b - \rho(t)]} = \frac{c}{a + b} + C_1 \exp\left[- \int_0^t \frac{R(t)(a + b)\, dt}{[a + \rho(t)][b - \rho(t)]}\right] \tag{135}$$

where C_1 is an integration constant. Since at $t = 0$, $z = 0$ and

$$\int_0^t \frac{R(t)(a + b)}{[a + \rho(t)][b - \rho(t)]}\, dt = 0$$

Eq. (132) becomes

$$\frac{z}{[b - \rho(t)]} = \frac{c}{a + b}\left\{1 - \exp\left[- \int_0^t \frac{R(t)(a + b)\, dt}{[a + \rho(t)][b - \rho(t)]}\right]\right\} \tag{136}$$

Since $z/[b - \rho(t)]$ is the specific activity of BX at time t and $c/(a + b)$ is the specific activity of the total X in the system at exchange equilibrium, Eq. (136) may be written as

$$1 - \frac{(S.A.)_t}{(S.A.)_\infty} = \exp\left[- \int_0^t \frac{R(t)(a + b)\, dt}{[a + \rho(t)][b - \rho(t)]}\right] \tag{137}$$

or

$$\log (1 - F) = - \int_0^t \frac{R(t)(a + b)\, dt}{[a + \rho(t)][b - \rho(t)]} \tag{138}$$

In general, the exchange rate can be expressed in the form of

$$R(t) = k_x[a + \rho(t)]^m[b - \rho(t)]^n \tag{139}$$

so that

$$\log (1 - F) = - \int_0^t k_x(a + b)[a + \rho(t)]^{m-1}[b - \rho(t)]^{n-1}\, dt \tag{140}$$

The behavior of $\log (1 - F)$ as a function of t will, then, depend upon the mathematical form of Eq. (139). The latter will reflect the mechanism by which exchange occurs. For the special (but not unusual) case in which $m = n = 1$,

$$\log (1 - F) = -k_x(a + b)t \tag{141}$$

An example of such a system is that of the chloroacetate-chloride system [39]. At temperatures of 70°C and above, an exchange of chloride,

$$ClCH_2COO^- + {}^*Cl \rightleftarrows {}^*ClCH_2COO^- + Cl^- \tag{142}$$

occurs simultaneously with hydrolysis of the chloroacetate ion

$$ClCH_2COO^- + H_2O \rightarrow HOCH_2COO^- + H^+ + Cl^- \tag{143}$$

Table 3 summarizes the results of one series of experiments with this system at 81.5°C. The consistency of the k_x values through at least 66% decomposition of the chloroacetate is evidence for the second-order nature of the exchange process.

TABLE 3

Simultaneous Chlorine Exchange and Hydrolysis of Chloroacetate Ion*
Total ^{36}Cl activity = 2632 cpm $(S.A.)_\infty = 4.35 \times 10^4$ cpm M^{-1}

Time (hr)	$ClCH_2COO^-$(M)	$[Cl^-]$(M)	$ClCH_2COO^-$ (cpm)	$ClCH_2COO^-$ (cpm M^{-1})	F	k_x (M^{-1} hr^{-1})
0	0.0399	0.0206	0.7	17.5	--	--
24.0	0.0324	0.0281	87.0	2,685	0.062	0.0439
48.0	0 0263	0.0341	138	5,247	0.121	0.0439
72.0	0.0203	0.0401	162	7,980	0.183	0.0465
96.0	0.0169	0.0436	171	10,120	0.233	0.0455
117.8	0.0134	0.0471	169	12,612	0.290	0.0479

*Reprinted with permission from *J. Phys. Chem.*, *67*:2812 (1963). Copyright by the American Chemical Society.

VI. HETEROGENEOUS EXCHANGE REACTIONS

Studies of isotopic exchange across solid-gas or solid-liquid interfaces have been used to obtain information about the surface characteristics of the solid and about molecular motions within its interior. The usual exchange behavior observed experimentally in such systems involves an initial, relatively fast exchange of molecules in or close to the surface of the solid, followed by a slower exchange of molecules within the solid.

A. Solid-solution Exchange

1. Estimation of Surface Areas

Isotopic exchange of molecules or groups at the surface with those in the saturated solution has been used in the estimation of effective surface areas of the solids. A varying exchangeability of the molecules in the solid with time reflects changes in particle size and structure during an aging process. Early work done by Kolthoff and others in this area has been summarized in Ref. 1. The exchange at the surface can be described in terms of a three-step process: (1) diffusion of solute to the solid-liquid interface, (2) a "bimolecular" exchange between adsorbed molecules or ions and those in the surface, and (3) a diffusion away from the surface of the exchanged molecules. Lieser [41] has suggested that in all aqueous solution-ionic solid systems, step 2 is rate-determining.

If the assumption is made that the initial fast exchange involves only a monomolecular layer of the solid, a measurement of the extent of this portion of the exchange allows the estimation of the surface area of the sample.

Let us consider an exchange of labeled X between solid AX and its saturated solution,

$$AX_{(s)} + A\overset{*}{X}_{(aq)} \rightleftarrows A\overset{*}{X}_{(s)} + AX_{(aq)} \tag{144}$$

If the solution phase, before exchange, contains a_1 μCi of $A\overset{*}{X}$ and n total moles of AX and if the number of gram atoms of readily exchangeable X in the surface layer of the solid is m_s, then the specific activity of AX in the solution (and in the surface layer) after the exchange is $S_x = a_1/(n + ms)$. Measurement of the specific activity of AX in the solution before, $S_0 = a_1/n$, and after the surface exchange permits the evaluation of m_s from

$$m_s = n\left[\frac{S_0}{S_x} - 1\right] \tag{145}$$

If only a monomolecular layer is involved in the exchange, this result, when combined with an effective cross-sectional area per AX unit in the surface, may be used to evaluate the total surface area of the solid.

An illustration of such a measurement is one in which a sample of solid $CaSO_4 \cdot 2H_2O$ was exposed to a saturated solution of $^{45}CaSO_4$. A 0.977 g sample of the solid was exposed to a saturated solution (1.72×10^{-5} mol ml^{-1}). A 1-ml sample of the solution phase gave 651 and 408 cpm before and after exchange, respectively. The number of moles of $CaSO_4$ in the surface layer is $m_s = 1.17 \times 10^{-4}$ mol g^{-1} solid. Assuming a value of 25 Å^2 for the effective cross-sectional area per Ca^{2+} in $CaSO_4 \cdot H_2O$, the corresponding surface area is 1.75×10^5 cm^2 g^{-1}.

The surface exchange technique as applied to the determination of surface areas has the obvious disadvantage of requiring the assumption that only one molecular layer is involved in the exchange. It may be useful, however, in situations where BET adsorption measurements are not feasible. The exchanges

$$AgX_{(s)} + \overset{*}{X}^-_{(aq)} \rightleftarrows Ag\overset{*}{X}_{(s)} + X^-_{(aq)} \tag{146}$$

with X = Cl, Br and I, for example, have been used to estimate surface areas of electrodeposited layers of silver chloride, bromide, and iodide [42]. For other examples the reader should refer to the article by Muramatsu [43] and to Chapter 13 of this book.

2. The Slower Exchange Process

In certain systems the heterogeneous exchange process can be analyzed into two component rates, the slower of which is characterized by a well-defined half-life. When the initial "surface phase" exchange is sufficiently fast, the slower exchange process, in such cases, may be described in terms of a modified first-order rate law that assumes an exchange equilibrium initially existing between the solution and the surface phase.

Van't Riet and Parcell [44] have used such a modification of the first-order law to describe the slower exchange process occurring between solid silver chromate and saturated solutions of the labeled ions,

$$Ag_2CrO_{4(s)} + {}^*Ag_2CrO_{4(aq)} \rightleftarrows {}^*Ag_2CrO_{4(s)} + Ag_2CrO_{4(aq)} \tag{147}$$

With a, b, and m_s representing the gram-atoms of exchangeable silver in the solution phase, total solid and surface layer, respectively, the exchange law becomes

$$\log (1 - F') = - \frac{R(a + b)t}{(a + m_s)(b - m_s)} \tag{148}$$

In Eq. (148), $F' = (S'_0 - S_t)/(S'_0 - S_\infty)$, where S'_0 is the initial specific activity of the solution assuming the surface phase to be initially in equilibrium with the solution phase. S_t and S_∞ are the specific activities of the solute in the liquid phase at time t and at exchange equilibrium, respectively. In the usual experimental situation, the mass of exchangeable material in the solid is much greater than in solution and $S_\infty << S'_0$ and $S_\infty << S_0$ where S_0 is the initial specific activity of the solution without correction for surface exchange. In this case,

$1 - F' \tilde{=} S_t/S'_0$ and $1 - F \tilde{=} S_t/S_0$, where F is the fractional exchange with no correction made for an initial surface exchange. Since both (1 - F') and (1 - F) vary linearly with S_t, the time dependence of either may be used for evaluation of the slower exchange rate. Some of the results obtained for the Ag_2CrO_4 system are shown in Fig. 5. The observed time dependence of log (1 - F) is linear for both Ag^+ and CrO_4^{2-} exchange and indicates an effectively constant rate for the slower process with a marked difference in exchangeability of the two ions. Silver ions exchange approximately eight times as fast as

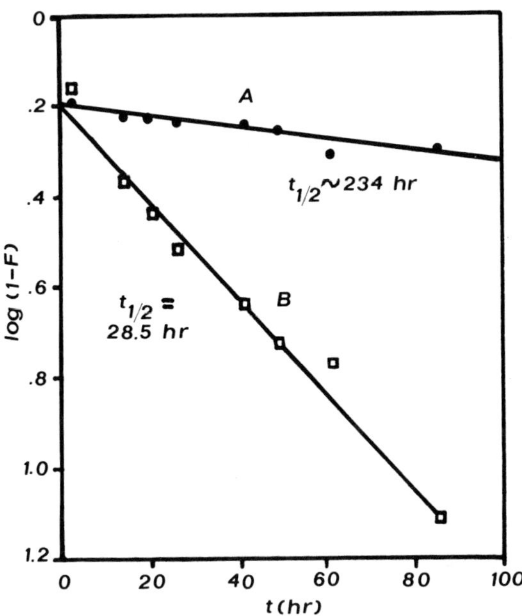

FIG. 5. The dependence of log (1 - F) upon t for the exchanges of ^{110}Ag and ^{51}Cr between solid Ag_2CrO_4 and 50% aqueous-ethanol solutions of Ag^+ and CrO_4^{2-}. The solutions were saturated with Ag_2CrO_4. In (A), 0.60×10^{-4} mol Ag_2CrO_4 was exposed to 10^{-5} mol $CrO_4^{2-}(^{51}Cr)$ in solution. In (B), 0.73×10^{-4} mol Ag_2CrO_4 was exposed to 10^{-5} mol $Ag^+(^{110}Ag)$ in the solution. (Adapted from Ref. 44.)

chromate ions, probably reflecting a preferential adsorption of silver ions at the particle-solution interface.

B. Models for Diffusion-controlled Exchange

When dissolution and recrystallization are not involved in an exchange between molecules of a solid and those in the surrounding medium, the slower exchange process is best described in terms of diffusion processes within the solid.

Interpretations of diffusion-controlled exchange reactions have been based upon simple models for the process. Let us consider an exchange geometry such as that indicated in Fig. 6. The rate of exchange of X between the solid and the surrounding medium is determined by interdiffusion of X between the interior and the surface of the solid. Molecules in a surface layer are at all times in isotopic equilibrium with the molecules in the fluid phase. The diffusion processes within the solid are assumed to be described by solutions of Fick's laws with appropriate boundary conditions. The results for the two most frequently encountered situations are summarized below [45,46].

1. The total number of moles, n, of tracer exchanged between an initially labeled solid phase and a fluid phase after a time t (see Fig. 6a) is given by

$$n = 2AC_0\left(\frac{Dt}{\pi}\right)^{1/2} \tag{149}$$

where C_0 is the molar concentration of labeled X per cubic centimeter present initially in the solid. D is the diffusion coefficient of the exchanging species in the solid and A is the surface area. In an experiment of this type, the measured quantity is the tracer radioactivity in a fixed volume of the fluid phase as a function of time.

When the direction of the exchange that is followed experimentally is from an initially labeled gas or liquid phase into the solid, C_0 represents the tracer concentration per unit

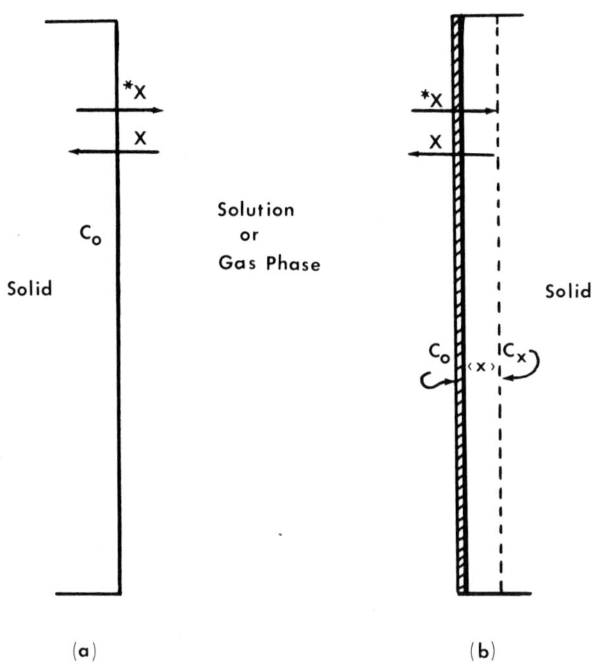

FIG. 6. (a) Exchange of X between an initially labeled solid and a fluid phase. C_0 is the initial tracer concentration in the solid and the number of moles of tracer transferred to the fluid phase at time t is given by Eq. (149). (b) Exchange of X between an initially labeled fluid and solid phase. Here, the dashed portion represents a surface layer in which the tracer concentration, C_0, is always in exchange equilibrium with that in the fluid. C_x is the tracer concentration at a depth x in the solid and is given by Eq. (152).

volume in a thin surface layer of the solid. The latter is assumed to be at all times in isotopic equilibrium with the fluid phase. In this case, the surface radioactivity of the solid is measured as a function of reaction time. In the usual experiments, the depth of penetration of the solid by the tracer is sufficiently small that corrections for self-absorption of the radiation are unnecessary. For whichever technique is used in determining n as a function of time, an allowance must be made for a surface exchange that is not described by diffusion within the solid. Sensui [47], for example, has studied the diffusion of chloride ions in solid metallic chlorides by measuring the exchange of ^{36}Cl between the labeled metal chlorides and gaseous Cl_2,

$$\overset{*}{M}Cl_{2(s)} + Cl_{2(g)} \rightleftarrows MCl_{2(s)} + \overset{*}{C}l_{2(g)} \tag{150}$$

The following modified form of Eq. (149) was used in treating the data,

$$B - B_0 = 2A\left(\frac{C_0}{\alpha}\right)\left(\frac{Dt}{\pi}\right)^{1/2} \tag{151}$$

Here, B represents the counting rate for a fixed volume of the gaseous Cl_2 at time t, and B_0 is the corresponding counting rate following equilibration with the surface layers of the solid. α is a detection coefficient for converting radioactivity to moles of tracer.

2. When the liquid or gas phase is initially labeled, the exchange may also be followed by measuring the tracer concentration, C_x, as a function of depth, x, in the solid. With this geometry (Fg. 6b), the tracer concentration at x at a time t is given by

$$C_x = C_0\left[1 - \operatorname{erf}\frac{x}{2(Dt)^{1/2}}\right] \tag{152}$$

with

$$\operatorname{erf}\frac{x}{2(Dt)^{1/2}} = \frac{2}{\sqrt{\pi}}\int_0^{x/2(Dt)^{1/2}} \exp(-z)^2\, dz \tag{153}$$

The quantity, C_0, the tracer concentration at the surface of the solid, may be obtained by extrapolation of C_x to zero thickness.

One example of such a study is that by King, Newman, and Suriani [48] in which the exchange

$$Cd_{(s)} + {}^{*}Cd^{2+}_{(aq)} \rightleftharpoons {}^{*}Cd_{(s)} + Cd^{2+}_{(aq)} \qquad (154)$$

was studied by the use of Cd discs exposed to solutions containing $^{115}Cd^{2+}$. After a given exposure time, microlayers of the surface were successively removed in an etching solution and the residual tracer concentration in the solid was determined by counting the radioactivity of the disc. In one experiment at 25°C, for a diffusion time of 5.9×10^5 sec, the ratio C_x/C_0 was found to be 0.41 for $x = 4 \times 10^{-5}$ cm. From tabulations of the error function, this gave $x/2(Dt)^{1/2} = 0.58$. The corresponding value of D was, based upon these data, 2×10^{-15} cm^2 sec^{-1}, a value in reasonable agreement with those obtained by other methods at higher temperatures and extrapolated to the temperature of the experiment. The results showed convincingly that the nature of the exchange involved a fast surface reaction accompanied by a diffusion-controlled exchange in the interior of the solid.

C. Other Heterogeneous Exchange Processes

1. Gas-Solid Exchange Reactions

The procedures summarized in Section VI B have been applied extensively to gas-solid systems in the determination of diffusion coefficients in the solid state. Secco and coworkers [49] have used the exchange reactions

$$ {}^{*}Zn_{(g)} + ZnX_{(s)} \rightleftharpoons Zn_{(g)} + {}^{*}ZnX_{(s)} \qquad (155)$$

to study the diffusion of Zn in ZnS, ZnSe, and ZnTe. Sensui [47] has studied the diffusion of chloride in nickel, magnesium, cobalt, and cadmium chlorides by means of the exchange reaction (150) of chlorine between the solid and gaseous Cl_2. In each case, meaningful diffusion coefficients have been evaluated by this technique.

2. Liquid-Liquid and Liquid-Gas Exchanges

Quantitative studies of exchange across liquid-liquid and gas-liquid interfaces are relatively rare. Filip and Mirnik [50] have measured the exchange rates of Zn across amalgam-solution interfaces

$$^{*}Zn(Hg) + Zn^{2+}_{(aq)} \rightleftarrows Zn(Hg) + {}^{*}Zn^{2+}_{(aq)} \qquad (156)$$

In the reference cited, these workers were studying the effects of absorbed layers of aliphatic amines at the amalgam-solution interface upon the rate of exchange of Zn.

Shah [51] has used NMR techniques to measure the rate of $H_2O(D_2O)$ exchange across the liquid-gas interface. In each study of such an interfacial exchange process, it is necessary to ensure that the interfacial exchange is truly the rate-limiting process.

Isotopic exchange techniques have been applied to studies of the nature of adsorbed films as well as insoluble monolayers at air-solution interfaces. For details see Ref. 43 and the discussion by Spink and Yates in Chapter 13.

REFERENCES

1. O. Myers and R. Prestwood, in Radioactivity Applied to Chemistry (A. C. Wahl and N. A. Bonner, eds.), Wiley, New York, 1951, p. 6.
2. M. Kahn et al., USAEC SCR-512 (1962).
3. D. R. Stranks and R. G. Wilkins, Chem. Rev., 57:743 (1957).
4. P. D. Boyer, Arch. Biochem. Biophys., 82:387 (1959).
5. H. A. C. McKay, Nature, 142:997 (1938).
6. J. F. Hinton, Ph.D. Dissertation, Univ. of Georgia, 1964.
7. F. J. Torre, M. Czuczak, and E. U. Monse, J. Chem. Phys., 58:1804 (1973).
8. T. F. Eriksen, Acta Chem. Scand., 26:573 (1972).
9. G. M. Harris, Trans. Faraday Soc., 47:716 (1951).
10. G. M. Fleck, J. Theoret. Biol., 34:509 (1972).
11. I. V. Vereschinskii and A. K. Pikaev, Introduction to Radiation Chemistry, Daniel Davey, New York, 1964, p. 29.
12. V. V. Gromov and V. I. Spitsyn, Dokl. Phys. Chem., 183:824 (1968).

13. P. Beronius, Acta Chem. Scand., 15:1151 (1961); Z. Phys. Chem., N. F., 47:246 (1965).

14. A. Gosman, Nature, 212:747 (1966); Z. Phys. Chem., 239:422 (1968); A. Gosman and I. Sedlacek, Ibid., 249:161 (1972); 241 (1973).

15. A. Bar-Nun and A. Lifshitz, J. Chem. Phys., 47:2878 (1967).

16. A. J. Kassman and D. S. Martin, Jr., J. Am. Chem. Soc., 91:6237 (1969).

17. E. Rona, J. Am. Chem. Soc., 72:4339 (1950).

18. B. J. Masters and L. L. Schwartz, J. Am. Chem. Soc., 83:2620 (1961).

19. G. B. Schmidt, Z. Phys. Chem. N.F., 50:222 (1966).

20. T. Shinohara, T. Yamada, N. Takebayashi, S. Hiraki, and A. Oyoshi, Bull. Chem. Soc. Japan, 45:3081 (1972).

21. Ersel A. Evans, J. L. Huston, and T. H. Norris, J. Am. Chem. Soc., 74:4985 (1952).

22. J. Hudis and A. C. Wahl, J. Am. Chem. Soc., 75:4153 (1953).

23. J. F. Hinton and F. J. Johnston, J. Phys. Chem., 67:2557 (1963).

24. W. E. Becker and R. E. Johnson, J. Am. Chem. Soc., 79:5157 (1957).

25. D. Clark, H. O. Pritchard, and A. F. Trotman-Dickenson, J. Chem. Soc., 2633 (1954).

26. F. B. Barker and M. Kahn, J. Am. Chem. Soc., 78:1317 (1956).

27. H. Gamsjäger, A. Grütter, and P. Baertschi, Helv. Chim. Acta, 55:781 (1972).

28. E. Ciuffarin and W. A. Pryor, J. Am. Chem. Soc., 86:3621 (1964).

29. G. Yagil and H. Hoberman, Biochemistry, 8:352 (1969).

30. J. Flossdorf and M. Kula, J. Biochem., 30:325 (1972).

31. E. C. Dinovo and P. D. Boyer, J. Biol. Chem., 246:4586 (1971).

32. A. Nath and S. Khorana, J. Chem. Phys., 46:2858 (1967).

33. A. Nath and M. P. Klein, Nature, 224:794 (1969).

34. D. Buntrock and K. Neumann, Z. Phys. Chem., N. F., 48:290 (1966).

35. L. H. McAmish, Ph.D. Dissertation, Univ. of Georgia, 1973.

36. L. H. McAmish and F. J. Johnston, Unpublished.

37. F. J. Johnston, J. Inorg. Nucl. Chem., 38:537 (1976).

38. C. P. Luehr, G. E. Challenger, and B. J. Masters, J. Am. Chem. Soc., 78:1314 (1956).

39. F. J. Johnston, J. Phys. Chem., 66:1719 (1962).

40. F. J. Johnston and J. F. Hinton, J. Phys. Chem., 67:2812 (1963).

41. K. H. Lieser, Monat. Chem., 101:1630 (1970).

42. F. J. Johnston, Int. J. Appl. Radiat. Isotopes, 18:435 (1967).

43. M. Maramatsu, in Surface and Colloid Science (E. Matijevic, ed.), Vol. 6, Wiley, New York, 1973, p. 101.

44. B. van't Riet and L. J. Parcell, Talanta, 11:321 (1964).

45. A. B. Lidiard and K. Tharmalingam, Disc. Faraday Soc., 28:64 (1959).

46. W. Jost, Diffusion in Solids, Liquids, Gases, Academic Press, New York, 1960, p. 256.

47. Y. Sensui, Bull. Chem. Soc. Japan, 46:3324 (1973); 45:2677 (1972).

48. C. V. King, D. S. Newman, and E. Suriani, J. Electrochem. Soc., 108:291 (1961).

49. E. A. Secco, Can. J. Chem., 42:1396 (1964); E. A. Secco and C.-H. Su, Can. J. Chem., 46:1621 (1968); E. A. Secco and R. S.-C. Yeo, Can. J. Chem., 49 (1971).

50. A. Filip and M. Mirnik, Z. Phys. Chem., N. F., 66:269 (1969).

51. D. O. Shah, J. Phys. Chem., 75:2694 (1971).

Chapter 11

SOLUTION PROPERTIES

Charles H. Bovington and Brian Dacre

Chemistry Department
Rutherford Laboratories
Royal Military College of Science
Shrivenham, Swindon
Wiltshire, England

I. GENERAL INTRODUCTION

For many years radioisotopes have been used as indicators in the study of chemical and biological solutions. However, much of the work has been of a qualitative nature in which the emphasis has been on detection of specified species (molecules, ions, etc.) rather than on quantitative measurements of physicochemical properties.

The use of radiotracers to obtain quantitative information is accepted as a useful technique but is regarded as yielding data of only modest precision. Recent work has, however, demonstrated that data of high precision can be obtained if sufficient attention is given to reducing errors other than that due to counting statistics; see particularly the work of Marx et al. quoted in Section VI of this chapter and Chapter 12 of this book.

In this chapter we shall consider the use of radiotracers to obtain quantitative information on equilibrium properties of solutions, e.g., solubility, complexation, transport numbers, etc. In addition, some discussion of physical rate processes will be included.

The use of radioisotopes as tracers for particular species, atoms, ions, or molecules in a system demands that certain assumptions be valid. These are listed as follows.

1. The radioisotopic species behaves exactly like the unlabeled species present and no change in isotopic ratios occurs during the physical or chemical process being studied. This assumption is generally correct within the normal precision of experiments of interest here. Important well-known exceptions occur in the use of hydrogen isotopes (see Chap. 9).
2. The radiation emitted from the radioactive tracer does not interfere with the system under investigation. It is well known that radiation interacts with materials and the problem of self-irradiation has been reviewed in Chapter 7.

Fundamental to the radiometric method is the assumption that the radioactivity introduced has no effect upon the physical and chemical properties of the substance under investigation. A series of papers by V. Spitsyn et al. [1] reported that labeling had quite considerable effects on a range of physical and chemical properties of solids. These effects, including considerably enhanced solubilities, were thought to be due to the residual charge on the surface when a β particle was emitted, and the consequent formation of an electrical double layer.

Much of the work of Spitsyn was carried out with crystals obtained by precipitation from very highly concentrated solutions. Under these conditions nucleation is rapid, and large numbers of minute crystallites are formed which would be expected to show enhanced solubilities.

Later studies by Ramette and Anderson [2], Bovington [3], and Gelsema, DeLigny, and Blijleren [4] largely contradict these findings, although electron microscopic examination of samples of $BaSO_4$ [3] which exhibited a 10% enhancement in solubility showed the presence of minute crystallites on the surface, which may have been due to self-irradiation damage.

In conclusion, it is probable that no abnormal effect on physical properties will be found when radiotracers are incorporated into a substance at the levels normally required in such experiments.

II. DISSOLUTION AND CRYSTALLIZATION

A. Introduction

Because of the ease with which extremely low concentrations of radioactive solutes may be determined, radioactive tracers are a valuable tool in the study of both solubilities and rates of solution of sparingly soluble substances. The radiotracer method is direct, unambiguous in interpretation and applicable to dissolution in any solvent irrespective of the degree of dissociation of the solvent or the degree of association of the solute in the particular solvent; the level of radioactivity is directly proportional to the mass of the particular element in the solution.

Very accurate determinations are possible provided that care is taken to ensure that equilibrium has been achieved by following the change in radioactivity of the solution with time, and that the solid in question is free from adsorbed species for reasons discussed later. For obvious reasons the half-life of the radioisotope chosen must be long compared to the time required for equilibrium to be achieved, the radiochemical purity of the isotope chosen must be high, since the impurity may be detected with greater efficiency than the isotope under study, and both the saturated solution and the samples taken during the equilibrium process must be free of colloidal or suspended material.

If measurements are carried out for solutions of different ionic strength at different temperatures, then heats of solution and solubility products can be measured accurately.

B. Theory of Solubility Product and Rates of Solution

1. Solubility Product and Ionic Strength

For sparingly soluble electrolytes an equilibrium constant may be expressed for the equilibrium between the crystal and the solution phase.

The solubility product K_{sp} is defined as

$$K_{sp} = a_+^{\nu_+} \cdot a_-^{\nu_-} \tag{1}$$

where ν_+ and ν_- are the numbers of positive and negative ions from dissociation of one molecule of electrolyte and a_+ and a_- are the activities of the cation and anion. If m is the concentration and γ the activity coefficient of the ions Eq. (1) becomes

$$K_{sp} = (\gamma_+ \cdot m_+)^{\nu_+} (\gamma_- \cdot m_-)^{\nu_-} = m_+^{\nu_+} \cdot m_-^{\nu_-} (\gamma_\pm)^{\nu_+ + \nu_-} \tag{2}$$

where $\gamma_\pm$ is the mean ionic activity coefficient and can be shown to be approximately the same for a given strong electrolyte in all solutions of the same ionic strength.

The ionic strength μ of an electrolyte solution is defined as

$$\mu = \frac{1}{2} \Sigma\, C_i Z_i^2 \tag{3}$$

where C_i and Z_i are the molarity and valence of each ion in solution. If this ionic strength is known, $\gamma_\pm$ for dilute solutions can be calculated by

$$-\log \gamma_\pm = A\sqrt{\mu} \tag{4}$$

where A depends on the solvent, temperature, etc. For a 1:1 electrolyte $m_+ = m_- = m$ (solubility) and $\nu_+ = \nu_- = 1$, hence combining Eqs. (2) and (4) and rearranging we obtain

$$\log m = A\sqrt{\mu} + \frac{1}{2} \log K_{sp} \tag{5}$$

A plot of $\log m$ against $\sqrt{\mu}$ permits calculation of K_{sp} from the intercept.

2. Variation of Solubility with Particle Size

It has long been generally assumed that solubility increases with decreasing particle size. There have been few satisfactory attempts to verify this assumption: if one can experimentally substantiate the Ostwald-Freundlich equation (5) the interfacial tension between solid and liquid can be estimated. If the minimum concentration for precipitate formation is known, the size of nucleus of solid for crystallization of the solid from solution may be estimated. The main difficulties are to obtain samples of particle size narrowly distributed in the submicrometer range, to clearly separate solid from solution, and to determine the particle size distribution of the solid in equilibrium with the separated solution.

Enustun and Turkevich [6] have used electron microscopy for measuring the particle size of strontium [^{35}S]sulfate in equilibrium with its saturated solution by stirring the thermostatted suspension. In Fig. 1, log (a/a_0) is plotted against the reciprocal of the minimum particle size, x in angstroms, where a and a_0 are the mean ionic activities for saturated solutions of a given and standard particle size, respectively. The results substantiate the Ostwald-Freundlich equation of solubility whereby the solubility is determined by the smallest particles present according to the following empirical formula:

$$\log \frac{a}{a_0} = \frac{16}{x} \tag{6}$$

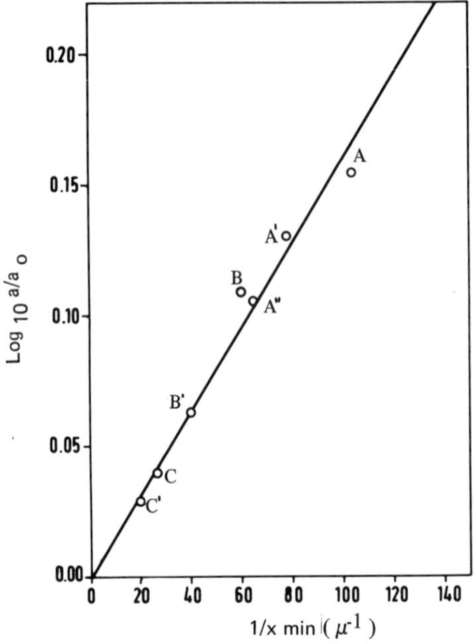

FIG. 1. Variation of aqueous solubility of $Sr^{35}SO_4$ with particle size at 25°C. x is the minimum particle size; a and a_0 are respectively, the mean ionic activities for saturated solutions of a given and standard particle size. (Reprinted from Ref. 6, p. 545, by courtesy IAEA.)

The interfacial tension was thus found to be 84 ± 8 erg/cm^2, a value lower than had been previously reported. The Ostwald-Freundlich equation was extended to crystals of any geometry and from this the interfacial tensions of the different faces of $Sr^{35}SO_4$ were calculated. The minimum $Sr^{35}SO_4$ concentration required to initiate precipitation (the critical supersaturation) was found to be 0.015 M. The difference between the critical supersaturation and the normal solubility was calculated and the Debye-Hückel equation used to determine the ratio $a/a_0 = 7.53$ [see Eq. (6)] whence $x = 18$ Å. The dimensions a, b, c, of the $Sr^{35}SO_4$ unit cell are 8.36, 5.36, 6.84 Å, respectively [7]; x is approximately 3 times the c dimension of each cell. Hence the critical nucleus is $3^3 = 27$ unit cells.

3. Kinetics of Dissolution

Any reaction between a solid and a solution resulting in soluble products may comprise as many as five primary steps:

1. Transport of solute materials to the interface
2. Adsorption at the surface
3. Reaction at the surface
4. Desorption of the products
5. Recession of the products from the interface

Of these, 2, 3, and 4 are characterized by an interaction between the solid and solute and may be termed chemical processes. Transport processes (1 and 5) involve mass transfer of a species by diffusion.

According to Nernst it is likely that the chemical processes at the interface are very much faster than one or other of the transport processes; unless there is a slow process occurring within the bulk of one of the phases the observed rate is transport-controlled.

In the dissolution of any substance the change in concentration C with time t can often be related to the solution concentration by

$$- \frac{dC}{dt} = kS(C - C_0)^n \tag{7}$$

where k is a rate constant, S the surface area, and C_0 is C at $t = \infty$. If the dissolution is transport-controlled, i.e., controlled by the rate at which solute molecules diffuse from a saturated layer at the interface into the bulk of the solution, then n will equal unity. If the diffusion coefficient D of the solute is known, the thickness δ of this diffusion layer can be calculated from

$$k_T = \frac{kV}{A} = \frac{D}{\delta} \tag{8}$$

where k_T is the velocity constant for the transport process, A the surface area, V the volume of solution and k the first-order rate constant $(= \frac{DA}{V\delta})$. Values of δ calculated from Eq. (8) are typically of the order of 0.03 mm, i.e., 50,000 molecules thick. This is, to say the least, unlikely. It is much more likely that fluid motion exists down to very short distances from the solid surface and that a gradient exists between this surface layer and the bulk solution. Experimentally, however, relations of the type $k_T \propto D^{0.70}$ instead of $k_T \propto D$ are usually observed. This implies that δ is a function of D and, because of the probable fluid flow within the Nernst Layer, probably suggests that δ is also a function of viscosity and temperature, although the dependence of δ on temperature will be slight.

An increase in stirring rate should be accompanied by an increase in the observed velocity, since the thickness of the liquid layer adhering to the surface will decrease. In addition, if δ is a function of the rate and type of stirring only, it should be temperature-invariant and k_T and D will have the same temperature coefficient. At 25°C the value of the activation energy for diffusion-controlled processes ranges between 11.7 and 28.5 kJ mol^{-1} throughout the fluid body and for a given surface area of solid the rate of dissolution will be inversely proportional to the volume of solution and the rate may be expressed as

$$\frac{dC}{dt} = \frac{k_c AC^n}{V} \tag{9}$$

where k_c is the chemical rate constant per unit volume and n is the order of the reaction. Heterogeneous reactions show a gradation between chemical and transport control as the rates of these processes approach the same order of magnitude and the observed rate is determined by a combination of these two. These may be termed as processes of an intermediate nature.

4. Kinetics of Crystal Growth

The processes which control the rate of crystal growth are analogous to those discussed in Section II B 3.

The rate of growth may be controlled by (a) diffusion in the surrounding liquid, (b) convection in the liquid, and (c) molecular processes at the surface such as adsorption, surface diffusion, desolvation, and the fitting of the units into the surface structure.

The mechanism of crystal growth has relevance to analytical chemistry, desalination, geochemistry, production, and dissolution of bone minerals. The growth of crystals proceeds in two steps: first, nucleation which corresponds to the production of new centers for spontaneous growth, and second, the deposition on the nuclei of material from the supersaturated solution. Investigation of the process of nucleation is complicated by the difficulty of detection of nuclei before they have grown to microscopic dimensions. The rate of this process determines the size and size distribution of the resulting crystals.

Attempts to understand the nucleation process were based on analogies with theories for the condensation of liquids. However, these theories are not simply applicable because of diffusion of solute ions, desolvation on crystallizing, interfacial adsorption, and double layer formation at the solid/solution interface. The process is further complicated by the almost impossible task of ensuring the absence of foreign bodies in the solution, since these act as nucleation sites. There is furthermore no universal agreement as to whether nucleation and growth occur simultaneously or consecutively.

In most electrolytes one observes a critical supersaturation above which instantaneous precipitation occurs and below which stable supersaturated solutions may exist for long periods. If these supersaturated solutions are seeded with crystals of the electrolyte the rate of growth of these seed crystals may be observed. This study of the growth from seed crystals in their supersaturated solution overcomes the difficulties associated with the study of spontaneous crystallization since reproducible results are easily obtained. Radiotracer techniques would appear to offer a number of advantages over other methods (conductivity, specific ion electrode potentiometric analysis, and chemical methods). The solute may be studied in the presence of higher concentrations of inert electrolyte to observe the effect of changing ionic strength. Furthermore, one is not limited to any particular range of solubilities or stable supersaturation concentration. The rate of crystallization is generally described by the equation $-dC/dt = kS(C - C_0)^n$ where n is the reaction order, S a function of the number of growth sites available, and k the rate constant. For a number of symmetrical electrolytes there is experimental evidence for $n = 2$. This is interpreted as being consistent with a surface-controlled mechanism; crystallization takes place via the simultaneous dehydration of pairs of oppositely charged ions in an adsorbed monolayer of hydrated ions.

To summarize, processes where the rate is determined by transport (diffusion) will obey first-order kinetics, will be susceptible to changes in stirring rate and will exhibit an energy of activation of between 11.7 and 28.5 kJ mol^{-1}. On the other hand, processes where the rate is determined by chemical factors will not obey first-order kinetics, will not be susceptible to changes in stirring rate and will probably have energies of activation differing from that of a diffusion-controlled process.

C. Experimental Methods

1. Solubility

Because of the marked influence (p. 462) of particle size on solubilities and rates of solution it is necessary to use some fractionating technique to obtain particles of uniform size. Methods based on simple sieving and sedimentation are not always satisfactory.

Koprda and Fojtik [8] describe a simple but effective apparatus for achieving fractionation by a hydraulic flow method. Crystals are separated according to their differing hydraulic properties of sedimentation in a counter flowing liquid. The particle size of the separated fraction is determined by the value of the linear flow rate of the liquid. The change of this linear flow rate is obtained by varying the cross section of a series of interconnected fractionating tubes at a constant flow rate of liquid. In this work a set of vertical tubes with stepwise increasing diameters connected in series was used. At a given volume flow rate the linear flow rate in these tubes is inversely proportional to the cross section. Different linear flow rates of the liquid operating in the individual tubes determine the hydrodynamic properties and produce the required fractionation of particles with differing sedimentation speeds.

Samples of labeled material thus fractionated are usually equilibrated with purified solvent in a thermostatted glass vessel. Samples of the solution are removed periodically and the radioactivity assayed until equilibrium has been achieved. The solution should be centrifuged through a filter tube to remove small crystallites whose presence would give rise to spuriously high results. Pipettes should be thermostatted prior to use to preclude crystallization. For determination at elevated temperatures, crystallization from the solution may occur while liquid and crystals are being separated by centrifuging. To avoid this a simple modified pipette of the type shown in Fig. 2 is useful.

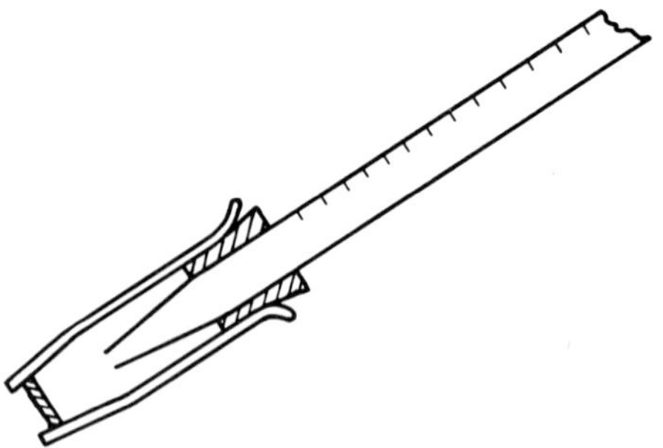

FIG. 2. Sample removal pipette.

The filter tube is connected to a pipette by a rubber bung. The thermostatted filter disc is immersed into the solution and the plunger of the pipette withdrawn to create a partial vacuum inside the filter tube. Crystal-free solution passes through the filter disc. In all these pipetting operations it must be remembered that adsorption from the solution will occur at the fresh liquid/glass interfaces and so the initial few pipetted portions should be discarded. This is especially important when we deal with extremely low solubility compounds. The apparatus is withdrawn, the filter tube disconnected from the pipette, and the portion of solution for assay removed with a second prethermostatted pipette. Specific activities of the solution are compared with those of a range of accurately prepared standards. This method of determining solubilities is simple, foolproof, and accurate. Care must be taken to ensure that equilibrium has been achieved and it is this which may lead to sources of error especially if the rate of solution is low. As will be discussed later, it is sometimes possible to determine the solubility from the rate of solution if the kinetics of dissolution is known. One interesting method of overcoming the problem of achieving equilibrium is the use of the packed column technique (Fig. 3) described by Kraus et al. [9]. Here at its simplest solvent flows slowly through a packed bed of labeled material. Provided that the amount of material forming

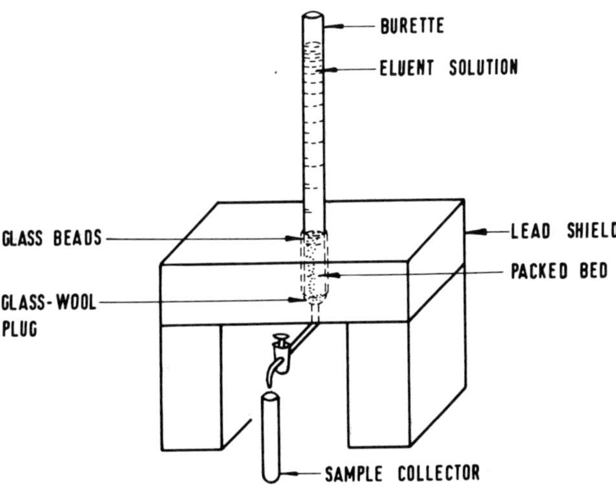

FIG. 3. Simple shielded packed column for solubility determination. Solvent flows through the packed column of labeled solute. Samples are collected and counted externally. (Reprinted from Ref. 9, p. 392, by courtesy IAEA.)

the bed is large relative to the solubility, thus avoiding the exhaustion of the bed or channeling of the solvent through the bed, equilibrium is established in only a few seconds and in the measurements reported by Kraus extremely satisfactory results were obtained.

Modifications of the simple apparatus (Fig. 4), allow measurements to be made at high temperatures and pressures. The effluent from the column passes through a long jacketed capillary. This capillary contains a large diameter portion before the exit end which serves as a counting cell. The concentration of radioactive tracer is ascertained with a scintillation counter located above the cell in a thermally insulated well. The crystal of the counter and the photomultiplier tube housing are water-cooled. Although a substantial distance separates the counter from the packed bed, capillary shielding is needed to lower the background. The counting cell must be calibrated by comparing its counting

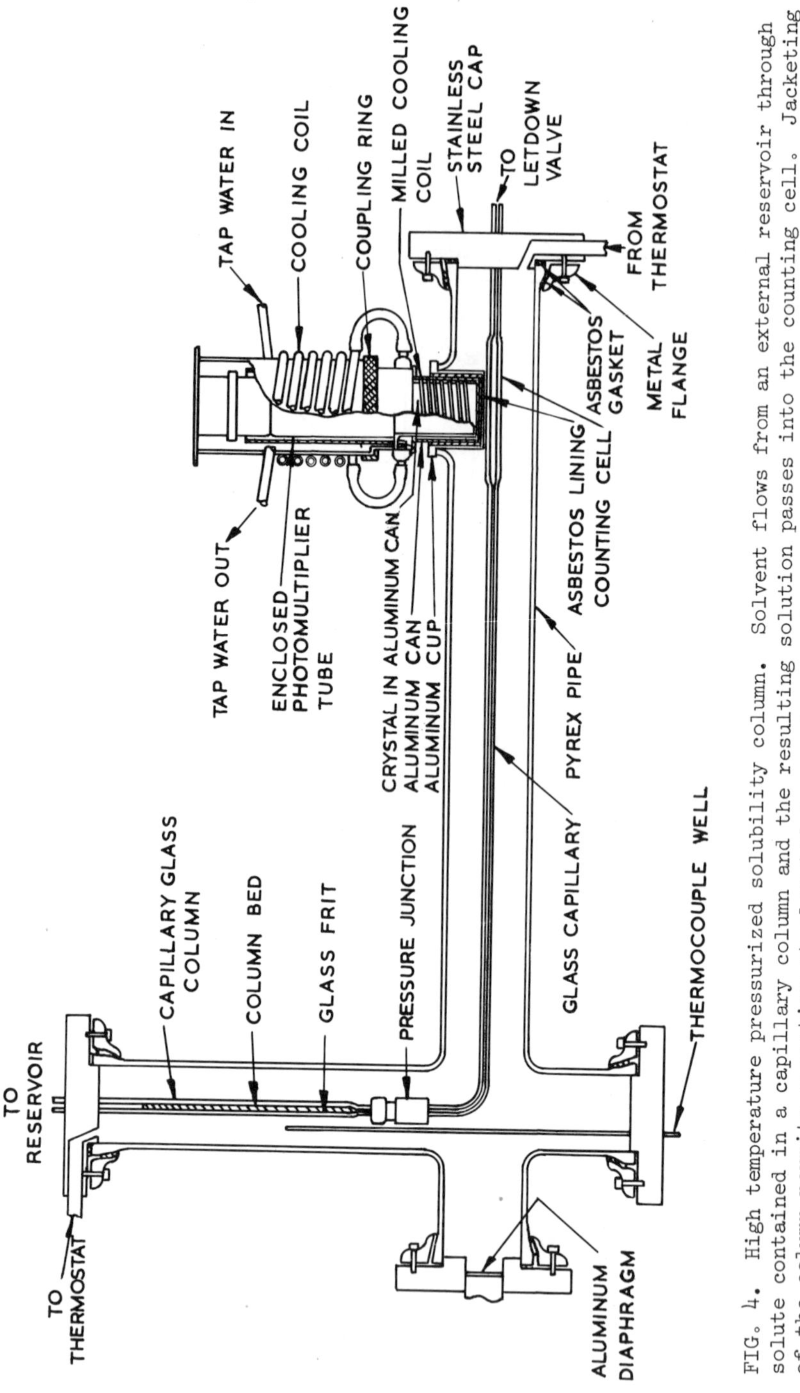

FIG. 4. High temperature pressurized solubility column. Solvent flows from an external reservoir through solute contained in a capillary column and the resulting solution passes into the counting cell. Jacketing of the column permits operation at elevated temperatures and pressures. Temperature is monitored by thermocouples. NaI detector is water-cooled. (Reprinted from [9], p. 393, by courtesy IAEA.)

rate under steady state conditions with that of effluent samples used for establishing the counting rate of the original sample. Provided this is done absolute measurements of solubilities may be made. Results were obtained for the solubility of $^{110m}AgCl$ in solutions of HCl at temperatures from 25°C to 200°C. Heats of solution thus obtained were consistent with previous results based on other techniques. The degree of complexing of AgCl by Cl^- was determined over a range of temperatures from the variation of solubility with temperatures at constant HCl concentration. This method seems potentially very useful especially when one considers its usefulness at elevated temperatures.

Applications of radiotracer techniques are by no means restricted to crystalline inorganic solids. Atkins [10] has studied atmospheric washout and deposition of labeled organic chlorine pesticides. ^{14}C-Labeled γ-benzene hexachloride, *p,p'*-DDT and Dieldrin (HEOD) were used to prepare aqueous solutions of concentrations varying from very dilute to greater than saturated (the excess pesticide is presumably present as a suspension). 20-40 liters of filtered, water-saturated air at 50 cm^3/min was passed through 100 cm^3 of these solutions (see Fig. 5); equilibrium was established rapidly between the solution and gas phase, and the gas phase pesticide was subsequently absorbed in a xylene-based scintillator. The concentration of pesticide in the contact bubblers at the start and finish of each experiment was measured and the partition coefficient p was determined by

$$p = \frac{\text{conc. pesticide/cm}^3 \text{ aqueous phase}}{\text{conc. pesticide/cm}^3 \text{ air}} = \frac{(C_1 + C_2)V}{2R} \tag{10}$$

where C_1 and C_2 are the radioactivity per unit volume (cpm/cm^3) at the initial and final pesticide concentration, V the volume (cm^3) of air passed, and R (cpm) is the pesticide absorbed in the scintillator in the Arnold bulbs. Figure 6 shows partition coefficients for all three pesticides plotted against concentration. A sharp increase in the apparent partition coefficient

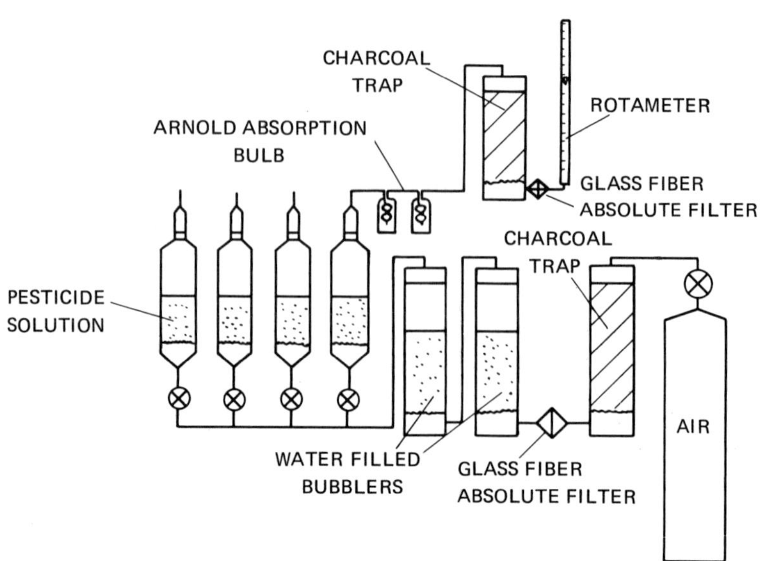

FIG. 5. Apparatus for determining the air-liquid distribution of pesticides. Water saturated air is bubbled through labeled pesticide solution. Pesticide in the air-water mixture is adsorbed for subsequent assay by liquid scintillation counting. (Reprinted from Ref. 10, p. 523, by courtesy of IAEA.)

occurs when the solubility is exceeded. The constant value below saturation indicates that Raoult's law is obeyed and indicates the high radiochemical purity of the pesticide.

The point of deviation allows values of solubility and hence saturation vapor pressure to be calculated as

$$p = \frac{S \times 10^{-6} \times 760 \times 2240 \times 293}{VP \times M \times 273} \qquad (11)$$

where p is the partition coefficient at 20°C, VP the saturation vapor pressure (torr) at 20°C, M the molecular weight, and S is the solubility in ppm at 20°C.

The solubility figures obtained (5.7 ppm γ-BHC; 9×10^{-3} ppm *p,p'*-DDT; 3.3×10^{-2} ppm Dieldrin) are considered more

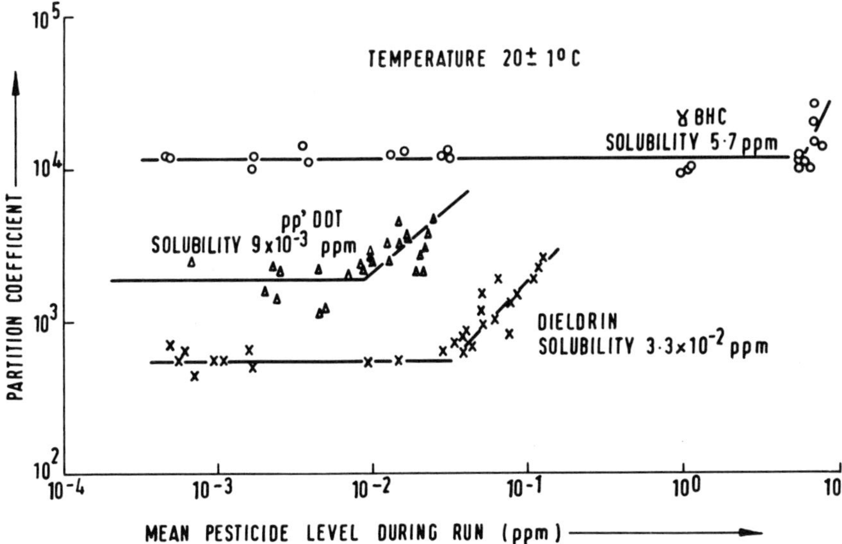

FIG. 6. Partition and aqueous solubility measurements on γ-BHC, *p,p'*-DDT and Dieldrin at 20°C. (Reprinted from Ref. 10, p. 525, by courtesy of IAEA.)

reliable than those obtained by previous workers using conventional weight loss techniques which give difficulty with separation of solution from solid phase or adsorption of solute at surfaces. Both problems are avoided in this radiometric method. This work should stimulate the application of radioisotopes to a large number of environmental pollution problems.

2. Kinetics of Dissolution

As is apparent from the theoretical introduction (Section II B) the rate of solution is critically affected by surface area, crystal geometry, stirrer geometry, solution total volume, and stirring rate.

Samples may be cylindrical and rotated at a controlled monitored rate as by Jones for $^{110m}AgCl$ [11] (a diffusion-controlled process), or partially embedded in an inert medium and the solution stirred by an efficient stirrer whose stirring rate is monitored. The partial embedding may be achieved by pressing heated crystals on to plates of high-purity polyethylene, or spread on a thin film of epoxy adhesive. Care must be taken, however, since certain polyethylenes are prepared with the use of catalysts which may leach out and give spurious results. Samples of potentially usable material should be immersed in high purity water for at least 1 week prior to use and should only be accepted if the conductivity of the water does not change significantly. Similarly not all commercial epoxy resins are completely stable to water over prolonged periods and caution must be exercised to ensure that a stable example is chosen. Sufficient sample must be mounted to ensure that no large change in surface area occurs during the dissolution process (one should remember that one of the great advantages of the radiotracer technique is that it allows dissolution to be followed over a wide range of concentration), and the position of the sample in the reaction vessel should be reproducible. Samples of the solution may be removed by pipetting for assay. Care must be taken to preclude the pickup of any minute crystals which might be accidentally removed from their mount and the use of a filter pipette illustrated in Fig. 2 is recommended. If samples are removed for assay, the volume of liquid in the reaction flask will change. This is not of practical significance if the change in volume during the experiment is only a few percent, but it means that sampling intervals must be carefully selected. In theory there is no reason why a flow cell should not be used to monitor the radioactivity of the solution. If this is chosen then the same considerations apply as with the AKUFVE (Section III D 2). The reaction may be allowed to continue until equilibrium has been established to determine C_0 (in cpm) for conversion to molarity by comparing the radioactivity of the solution with that of standard solutions.

Certain graphical methods of analysis of data allow the establishment of the rate law and the measurement of the solubility without equilibrium being achieved. For a second-order, chemically controlled dissolution process a plot of C/t against C will be linear (see Fig. 7 for $Ba^{35}SO_4$) and the intercept of C at C/t = 0 is C_0 [2]. In the case of a first-order (transport-controlled) process like AgCl log (C_0 - C) was linear with time, and the energy of activation, measured by a rapid change of temperature after 40% dissolution, was comparable to that for diffusion [11]. This temperature change method ensures that no unusual change of surface area occurs which would invalidate comparisons achieved with runs carried out at different temperatures on different occasions. In the case of the second-order processes C/t was linear with C [11] (in addition to $(C_0 - C)^{-1}$ with t), and the energy of activation measured by the same temperature change method was higher than that for diffusion. The rate of dissolution of ^{110m}AgCl depended on the

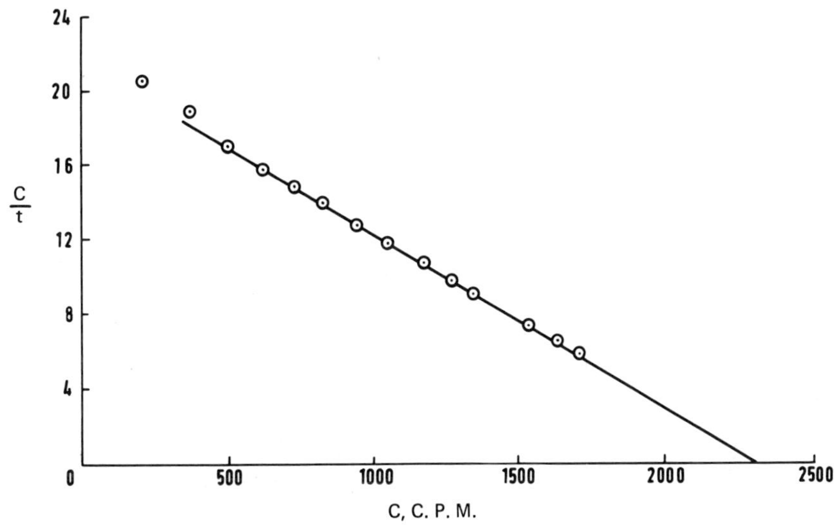

FIG. 7. Determination of the aqueous solubility of $Ba^{35}SO_4$ at 25°C from the rate of solution for a second-order process. C, radioactivity of solution (counts per minute); t, time. (Reprinted from Ref. 3, p. 1977, by courtesy of Pergamon Press.)

stirring rate [11], whereas that for $BaSO_4$ was unaffected by changes in stirring rate [12].

$PbSO_4$ [13] provides an interesting example of intermediate behavior; at low temperatures and low stirring rate the dissolution is influenced significantly by transport processes, whereas at high stirring rates or temperatures, chemical control is observed. Similar behavior has been observed by Delmas in these laboratories, who used the cell shown in Fig. 8 for studying the dissolution kinetics of $SrSO_4$ [14].

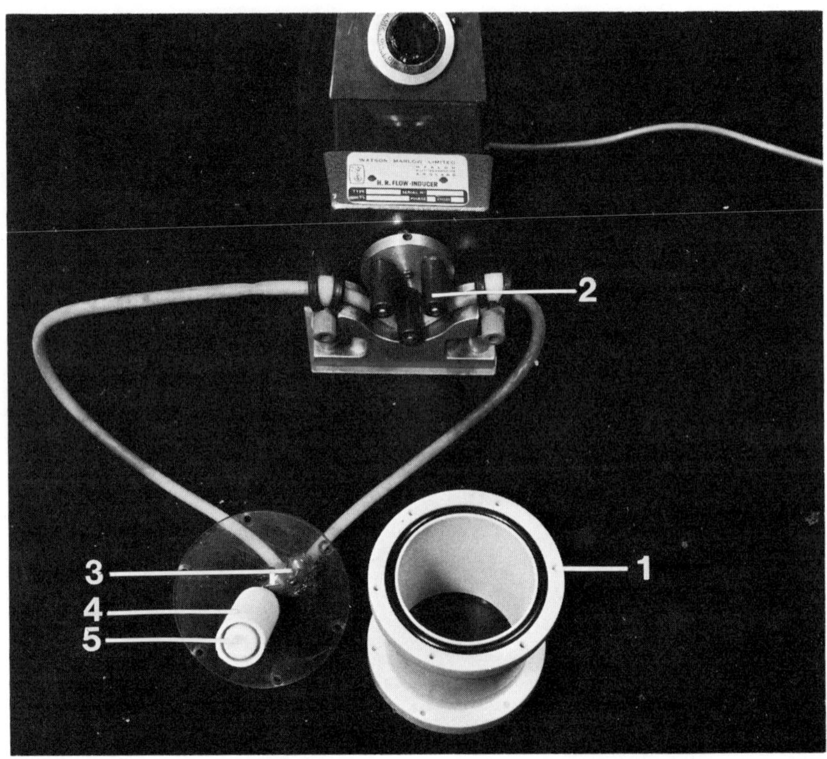

FIG. 8. Cell for crystal dissolution studies: (1) cell constructed from PVC (polyvinylchloride); (2) peristaltic pump for solution circulation; (3) syringe inlet and outlet; (4) PVC cylinder coated with araldite, and covered with $Sr^{35}SO_4$; (5) perforated diffuser to minimize turbulence due to solution circulation.

3. Kinetics of Growth

Reddy and Nancollas [15] studied the kinetics of crystal growth of calcite using ^{45}Ca as a tracer. Inactive seed crystals of $CaCO_3$ were added to stable supersaturated solutions of $^{45}CaCO_3$, and the kinetics of growth have been studied with consideration of the isotopic exchange between inactive crystals and active solution. The results indicated that crystal growth commences immediately upon the addition of seed crystals. There is little or no exchange between supersaturated solution and crystal surface showing that the new solid phase is in isotopic equilibrium with the solution from which it has grown rather than with the underlying seed crystal. The rate equation for growth obtained was second order with respect to concentration and not diffusion-controlled, stirring rate having no effect on dissolution. Surface control of growth was proposed.

There is obviously ample potential for further applications of radiotracer techniques to growth kinetics including the use of labeled adsorbants to demonstrate the effects of crystal imperfections on growth.

D. Radiometric Titrations and the Determination of Solubility Products

1. Endpoint Determination by Precipitation

Radiometric endpoint determination was first reported by Langer [16] as long ago as 1940. He describes the principle of radiometric titrations based on precipitate formation and an instrument using separation by filtration. Langer [17,18] has described titrations based on the precipitation of phosphate labeled with ^{32}P and with ^{110m}Ag as the indicator in radioargentimetry. The technique was extended by Duncan and Thomas [19] and Korenman et al. [20] to complex formation with separation by solvent extraction. Solid indicators were used by Braun et al. [21] to investigate complex formation (radiocomplexometry).

In radiometric titrations based on precipitate formation the radioactive indicator is precipitated during the titration. The endpoint is detected by following the change in the radioactivity of the liquid phase. If $^{110m}Ag^+$ is titrated with iodide, the endpoint may be detected by the fall in the solution activity. If, however, the titrant is labeled with ^{125}I the endpoint is determined by the increase in solution activity. Ideally if continuous activity measurements are made on the solution during the titration, one will observe either a falling straight line before or a rising straight line after the endpoint, and the point at which this line intersects the constant activity axis is the endpoint of the titration. The level of activity of this constant axis is determined by the solubility of the precipitate. If solution and titrant are labeled with different isotopes separately, the endpoint is given as the point at which the two straight lines of positive and negative slope intersect. Such a dual labeling method would allow a more accurate estimate of the endpoint than in the case of single labeling.

The definition of the endpoint may be blurred by incomplete precipitation and the accuracies obtained have been rather low, partially due to the uncertainty of particle size and distribution.

2. Principle of Solubility Product Determination by Radiometric Titrations

Radiometric titration was first used by Duncan in 1959 [22] to determine the solubility product of PbI_2 by titrating 5 ml of 0.01 M $K^{131}I$ with 0.005 M $Pb(NO_3)_2$. If a volume V of a solution of the ion N^- at concentration $[N^-]_0$ is added to a solution volume V_0 of the labeled ion $\dot{M}^+$ at a concentration $[M^+]_0$ and an activity A_0 cpm and if the solubility product $S = [M^+][N^-]$ where $[M^+]$ and $[N^-]$ are the concentrations of M^+ and N^- in equilibrium with any precipitated solid MN, then

$$SV^2 + VV_0\{2S - [N^-]_0[M^+]_0R\} + V_0^2\{S + [M^+]_0^2R(1 - R)\} = 0 \tag{12}$$

with $R = A/A_0$, A is the activity of M^+ remaining in solution after precipitation. Generally the variation of R with V has a curvature, as shown in Fig. 9 for PbI_2. From each point on the graph S may be determined by the use of Eq. (12). For more generalized reactions such as

$$nM^{m+} + mN^{n-} = M_nN_m$$

a similar calculation leads to

$$S = [M^{m+}]^n[N^{n-}]^m$$

$$\left\{\frac{[M^{m+}]_0RV_0}{V + V_0}\right\}^n V \left\{\frac{[N^{n-}]_0 - (m/n)[M^{m+}]_0V_0(1 - R)}{V + V_0}\right\}^m \tag{13}$$

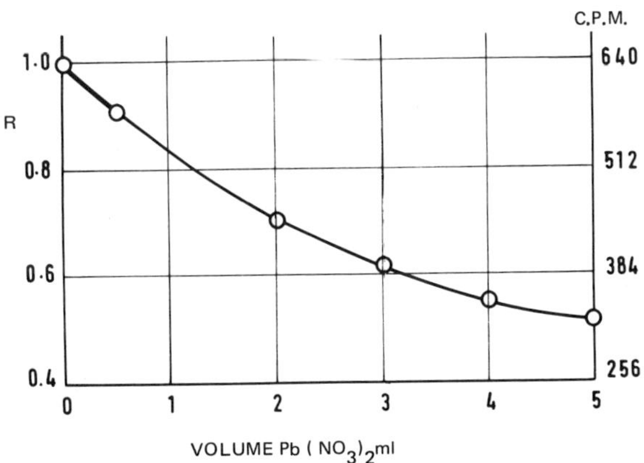

FIG. 9. Determination of the solubility of $Pb^{131}I_2$ by radiometric titration. Titration of 5 ml of 0.01 M $K^{131}I$, 0.005 M $Pb(NO_3)_2$ solution as titrant at 18°C. Solubility product determined as $(0.88 \pm 0.2) \times 10^{-18}$ M^3. (Reprinted from Ref. 22, p. 162, by courtesy of Pergamon Press.)

Assuming $[M^{m+}]_0 = k[N^{n-}]_0$ and equivalent mixing, i.e., $V = kV_0 m/n$, we obtain

$$S = \left(\frac{m}{n}\right)^m \left\{\frac{R[N^{n-}]_0}{m/n + 1/k}\right\}^{m+n} \tag{14}$$

Equation (14) is further simplified to

$$S = \frac{R^2[M^+]_0^2}{4} \tag{15}$$

for a compound of MN type, or to

$$S = \frac{R^3[N^-]_0^3}{16} = \frac{R^3[M^{2+}]_0^3}{2} \tag{16}$$

for MN_2 compound. Equations (15) and (16) allow the most direct experimental procedure, since it is necessary only to mix equivalent amounts of cation and anion and to determine the fraction of labeled species remaining in solution.

The usual care is necessary to avoid the picking up of small crystallites in the solution for assay and the results obtained by Duncan were in good agreement with the literature. The accuracy, ±20%, could be improved by the use of higher specific activities. More reliable results would have been obtained if less concentrated solutions had been used, although this would have increased the time scale of the measurement greatly.

Similar accuracies were obtained by Casey and Robb [23] who, using $^{137}CsCl$ as titrant, obtained a value of $(3.51 \pm 1.09) \times 10^{-12}$ mol^3 $liter^{-3}$ for the solubility product of Cs_2PtCl_6 in water. Generally this method can be satisfactory and reliable and is particularly useful when a long time is required for equilibrium between precipitate and solvent.

III. LIQUID-LIQUID DISTRIBUTION

A. Introduction

The extensive application of liquid-liquid extraction processes in the nuclear science industry both as a means of processing fuels and as a separation method preceding analysis has led to an abundance of papers on this topic during the past 20 years or so. Only a small proportion of the work, however, has made specific use of radioactivity of the dissolved materials. The application of liquid-liquid extraction extends wider and it has been usefully employed in the study of solution equilibrium for its own sake, e.g., simple distribution experiments can provide value of the Gibbs free energies of transfer [24] and ion association and complexation constants. The principles of liquid-liquid extraction have been dealt with in detail elsewhere [25] and only an outline will be given here.

B. Principles of Liquid-liquid Distribution

Consider a system of two immiscible solvents 1 and 2 and a solute Y which contains a radiometrically detectable group or atom X. If the concentration of Y is $[Y]_1$ and $[Y]_2$ in the solvents 1 and 2, respectively, the distribution coefficient K_D can be written by

$$K_D = \frac{[Y]_2}{[Y]_1} = \frac{[X]_2}{[X]_1} \tag{17}$$

For equilibrium distribution of Y between the two phases we can show that

$$K_D = \frac{\gamma_1}{\gamma_2} K \tag{18}$$

where K is the true thermodynamic equilibrium constant and γ_1 and γ_2 are the activity coefficients of Y in solvents 1 and 2, respectively, referred to the same standard state. The standard state may, for example, be infinite dilution in solvent 1 or a 1 M solution in solvent 1, etc.

Variations in the distribution coefficient K_D will clearly occur for a system in which γ_1 and γ_2 show different concentration dependence. Also systems in which the mutual solubilities of 1 and 2 change with the concentration of Y will also show a concentration dependence of K_D. These latter effects are always present, but a much more important factor affecting the distribution of X is self-association or dissociation in either solvent or chemical reaction with other components present in each phase. In such situations the ratio $[X]_2/[X]_1$ is no longer equal to K_D. It is then more practical to use the distribution ratio D defined as

$$D = \frac{\text{Total concentration of X in all species present in phase 2}}{\text{Total concentration of X in all species present in phase 1}} \tag{19}$$

which equals K_D when no chemical interaction takes place in the system. The effect of chemical interaction may produce gross changes in D as concentration is varied. From studies of the factors affecting D the nature and magnitude of the interactions can often be elucidated [25]. Extensive use of liquid-liquid distribution has been made in the study of inorganic complexes [26,27].

C. Tracer Application

Whatever application is made to liquid-liquid distribution, experimental work must always involve determination of the total concentration of X in each phase. The usefulness of the conventional distribution method has been greatly extended by the wide range of radioisotopes now available. Analysis of a single component in a complex mixture can be achieved from direct measurements on the phases since there is no interference from other components. Because of the sensitivity of detection and measurement, the determination of the distribution ratio for a component which is overwhelmingly present in one phase is also facilitated. For the same reason accurate measurements can be made on very dilute solutions, this is particularly advantageous if species are only slightly soluble or if polynuclear complexes are formed at higher concentrations. Examples of the latter will be discussed later.

A number of precautions must be taken to ensure the reliability of distribution ratios. Radiochemical impurities will be present in each phase according to their distribution ratios, hence a high degree of radiochemical purity may be necessary. This will be particularly important in measurements on systems with very large (or small) distribution ratios. Let us consider a mixture of the main component $(D = 10^{-4})$, and an impurity $(D = 1)$ in the ratio of 10^4:1. The measured D will be apparently 1.5×10^{-4}. If the ratio is 10^3:1, then the measured D is 6×10^{-4}. In any case radiochemical purity should be carefully checked for all phases being analyzed. The presence of impurities may increase or decrease the measured distribution ratio, depending on the balance of their interactions with the components in both phases. The tracer should also be in the same chemical form as the species being studied; e.g., addition of $^{63}NiCl_2$ to a solution of a nickel complex will not be of use to trace the movement of the complex, unless the exchange between the two compounds is rapid.

If phase equilibrium is not established rapidly, then D will depend upon the time of contact between phases [25]. In preliminary studies wherever possible, the equilibrium should be approached from both sides, i.e., solutes dissolved in solvent 1 are then mixed with solvent 2 and vice versa, providing solubilities are sufficiently great in the two solvents.

D. Experimental Technique

The essential requirements for liquid-liquid extraction work are good mixing, to provide large areas of contact between phases in order to establish rapid equilibration, followed by a clean separation of the phases. The clean separation is very important in cases where the distribution ratio is very large or very small, as is well illustrated as follows. For a system in which the distribution ratio is 10^{-6}, inclusion of one part aqueous phase in 10^4 of the organic phase will lead to a measured value of D, 100 times too large. For D values of 10^{-5} and 10^{-4} the apparent values would be approximately 10 times and twice too large.

1. Batch Method

Conventional laboratory apparatus has been extensively used, e.g., see Rydberg [28] and Marcus [29]. Preliminary experiments are necessary to establish the time to reach equilibrium. Phases often separate rapidly under gravity, but a quicker separation is achieved by centrifugation. A very simple device which can be used for sampling the lower layer and which effectively prevents cross-contamination is shown in Fig. 10. To sample the lower phase the glass tube b is inserted into the flask or centrifuge tube, maintaining a slightly positive pressure by pinching the rubber bulb. The bulb is then removed from the tube which provides contamination-free access. Such equipment is inexpensive and simple to use, but for a large program of experiments the batch method is tedious and time-consuming, particularly if equilibration is slow.

2. Continuous Methods: The AKUFVE

A system [30] which is capable of continuous monitoring of phase composition and which can give rapid acquisition of data has recently been developed. The apparatus is known as AKUFVE, which

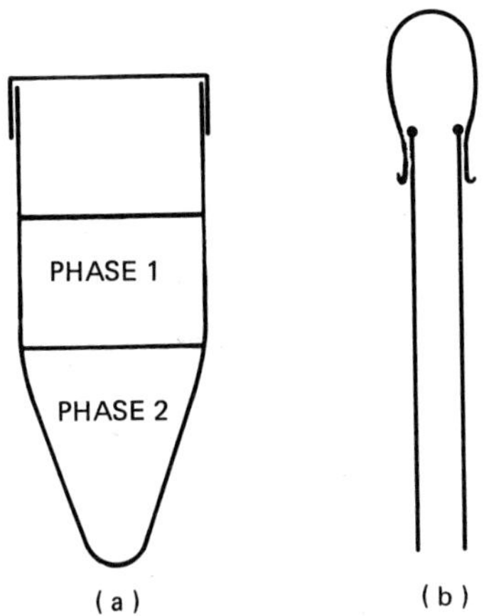

FIG. 10. Phase sampling tube.

is the Swedish abbreviation for "apparatus for continuous measurement of distribution factors in solvent extraction." An outline diagram of the apparatus is shown in Fig. 11.

The two liquid phases are stirred in the mixing chamber to which further additions of reagents, solvents, etc. can be made. The mixture then flows, via 1, into the continuous centrifuge for quick phase separation. The two phases then pass, via 2 and 3, through suitable devices for continuous measurement of flow, temperature, and concentration and then flow either back into the mixing chamber in a closed cycle or out into a collector. About 450 ml of each phase is added to the mixing chamber.

Flow rates and centrifuge speed are adjusted until pure phases are obtained. The importance of absolute phase separation has already been discussed and a means of checking that this has been achieved is discussed in detail by Reinhardt and Rydberg [31]. In the AKUFVE apparatus used by Anderson et al. [32] the radioactivities R_m and R_0 for the sample and total background, respectively, are measured in a flow cell placed in a well-type NaI (Tl) scintillation crystal (A and B in Fig. 11) enclosed in a

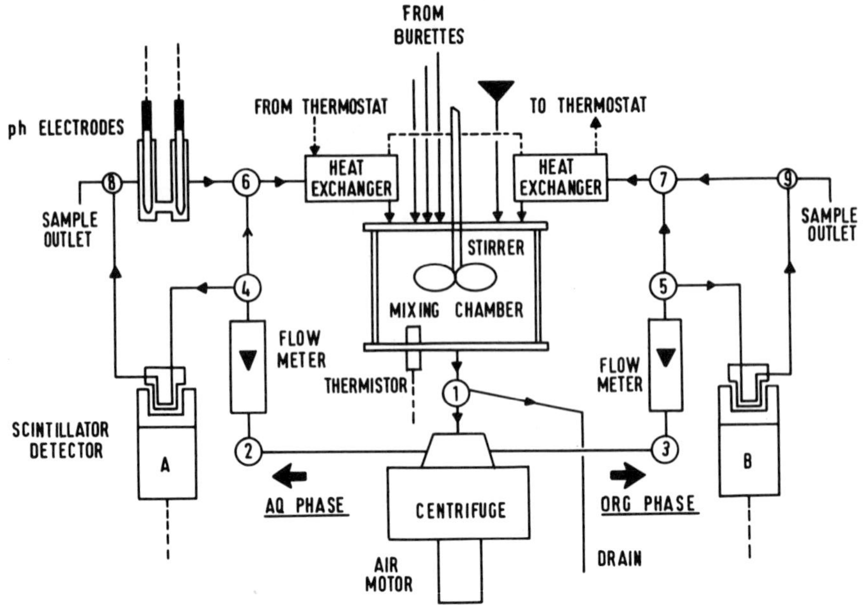

FIG. 11. AKUFVE apparatus. Parts 1-9 are valves, A and B are well-type NaI (Tl) scintillation detectors into which are placed liquid flow cells. (Reprinted from Ref. 31, p. 2783, by courtesy of *Acta Chimica Scandinavica*.)

7 cm-thick lead shield. The distribution ratio D is given by

$$D = \frac{(R_{m(org)} - R_{0(org)})V_{aq}\Phi_{aq}}{(R_{m(aq)} - R_{0(aq)})V_{org}\Phi_{org}} \tag{20}$$

in terms of the detector cell volume V and counting efficiency Φ. The background $R_{0(org)}$ or $R_{0(aq)}$ arises, as discussed by Reinhardt and Rydberg [31] from such sources, as natural background, radiation from the parts other than the flow cell and radioisotopes adsorbed on the walls of the flow cells. Because of the importance of these corrections, particularly in the cases of high and low D values, background radioactivity for each phase should be determined regularly during experiments. Anderson et al. [32] have suggested two ways of doing this. In the first equal volumes of samples are taken from the phases and measured in an external counter under the same conditions. For such measurements ϕ is constant. If $r_{m(org)}$ and $r_{m(aq)}$ are the externally measured radioactivities of the two phases and r_0 the background, then the relationship between the AKUFVE measurements and the externally measured activities is:

$$\frac{(R_{m(org)} - R_{0(org)})\phi_{aq}V_{aq}}{(R_{m(aq)} - R_{0(aq)})\phi_{org}V_{org}} = \frac{r_{m(org)} - r_{0(org)}}{r_{m(aq)} - r_{0(aq)}} = D \qquad (21)$$

For very low D values $R_{m(aq)} >> R_{0(aq)}$ so $R_{0(aq)}$ can be neglected and $R_{0(org)}$ calculated. Similarly for high D values $R_{0(aq)}$ can be calculated.

In the second method R_0 is obtained directly by flushing the cell with an inactive solution of identical composition and is based on the uncertain assumption that the amount of tracer adsorbed and trapped in the cell remains unchanged.

Anderson [33] in experiments with tri-n-butyl phosphate has shown that this compound penetrates into the Viton and Teflon tubing used in the AKUFVE cell construction. In subsequent experiments, e.g., using metals at trace concentrations, the tri-n-butyl phosphate in the tubing extracts metal ions from the solution phase producing a change in the radioactivity.

Ottertun [34] has described a flow cell in which contact between the solution and the tube surface is avoided by creating a liquid jet which passes down the center of the tube. This technique, used in conjunction with the AKUFVE system for solvent extraction studies, has been shown to be reliable.

Anderson et al. [32] describe a data collection and processing system which allows full advantage to be taken of the high data output from the AKUFVE. It is based on a data logging unit, produced by SAAB Electronics AB, which accepts digital information from the detectors, pH electrodes, etc., and punches all readings on paper tape. They suggest that by connecting a small computer on-line to the data logging unit evaluation of measurements can be made continuously during an experiment. Such a procedure permits rapid calculation of D values which may suggest what concentration changes should be made in the system. All reagent additions can be accurately calculated and automatically performed under control of the computer.

E. Example of Application to Complex Equilibria

Manning and Monk [35,36] used the batch technique for studying association constants of ^{60}Co(II) with dicarboxylates and (^{154}Eu + ^{152}Eu)(III) with acetate, glycollate, sulfate, and tartrate. Because of the low solubility of the Co(II) dicarboxylates in the organic phase, an auxiliary extractant, oxine (HO), was added to produce an extractable uncharged complex MO_2. For the mixture of oxine and Co(II) the equilibria existing in the organic and aqueous phase are expressed as follows:

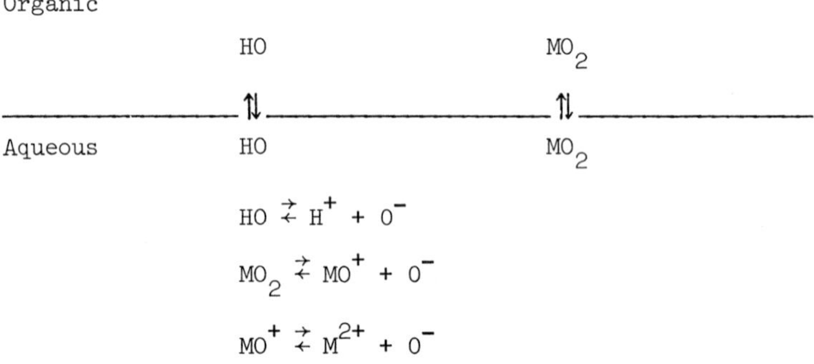

In the presence of dicarboxylate ligand (L^{2-}) the equilibria can be represented by:

Organic

$$\begin{array}{lccc} \text{Organic} & HO & MO_2 & ML \\ & \rightleftarrows & \rightleftarrows & \rightleftarrows \\ \text{Aqueous} & HO & MO_2 & ML \end{array}$$

$$HO \rightleftarrows H^+ + O^- \qquad ML \rightleftarrows M^{2+} + L^{2-}$$

$$MO_2 \rightleftarrows MO^+ + O^- \qquad ML + L^{2-} \rightleftarrows (ML_2)^{2-}$$

$$MO^+ \rightleftarrows M^{2+} + O^-$$

Higher and mixed complexes are also possible with ligands. The measured distribution ratio D is given by

$$D = \frac{\text{total Co in the organic phase}}{\text{total Co in the aqueous phase}}$$

$$= \frac{(C_1 - C_2)V_{aq}}{C_2 V_{org}} \qquad (22)$$

where C_1 and C_2 are the count rates of equal volumes of aqueous phases before and after equilibration and V_{aq} and V_{org} are the volumes of equilibrated aqueous and organic phases, respectively. Consequently,

$$D = \frac{[MO_2]_{org} + [ML]_{org}}{[M^{2+}]_{aq} + [MO^+]_{aq} + [MO_2]_{aq} + [ML]_{aq} + [ML_2^{2-}]_{aq}} \qquad (23)$$

and if the complex ML has low solubility in the organic phase,

$$D = \frac{[MO_2]_{org}}{[M^{2+}]_{aq} + [MO^+]_{aq} + [MO_2]_{aq} + [ML]_{aq} + [ML_2^{2-}]_{aq}} \qquad (24)$$

Manning and Monk [36] measured the distribution of tracer ^{60}Co as $CoCl_2$ ($\approx 10^{-6}$ M) between oxine in chloroform and aqueous sodium dicarboxylate. The aqueous solutions were made up to specific ionic strengths with sodium chloride. The distribution coefficient K_D in the sense of Eq. (17) for the complex MO_2 is given by

$$K_D = \frac{[MO_2]_{org}}{[MO_2]_{aq}} \tag{25}$$

Combination of Eq. (25) with Eq. (24) yields

$$D = \frac{K_D[MO_2]_{aq}}{[M^{2+}]_{aq} + [MO^+]_{aq} + [MO_2]_{aq} + [ML]_{aq} + [ML_2^{2-}]_{aq}} \tag{26}$$

Putting

$$\beta_1(o) = \frac{[MO^+]_{aq}}{[M^{2+}]_{aq}[O^-]_{aq}} \qquad \beta_2(o) = \frac{[MO_2]_{aq}}{[M^{2+}]_{aq}[O^-]_{aq}^2}$$

$$\beta_1(L) = \frac{[ML]_{aq}}{[M^{2+}]_{aq}[L^{2-}]_{aq}} \qquad \beta_2(L) = \frac{[ML_2^{2-}]_{aq}}{[M^{2+}]_{aq}[L^{2-}]_{aq}^2}$$

we have

$$D = \frac{K_D\beta_2(o)[O^-]_{aq}^2}{[M^+]_{aq}\{1 + \beta_1(o)[O^-]_{aq} + \beta_2(o)[O^-]_{aq}^2 + \beta_1(L)[L^{2-}]_{aq} + \beta_2(L)[L^{2-}]_{aq}^2\}} \tag{27}$$

For a constant concentration of auxiliary extractant the second and third terms in the denominator are constant, so that

$$D = \frac{K_D\beta_2(o)[O^-]_{aq}^2}{[M^+]_{aq}\{1 + \text{constant} + \beta_1(L)[L^{2-}]_{aq} + \beta_2(L)[L^{2-}]_{aq}^2\}} \tag{28}$$

If D^o is the value of D at $[L^{2-}] = 0$, then

$$\frac{D^o}{D} = 1 + \frac{\{\beta_1(L)[L^{2-}]_{aq} + \beta_2(L)[L^{2-}]^2_{aq}\}}{\{1 + \text{constant}\}} \tag{29}$$

If it is further assumed that the aqueous solubility of MO^+ and MO_2 are small, the constant can be put equal to zero. A plot of 1/D against $[L^{2-}]$ is linear at low $[L^{2-}]$ hence $\beta_1(L)$ can be calculated. From the variation of $\beta_1(L)$ with ionic strength, a value of 2.3×10^{-5} mol liter^{-1} is derived for the thermodynamic dissociation constant. This is rather higher than the value of 1.6×10^{-5} mol liter^{-1} from emf measurements [37]. Prue [38] has shown that the agreement is improved if some account is taken of the constant in the denominator of Eq. (29). The distribution of auxiliary extractant and its dependence on ionic strength could of course be studied by labeling.

Other more recent work has included determinations of activity coefficients of electrolytes [39] and further studies on complex formation [40].

IV. ION EXCHANGE

A. Principles of Ion Exchange

In the preceding section we showed how radiotracers are useful for studying complexation in solution. Ion exchange resins can be used in much the same way to provide a second phase.

A typical ion exchange resin consists of a three-dimensional polymer network containing ionizable groups. For a resin of the strong acid type the anion sites are fixed within the network, whereas the hydrogen ions can undergo exchange with other cations. When a resin is put in contact with pure water, the water shows a tendency to enter the network thus causing the resin to swell. This produces a solution within the three-dimensional framework of the resin. Since electrical neutrality must be preserved, this solution will contain a cation concentration equivalent to the rigidly held anion concentrations. For strong resins, ion concentrations within the resin phase are usually high (5-10 M). At equilibrium ionic species are partitioned between the solution and resin phases. If a 1:1 salt M_1X is present in an aqueous solution in contact with a 1:1 cation exchange resin RM_2, then the exchange can be expressed as

$$M_1X + RM_2 \rightleftarrows M_2X + RM_1 \tag{30}$$

or more briefly as

$$M_1^+ + \overline{M}_2^+ \rightleftarrows \overline{M}_1^+ + M_2^+ \tag{31}$$

where the bar denotes the exchanger phase. The thermodynamic equilibrium constant is given by

$$K = \frac{a_{\overline{M}_1^+} \cdot a_{M_2^+}}{a_{M_1^+} \cdot a_{\overline{M}_2^+}} \tag{32}$$

where a denotes the thermodynamic activity. Addition to the aqueous solution, of species which form a complex with the "free" ion M_1^+ (or M_2^+), will clearly disturb the exchanger/solution equilibrium by reducing the "free" ion concentration. Hence transfer of M_1^+ (or M_2^+) will occur from the resin to the solution in order to reestablish this equilibrium. A study of the way in which the equilibrium is affected by added substances can in principle be used to study interactions in solution. We rewrite Eq. (32) as follows:

$$K = \frac{[\overline{M}_1^+]}{[M_1^+]} \frac{[M_2^+]}{[\overline{M}_2^+]} \frac{f_{\overline{M}_1^+}}{f_{M_1^+}} \frac{f_{M_2^+}}{f_{\overline{M}_2^+}} \tag{33}$$

The quantities, $[\overline{M}_1^+]/[M_1^+]$ and $[\overline{M}_2^+]/[M_2^+]$, are called the distribution coefficients $K_{D(1)}$ and $K_{D(2)}$, respectively, for the ions 1 and 2 and expressed as

$$K_{D(1)} = \frac{\text{amount of solute 1 per gram of resin}}{\text{amount of solute 1 per milliliter of solution}} \tag{34}$$

etc. Hence

$$[M_1^+] = \left(\frac{1}{K} \times \frac{1}{K_{D(2)}} \times \frac{f_{M_2^+}}{f_{M_1^+}} \frac{f_{\overline{M}_1^+}}{f_{\overline{M}_2^+}} \right) \times [\overline{M}^+] \tag{35}$$

In general the quantity in the bracket will depend on concentration, with $f_{M_2^+}$ and $f_{M_1^+}$ being usually <1, and $f_{\overline{M}_1^+}$ and $f_{\overline{M}_2^+} < 1$ at the high internal concentration existing in the exchanger.

B. Experimental Techniques for Tracer Application

The use of radiotracers permits establishment of conditions for which the quantity in parentheses, Eq. (35), is a constant. This is achieved by employing trace concentrations of M_1^+ such that in each phase $[M_2^+] \gg [M_1^+]$. The concentration of M_2^+ in each phase is therefore negligibly altered by the exchange process and the activity coefficients which are then determined by $[M_2^+]$ will also remain constant so that

$$[M_1^+] = A[\overline{M}_1^+] \tag{36}$$

We now discuss experimental techniques before applying the method to specific examples. For details such as selection, storage, and conditioning of resins readers should consult Ref. 41. To achieve ion exchange equilibrium with the solution the following two methods [42,43] are currently used.

1. Column Equilibration

Some variants on this method have been described by Kraus et al. [44,45]. The particular choice depends primarily on the magnitude of the distribution coefficient K_D. For a small K_D value the solution containing tracer may be passed through the resin bed. K_D can be calculated provided that the breakthrough curve, i.e., concentration of tracer in effluent against effluent volume, approximates to the ideal one, i.e., there are no large departures from equilibrium.

Alternatively small K_D values ($K_D < 1$) can be measured using a retention method. In this case sufficient solution is passed through the column until inflowing and outflowing solutions have identical composition. Both solution and resin are now analyzed. The resin-adhering solution can be removed by low-speed centrifuging or mild suction.

For large K_D values ($K_D > 10^3$) Kraus et al. [44] have developed the preloaded column technique in which the resin is initially equilibrated with tracer before being placed in the column. The solution for which K_D is to be measured is then allowed to flow through the column. Distribution coefficients are calculated from the tracer concentrations in the resin and solution.

2. Batch Equilibration

Small measured amounts of resins and solution are simply shaken for sufficient time to establish equilibrium. The shaking time necessary will depend on a number of factors, such as mesh size of the resin, degree of molecular crosslinking, and its acidity or basicity. In principle an analysis of the solution phase before and after equilibration is sufficient to determine the distribution coefficients, but it may be necessary to analyze the resin when K_D is small, i.e., most of the tracer remains in the solution.

C. Experimental Results

1. Distribution Coefficients and Complexation

Kraus and coworkers have made much use of column techniques in their extensive work on anion exchangers [46,47]. Their principal aim has been to provide data which can be used as a basis for the design of separation methods.

Schubert et al. [48,49] have used the batch method to study complexation, in particular the dissociation of strontium citrate and tartrate. Shaking periods of up to 3 hr were found to be necessary to achieve equilibration in their solutions which contained resin in the ammonium form. Measurements were made of the distribution of tracer as a function of tartaric acid and citric acid concentrations. The concentration quotient defined by

$$K_c = \frac{[Sr^{2+}][tartrate^{2-}]}{[strontium\ tartrate]} \tag{37}$$

was found to be 2.02×10^{-2} ($pK_c = 1.69$) at 25°C, pH = 7 and an ionic strength of 0.2, the value being in fair agreement with $pK_c = 1.65$ obtained by potentiometric titration [50]. For strontium citrate, under similar conditions of pH and ionic strength, pK_c was found to be 2.81 compared with $pK_c = 2.70$ [51] and 2.92 [52], which have been reported previously. Similarly the results on the dissociation of cobaltous oxalate [53] in aqueous sodium chloride were comparable with those obtained from liquid-liquid distribution studies [38]. The calculations made by Schubert are based on the assumption that only complexes with zero or negative charge are formed. A method of greater generality has been developed by Fronaeous [54,55].

2. Calculation of Relative Activity Coefficients

Vanselow [56] suggested a method, based on ion exchange, for the measurement of relative activity coefficients in mixed electrolyte solutions. His method is explained by reference to Eq. (32), viz.,

$$K = (a_{M_2^+} \cdot a_{\overline{M}_1^+})/(a_{M_1^+} \cdot a_{\overline{M}_2^+})$$

Vanselow, using as a basis his work on aluminosilicates, argued that partial substitution of one cation for another results in the formation of mixed crystals so that the solid phases can be thought of as consisting of solid solutions of RM_1 and RM_2. He then assumed such solutions are ideal so that the activities of RM_1 and RM_2 are given by their respective mole fractions. Equation (32) now becomes

$$K = \frac{a_{M_2^+} \; n_{\overline{M}_1^+}}{K_{M_1^+} \; n_{\overline{M}_2^+}} \tag{38}$$

or

$$K = \frac{f_{M_2^+}}{f_{M_1^+}} \; \frac{[M_2^+]}{[M_1^+]} \; \frac{n_{\overline{M}_1^+}}{n_{\overline{M}_2^+}} \tag{39}$$

where $n_{\overline{M}_1^+}$ and $n_{\overline{M}_2^+}$ are the numbers of moles of M_1^+ and M_2^+ in the solid, respectively.

Evaluation of K is as follows: A series of equilibrium measurements are made over a range of concentrations and the values of K^1 defined as

$$K^1 = \frac{[M_2^+]}{[M_1^+]} \; \frac{n_{\overline{M}_1^+}}{n_{\overline{M}_2^+}} \tag{40}$$

are plotted against some function of ionic strength. Extrapolation to infinite dilution, at which $f_{M_2^+}/f_{M_1^+} = 1$, yields a value for K. It follows that the ratio of the activity coefficients of the two ions at a total concentration C is

$$\frac{f_{M_2^+}}{f_{M_1^+}} = \frac{K}{K_c^1} \qquad (41)$$

Schubert [57] points out that for trace concentration of a metal ion M_1^+, e.g., $[M_1^+]$ is effectively zero and the activity coefficient of M_2^+ is that of its pure solution. Hence

$$f_{M_1^+} = \frac{K_c^1 \cdot f_{M_2^+}}{K} \qquad (42)$$

The underlying assumption of solid solution ideality is, however, incorrect [58] and a better approach is that described by Helfferich [58].

V. DETERMINATION OF COMPOSITION OF COMPOUNDS AND COMPLEXES

Several examples exist of the use of radioisotope techniques to investigate the composition of compounds and complexes formed during titrations.

The method used is that of radiometric titration, followed either by solvent extraction of a labeled product or by filtration of a sparingly soluble precipitate.

1. Study of Precipitates

The supernatant liquor may be removed to a counter by filtering through sintered glass, the level of activity is determined and the liquid is then returned to the reaction cell, as shown in Fig. 12 [59]. The measured activity must be corrected, for the change in total volume during the titration by a factor, $K = (V_0 + V)/V_0$ with V_0 = initial volume and V = volume of

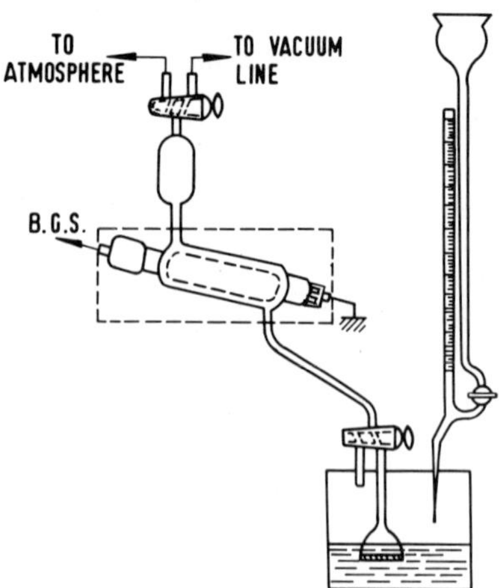

FIG. 12. Apparatus for radiometric titration. Dashed line represents lead shielding which surrounds a dip-type GM tube. The taps are for sampling of the filtered solution. (Reprinted from Ref. 59, p. 118, by courtesy of Pergamon Press.)

reagent added. The titration proceeds in the normal way and at the equivalence point a small addition of reagent results in a sharp increase in the radioactivity of the solution.

The above work by Alimarin et al. [59] produced typical results. In the titration of 3 vol $BeSO_4$ and 2 vol $(NH_4)_2\ H^{32}PO_4$ in equimolar concentrations (0.1 M, 0.05 M, 0.01 M), it was found that beryllium orthophosphate and not beryllium ammonium phosphate is formed by the following equation.

$$3\ BeSO_4 + 2\ (NH_4)_2H^{32}PO_4 = Be_3(^{32}PO_4)_2 + (NH_4)_2SO_4 + H_2SO_4 \tag{43}$$

Similarly the reaction of zirconyl chloride and $H_3{}^{32}PO_4$ in equimolar amounts showed the precipitate to be $ZrOH^{32}PO_4$. Such a technique has also been used by Shinagawa et al. [60], Bradhurst, Calles, and Duncan [61], and Basinski et al. [62].

2. Use of Liquid-Liquid Extraction

Duncan and Thomas [63] investigated the complex formed by dithizone with $^{60}Co^{2+}$. Aqueous cobalt nitrate in an acetate buffer solution was titrated with a solution of dithizone in CCl_4, shaken to extract the complex and the activity of the aqueous solution measured after each addition of the complexing agent. Figure 13 shows a typical titration curve. Investigations using standard solutions of reagents showed a cobalt to dithizone ratio of 1:3. References 64-67 list other similar applications. The general theoretical background to this method is covered in Section III of this chapter.

3. Ion Exchange Separation

Standard solutions of metal ions labeled with an appropriate isotope may be titrated with a chelating agent (typically EDTA) and the resulting solution passed through an anion exchange column where the uncomplexed ion will be retained. The residual radioactivity in the eluate is thus due to the complexed form of the ion. Sequential measurements may be carried out with

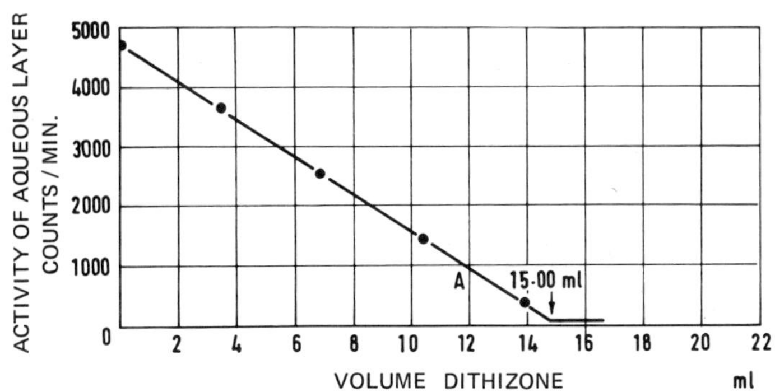

FIG. 13. Radiometric titration of 10.7 μg $^{60}Co^{2+}$ with 3.62×10^{-5} M dithizone in CCl_4. (Reprinted from Ref. 63, p. 377, by courtesy of Pergamon Press.)

increasing amounts of chelating agent and the end point for the reaction obtained. References 68-70 list typical applications, all involving chelation with EDTA.

4. Focusing Ion-Exchange Separation

This method developed by Schumacher [71,69] is illustrated in Fig. 14. A drop of a dilute solution of cation $M^{\nu+}$ is placed in the middle of a strip of chromatographic paper which is passed through the coolant bath of CCl_4 and the cathodic compartment contains a solution of complexing agent $L^{\nu-}$, e.g., acetate which forms an anionic complex with $M^{\nu+}$. The anodic compartment is filled with mineral acid. When by capillary migration these solutions reach the spot of $M^{\nu+}$ a potential of 300-500 V is applied. The protons and complexing anions migrate toward each other and react on meeting to form a pH-pL gradient along the paper strip. M may be present in this gradient as $M^{\nu+}$ or as a complex anion depending on the concentration of the free ligand $L^{\nu-}$. Figure 15 illustrates various stages of the process. The

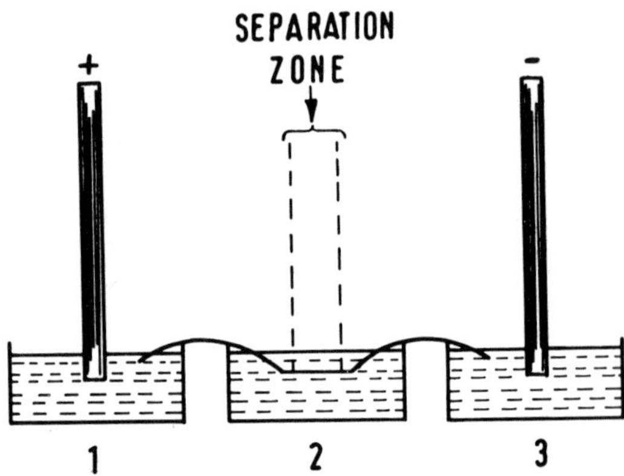

FIG. 14. Focused ion exchange separation apparatus: (1) anode compartment containing mineral acid; (2) cooling compartment containing carbon tetrachloride; (3) cathode compartment containing complexing agent. (Reprinted from Ref. 71, p. 229, by courtesy Helvetica Chimica Acta.)

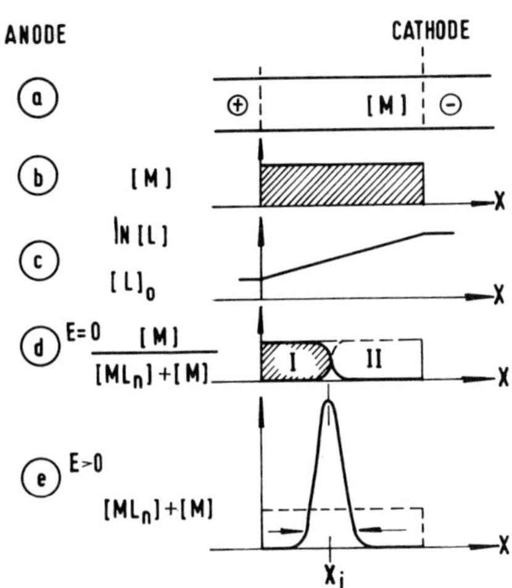

FIG. 15. Schematic sketch of separation by focusing ion exchange. Charges on cation anion and complex are omitted for clarity. Zone I, cation; zone II, complex formed on mixing of cation and ligand; E, applied potential. (Reprinted from Ref. 71, p. 235, by courtesy of *Helvetica Chimica Acta*.)

effect of the applied potential is to cause focusing of M in a narrow region of the strip. Schumacher showed that the position of this focus depends on the stability constant of the complex $[ML_n]^{-n\nu_-+\nu_++\nu}$. If the migration velocities of $M^{\nu+}$ and complex ion are similar, we obtain

$$x = \frac{-\log(K[L^{\nu-}]^n_{anode})}{\alpha n} \tag{44}$$

where x is the position of the focus, n the number of ligands in the complex, $\alpha = d\log[L^{\nu-}]/dx$, K the stability constant of $[ML_n]^{-n\nu+\nu}$ and $[L^{\nu-}]_{anode}$ is the concentration of L^{ν} in the anodic part of the separation zone. Schumacher [69] showed that when two ligands are present in the system, the second $M^{\nu+}$ focus the intensity of which is proportional to the amount of EDTA added to the spot, appears on the strongly acidic side of the pH gradient. This formation of two foci in the presence of two ligands that form complexes of different stabilities is called double focusing. The distribution ratio of M, i.e., $[M_n]_{right\ focus}/[M_n]_{left\ focus}$ is constant and depends only on the amount of EDTA added to the spot.

The above technique was exemplified by titration of $^{90}Y^{3+}$ with EDTA and separation of complexed and uncomplexed $^{90}Y^{3+}$ with acetate as complexing agent. Known volumes of EDTA (0.0954 M) were added to 2 ml aliquots of the $^{90}Y^{3+}$ solution. Samples were withdrawn from the solution and complexed and uncomplexed $^{90}Y^{3+}$ separated by focusing ion exchange with acetate as the separating ligand. The activity along the strip was subsequently measured and the $^{90}Y^{3+}$ which had reacted with the EDTA determined.

Figures 16 and 17 show, respectively, the foci obtained as a function of EDTA added to the initial solution and the ratio

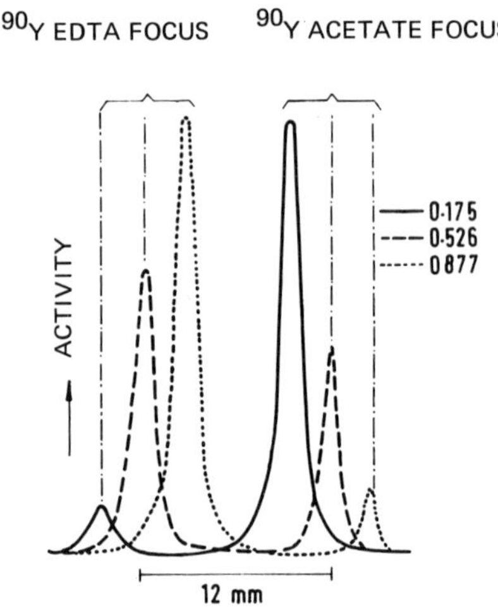

FIG. 16. Radiograms of foci obtained in the titration of yttrium at various degrees of titration (0.175, 0.526, 0.877). (Reprinted from Ref. 69, p. 1776, by courtesy of *Helvetica Chimica Acta*.)

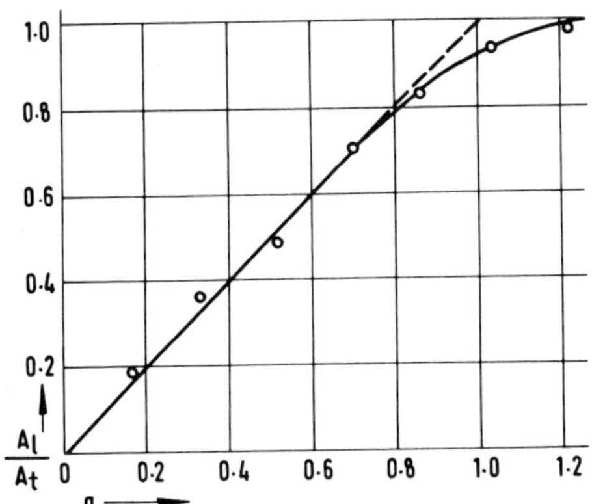

FIG. 17. Radiometric titration of yttrium with EDTA. A_ℓ, radioactivity of the left focus; A_t, total radioactivity; EDTA concentration 0.0954 M. a = degree of titration. (Reprinted from Ref. 69, p. 1776, by courtesy of *Helvetica Chimica Acta*.)

of the radioactivity of the EDTA focus to total radioactivity as a function of the degree of titration.

5. Paper Chromatographic Separations for Endpoint Determination

Schumacher and Friedli [68] developed a simple form of radiometric titration based on paper chromatographic separation of the complex formed for endpoint detection. Labeled cation M of initial concentration $[M]_t$ is titrated with complexing agent L of initial concentration $[L]_t$. During the titration samples are removed and M and ML (the complex) separated by paper chromatography. The radioactivities of the M and ML spots are assayed and if standard solutions of M and L are used the endpoint of the titration and hence the composition of the complex may be determined.

The usual titrant is a multidentate-ligand such as the aminocarboxylic acids. Schumacher and Friedli illustrate the technique for titrating traces of cobalt with EDTA with paper chromatographic separation of complex and uncomplexed ^{60}Co followed by a radiochromatogram scanning technique for evaluating the ratio of free to complexed ^{60}Co. If standard solutions of Co(III) and titrant EDTA are utilized, then provided all of the ligand is consumed, the composition of the complex can be calculated.

VI. TRANSPORT NUMBERS

A. Introduction

When electricity is passed through an electrolyte solution, the current is carried by countercharged ions moving in opposite directions under the applied potential. The fraction of the total current carried by an ionic species is called the transport number of that ion.

Precise measurements of transport numbers have been used to test interionic attraction theory. Extensive data are available on the concentration dependence of transport numbers for univalent electrolytes in water, for which there is good quantitative agreement with theory [72]. Much less information is available for higher electrolytes in water and for all electrolytes in nonaqueous solvents.

The magnitude of transport numbers often reflects the importance of ion-solvent interaction and may also provide information on complex ion formation. The application of radiotracers to transport number determination has received far less attention than it deserves.

All methods for measurement of transport numbers fall into three categories: (a) Hittorf method, (b) Moving boundary method, and (c) the method based on emf measurements of cells with and without a liquid junction. We shall consider the basic principles of methods (a) and (b) and then show how radioactive tracers can obviate certain difficulties associated with these methods.

B. The Hittorf Method

In the Hittorf method a measured quantity of electricity is made to flow through an electrolyte solution of initially known concentration. After passage of current the concentration is redetermined in either cathode or anode compartment. For high accuracy at least one of the electrodes must be reversible [72] so that no unwanted species, which may affect subsequent analyses, are produced by electrode reactions. In addition, a precise analytical method is required to determine the change in concentration which may amount to only 20% of the initial value. A number of experimental cells have been described [72]. The analytical problem is reduced if sufficient electricity is passed to produce a large concentration change. Under such conditions, however, the electrode may not continue to operate reversibly. Other problems may arise from how the electricity is passed through the cell. A high current passing for a short period will cause heating which may cause turbulent mixing. For a low current passing for a long period diffusion may result in some mixing of anode and cathode solutions.

1. Tracer Application

We examine first the situation in which the radioisotope method serves simply as a convenient and rapid analytical method. Consider the Hittorf-type cell shown schematically in Fig. 18. We regard the cathode and anode compartments as the spaces to the left and to the right, respectively, of the dashed line which

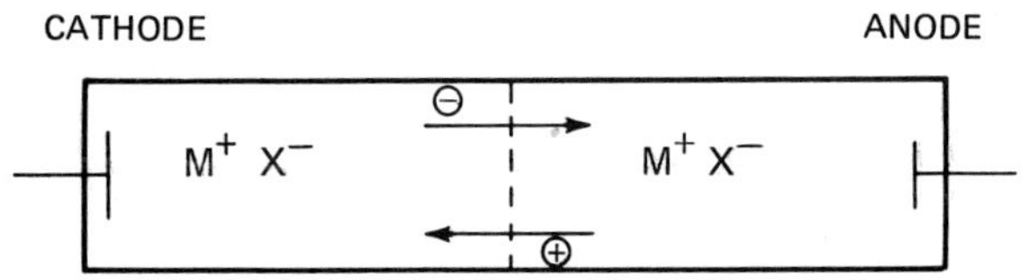

FIG. 18. Schematic representation of a Hittorf cell.

represents a tap or a porous membrane. Initially the electrolyte concentration is uniform throughout. If we assume that both cathode and anode are reversible, then the overall concentration remains unchanged during the experiment, since the increase (decrease) in concentration in the cathode (anode) compartment equals the decrease (increase) in the anode (cathode) compartment [73]. Let the volume of the cathode solution be V_c and of the anode solution be V_a. Suppose we add tracer for the cation. This can be added to one compartment only, say anode, and its transfer to the cathode measured, or can be uniformly distributed throughout the cell and the change in concentration of tracer measured after passage of current.

a. Mode of application. *Method 1.* Consider the former case in which initially the concentration C of electrolyte is uniform throughout, but cation tracer is present only in the anode compartment. After a quantity of electricity Q has been passed, the radioactivity S_1 per unit volume of cathode compartment is measured. The final total activity in the cathode section is given by S_1V_c. At the same time, the radioactivity transferred from the anode is expressed as $(Qt_+S_0)/\underline{F}C)$, where t_+ is the transport number of the cation, S_0 the radioactivity per unit volume of the anode compartment and $\underline{F}$ the Faraday constant. This will be true as long as the active material in the vicinity of the connection is not allowed to mix with inactive salt from the anode. We thus have

$$\frac{S_0Qt_+}{C\underline{F}} = S_1V_c \tag{45}$$

or

$$t_+ = \frac{S_1\underline{F}V_cC}{S_0Q} \tag{46}$$

Method 2a. We now consider the alternative method in which the tracer is uniformly distributed in the cell. The initial activity in the anode compartment is S_oV_a and in the cathode compartment is S_oV_c. After passing a quantity of electricity Q the radioactivity transferred from anode to cathode section is $(Qt_+S_o)/(\underline{F}C)$ and the total activity in the cathode section is $S_oV_c + (Qt_+S_o)/(\underline{F}C)$. We note here that if the cathode reaction removes cation from the solution, e.g., formation of an insoluble salt, then there is a loss of radioactivity. If such an electrode is reversible, an additional term due to this loss is $QS_o/(\underline{F}C)$, so that the final total activity becomes $S_oV_c + (Qt_+S_o)/(\underline{F}C) - (QS_o)/(\underline{F}C)$. Comparing the activities A_1 and A_f of equal volumes of cathode solution before and after the experiment, we obtain

$$\frac{A_f}{A_1} = \frac{S_oV_c + \frac{Qt_+S_o}{\underline{F}C} - \frac{QS_o}{\underline{F}C}}{V_cS_o} \tag{47}$$

or

$$\frac{C\underline{F}V_c}{Q}\left(\frac{A_f}{A_1} - 1\right) = -(1 - t_+) \tag{48}$$

Because of $1 - t_+ = t_-$

then $$t_- = \frac{C\underline{F}V_c}{Q}\left(1 - \frac{A_f}{A_1}\right) \tag{49}$$

Alternatively, if there is no loss of cation at the cathode, e.g., if the cathode is Ag/AgCl and the cation is an alkali metal, the electrode reaction produces soluble alkali iodide and no radioactivity is lost. In this case Eq. (48) reduces to

$$t_+ = \frac{C\underline{F}V_c}{Q}\left(\frac{A_f}{A_1} - 1\right) \tag{50}$$

because $QS_o/\underline{F}C = 0$.

Method 2b. If the analyses are done in the anode compartment initial and final total activities are given by S_oV_a and

$S_oV_a - (Qt_+S_o)/(\underline{F}C)$, respectively. Changes in the anion concentration due to the anode reaction will not affect the amount of radioactivity in solution. In the absence of other reactions which may affect the cation concentration we can easily obtain

$$t_+ = \frac{\underline{F}CV_a}{Q}\left(1 - \frac{A_f}{A_1}\right) \quad (51)$$

Since methods 2a and 2b depend on the difference between two measured activities, A_1 and A_f, whereas method 1 depends on the ratio of activities, obviously, the error in the latter will be less. There are situations, however, where the choice of method is determined by the nature of the transported ions and where the "difference" method is to be preferred.

b. Factors affecting the choice of method. When the electrolyte produces simple species only, e.g., KCl, either method can be used. However, many electrolytes, especially concentrated solutions of certain transition metal halides, give rise to complex ions for which cations and anions contain a common element; e.g., aqueous zinc chloride is thought to contain Zn^{2+}, $ZnCl^+$, $ZnCl_2$, $ZnCl_3^-$, and $ZnCl_4^-$ [74,72]. Similar species are observed for mercury halides.

Micellar solutions of ionic surfactant may contain an ion aggregate (micelle) plus tightly bound counterions [75]. Let MN_n be such an ionic micelle which dissociates as

$$MN_n = zN^+ + \{M(N)_{n-z}\}^{z-} \quad (52)$$

If we now add tracer for N to a solution of MN_n exchange will occur between "free" and bound N until isotopic equilibrium is established. Huizenga et al. [76] have discussed the important consequences which follow if exchange rates are zero, finite, or infinite. If rates are finite then method 1 cannot be used, since transfer of activity is determined not only by the transport number but also by exchange rate. In the absence of information on rates, method 2 is to be preferred.

2. Experimental Technique

The apparatus devised by Brady [75] is shown in Fig. 19. His method can be regarded as a combination of the Hittorf and moving boundary methods and has been called the analytical boundary method. The experimental cell consists of a U-tube divided into two compartments, A and B, by a sintered glass disc (Pyrex medium

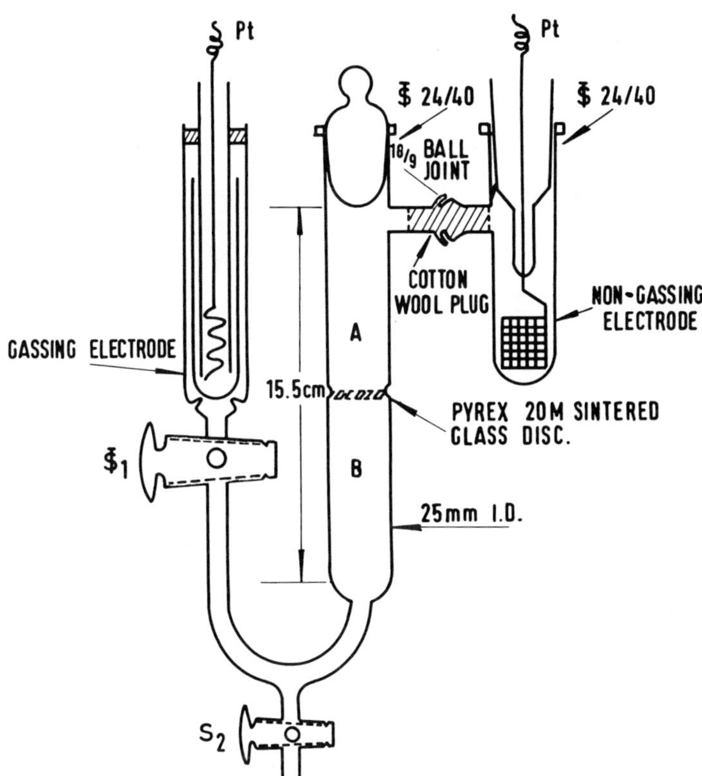

FIG. 19. Analytical boundary apparatus: (A) compartment containing the leading electrolyte; (B) compartment containing the solution under investigation; S_1 and S_2 are stopcocks. (Reprinted from Ref. 75, p. 911, by courtesy of American Chemical Society.)

grade). B contains the ion under investigation and A contains a reference solution. The method consists of an analysis of the amount of ion initially in B that has been transferred electrically to A. The left-hand side of the U-tube, the whole of B, and the sintered glass disc are filled with the labeled solution. Compartment B may be rinsed via the stopcock S_2; stopcock S_1 is present so that the tube can be closed to prevent streaming through the disc while A is rinsed and filled. The cell is thermostatted during an experiment. Current can be supplied from storage batteries and in the original work a milliammeter and voltmeter were used to check the approximate wattage; the current was determined periodically by measuring the voltage drop across a known resistance in series with the circuit, using a Leeds and Northrop type K-2 potentiometer. Duration of experiments varied between 30 and 90 min, after which the contents of compartment A were made up to a specified volume. The radioactive solutions were counted in a glass-jacketed GM counter. Table 1 allows a comparison of the results from this method with those from other methods. The accuracy is comparable to that obtained in a reasonably careful conventional Hittorf determination. This

TABLE 1

Comparison of Analytical Boundary Results with Those from Other Methods*

		Cation transport number		
Solution	Concentration (M)	Analytical boundary	Moving boundary	Hittorf**
$AgNO_3$	0.1	0.473	0.468	0.474
	0.01	0.469	0.465	0.474
KBr	0.1	0.481	0.485	--
	0.01	0.471	0.483	0.496
NaCl	0.05	0.399	0.390	0.396
	0.02	0.396	0.389	0.396

*Reproduced from Ref. 75, p. 914, by courtesy of the American Chemical Society.

**At 18°C.

technique has also been used to study the transport numbers of small ions in colloidal electrolytes [77]. The data in conjunction with conductance data were used to estimate the number of counterions associated with the anionic micelles. These authors do not appear to have considered the importance of rates of exchange of free ions with bound ions.

Huizenga et al. [76] have used a cell of the type designed by Brady, to measure transport numbers of sodium in aqueous solutions of polyacrylic acid and sodium hydroxide. They state that the rate of exchange of sodium ions between the polyacrylate-sodium complex anion is finite and made their measurements on a cell in which the initial tracer as well as the stoichiometric concentrations of polyacrylic acid and sodium hydroxide were uniform throughout.

The cell shown in Fig. 20 has been used by Wall et al. [78] to study the transport numbers of iodide ions in aqueous solutions of polyvinylpyridine neutralized with hydriodic acid. They were able, by combining their transport data with measurements of hydrogen ion concentrations and conductances, to calculate the fraction of iodide ions associated with the polyanion and a figure for the overall degree of ionization of the polymer. The cell consisted of a horizontal glass tube divided by sintered

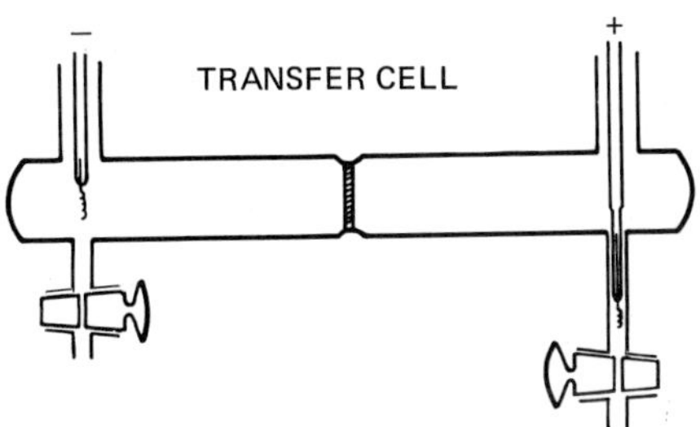

FIG. 20. Cell for measurement of the transference numbers of polyvinylpyridine partially neutralized with hydrogen iodide. (Reprinted from Ref. 78, p. 2823, by courtesy of American Chemical Society.)

glass disc into anode and cathode sections. The electrodes were inserted through the vertical glass tubes which also served as filling tubes. Two glass tubes sealed beneath the cell permitted drainage of the two sections. To localize the products of the anode reaction the anode drainage tube was made longer than the cathode tube. This apparatus has been used by Wall and Eitel [79,80] to measure transport of strontium in the presence of polymeric acid anions. From these data extents of association (binding) have been calculated. Mysels et al. [81-83] have measured electrophoretic mobilities of human serum albumin labeled with ^{131}I. They have also examined mobility of Na^+ in sodium dodecyl sulfate, using the apparatus illustrated in Fig. 21. The horizontal capillary is filled with tracer solution via the downward arms of the three-way stopcocks (1) and (2) by applying suction at (5). Electrode compartments are filled with concentrated sodium chloride solution via the funnels and the arching tubes filled with untagged solution to form a liquid junction at (4). The apparatus is allowed to reach temperature

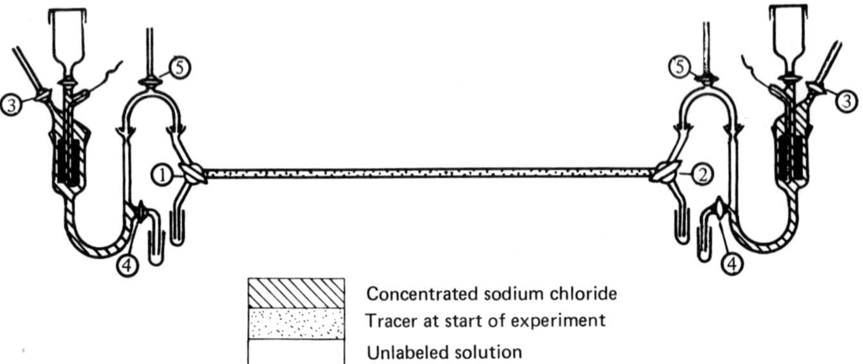

FIG. 21. Free liquid tracer electrophoresis cell. Parts 1-5 are stopcocks. The radioactivity of the capillary section is measured prior to and after the passage of current. (Reprinted from Ref. 81, p. 386, by courtesy of American Chemical Society.)

equilibrium with stopcocks (5) open; movement of the liquid level in (5) shows when equilibrium is achieved. Upon completion of an experiment measurements are made of the radioactivity remaining in the central capillary.

The normal method of measuring electrophoretic mobilities of dissolved material is a moving boundary method based on the work of Tiselius [84]. This method has been used as a means of separation, identification, and measurement of components in multicomponent systems in which the components have different electrophoretic mobilities. The latter method provides accurate values of electrophoretic mobilities, but unfortunately requires an ionic strength greater than 0.01 in order to achieve satisfactory boundaries. An important reason for this constraint is the difference in electrical conductivity between dialyzed protein/buffer and the pure buffer solution [85]. This can cause the boundary to be ill-defined. Salt boundaries are also present, in each limb of the Tiselius cell, due to differences in buffer concentrations above and below the original boundaries.

The importance of such effects can be reduced by working at high salt/protein ratios so that the protein has a negligible effect on the conductivity. In a conventional application of the method using an optical system to determine concentration, the sensitivity of the system does not permit protein concentrations of much lower than 1% to be studied. An additional restriction applies to solutions at higher concentrations, viz., solution concentrations should be insufficient to cause unacceptable joule heating. The use of labeled proteins allows measurements to be extended to extremely low protein concentration at low ionic strengths, e.g., 0.001% protein and 0.001 M buffer [83].

Zhitomirski et al. [86] have used a cell comprised of three spherical compartments connected through porous glass membranes. The solution composition is made uniform throughout the cell and tracer is added to the middle section. The authors point out that such an arrangement should allow measurement of transport numbers "in those very common cases where the element being determined forms part of an ion of opposite charge." Radiometric analyses of the cathode and anode solutions after electrolysis then allows calculation of cation and anion transport numbers. However, in their check on the cell, measurements were made on aqueous solutions of sulfuric acid labeled with ^{35}S; such solutions contain both HSO_4^- and SO_4^{2-}, the value found for the anion transport number, 0.177 ± 0.006, is not of high precision and there is no discussion on sources of error. Inspection of their data (especially Table 1), shows that the low specific activity is no doubt responsible for much of the error. The radiometric assay involved conversion of the sulfate to barium sulfate. No details are given on the procedure for preparing reproducible $BaSO_4$ samples.

The technique was then applied to ion migration in absolute sulfuric acid. No data are given but simply a statement that "radioactivity moves only into the cathode section and hence there are no anions containing sulfur in absolute sulfuric acid." It has been well established [87], however, that the ions present in absolute sulfuric acid are $H_3SO_4^+$, HSO_4^-, H_3O^+, and $HS_2O_7^-$ and that the first two $H_3SO_4^+$ and HSO_4^- have abnormal mobilities. Such abnormal mobilities arising, as they apparently do, from a chain mechanism should not cause transport of sulfur, but there will be a contribution to the total mobility from the normal movement of these ions in an electric field. The latter process should cause transport of sulfur. The absence of transport into the anode section is puzzling, since HSO_4^- might be expected to have a greater normal mobility than $H_3SO_4^+$. In addition, some transport of sulfur should occur via the ion $HS_2O_7^-$. More information is obviously desirable.

C. The Moving Boundary Method

In this method transport numbers are obtained by direct observation of the rate at which a boundary between two electrolyte solutions moves along a tube. We now briefly describe how transport numbers are derived from such observations. Two salts AX and BX having the common ion X are made to form a boundary ab (Fig. 22). Let current be passed such that the cathode is at the top of the tube, i.e., cations A^+ and B^+ move upward and the anion X^- moves downward. The sharpness of the boundary can be maintained if the conditions are properly chosen, and its boundary movement from ab to cd during time t can be closely followed. All the cations A^+ originally present in the volume V bounded by the planes ab and cd must have crossed the plane cd. If the total quantity of electricity passed is Q and the concentration of AX is C_{AX} then the amount of A^+ moving upward is $C_{AX}V$, and

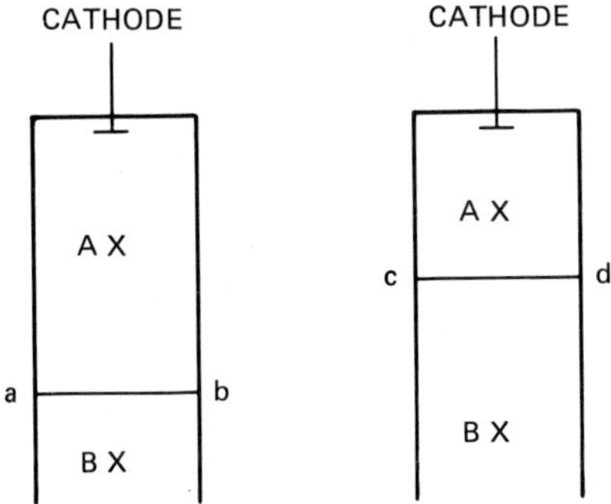

FIG. 22. Schematic representation of the moving boundary method.

$$\frac{t_{A^+} \cdot Q}{\underline{F}} = C_{AX}V \tag{52}$$

where t_{A^+} is the transport number of cation A^+. Hence

$$t_{A^+} = \frac{\underline{F}C_{AX}V}{Q} \tag{53}$$

Similarly from the amount of B^+ transported across the plane ab into the volume V we have

$$t_{B^+} = \frac{\underline{F}C_{BX}V}{Q} \tag{54}$$

where t_{B^+} and C_{BX} are the transport number of cation B^+ and the concentration of BX, respectively, that

$$\frac{t_{A^+}}{t_{B^+}} = \frac{C_{AX}}{C_{BX}} \tag{55}$$

This condition, sometimes called the Kohlrausch regulating condition, is one which must be approximately satisfied in order to achieve a stable boundary. Other requirements are (i) the boundary ab must initially be sharp, (ii) the solution of lower density should be on top, (iii) the conductance of the indicator ion, which in our illustration is B^+ must be lower than that of the leading ion A^+, and (iv) the two electrolytes must have a common ion. Methods of producing the initial sharp boundary are discussed in most texts of physical chemistry.

1. Tracer Application

The original idea of following the moving boundary between radioactive and inactive solutions appears to be due to Schiff [88], although no measurements were made by him. The method has been developed by Marx et al. [89-96] who measured the time t for the active-inactive boundary to pass two detectors placed a known distance apart along the flowing direction. The transport number $t_{\pm}$ can then be calculated from

$$t_{\pm} = \frac{CV\underline{F}}{It} \tag{56}$$

where C is the leading solution concentration, V the volume swept out, $\underline{F}$ the Faraday constant, and I the current.

2. Experimental Technique

a. The electrolyte cell. A diagram of the apparatus described in Ref. 89 is shown in Fig. 23. The cell made of Jena glass consists of two electrode compartments A and B connected by a U-tube which incorporates a capillary tube M above the two-way tap H_1. The cathode compartment B is at right angles to the plane of the paper so as to allow the detectors to be positioned to the right of M. The whole cell must be vibration-free and M must be vertical.

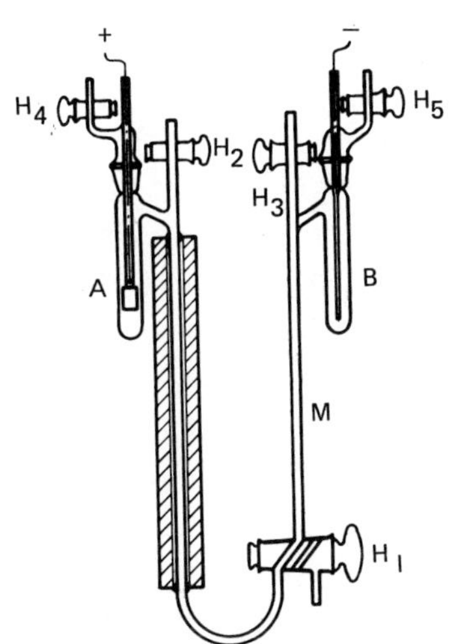

FIG. 23. Moving boundary cell: (A) compartment containing high purity Cd anode; (B) compartment for Ag/AgCl cathode; (H_1) two-way tap; (H_2-H_5) stopcock; (M) capillary tube of 0.8 mm-thick wall. The hatched portion represents a 1.5 cm-thick Perspex radiation shield. (Reprinted from Ref. 89, p. 12, by courtesy of Academic Press.)

To fill the cell the joints containing the electrodes are removed and tap H_1 opened. ^{42}KCl solution is then introduced through tap H_2 until a counting tube held just above H_1 registers radioactivity present in M. H_1 is then closed across. With the anode compartment stoppered more active solution is introduced through H_2 until the level is just above the tap H_2 which is then closed. With the stopper then removed the anode space is filled to the highest point with inactive KCl solution so that any spillage resulting from the next step, i.e., insertion of the electrode joint, does not cause contamination. Filling of the anode compartment is completed through H_4. A small amount of active solution still remains in M and this is rinsed out through the second opening of H_1. M and B can then be filled with the leading solution 0.05 M HCl. The cathode is inserted and the filling operation is completed through H_5. At the start of the electrolysis H_1 is turned in the direction shown in Fig. 23.

b. Radioactivity measurement. Each of two end-window GM tubes of 1.2 mg/cm^2-thick window was completely covered with a 6 mm-thick Perspex plate having a 1 mm × 30 mm slit. The distance between the centers of the slits was 99.94 ± 0.02 mm, giving a volume of 1.2554 ± 0.005 ml swept out V between the slits. Alignment of the slit with a mark on M was accomplished with a cathetometer. The signals from the two counting tubes were fed to a recorder to give a trace as shown in Fig. 24. A graphical means of measuring the time difference is described. All experiments were performed in an air thermostat at 25°C and the transport number of the hydrogen ion thus obtained $t_{H^+}^{HCl}$ (25°C) = 0.8295 ± 0.0011, compared well with the accepted literature value [97] of 0.8292.

More recently Marx and Hentschel [93] have used γ-emitting ^{24}Na for monitoring the movement of sodium ion in a 1×10^{-2} M solution of sodium chloride in absolute methanol. A value of 0.4584 ± 0.004 for the transport number is in good agreement with the value of 0.4585 ± 0.0001 obtained by Davies et al.

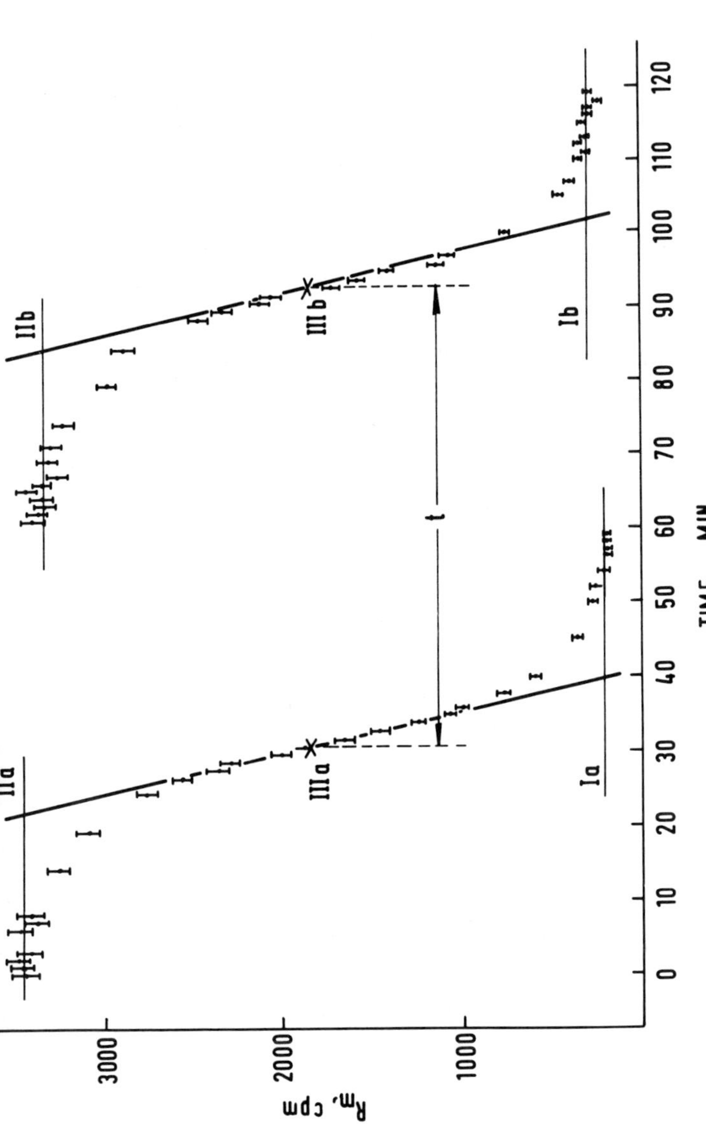

FIG. 24. Measurement of the time taken for the active/inactive boundary to pass between two detectors: R_m, observed radioactivity; IIa and Ia are the constant count rates, respectively, before and after the boundary passes the first detector and IIb and Ib are the corresponding count rates for the second detector; t, time for boundary to pass between the detectors. (Reprinted from Ref. 93, p. 1164, by courtesy of Pergamon Press.)

[98]. Measurements have been extended to nonaqueous solutions [96].

D. General Comments

We have described how radiotracers have been used as a convenient analytical tool in modifications of the Hittorf method (and their use has enabled measurements to be extended to low concentrations). The use of tracers in this way has not so far produced data of precision better than about ±1% but the method is quicker than the conventional Hittorf technique. If the precision can be increased, without undue sophistication and cost, the method may become more generally used. There are a number of obvious ways in which this could be achieved; these can be illustrated by taking method 1 in VI B 1 a as an example. The major counting error in this method arises in the determination of S_1. If the activity transferred from the active to inactive compartment can be confined to a smaller known volume then the activity per unit volume can be made larger. The error in the volume measurement will now be larger, but since we can measure volume more precisely than radioactivity, perhaps a better compromise can be reached than in the techniques employed to date. This will necessitate a division of the cell into isolable sections. A means of mixing the solution within a compartment, after an experiment, is needed to produce a completely uniform tracer concentration prior to sampling.

The use of liquid scintillation counting greatly increases the range of easily detectable radioisotopes. In addition the simultaneous use of several tracers, e.g., labeling of a polyanion as well as counter ions, may help to simplify interpretation of results from complex systems.

Transport numbers in fused salts have been measured using the Hittorf method [99].

VII. EFFICIENCY OF SEPARATION PROCESSES

Weiss and Shipman [100] have studied methods for separation of trace quantities of a number of elements such as ^{59}Fe, ^{60}Co, ^{95}Zr, etc. from solution by coprecipitation with 8-quinolinol formed homogeneously in situ by hydrolysis of 8-acetoxyquinoline. Purity checks on the radiotracers were made by γ-ray spectrometry. Similar experiments [101] on collection of such elements as ^{195}Au, ^{191}Os, ^{182}Ta, ^{144}In, and ^{110m}Ag from seawater have also been carried out to determine the optimum conditions for efficient cocrystallization with water-insoluble thionalide (α-mercapto-N,2-naphthylacetamide). Further examples are listed in Ref. 102.

Recent interest in reverse osmosis as a separation technique has centered principally around the use of cellulose acetate semipermeable membranes to separate both electrolytes and nonelectrolytes from aqueous solutions [103]. In the majority of these solutions, "conventional" analytical techniques have been used to measure fairly high concentrations. Recently, during a study of radioactive fallout in natural water supplies, radiotracers have been used to assess the ability of commercial equipment to remove ions at *very low concentrations* [104]. The equipment was a tubular unit manufactured by Paterson Candy International and consisted of B-type modules. The membranes, T1/12, were the tightest available cellulose acetate membranes and had a maximum working pressure of 1200 psi ($\sim$83 atm). A standard unit, consisting of 27 modules can produce up to 23 m^3 day^{-1} of purified water. In order to reduce the solution throughput to levels acceptable for small scale testing and to avoid the use of large quantities of radiotracer the equipment was modified in order to recycle much of the solution. The arrangement used for experimentation is shown schematically in Fig. 25. As a measure of the performance, the membrane leakage factor K = (ion concentration in permeate)/(ion concentration in feed) has been measured for a number of ions.

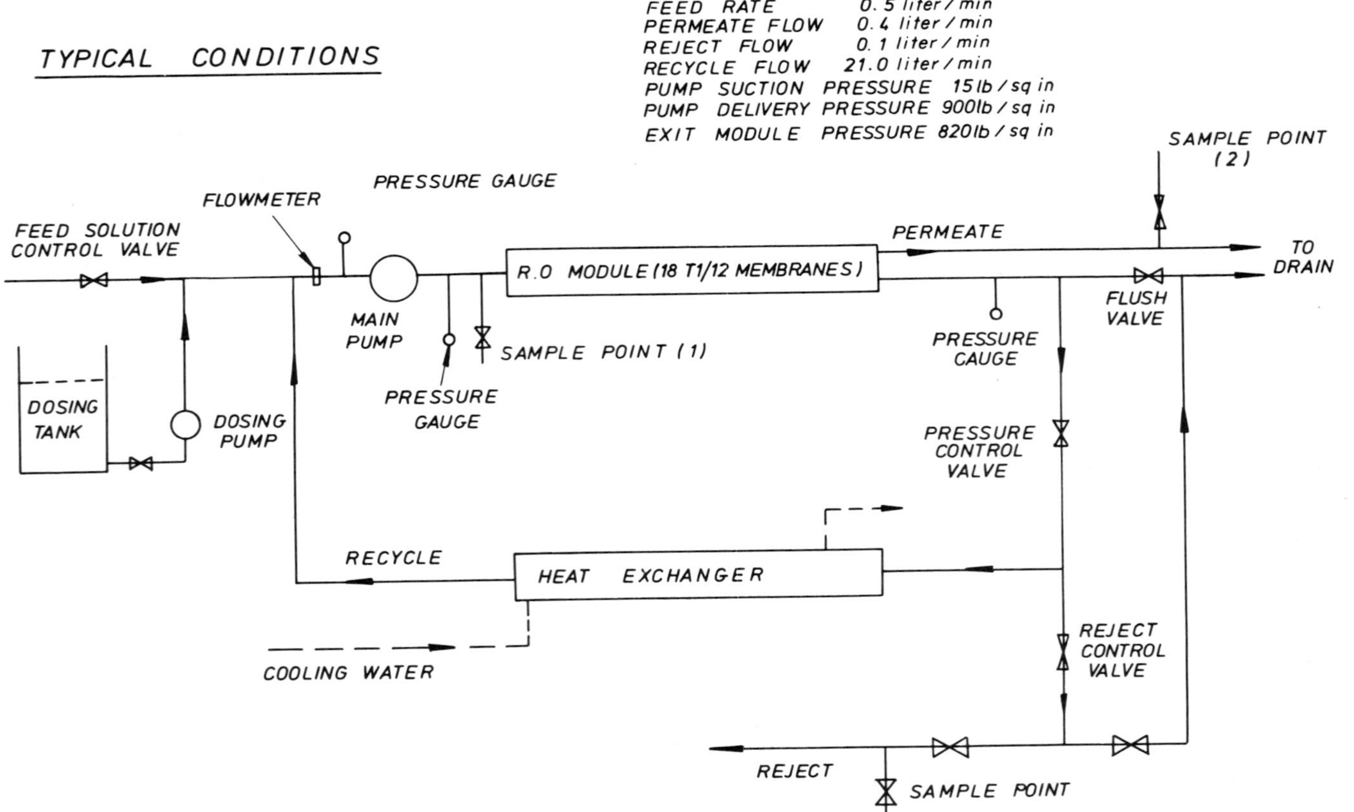

FIG. 25. Schematic diagram of an experimental arrangement for reverse osmosis.

In normal "straight through" operation, i.e., with no recycle, the permeate and feed concentrations vary along the length of the module. In the experimental arrangement, however, with high recycle, the concentration of both permeate and feed remain essentially constant throughout the module and performance of the equipment under different conditions of recycle can be predicted by making certain assumptions about the concentration variation along the module length. The K values, calculated from samples taken at points (1) and (2) shown in Fig. 25, are listed in Table 2. The order of separation $Pm^{3+} \sim Y^{3+} > Sr^{2+} > Ca^{2+} > I^-$ is in accord with previous data [103] at *high* salt concentrations.

TABLE 2

Membrane Leakage Factors, K, at 20°C

Ion	Estimated range of feed concentrations (M)	K
Pm^{3+}	$<10^{-9}$	5.7×10^{-4}
Y^{3+}	$<10^{-11}$	8.1×10^{-4}
Sr^{2+}	$<10^{-5}$	1.6×10^{-3}
Ca^{2+}	10^{-3}-10^{-2}	2.5×10^{-3}*
I^-	$<10^{-5}$	4.8×10^{-2}

*Measured by flame photometry.

REFERENCES

1. V. Spitsyn and I. Mikhailenko, Dokl. Akad. Nauk SSSR, 121:319 (1958)
2. R. Ramette and O. Anderson, J. Inorg. Nucl. Chem., 25:763 (1963).
3. C. Bovington, J. Inorg. Nucl. Chem., 27:1975 (1965).

4. W. Gelsema, C. L. DeLigny, and H. Blijleren, Rec. Trav. Chim., 86:1372 (1967).

5. H. Freundlich, Colloid and Capillary Chemistry, Dutton, New York, 1922, p. 155.

6. B. Enustun and J. Turkevich, Radioisotopes in the Physical Sciences and Industry, Vol. III, IAEA, Vienna, 1962, p. 531.

7. R. Wyckoff, Crystal Structure, Vol. 2, Wiley-Interscience, New York, 1951, Chap. 8, Table, p. 21.

8. V. Koprda and M. Fojtik, Chem. Listy Svazek, 62:679 (1968).

9. K. Krause, H. Phillips, and F. Nelson, Radioisotopes in the Physical Sciences and Industry, Vol. III, IAEA, Vienna, 1962, p. 387.

10. D. Atkins and A. Eggleton, Nuclear Techniques in Environmental Pollution, IAEA, Vienna, 1971, p. 521.

11. A. Jones, Trans. Faraday Soc., 59:2355 (1963).

12. C. Bovington and A. Jones, Trans. Faraday Soc., 66:764 (1970).

13. C. Bovington and A. Jones, Trans. Faraday Soc., 66:2088 (1970).

14. C. Bovington and A. Delmas, Unpublished work.

15. M. Reddy and G. Nancollas, J. Colloid Interface Sci., 36:166 (1971).

16. A. Langer, Radiometric Titration Method, U. S. Pat. 2,367, 949, 23, Nov. 1940.

17. A. Langer, J. Phys. Chem., 45:639 (1941).

18. A. Langer, Anal. Chem., 22:1288 (1950).

19. J. Duncan and F. Thomas, Proc. Australian Atomic Energy Symp. Radioisotopes in the Physical Sciences, 1958, p. 617.

20. I. Korenman, F. Seganova, H. Mezina, and M. Ostaseva, Zhur. Anal. Khim., 12:48 (1957).

21. T. Braun, I. Maxim, and I. Galateanu, Nature, 183:936 (1958).

22. J. Duncan, J. Inorg. Nucl. Chem., 11:161 (1959).

23. A. Casey and W. Robb, Nature, 198:581 (1963).

24. A. Parker, Chem. Rev., 69:1 (1969).

25. G. Morrison and H. Freiser, Solvent Extraction in Analytical Chemistry, Wiley, New York, 1957.

26. F. Rossotti and H. Rossotti, The Determination of Stability Constants, McGraw-Hill, New York, 1961.

27. D. Peppard, Ann. Rev. Nucl. Sci., 21:365 (1971).

28. J. Rydberg, Acta Chem. Scand., 4:1503 (1950).

29. Y. Marcus, Acta Chem. Scand., 11:329 (1957).

30. J. Rydberg, Acta Chem. Scand., 23:647 (1969).

31. H. Reinhardt and J. Rydberg, Acta Chem. Scand., 23:2773 (1969).

32. C. Andersson, S. Sanderson, J. Liljenzin, H. Reinhardt, and J. Rydberg, Acta Chem. Scand., 23:2781 (1969).

33. S. Andersson, private communication.

34. H. Ottertun, Int. J. Appl. Rad. Isotopes, 23:13 (1972).

35. P. Manning and C. Monk, Trans. Faraday Soc., 58:938 (1962).

36. P. Manning and C. Monk, Trans. Faraday Soc., 57:1996 (1961).

37. A. McAuley and G. Nancollas, J. Chem. Soc., 4367 (1961).

38. J. Prue, International Encyclopedia of Physics and Chemistry, Vol. 3, Pergamon, London, 1966, Chap. 5.

39. G. Scibona, P. Danesi, F. Orlandini, and C. Coryell, J. Phys. Chem., 70:141 (1966).

40. G. Scibona, R. Nathan, A. Kertes, and J. Irvine Jr., J. Phys. Chem., 70:375 (1966).

41. J. Salmon and D. Hale, Ion Exchange--A Laboratory Manual, Butterworths, London, 1959.

42. J. Duncan and B. Lister, Disc. Faraday Soc., No. 7, 104 (1949).

43. J. Cosgrove and J. Strickland, J. Chem. Soc., 1845 (1950).

44. K. Kraus and F. Nelson, American Society for Testing Materials, Special Technical Publication No. 195, 27, 1958.

45. K. Kraus, H. Phillips, and F. Nelson, Radioisotopes in the Physical Sciences and Industry, Vol. III, IAEA, Vienna, 387, 1962.

46. F. Nelson, T. Murase, and K. Kraus, J. Chromatog., 13:503 (1964).

47. K. Kraus and R. Raridon, J. Am. Chem. Soc., 82:3271 (1960), and earlier papers.

48. J. Schubert, J. Phys. Colloid Chem., 52:340 (1948).

49. J. Schubert and J. Richter, J. Phys. Colloid Chem., 52:350 (1948).

50. R. Cannan and A. Kilbrick, J. Am. Chem. Soc., 60:2314 (1938).

51. A. Hastings, F. Mclean, L. Eichelberger, J. Hall, and E. DaCosta, J. Biol. Chem., 107:351 (1934).

52. N. Joseph, J. Biol. Chem., 164:529 (1946).

53. J. Schubert, E. L. Lind, W. M. Westfall, R. Pfleger, and N. C. Li, J. Am. Chem. Soc., 80:4799 (1958).

54. S. Fronaeous, Acta Chem. Scand., 5:859 (1951).

55. S. Fronaeous, Acta Chem. Scand., 6:1200 (1952).

56. A. Vanselow, J. Am. Chem. Soc., 54:1307 (1932).

57. J. Schubert, J. Phys. Chem., 52:340 (1948).

58. F. Helfferich, Ion Exchange, McGraw-Hill, New York, 1962, p. 183.

59. I. Alimarin, I. Gibala, and I. Sirotina, Int. J. Appl. Radiation Isotopes, 2:117 (1957).

60. M. Shinagawa, H. Matsuo, and M. Yoshida, Japan. Analyst, 4:139 (1955).

61. D. Bradhurst, B. Colles, and J. Duncan, J. Inorg. Nucl. Chem., 4:379 (1957).

62. A. Basinski, W. Szymanski, A. Krygier, and G. Zapalowska, Roczniki Chem., 37:1345 (1963).

63. J. Duncan and F. Thomas, J. Inorg. Nucl. Chem., 4:376 (1957).

64. H. Spitzy, Microchim Acta, 789 (1960).

65. I. Korenman, F. Shayanova, H. Mezina, and M. Ostasheva, Zhur. Anal. Khim., 12:48 (1957).

66. P. Spacu and V. Boicu, Acad. Rep. Pop. Rom. Stud. Cercet. Chim., 10:305 (1962).

67. S. Ionescu and C. Grigonescu-Sabau, Proc. Int. Conf. Peaceful Uses Atomic Energy, IAEA, Geneva, 28, 148 (1958).

68. E. Schumacher and W. Friedli, Helv. Chim. Acta, 43:1013 (1960).

69. E. Schumacher and H. Streiff, Helv. Chim. Acta, 41:1771 (1958).

70. J. Stary, J. Ruzicka, and A. Zeman, Talanta, 11:481 (1964).

71. E. Schumacher, Helv. Chim. Acta, 40:228 (1957).

72. R. Robinson and R. Stokes, Electrolyte Solutions, Butterworths, London, 1967, Chap. 7.

73. D. MacInnes and M. Dole, J. Am. Chem. Soc., 53:1361 (1931).

74. A. Harris and H. Parker, Trans. Faraday Soc., 36:1149 (1940).

75. A. Brady, J. Am. Chem. Soc., 70:911 (1948).

76. J. Huizenga, P. Grieger, and F. Wall, J. Am. Chem. Soc., 72:2636 (1950).

77. A. Brady and D. Salley, J. Am. Chem. Soc., 70:914 (1948).

78. F. Wall, J. Ondrejcin, and M. Pikramenow, J. Am. Chem. Soc., 73:2821 (1951).

79. F. Wall and M. Eitel, J. Am. Chem. Soc., 79:1550 (1957).

80. F. Wall and M. Eitel, J. Am. Chem. Soc., 79:1556 (1957).

81. H. W. Hoyer, K. J. Mysels and D. Stigter, J. Phys. Chem., 58: 385 (1954).

82. K. J. Mysels and C. I. Dulin, J. Colloid Sci., 10:461 (1955).

83. E. K. Mysels and K. J. Mysels, J. Am. Chem. Soc., 83:2049 (1961).

84. A. Tiselius, Trans. Faraday Soc., 33:524 (1937).

85. A. Tiselius and H. Svensson, Trans. Faraday Soc., 36:16 (1940).

86. A. Zitomirski, O. Kudra, and Y. Fialkov, Dokl. Acad. Nauk Tadzhik., SSR, No. 5, 15 (1962).

87. R. Gillespie and S. Wasif, J. Chem. Soc., 946 (1953), and later papers.

88. H. Schiff, Ph.D. Thesis, University of Toronto, 1948.

89. G. Marx, L. Fischer, and W. Schultz, Radiochim. Acta, 2:9 (1963).

90. G. Marx and L. Fischer, Z. Physik. Chem. Neue Folge, 41:315 (1963).

91. G. Marx and W. Schulze, Kerntechnik, 7:13 (1965).

92. W. Schultz, M. Hornig, and G. Marx, Z. Physik. Chem., Neue Folge, 53:106 (1967).

93. G. Marx and D. Hentschel, Talanta, 16:1159 (1969).

94. G. Marx, W. Riedel, and J. Vehlow, Ber. Bunsenges., 73:74 (1969).

95. J. Vehlow and G. Marx, Naturwissenschaften, 58:320 (1971).

96. L. Strauss and G. Marx, Z. Naturforsch., 28:543 (1973).

97. L. Longsworth, J. Am. Chem. Soc., 54:2741 (1932).

98. J. A. Davies, R. L. Kay, and A. R. Gordon, J. Chem. Phys., 19:749 (1951).

99. H. Jonassen and A. Weissberger (eds.), Technique of Inorganic Chemistry, Wiley, New York, 1963, p. 139.

100. H. Weiss and W. Shipman, Anal. Chem., 34:1010 (1962).

101. M. Lai and H. Weiss, Anal. Chem., 34:1012 (1962).

102. G. Hevesy, Radioactive Indicators, Wiley-Interscience, New York, 1948.

103. S. Sourirajan, Reverse Osmosis, Lagos Press, London, 1971, Chap. 5.

104. A. Garratt and B. Dacre, Unpublished work.

Chapter 12

DIFFUSION

Peter J. F. Griffiths and Geoffrey S. Park

Department of Chemistry
Institute of Science and Technology
University of Wales
Cardiff, Wales

I. INTRODUCTION

Essentially diffusion may be defined as the reduction of concentration differences in a system by random molecular motion. The rate constant for this mixing process is the chemical diffusion coefficient, D, which is given by the ratio of the flux, F, of the diffusant per unit time through unit area normal to the direction of flow, to the concentration gradient, dC/dx, of the diffusant in the given direction,

$$F = - D \frac{dC}{dx} \tag{1}$$

This is often called Fick's first law and leads to the second law which relates the rate of change of concentration, $\partial C/\partial t$, to the concentration gradient. For diffusion in one dimension the second law is

$$\frac{\partial C}{\partial t} = \frac{\partial}{\partial x} (D \frac{\partial C}{\partial x}) \tag{2}$$

while in three dimensions it has the form

$$\frac{\partial C}{\partial t} = \text{div } (D \text{ grad } C) \tag{3}$$

The diffusion coefficient is of considerable importance to both the chemical engineer and the fundamental scientist since it measures rates of transport and changes of concentration on the macroscopic level, while on a molecular level the rates of molecular motion are obtained. However, a coefficient that is of even greater importance from a fundamental point of view is the self-diffusion coefficient, D^*, which measures the flux of one set of molecules into their exact counterparts. In a one-component system D^* is defined by Eq. (1) in which both F and C refer to a set of molecules that differs from the rest by some label that introduces no significant change in molecular motion or equilibrium. In multicomponent systems there are self-diffusion coefficients, sometimes called tracer diffusion coefficients or intradiffusion coefficients, for each component and these measure the ease of molecular motion of the component in the absence of any overall concentration gradient.

Chemical diffusion coefficients can be determined by any of the many analytical and physical processes that measure concentration and changes therein. One such process involves isotopic labeling and radioassay.

An interesting example is the measurement of the diffusion of oxygen in single-crystal CoO. Crystals of the oxide were annealed in oxygen enriched with ^{18}O at 1000-1500°C and the diffused ^{18}O was activated by bombardment with 2.8 MeV protons to produce the positron emitter ^{18}F. Autoradiographs of sections yielded penetration profiles [1]. For self-diffusion studies, however, the radiotracer technique is unique in that it permits labeling of some of the molecules, with negligible change in properties. In general, chemical labeling produces too great a change in behavior of the diffusing molecules. Characterization in terms of nuclear magnetic spin produces even less change in properties than isotopic labeling but is often not very accurate and, furthermore, uses the laborious and expensive technique of spin-echo NMR [2].

Basically, techniques for measuring diffusion coefficients fall into two categories. The first of these follows directly from the definition in Eq. (1) and is based on steady-state conditions.

The second set of methods is based on the transient state and in the mathematically simplest case D is obtained from independent measurements of dC/dt and dC/dx followed by substitution in Eq. (2) or (3). Usually, methods involving the transient state depend primarily on the mathematical integration of Eqs. (2) and (3) for appropriate boundary conditions to give, for instance, equations describing the rate of sorption of diffusants by media of known geometrical shape or the rate of approach to steady state in permeation experiments.

It is appropriate to classify diffusion systems as solid or fluid. In solids diffusion is the only means of mixing and this, coupled with the dimensional stability of solid systems, means that the diffusion coefficients are more easily obtainable than in fluid systems where convective mixing and flow occur readily.

Important solid systems include metals, metallic oxides, ionic crystals, and organic crystals. Many polymeric systems can also be regarded as solid and so the techniques of measurement considered in the next section are applicable to gases, vapors, liquids, dyestuffs, and other materials dissolved in polymers, as well as to crystalline materials.

Liquid systems have been considered separately since convective mixing and flow create problems in measuring diffusion coefficients.

II. DIFFUSION IN SOLIDS

A. Steady-state Transmission Rates

This technique is mainly applicable to the diffusion in a solid membrane of materials that are normally fluid (gases or liquids) but can be used for the diffusion of a solid solute when this can be presented to the membrane surface as a solution in a suitable fluid. It has been used, for instance, for diffusion measurements on sodium chloride in cellulosic membranes [3] as well as for the diffusion of permanent gases or vapors in polymers. Steady-state diffusion measurements across a membrane using diffusants that are dissolved in a fluid on either side of the membrane have often been made using labeled diffusants. Examples are the diffusion of both elementary anions (Br^-) and dye anions (orange II) in films of horn keratin [4]. In another application the fluid on one side of the membrane is also a scintillator solution so that, while diffusion occurs, the activity may be monitored continuously by scintillation counting. This method has been used to measure the diffusion of n-decane in block copolymers [5]. The technique can, of course, be used for measuring the chemical diffusion coefficient or the self-diffusion coefficient. For the latter measurements, although the concentration of diffusant is the same on both sides of the membrane throughout the experiment, the diffusant on one side of the membrane is radioactively labeled and the course of diffusion is followed by measuring the increase of radioactivity on the opposite side. The apparatus designed by Meares and his coworkers [6] for studying self-diffusion in ion-exchange membranes is shown in Fig. 1 as an example of this method. The sealing of the membrane is very important in all permeation experiments, and in this apparatus the seal is accomplished by gripping the two faces of the membrane between the faces of similar Perspex cells using a constraining brass frame and an annular Perspex spacing ring which is slightly thinner than the membrane itself. Circulation is maintained by Perspex centrifugal pumps, F, and the formation of a stagnant diffusion layer at the surfaces of the membrane is minimized by directing the circulating solutions to flow rapidly over the membrane faces. Initially, the membrane is allowed to come into equilibrium with

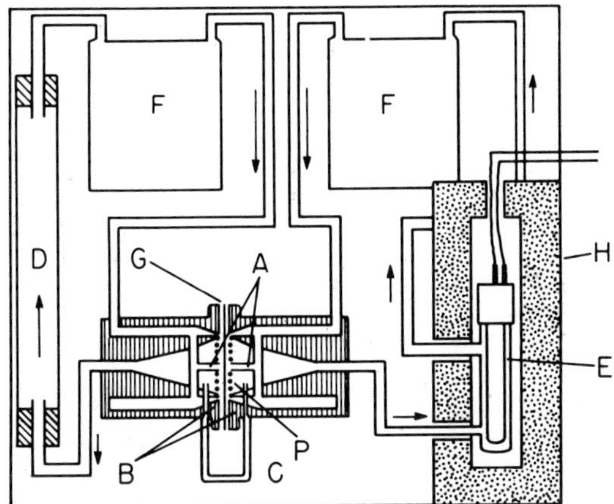

FIG. 1. Continuously recording membrane-diffusion apparatus. (A) central solution jet; (B) annular solution jet; (C) differential manometer; (D) flowmeter; (E) continuous flow Geiger tube; (F) Perspex centrifugal pump; (G) membrane; (P) perforated baffle plates; (H) lead castle. (Reproduced with permission from Ref. 6.)

inactive solutions of equal concentration circulating on each of its sides. The radiotracer is added to one side of the membrane and the transmission of activity to the other side is monitored continuously using the liquid-counting Geiger tube, E. Meares [6] has discussed corrections for the boundary films that occur on the membrane surfaces, for the edge effect due to the flow of diffusant into and out of the portion of the membrane clamped between the Perspex surfaces, and for the departure from true steady-state conditions that occurs due to the slowly increasing concentration of radiotracer on the counting side of the apparatus.

B. Permeation Time Lag

Although the concentration of diffusant in the membrane surface must be known before the diffusion coefficient can be determined from steady-state experiments it is possible to obtain the diffusion coefficient directly from the rate of approach to steady-state conditions. Figure 2 shows how the amount of radioactivity transmitted through a membrane that is free initially of diffusing radiotracer changes with time (t) when a constant concentration of radiotracer is introduced to one side of the membrane (at $t = 0$). By extrapolation of the steady-state rate of permeation to the time axis the time lag, θ, is obtained. Solution of Fick's second law, Eq. (2), then gives the relationship

$$D = \frac{\ell^2}{6\theta} \tag{4}$$

where ℓ is the thickness of the membrane. Most techniques for determination of the steady-state rate of permeation can also be used for obtaining the time lag [47]. Equation (4) is applicable only when there are no extra boundary layers on the surfaces of the membrane contributing to the diffusional resistance. When such layers do occur the relationship given by Barrie et al. [7] can be used to make corrections. In an interesting time-lag

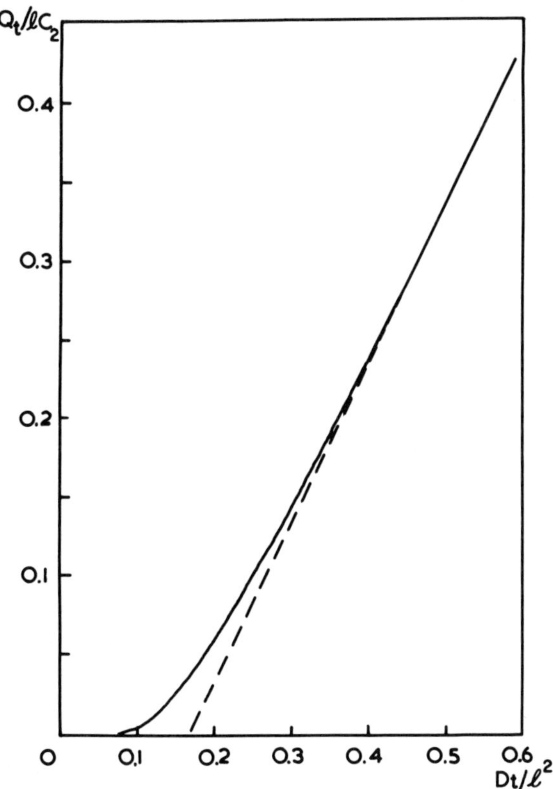

FIG. 2. Diffusional time lag for permeation through a plane sheet. Q_t, amount permeated in time t; ℓ, sheet thickness; C_2 surface concentration. (Reproduced with permission from Ref. 10, p. 48.)

technique, used by Haefelfinger and Grün [8] for labeled compounds at very low concentration in natural rubber, a constant concentration of diffusant was maintained on one side of the membrane, while all the activity which diffused out on the other face was completely removed by a rapid current of air. Measurement of the β-particle emission from the face of the membrane at zero concentration revealed that when a steady state had been reached the plot of total β-particle count against time increased linearly. Extrapolation similar to that in Fig. 2 yielded a time lag which could be used in Eq. (4) for thick membranes since β particles from the high concentration side of the membrane have insufficient energy to escape into the detector. For thin membranes, however, this is no longer true and Haefelfinger and Grün have shown that the relationship becomes

$$D = \frac{\ell^2}{6\theta}\,\frac{1 - 6\exp\mu\ell - 6 - 6\mu\ell - 3\mu^2\ell^2 - \mu^3\ell^3}{\mu^2\ell^2(\exp\mu\ell - 1 - \mu\ell)} \tag{5}$$

in which μ is the exponential absorption coefficient for β particles.

C. Diffusant Absorption Measurements

Rate measurements of sorption or desorption of gases, vapors, and liquids have been extensively used for obtaining diffusion coefficients in solids. The technique has been particularly important in studying the diffusion of organic molecules in polymers, and has usually involved absorption by a thin sheet of material in an atmosphere containing a constant pressure of diffusant. In this method the sorbing material contains a uniformly constant concentration of diffusant until the start of the experiment when the pressure of diffusant is suddenly changed. At this moment the surface concentration of diffusant makes a discontinuous change from C_1 to C_2 and the concentration change spreads subsequently throughout the whole of the material by diffusion. The solution of Eq. (2) for the appropriate boundary conditions gives the amount sorbed, or desorbed, M_t, as a function of time, t. Equations (6) and (7) are alternative forms of the solution

$$\frac{M_t}{M_\infty} = \left(\frac{Dt}{\ell^2}\right)^{1/2}\left(\frac{1}{\pi^2} + 2\sum_{n=0}^{\infty}(-1)^n \operatorname{ierfc}\frac{n\ell}{2(Dt)^{1/2}}\right) \tag{6}$$

$$\frac{M_t}{M_\infty} = 1 - \frac{8}{\pi^2}\sum_{m=0}^{\infty}\frac{1}{(2m+1)^2}\exp\left(-\frac{D(2m+1)^2\pi^2 t}{\ell^2}\right) \tag{7}$$

In these relationships n and m are integers and ierfc x is

the error function complement $\int_x^\infty (1 - \mathrm{erf}\, Z)\, dZ$. Equation (6) is the appropriate form for relatively short times in which the amount absorbed is less than, or not much greater than, one-half of the total equilibrium sorption. Equation (7), on the other hand, is most appropriate for long times in which M_t/M_∞ is greater than 0.5. Sorption and desorption kinetic experiments on polymer systems have shown that frequently the diffusion coefficient is dependent upon concentration so that special procedures had to be devised to use Eqs. (6) and (7) which strictly refer only to constant diffusion coefficients. To obtain self-diffusion coefficients by sorption the polymer is equilibrated with radioactive diffusant at a fixed vapor pressure and then transferred to an atmosphere containing nonradioactive vapor at the same vapor pressure. The rate of increase of radioactivity in the initially nonradioactive vapor is a measure of the amount sorbed or desorbed and can be used in Eqs. (6) and (7) to calculate the self-diffusion coefficient. Since the overall concentration of diffusant is constant the diffusion coefficient is constant in any one experiment so that Eqs. (6) and (7) apply strictly. In such self-diffusion measurements errors are likely to occur due to self-diffusion in the vapor phase, but this can be eliminated by rapid rotational stirring of the polymer specimen [9]. The relationships (6) and (7) and the further approximations to these given in Eqs. (8) and (9)

$$\frac{M_t}{M_\infty} = \left(\frac{16Dt}{\pi \ell^2}\right)^{1/2} \tag{8}$$

$$\frac{M_t}{M_\infty} = 1 - \frac{8}{\pi^2} \exp\left(-\frac{\pi^2 Dt}{\ell^2}\right) \tag{9}$$

for long and short times, respectively, would enable the diffusion coefficient, D, to be evaluated from the amount, M_t, of radioactive material desorbed at time t provided the volume of vapor phase was large enough to satisfy the assumption that the concentration of activity at the surface stayed at zero throughout the desorption. In practice this is not so and, although the surface concentration at the beginning of the experiment drops to zero, it then increases slowly with time. However, due to the finite volume of vapor the solutions appropriate to a finite bath of diffusant [10] have then to be used.

A method analogous to the vapor sorption technique has been developed for solid diffusants [94], and used to study the self-diffusion at 230-380°C of silver using ^{110}Ag. In this method silver foils were maintained in contact with molten radioactive silver nitrate for various times and the activity taken up by each foil was measured.

Sorption and desorption kinetics for obtaining the diffusion coefficient can be used for studying self-diffusion in textile polymers [11] and have been used extensively for studying diffusion in ion-exchange resins. In one technique, beads of ion-exchange resin are equilibrated with the radioactive ion or counter ion in solution. The beads are then transferred to a nonradioactive solution at the same overall concentration and the rate at which the solution becomes radioactive is measured. The use of relatively large beads decreases the barrier due to the layer of stagnant solution on the bead surfaces. With smaller beads attempts have been made to remove this layer by rapidly stirring the solution containing the freely suspended beads. In another method the beads are spread on a platinum screen and a solution is forced at high speed through the bed of resin particles. Using the stirring technique, stagnant layers as thick as 10 μm were found even at a stirring rate of 750 rpm, while the shallow bead technique reduces the thickness of the layer to about 1.3 μm with a linear flow rate of the solution of about 200 cm sec^{-1} [12]. Similar methods can be used for studying tracer sorption or desorption from rods, strips, or sheets of ion exchange material. In one method a long rod (about 20 cm) of ion-exchange resin that has been equilibrated with labeled diffusant is then allowed to desorb by placing the end of the rod only in contact with the same concentration of nonradioactive diffusant. Richman and Thomas [13] have used this method to study the self-diffusion of sodium ions in ion exchange material by circulating the originally nonradioactive salt solution through a flow-type Geiger counter. In a variant of this technique the total activity in a much shorter length of ion exchange resin is measured by using a gamma-active tracer such as ^{60}Co [14].

Elleman et al. [92], in a study of the diffusion of xenon in single crystals of CsI, used two techniques to introduce ^{133}Xe. One was a homogeneous labeling method in which ^{133}Xe was introduced by the decay of homogeneously distributed ^{133}I. In the other, ^{133}Xe was recoiled into the surface layers (10-15 μm) of the CsI specimen from a surrounding enriched uranium foil by irradiation with thermal neutrons. The diffusion coefficient was obtained from the rate of release of ^{133}Xe.

D. Techniques Dependent on Concentration-Distance Relationships

Permeation and sorption techniques for measuring the diffusion coefficient can be used only when it is possible to maintain the diffusant at a known or constant concentration at the surface of the solid. This is possible only when the diffusant is a gas or a liquid or can be dissolved in a liquid. The study of the interdiffusion of solids requires techniques that depend essentially on measurements of rates of mixing. The simplest system that can be considered is one in which two semiinfinite slabs are joined together so that diffusion from one slab to the other is possible, and is followed by finding the distribution of diffusant after a given time. When this technique is used for obtaining the chemical diffusion coefficient, for instance, for pure silver diffusing into pure lead [19] or for a plasticizer diffusing into an unplasticized polymer [16], the analysis of the system is complicated by the fact that the diffusion coefficient is concentration-dependent. Matano [17], however, has given a mathematical analysis of this situation which enables the diffusion coefficient at any concentration C' to be obtained from the relationship

$$D_{C=C'} = -\frac{1}{2t} \cdot \frac{dx}{dC} \int_0^{C'} x \, dC \tag{10}$$

where x is the distance from a specially defined plane. Labeling of one of the diffusants followed by radioassay has not been a common method for obtaining the chemical diffusion coefficient, but Kidson [19], for instance, used ^{198}Au as a tracer for the determination of the diffusion coefficient of gold in lead. Tracer techniques are more usually used to obtain self-diffusion coefficients. Since the overall composition is then constant the diffusion coefficient is constant throughout the system and instead of Eq. (10), the analytical solution

$$2\frac{C}{C_0} = \text{erfc}\,\frac{x}{(4Dt)^{1/2}} \tag{11}$$

can be used. This relationship has been used for measuring the self-diffusion coefficient of plasticizer in polyvinyl chloride [20]. A button of polyvinyl chloride plasticized with ^{14}C-labeled plasticizer is pressed on to a button of plasticized polymer in which the plasticizer contains no ^{14}C. Since the two buttons of material can be separated easily, the total gain in activity of the originally unlabeled button can be found as a function of time and the diffusion coefficient calculated from Eq. (12), the integrated form of Eq. (11)

$$\sigma = \sigma_s A \left(\frac{Dt}{\pi}\right)^{1/2} \tag{12}$$

where σ_s is the specific activity of the original ^{14}C-labeled plasticized button and A is the cross-sectional area. For metal systems the most common method has been to use an infinitely thin layer of radioactive material at the surface of an effectively infinite layer of unlabeled metal or sandwiched between two such layers. Uniformity of the layer is not essential since the criterion is that the layer must be very thin in comparison with diffusion distances. The infinitely thin radioactive layer can be applied by an electroplating technique [21] or vacuum evaporation of the tracer on to the surface [22]. Among other methods the welding of a thin metallic foil on to the surface and the production of radioactive atoms on the surface by ion bombardment [23] have been used. The relationship between concentration, C, and distance for one infinite layer is given by Eq. (13) in which Q_0 is the initial surface concentration.

$$\frac{C}{Q_0} = [(\pi Dt)^{1/2} \exp \frac{x^2}{4Dt}]^{-1} \tag{13}$$

A plot of log C against x^2 gives a straight line from which the diffusion coefficient can be computed.

Various methods have been used for obtaining the concentration/distance curve. In one of these the diffusion specimen is cut into thin sections (of the order of 10 μm) [25] and the activity of each section is then determined. In another method the specimen is progressively etched layer by layer and the activity of each dissolved layer is then found [26]. A grinding technique has also been used to remove one radioactive layer after another and the residual activity of the ground down specimen is found [27]. In this technique the apparent activity of the surface of the specimen is influenced by activity from the lower depths of the material. The diffusion coefficient is then obtained [27] from the relationship

$$a_x = \frac{a(1 - \text{erf}[x/2(Dt)^{1/2} + \mu(Dt)^{1/2}])}{1 - \text{erf}\ \mu(Dt)^{1/2}} \tag{14}$$

where μ is the exponential absorption coefficient for β particles, a is the count rate after time t before grinding, and a_x after grinding to a depth x. When finite layers of thickness h and ℓ of radioactive (activity a_0) and nonradioactive material are used the diffusion coefficient has to be obtained [24] from the relationship

$$\frac{a}{2a_0} = \sum_{n=-\infty}^{n=\infty} \left(\text{erf} \frac{h + 2n(h + \ell) - x}{(4Dt)^{1/2}} + \text{erf} \frac{h - 2n(h + \ell) + x}{(4Dt)^{1/2}} \right) \tag{15}$$

Yet another method of obtaining the concentration/distance curve uses autoradiography [28].

E. Surface Concentration and Radiation Absorption Techniques

1. Measurements at the Active Surface

The earliest determination of diffusion coefficients using a radiotracer were due to Hevesey and Obrutscheva [29] who joined a thin plate of radioactive lead to a nonactive one and measured the course of diffusion from the decreasing intensity of radiation at the face of the composite. The technique was later applied to diffusion in lead iodide by Hevesey and Seith [30] and has since been extensively applied to many metallic systems, to solid metallic oxides and salts, and also to diffusion in polymer systems.

Probably the most common method dependent on a measurement of concentration/distance curves is the technique of joining an infinitely thin layer of radioactive material to an infinitely thick layer of inactive substance. This technique is also the one that has been mainly used in obtaining the diffusion coefficient from surface activity measurements. When the radioactive material is an emitter of fairly low-energy particles, the radiation escaping from the face of the composite slab is controlled by the concentration distribution of the radiotracer close to the surface. At sufficiently low values of Dt the measured activity,

a_t, is proportional to the actual surface concentration as given by Eq. (13) and it can be shown that

$$a_t = \frac{\eta a_0}{(\pi Dt)^{1/2}} \tag{16}$$

where a_0 is the initial measured activity and η is a parameter depending on the relationship between the absorption of radiation by the diffusion medium and the absorber thickness. For α particles η is equal to one-half, $R/2$, of the range, while for γ radiation or for β particles that obey the exponential absorption law, $a_x = a\,\exp(-\mu x)$, η is equal to $1/\mu$. It is not often that this limiting law, Eq. (16), is reached in practice and particularly for very low diffusion coefficients it is usually necessary to use Eq. (17).

$$a_t = \int_0^\lambda \psi(x)\phi(Dt,x)\,dx \tag{17}$$

where λ is the thickness of the slab, $\psi(x)$ expresses the radiation absorption law and $\phi(Dt,x)$ is the concentration/distance relationship at time t. Hevesey and Seith's treatment of the diffusion of lead, ^{212}Pb, in lead iodide is an example of the application of Eq. (17), using the absorption of the α emission from the polonium daughter product of lead [30].

A relatively simple solution of Eq. (17) occurs when the emitted particles obey an exponential absorption law so that

$$\psi(x) = \exp(-\mu x) \tag{18}$$

as occurs for γ-rays and approximately at least for β particles from some sources. Equation (17) now becomes

$$a_t = a_0 \int_0^\infty (\pi Dt)^{-1/2} \exp\,(-\mu x)\, \exp\left(-\frac{x^2}{4Dt}\right) dx \tag{19}$$

and so on integration

$$a_t = a_0 \cdot \exp \mu^2 Dt \cdot \text{erfc}\,[\mu(Dt)^{1/2}] \qquad (20)$$

This relationship was used by Steigman et al. [31] in a study of self-diffusion in copper. Since the equations obtained by Hevesey and Seith and Eq. (20) enable the diffusion coefficient to be obtained from the dimensionless parameters Dt/R^2 or $\mu^2 Dt$ it is possible to use this technique to measure very low diffusion coefficients in a reasonable length of time provided that the range R or reciprocal absorption coefficient μ^{-1} of the emitted particles is sufficiently small. Table 1 gives the order of magnitude of some diffusion coefficients that have been measured using different radionuclides. Perhaps the most interesting instance is the one that makes use of the recoil atoms ^{208}Th ("ThC") from the disintegration of ^{212}Bi(ThC) in the series starting from ^{212}Pb(ThB). Diffusion coefficients as low as 10^{-21} m^2 sec^{-1} were obtained using this system [32].

The diffusion of α-active elements in nuclear fuels has been studied by the α-energy degradation method [93]. In this technique the distribution of the tracer nuclide is determined from changes in the shape of the energy spectrum of the escaping α particles.

2. Measurements at the Inactive Surface

The surface radioactivity technique has been used chiefly to obtain self-diffusion coefficients in metal systems or in simple salts. It has, however, been applied to polymer systems and has been used for instance to measure the self-diffusion coefficient of polystyrene and of polybutylmethacrylate using ^{14}C-labeled polymer [33] and in natural rubber using tritium-labeled polyisoprene [34].

For the diffusion of antioxidants, plasticizers, insecticides, and other low molecular weight materials in polymers, however, a more common technique involves the use of activity measurements at the originally inactive surface, while sometimes

TABLE 1

Some Examples of the Surface Activity Method for Self-Diffusion

Radionuclide	Emitted particle	Range R (mm)	Reciprocal absorption coefficient (μ^{-1})(mm)	System	Minimum D measured $(m^2\ s^{-1})$	Ref.
^{65}Zn	325 keV β^+	--		Zn metal	10^{-16}	32
^{60}Co	315 keV β^-	--		Co metal	10^{-16}	91
^{14}C	155 keV β^-	--	3.3×10^{-2}	Polybutylacrylate	10^{-16}	33
^{3}H	18 keV β^-	--	1.3×10^{-3}	Polyisoprene	10^{-18}	34
^{212}Pb ($^{212}P_o$)	8.6 MeV α	9.8×10^{-2}	--	PbI_2	10^{-16}	30
^{212}Pb (^{212}Bi)	$^{208}T\ell$ recoil	7.5×10^{-5}	--	$PbCl_2$	10^{-21}	30

quite complicated sandwich arrangements of polymers have also been used [35]. Table 2 summarizes some of the systems that have been used. Perhaps the most useful and easily applicable arrangement is the smear method used by Ferry and collaborators [36]. In this technique a very thin smear of diffusant of negligible mass but very high specific activity is spread on the face of a polymer disc away from the counting device. This is used to give the diffusion coefficient at zero concentration of diffusant. If the self-diffusion coefficient at a non-zero concentration is required then the polymer disc is replaced by a disc containing polymer and inactive diffusant at the required concentration. For a slab of thickness ℓ, which is large compared with the range of the β particles, it is readily shown that

$$a_t = a_\infty[1 + 2 \sum_{n=1}^{n=\infty} (-1)^n \exp(-n^2\pi^2Dt/\ell^2)] \tag{21}$$

where a_∞ is the radioactivity reached after infinite time. Since this relationship does not involve μ or R, a possible source of error in the diffusion coefficient is removed. For thinner slabs Chen and Ferry [36] have shown that

$$a_t = a_\infty[1 + 2 \sum_{n=1}^{n=\infty} (-1)^n[\exp(-n^2\pi^2Dt/\ell^2)](1 + n^2\pi^2\mu^{-2}\ell^{-2})^{-1}] \tag{22}$$

and used a curve-fitting procedure for obtaining diffusion coefficients from their data. This procedure is not necessary [38] provided that $\frac{a_\infty - a_t}{a_\infty}$ is larger than about 0.7 and that $\mu\ell$ is reasonably great. Under these circumstances both equations (21) and (22) reduce to the relationship

$$\ell n(a_\infty - a_t) = \text{Constant} - \pi^2Dt/\ell^2 \tag{23}$$

so that the diffusion coefficient can readily be determined from a simple linear plot. Instead of a thin smear, several investigators [39] have used a finite thickness of radioactive material.

TABLE 2

A Variety of Arrangements for Measuring Diffusion in Polymers Using β-Particle Absorption Techniques

Ref.	Isotope	Diffusant	Polymer	Layer facing detector	Thickness ratio: Inactive layer / Active layer	Range of − log D
33	^{14}C	Polystyrene Polybutylacrylate	Polystyrene Polybutylacrylate	Active	∞	10-12
34	^{3}H	Polyisoprene	Natural rubber	Active	∞	12-14
42	^{32}P	Tricresyl phosphate	Polystyrene	Inactive	10	9-10
36	^{14}C	Cetane	Polyisobutene Polymethacrylate	Inactive	∞ 1	6-9
18	^{14}C	Benzene	Natural rubber	Inactive Inactive	∞ 1	5-6 6-7
37	^{35}S	Sulfur	Natural rubber, GRS	Inactive	∞*	7
39	^{14}C	Octadecane Octadecanol Octadecyl stearate Stearic acid	Many rubbers Polyethylene Balata Polybutadiene	Inactive	1	7-9
15	^{14}C ^{35}S	Ionol Phenothiazine	Polypropylene Polyformaldehyde	Active	∞	7-9
35	^{14}C	Complex antioxidants (MW ∼ 350)	Polyethylene Polypropylene Poly-4-methylpentene	Inactive	100**	7-10

*Active layer is kept saturated with sulfur by contact with a layer of pure sulfur.

**1.5 mm of polymer is coated with 10 μm thickness of a 1:1 mixture of polyisobutene and diffusant covered with 1 μm thickness of the polymer.

The equations for this situation were given by Zimen et al [40] and for equal layers of active and inactive material by Timmerhaus and Drickammer [41]. At a reasonably long time all these relationships reduce to equation (23) in which ℓ is the total thickness of the slab, provided that this is large compared with the range of the β-particles. This relationship is applicable even when the active and inactive layers are of comparable thicknesses.

It is possible to apply the technique involving a finite labeled layer to a system in which the polymer specimen is thin in comparison with the range of the β particles. An example of this is given by the measurement of the self-diffusion coefficient of tricresyl [^{32}P]phosphate in polystyrene [42]. From the known and rather complicated relationship between measured activity and absorber thickness it was possible to compute activity against Dt plots from the known thicknesses of inactive and active layers. By comparison with measured activity/time plots diffusion coefficients could be evaluated even though the measured activities only increased by 10-20% during the course of an experiment.

III. DIFFUSION IN FLUIDS

In principle, all the methods that can be used for measuring the diffusion coefficient in solids can also be used for fluid systems. There are, however, two added complications with fluids. First, it is usually necessary to contain the fluid in a vessel and so there may be problems in approaching the fluid surfaces to measure surface activity. The second problem is much greater and is concerned with nondiffusional bulk convective mixing. If the conditions are not well chosen, convection currents in the system will lead to a much more rapid mixing than is obtained by diffusion. To overcome this problem in diffusion down a concentration gradient, the denser component is normally located at the bottom of the tube. Uniformity of temperature is essential if thermal convection is to be avoided. It is surprising that a very simple method for obtaining concentration ∿ distance curves by radioactivity measurement using a normal diameter test tube and a slit collimator for the β particles [43] has been reasonably effective. Walker's results on the phosphoric acid/water system [43] using the apparatus in Fig. 3 gives diffusion coefficients agreeing with published values to within a few percent! It could be that Walker's method worked because he was measuring diffusion down a concentration gradient and the denser solution at the bottom of the tube was therefore unlikely to mix convectively with the less dense solution at the top. In self-diffusion

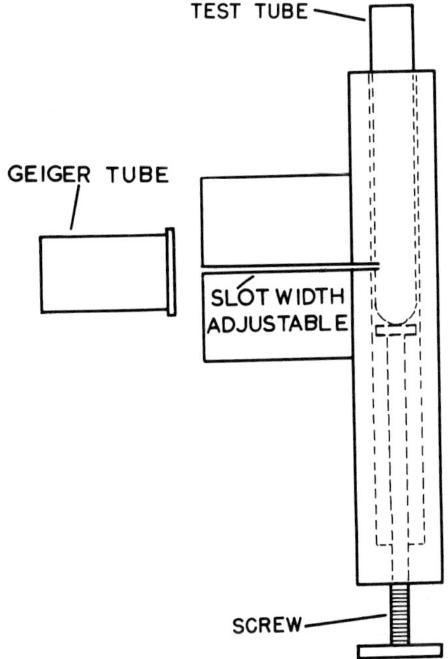

FIG. 3. A simple test-tube apparatus for obtaining the diffusion coefficient from the distribution of labeled material. (Reproduced with permission from Ref. 43.)

measurements where there is no density difference and with reasonably mobile liquids difficulties might be experienced using a diffusion tube as wide as the one used by Walker.

A fairly general method of overcoming the problems due to convective mixing is to contain the diffusing material in a fine capillary or in a system of fine capillaries. Perhaps the most publicized method using radiotracers for obtaining self-diffusion coefficients in liquid systems is the closed capillary technique of Anderson and Saddington [44].

A. Closed Capillary Technique

In this method a radioactively labeled sample of fluid is contained in a capillary of accurately known length which is closed at one end. The diameter of the capillary is usually between 0.2 and 1.0 mm and the length of the capillary is within the range 1 to 6 cm. The tube containing the labeled fluid, which may be either a pure liquid or a solution, is immersed in a large bath of nonlabeled fluid of the same composition. The active molecules then diffuse out of the capillary and the total amount, M_t, of labeled material that diffuses out in time t is given by Eqs. (6) or (7) in which M_∞ is the original total amount of activity in the tube and the length of the capillary is $\ell/2$. In practice the capillary is withdrawn after a given time and the total contents are counted. The application of Eqs. (6) or (7) to this diffusion situation depends upon the concentration of radioactive material being maintained at zero at the mouth of the capillary at all times. If a stagnant immersion fluid is used for the capillary this will not be true and the capillary will have an effective length somewhat greater than its actual length. Alternatively, vigorous stirring depletes some of the active solution from the top part of the capillary and the effective length will be less. In a study of the self-diffusion of water and of the diffusion of Na^+ in potassium chloride solution Wang [45] has shown how corrections for the uncertain effective length of the capillary can be made by carrying out parallel experiments with capillaries of different lengths. In his experiments he found that no correction was required when mild stirring was used. Further investigations of the effect of stirring have been made by Bambynek and Freise in their study of the self-diffusion coefficient of tetramethyl tin [46]. They have shown that when Eq. (7) is used for obtaining the diffusion coefficient at sufficiently high values of t so that only one term is required, Eq. (24) gives the relationship between the true value of the diffusion coefficient D and the value D_0 actually obtained from the experiment.

$$D = D_0[1 + 2 \frac{\Delta \ell}{\ell}] + 4 \frac{\ell \, \Delta \ell}{\pi^2 t} \qquad (24)$$

Here, $\Delta \ell$ is the difference between the actual length of the capillary and the virtual length. By determining D_0 for two different values of ℓ, good values for both D and $\Delta \ell$ can be obtained. Bambynek and Freise fill and empty the measuring capillary by inserting a much finer capillary into it and withdrawing or inserting fluid with a Rekord glass pump and a micromanipulator.

By setting the capillary in a scintillator block Mills and Godbole [89] were able to monitor the activity in the capillary over a considerable time, while Monk and collaborators [62] did this by taking samples of the immersion liquid.

Constant bore tubing can easily be obtained for measurements near room temperature but for the self-diffusion of molten salts at high temperatures silica capillaries which are usually slightly tapered have been used. The errors thus introduced have been considered by Talbot and Kitchener [48].

A modification of the closed capillary technique in which a 6-mm diameter rod contains 5000 parallel holes of 50 μm diameter has been used by Thomas and Ku [64], who eliminated stirring errors by covering the open ends of the capillaries with a thin cellophane membrane.

B. Diaphragm Cell

This technique of Northrop and Anson [49] is a very useful one for obtaining diffusion coefficients from radiotracer measurements. The cell consists essentially of two large chambers separated by a porous sintered frit that makes up the diffusion cell. For investigation of diffusion down a concentration gradient a horizontal sintered frit is used and the lower chamber is filled with the denser solution. The change of concentration of a diffusant with time is measured in one or both chambers which are assumed to be well mixed. Thorough mixing is important and can be achieved by using magnetic stirrers. A typical diaphragm cell for obtaining self-diffusion coefficients using labeled materials is shown in Fig. 4. One problem associated with

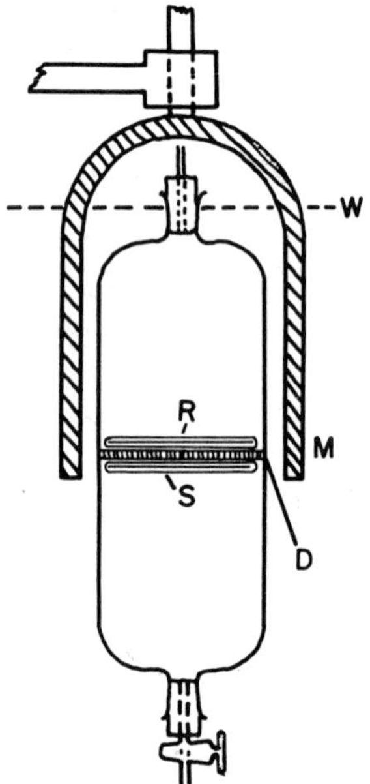

FIG. 4. Stokes diaphragm cell. (M) magnet; (D) porous diaphragm; (R) and (S) magnetic stirrers; (W) level of thermostat water. (Reproduced with permission from Ref. 95.)

this technique is that there is no direct measure of the cross-sectional area or of the thickness of the zone through which the diffusion is occurring. The ratio of these two quantities is determined by calibration with a material of known diffusion coefficient such as potassium chloride solution [66]. Johnson and Babb [50] mentioned in 1956 that two of the recent determinations of the diffusion coefficients in the ethanol/water system disagreed by as much as a factor of 2 even though each set of investigators considered their results to be accurate to between 1 and 3%. They suggested that to some extent at least this is a result of the calibration problem, since air or vapor trapped in the sintered frits could seriously change the effective cross-sectional area available for diffusion. More recently, Mills [51] pointed out that for solutions of electrolytes, very high reproducibility (±0.1% for KCl solution) is found and yet a 6% error for benzene is typical for organic diffusants. In a study of self-diffusion in benzene Mills found that radioactive impurities and impurity in the inactive material led to changes of about 1% in the diffusion coefficients, and that negligible error seems to arise from the radioactive analysis. An important source of error peculiar to organic liquid systems arises from the large expansion coefficient of these substances relative to water. In one diaphragm cell technique the bottom compartment has a fixed volume and so expansion of the liquid leads to forced flow through the membrane. Mills considers that short-term temperature variations are not major sources of error but long-term drifts in experiments that last for several days can lead to an error of 8% or higher. A real source of error arises from removal of the cell from the thermostat before sampling. In Mills' experiments this could lead to an error of almost 2%. Mills, Woolf, and Watts [52] discuss three methods for filling the diaphragm cell and Mills found that filling an evacuated cell with the solvent followed by the addition of labeled material after equilibration was the best method. Descriptions of the different kinds of diaphragm cells available have been given by Mills and Woolf [53].

C. Other Capillary Methods

In addition to the diaphragm cell and the closed capillary techniques other methods have been used in which the diffusing fluid is immobilized in a capillary or in a system of fine capillaries. These methods have been used mainly with molten salts or in high-pressure situations.

1. Surface Radioactivity Methods

Jeanneret and Grün [54] found that it was possible to obtain the diffusion coefficient of tristearin and triolein by depositing a layer of the inactive material 0.5 mm thick on a very thin layer of the ^{14}C-labeled diffusant and monitoring the β particles escaping from the surface. The method is successful for systems of high viscosity but for less viscous materials containment in a system of capillaries is necessary. In an investigation of the effect of pressure on the diffusion coefficient of gaseous carbon dioxide Drickammer and collaborators used a system in which the diffusion cell consisted of two equal lengths of fritted steel or parallel glass capillaries containing a uniform concentration of diffusant labeled with a weak β emitter in a one half and unlabeled material at the same concentration in the other half [41]. The activity at the face of the originally unlabeled material was measured by scintillation counting. They used a similar method to study the effect of pressure on the self-diffusion coefficient of carbon disulfide, water, and various ions in salt solutions [55].

2. Concentration-Distance Curves

Errors due to convective mixing in the wide diameter tubes used by Walker [43] are minimized when a long narrow capillary is used. Brown and Tuck [56] used a 30 cm-long capillary of 0.75 mm diameter closed at one end to obtain the self-diffusion coefficient of liquid mercury from the concentration/distance relationship of the γ-emitting ^{203}Hg isotope. The tube was half-filled with radioactive mercury and then with ordinary mercury, and the diffusion coefficient was determined from the activity/distance relationship measured with an accurately positioned collimated sodium iodide crystal scintillation counter. Another technique [57] used a system of rotating discs that enabled the capillary to be split into four equal parts. At the start of the experiment three sections of the capillary were filled with unlabeled mercury and the fourth one with labeled material. After aligning the four sections and allowing diffusion to proceed for a measured time, the sections were separated and the activities determined in each one. This method has also been used for determining the diffusion coefficient of liquid gallium [96]. Broome and Walls [70] have used a capillary for mercury comprising only two segments.

With high melting point liquids the concentration/distance curves can be obtained by sectioning after cooling to solidify the contents of the capillary. This was a method used for instance by Morgan and Kitchener [58] to obtain the diffusion coefficient of cobalt and carbon in molten iron, and, by substituting a sintered frit for the capillary, Saxton and Drickammer obtained the self-diffusion coefficient of sulfur. The activity/distance curve for ^{35}S was determined by sectioning after the sulfur in the frit had been solidified [59]. This technique was also used for measuring self-diffusion coefficients of indium and thallium [60].

Ketelaar and Honig [61] used a particularly simple technique for obtaining the self-diffusion coefficient of metallic ions in molten nitrates. They saturated a strip of glass fiber paper with the molten unlabeled salt, applied a minute quantity of labeled material to the center of the strip and allowed diffusive mixing to occur. Afterward they determined the variation of concentration with distance along the strip and obtained the diffusion coefficient from the resulting distribution.

3. The Open Capillary and Sintered Frits Techniques

When the closed capillary method described in III A is replaced by a capillary open at both ends the possibility of convective flow through the capillary is very great. If an open capillary is to be used, it must be of extremely narrow bore. Consequently the amount of tracer that can be used is so small that the method will only work when several capillaries are used in parallel. For this practical reason the main open-ended tube technique employs a block of finely sintered material. Since the capillary length is not easily defined in such a system the method is a comparative one which may be calibrated using a material of known diffusion coefficient. The method has been used to obtain the diffusion coefficient of polymers in solution and the ionic self-diffusion coefficient of Na^+ and CO_3^{2-} in molten sodium carbonate [63]. Sintered alumina discs of 1.5 cm diameter, 3 cm length, and having a mean pore size of 5-10 μm were used for the molten salt and in this work the effective length of the pores was obtained by measuring the conductivity of the sintered aluminum discs when saturated with a solution of known specific conductance. Use of the γ emitting ^{22}Na and the weak β emitter ^{14}C enabled the simultaneous determination of the ionic diffusion coefficients of both sodium and carbonate to be made from the rate of desorption into the molten salt.

D. Gel Immobilization

The addition of quite small amounts of a cross bonding, high molecular weight additive is sufficient to turn many liquids into gels in which convective flow is completely prevented. In systems where the gelating agent has little effect on the diffusion coefficient conversion to a gel facilities determination of the diffusion coefficient by techniques similar to those used for solids.

Typical of the gel techniques is that due to Spalding [67] who filled cylindrical stainless cells (3 mm deep) with Agarose gels containing $^{22}NaCl$ and then brought these into contact with similar cells containing unlabeled sodium chloride gels. The diffusion coefficients were obtained from radioassay of the separated cells after a suitable period of standing and were calculated from the relationship

$$\frac{\Sigma Q^0 - 2Q_i^t}{\Delta Q^0} = \frac{\pi^2}{8} \sum_{n=0}^{\infty} \frac{1}{(2n + 1)^2} \exp\left[-\frac{(2n + 1)^2 \pi^2 Dt}{4\ell^2}\right] \tag{25}$$

Here, ΣQ^0 and ΔQ^0 are, respectively, the sum and difference of the initial activities of the two cells, Q_i^t is the activity of the initially "inactive" cell after time t, and ℓ is the depth of the gel in each cell. By calibrating with a salt of known diffusion coefficient Spalding showed that in his determinations the sole effect of the Agarose was to increase the effective value of ℓ by about 4%. When he allowed for this he obtained the very good agreement shown in Fig. 5 between his results for the self-diffusion of Na^+ and the free solution ones obtained by Mills [68]. With 0.1 M Ba^{2+}, however, the errors were larger suggesting a specific interaction between these doubly charged ions and the Agarose.

In the determination of the self-diffusion coefficient of water using tritiated water Nakayama and Jackson [69] used a 12 cm-long cylinder of Agarose gel in a 1.9 cm diameter tube. This cylinder was cut into 1 cm lengths and assayed for tritium after it had been left for 48 hr in contact with 1 μCi of tritium on its upper surface. The diffusion coefficient was then obtained from Eq. (26) where a_x is the total activity from the top of the tube to a point distance x down the tube and a_T is the total of all the activity in the tube.

$$\frac{a_x}{a_T} = \operatorname{erf} \frac{x}{(4Dt)^{1/2}} \tag{26}$$

In a technique developed by Thomas and his collaborators [65] for determining the self-diffusion coefficient of ions in

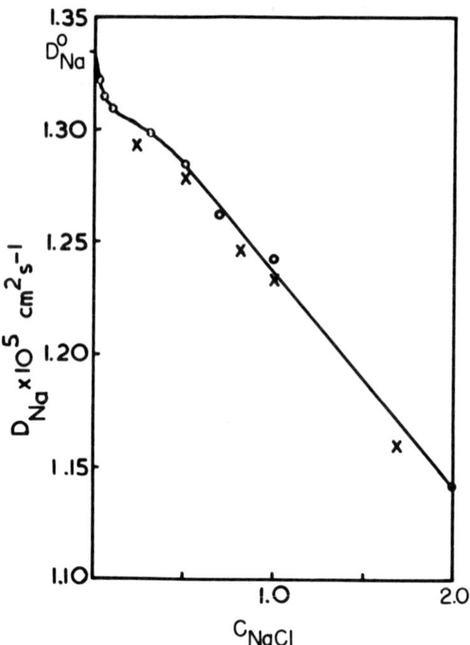

FIG. 5. Tracer diffusion coefficient for Na^+ in aqueous NaCl. (x) diaphragm cell results; (o) Agarose gel technique. (Reproduced with permission from Ref. 67.)

aqueous solution, a disc of gel 3 or 6 mm-thick is contacted on one face through a thin cellophane sheet with well-stirred solution. The decreasing tracer content of the gel is monitored by γ-ray scintillation counting. The thin cellophane membrane allows quite rapid stirring to be used without distorting even quite weak gels, but corrections have to be made for the barrier effect of this layer. The validity of these corrections was shown by the coincidence of the values obtained with cellophane films having different thicknesses. In another technique Langdon and Thomas [71] contained the gel in a capillary open at both ends. Gelation prevented flow through the capillary and also permitted rapid stirring without decrease of the effective capillary length by convective mixing. The blocking effect of the gel molecules has to be taken into account before measurements in gels can be identified with free solution values. Calibration with a salt of known diffusion coefficient was used for this purpose by Spalding [67], while Nakayama and Jackson [69] obtained a correct value for the self-diffusion coefficient of water by extrapolation to zero concentration of Agarose. Thomas and collaborators, on the other hand [71], found that, although extrapolation was satisfactory for the self-diffusion of anions in Agarose gels and for high concentrations of cations, specific absorption prevented its use at low concentrations, even with sodium ions, as is clearly shown in Fig. 6.

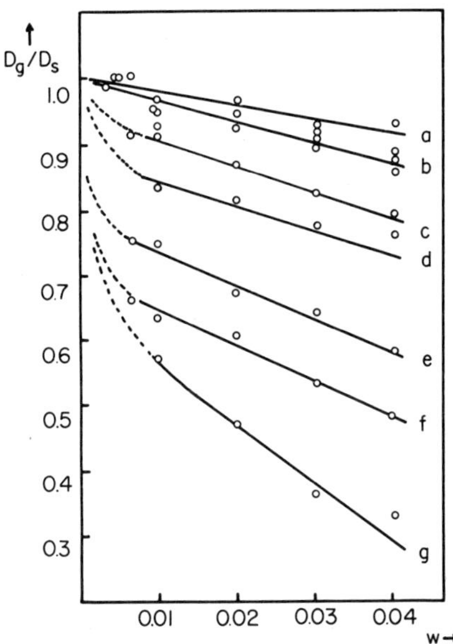

FIG. 6. Tracer diffusion coefficient for Na^+ in Agarose gels. D_g, diffusion coefficient in gel; D_s, diffusion coefficient in aqueous solution; w, concentration of gel (g cm^{-3}); NaCl concentrations (mol $liter^{-1}$), (a) 2.0; (b) 0.5, 0.1, 0.056; (c) 0.01; (d) 0.001; (e) 0.0004; (f) 0.0001; (g) 0.00001. (Reproduced with permission from Ref. 71.)

E. Gases

Although low viscosity and high expansion coefficients make thermal convective mixing far more important in gases than in condensed fluids the high value of the diffusion coefficient in gases ($\sim 10^{-5}$ m^2 sec^{-1}) means that this can be determined with relatively few precautions.

The porous frit method of Timmerhaus and Drickammer [41] for carbon dioxide has already been mentioned. In general the use of capillaries is not necessary in gaseous systems as is illustrated by the particularly simple method used by Hutchinson [72] for obtaining the self-diffusion coefficient of argon. A 45 cm-long tube of 1.2 cm diameter filled with ^{41}Ar was brought into line with a similar tube of inactive argon. The change in activity of the two tubes was measured as a function of time and the diffusion coefficient was calculated from a relationship similar to Eq. (25).

In another technique due to Ney and Armistead [73] a radioactive diffusant in a relatively large bulb is allowed to migrate down a connecting tube to a second bulb and the diffusion coefficient is obtained from the rate of increase of activity in the second bulb. Miller and Carman [74] used bulbs of about 700 cm^3 capacity and a 10 cm-long 1.1 cm diameter connecting tube. They obtained the binary diffusion coefficients with hydrogen and the self-diffusion coefficients for $^{14}CO_2$, ^{85}Kr, and $CF_2{}^{36}Cl_2$. Because of the low energy of the β radiation from ^{14}C, direct counting of $^{14}CO_2$ in continuous experiments was not possible since it was necessary to count the $^{14}CO_2$ as barium [^{14}C]-carbonate. For ^{85}Kr and ^{36}Cl the activity could be monitored continuously by a thin end-window Geiger tube incorporated into one of the bulbs. In similar experiments, tritium was monitored by using the lower bulb as an ionization chamber [75]. In these latter determinations they found it convenient to oppose convective mixing by maintaining a small temperature gradient from the bottom to the top of the apparatus. The increase in mass on substituting H^3H for H_2 means that the measured diffusion coefficient differs somewhat from that for normal hydrogen. The binary diffusion coefficient D_ℓ of a labeled molecule of molecular weight M_ℓ in an unlabeled one of M_u is related to the self-diffusion coefficient D^* of the unlabeled material by the relationship [75]

$$D^* = \left(\frac{2}{1 + M_u/M_\ell}\right)^{1/2} D_\ell \tag{27}$$

while the binary diffusion coefficient D_{ux} of the unlabeled material into material of molecular weight M_x is related to the binary diffusion coefficient $D_{\ell x}$ of the labeled material [75] by

$$D_{ux} = D_{\ell x}\left[\frac{1 + \dfrac{M_x}{M_u}}{1 + \dfrac{M_x}{M_\ell}}\right]^{1/2} \tag{28}$$

When ^{14}C is used as a label for carbon dioxide Eq. (27) leads only to a correction of 1%, while (28) gives a maximum correction of about 2%. When $H \cdot {}^3H$ is used as a label for H_2, however, these corrections amount to 15 and 40%, respectively.

Knowledge of the self-diffusion coefficient of gaseous uranium hexafluoride is important in the gaseous separation methods for uranium. In a modification [76] of Ney and Armistead's method, volumes of 7 liters and 2 liters joined by a 20 cm-long tube were used to measure the intermixing of ^{235}U with ^{238}U (see Fig. 7). The change in ^{235}U content was monitored continuously in the smaller bulb with a γ-ray spectrometer set for 185 keV γ-rays. In these methods the diffusion coefficient is obtained [76] from the relationship

$$D = -\frac{(\ell + 2r)}{\pi r^2}\left[\frac{1}{(V_1 + \frac{1}{2}v)} + \frac{1}{(V_2 + \frac{1}{2}v)}\right]^{-1} \frac{d \log (\Delta a)}{dt} \tag{29}$$

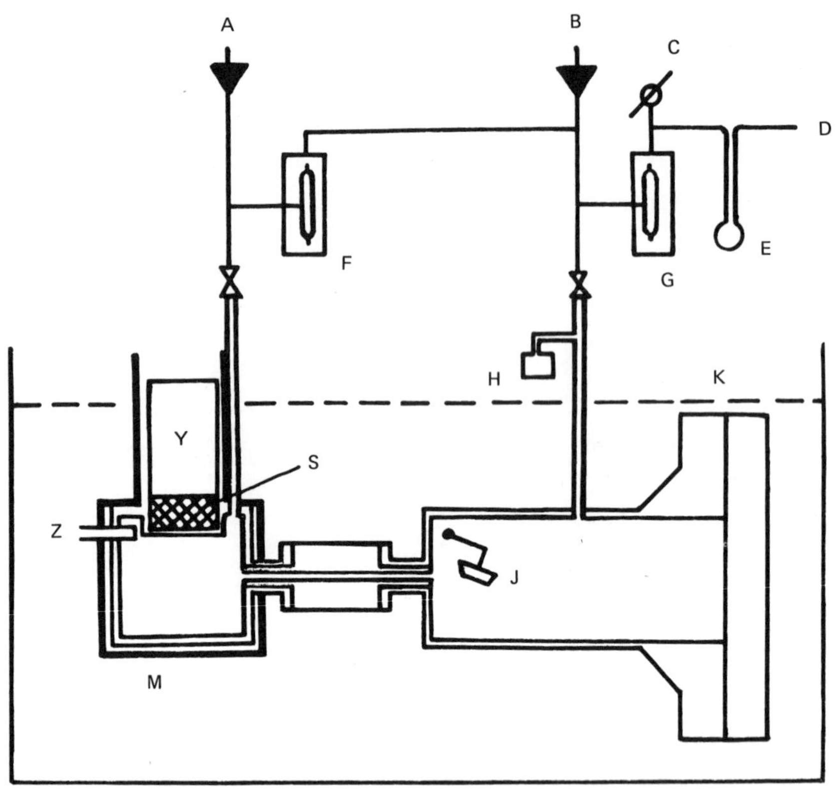

FIG. 7. Apparatus used for determination of the self-diffusion coefficient of UF_6. (A) enriched gas supply; (B) depleted gas supply; (C) air; (D) vacuum pump; (E) dibutylphthalate manometer; (F) pressure difference micromanometer; (G) pressure-measuring micromanometer; (H) trap; (J) valve; (K) water level; (M) lead shield; (S) NaI crystal; (Y) photomultiplier; (Z) standard finger. (Reproduced with permission from Ref. 76.)

Here ℓ, r, and v are the length, radius, and volume of the diffusion tube, V_1 and V_2 are the volumes of the two bulbs, and Δa is the difference between the activity in one of the bulbs at time t and the activity after infinite times. Brown and Murphy [76] discuss corrections to Eq. (29) that enable the finite rate of diffusional mixing in the two bulbs to be taken into account. This leads to an increase of about 1% above the measured value of the diffusion coefficient.

F. Relationships between the Diffusion Coefficients in Two-component Systems

Many liquid systems contain several components for which a large number of diffusion coefficients can be defined. Onsager's reciprocal relations [88] show that many of these coefficients are not independent and for a two-component system in which labeled molecules are also present only three basic diffusion coefficients need to be defined. For a system containing molecules of type A and of type B there is one mutual-diffusion coefficient, D_{AB}, and two self-diffusion coefficients or, as they are sometimes called, tracer diffusion coefficients or intradiffusion coefficients, D_A^*, D_B^*. Several attempts have been made to produce theories correlating these three diffusion coefficients. The feature which these relationships have in common is that the limiting values of the mutual diffusion coefficients reduce to the two intradiffusion coefficients at zero concentrations, namely,

$$D_{AB} = D_A^* \quad \text{at} \quad [A] = 0 \tag{30}$$

$$D_{AB} = D_B^* \quad \text{at} \quad [B] = 0 \tag{31}$$

At intermediate concentrations, however, the predictions are not so certain.

Possibly the simplest approach to the problem is that of Darken [77] and of Hartley and Crank [78] which starts with the basic assumption that the diffusional motion of the A molecules is independent of the motion of the B molecules and vice versa. In general these two processes will not be equal so that there is an excess accumulation of matter toward either the high or low end of the concentration gradient which is relieved by a mass flow of the system. When this concept is combined with the idea that the gradient of chemical potential of each species provides the driving force for the diffusion of that species, the relationships (32) or (33) are obtained.

$$D_{AB} = \phi_B D_A^* \frac{\partial \log \lambda_A}{\partial \log C_A} + \phi_A D_B^* \frac{\partial \log \lambda_B}{\partial \log C_B} \tag{32}$$

$$D_{AB} = N_B D_A^* \frac{\partial \log \lambda_A}{\partial \log N_A} + N_A D_B^* \frac{\partial \log \lambda_B}{\partial \log N_B} \tag{33}$$

Here, ϕ_A, C_A, N_A, and λ_A are the volume fraction, concentration per unit volume, mole fraction, and thermodynamic activity of species A. The same symbols with subscript B refer to species B.

A different approach to this problem is a kinetic one that considers the relative velocity of molecules and the frictional resistances between them. In a two-component system there are three frictional resistances, namely, those between A molecules and A molecules, r_{AA}; those between B molecules and B molecules, r_{BB}; and those between A molecules and B molecules, r_{AB}. This kind of treatment has been considered by Lamm [79] and Laity [80]. Since the diffusion coefficient D_A^* involves r_{AA} and r_{AB}, while D_B^* involves r_{BB} and r_{AB} and D_{AB} involves r_{AB} only, it is possible to eliminate r_{AB} and hence Eq. (34) is obtained [80]. Further progress is impossible without additional assumptions.

$$D_{AB}\left(2 - \frac{r_{AA}}{RT} N_A D_A^* - \frac{r_{BB}}{RT} N_B D_B^*\right) = N_B D_A^* \frac{\partial \log \lambda_A}{\partial \log N_A} + N_A D_B^* \frac{\partial \log \lambda_B}{\partial \log N_B} \tag{34}$$

The simplest assumption is that r_{AA} and r_{BB} are independent of concentration. It then follows that

$$D_{AB}\left(\frac{D_A^{*0} - N_A D_A^*}{D_A^{*0}} + \frac{D_B^{*0} - N_B D_B^*}{D_B^{*0}}\right) = N_B D_A^* \frac{\partial \log \lambda_A}{\partial \log N_A} + N_A D_B^* \frac{\partial \log \lambda_B}{\partial \log N_B} \tag{35}$$

where D_A^{*0} and D_B^{*0} are the self-diffusion coefficients in pure A and in pure B, respectively. Equation (33) was first tested for interdiffusion in a 50/50 gold/silver alloy [81]. The relationship appeared to be satisfactory for the solid system. The evidence from polymer systems is less certain. Fair agreement was obtained in the semisolid isopentane/polyisobutene system [90] at isopentane concentrations up to about 15%. However, for solutions of polystyrene in toluene Eq. (32) gave values of D_{AB} which were less than half the measured values at polymer concentrations of 9% [87]. In these systems the thermodynamic terms are very large and so it is of interest to examine simple fluid systems. A series of simple and ideal two-component systems are those involving gases. Miller and Carman [74] found that, for the interdiffusion of hydrogen with carbon dioxide, krypton, and dichlorodifluoromethane, Eq. (33) is not adequate and the application of kinetic theory in which mass flow occurs as a result of the conservation of momentum leads to Eq. (36) which gives a good fit to the results.

$$\frac{1}{D_A^*} = \frac{N_A}{D_A^{*0}} + \frac{N_B}{D_{AB}} \tag{36}$$

On the assumption that a similar relationship might hold in liquid systems an equation similar to (36) has been derived by Lamm [79]. This is

$$\frac{1}{D_A^*} = \frac{\phi_A}{D_A^{*0}} + \frac{N_B}{D_{AB}} \frac{\partial \ln \lambda_A}{\partial \ln N_A} \tag{37}$$

Numerous tests of these relationships have been made in liquid systems. Miller and Carman [82] have looked at a 50/50 mixture of heptane as molecule B and cetane as molecule A. At 20°C they obtained the values in Table 3. Values for D_{AB} calculated from Eqs. (33) and (35) agree with each other within experimental error. Values for D_A^* and D_B^* calculated from Eq. (37), on the other hand, are not very good values and a further hypothesis

TABLE 3

Observed and Calculated Diffusion Coefficients for Cetane (A)/Heptane (B) Mixtures at Mole Fractions of 0.5

$$N_A = N_B = 0.5$$

$$\frac{\partial \log \lambda_A}{\partial \log N_A} = \frac{\partial \log \lambda_B}{\partial \log N_B} = 1.054$$

Source of diffusion coefficient values	Diffusion coefficients × 10^9 cm^2 sec^{-1}		
	Self-diffusion		Mutual
	Cetane D_A^* and D_A^{*0}	Heptane D_B^* and D_B^{*0}	D_{AB}
Measured values at $N_A = N_B = 0.5$	0.54 ± 0.06	1.18 ± 0.10	0.90 ± 0.08
Measured values in pure cetane D_A^{*0} and pure heptane D_B^{*0}	0.32 ± 0.04	2.86	--
Calculated values from			
Eq. (33)	--	--	0.89 ± 0.08
Eq. (35)	--	--	0.94 ± 0.27
Eq. (37)	0.38	1.43	--
Eq. (38)	0.65	1.39	--

in which a direct place exchange of A molecules and B molecules was postulated giving the relationship

$$D_A^* = \phi_A D_A^{*0} + N_B D_{AB} \frac{\partial \log N_A}{\partial \log \lambda_A} \tag{38}$$

yields values for the self-diffusion coefficients which are no better. From this work it would appear that Eq. (33) gives a reasonable representation of the situation. A better test, however, would be to compare the diffusion coefficients over a whole range of concentrations. Among the early attempts to do this are the comparisons of nonideal systems involving ethanol or methanol in benzene and ideal ones using benzene/carbon tetrachloride mixtures [83]. Although the ideal system could be represented by Eq. (38), considerable deviations were found for the nonideal systems and this was attributed to association of the alcohols to form polymer aggregates with quite different diffusional properties. In the nonideal systems involving nitromethane with benzene or carbon tetrachloride, mutual diffusion coefficients calculated from Eq. (38) were again appreciably smaller than the measured values, and it was suggested that the kinetic units might involve associated species formed by the interaction between the two components [84].

Harris, Pua, and Dunlop [85] have presented data for three nearly ideal systems (benzene/chlorobenzene, n-dodecane/n-hexane, n-dodecane/n-octane) and for three nonideal systems (benzene/n-hexane, benzene/n-heptane, benzene/cyclohexane). They have found that for systems which show positive excess Gibbs free energy the use of Eq. (38) predicts mutual-diffusion coefficients which are much too low. This is shown very clearly in their data for the benzene/n-heptane system illustrated in Fig. 8. Bearman and collaborators [86] have developed a statistical mechanical theory of transport which for the simple case of regular binary mixtures predicts that Eq. (33) should be approximately true. In these systems the frictional coefficients are related to each other through the equation

$$r_{AB}^{2} = r_{AA} r_{BB} \tag{39}$$

whereas for systems having positive excess Gibbs free energy $r_{AB}^{2}/r_{AA}r_{BB}$ is much less than unity as has been shown by Harris, Pua, and Dunlop [85]. It would be expected that systems having negative excess Gibbs free energy should have values of this ratio that are greater than 1.

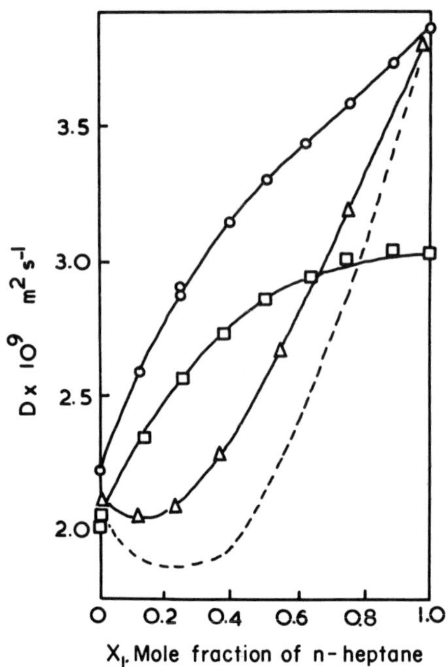

FIG. 8. Diffusion coefficients for the benzene/n-heptane system (○) Tracer diffusion coefficient for benzene; (□) Tracer diffusion coefficient for n-heptane; (△) measured mutual diffusion coefficient; (---) mutual diffusion coefficient calculated from Eq. (38). (Reproduced with permission from Ref. 85.)

REFERENCES

1. J. B. Holt, Proc. Brit. Ceram. Soc., 9:157 (1967).
2. D. C. Douglass and D. W. McCall, J. Phys. Chem., 62:1102 (1958).
3. D. L. Michelsen and P. Harriot, Appl. Polym. Symp., 13:27 (1970).
4. M. L. Wright, Disc. Faraday Soc., 16:58 (1954).
5. H. K. Tok, A. R. Ware, and R. E. Wetton, Nucl. Instrum. Methods, 103:93 (1972).
6. P. Meares, in Diffusion in Polymers (J. Crank and G. S. Park, eds.) Academic Press, London, 1968, p. 398.

7. J. A. Barrie, J. D. Levine, A. S. Michaels, and P. Wong, Trans. Faraday Soc., 59:869 (1963).

8. P. Haefelfinger and F. Grün, Helv. Chim. Acta, 43:529 (1960).

9. G. S. Park, Trans. Faraday Soc., 57:2314 (1961).

10. J. Crank, Mathematics of Diffusion, Clarendon, 1956, p. 55.

11. J. A. Medley, Trans. Faraday Soc., 53:1380 (1957).

12. M. Tetenbaum and H. P. Gregor, J. Phys. Chem., 58:1156 (1954).

13. D. Richman and H. C. Thomas, J. Phys. Chem., 60:237 (1956).

14. G. F. Allen, H. Schurig, L. Slade, and H. C. Thomas, J. Phys. Chem., 67:1402 (1963).

15. B. A. Gromov, B. V. Miller, M. B. Neiman, and Yu. A. Shlyapnikov, Int. J. Appl. Radiat. Isotope, 13:281 (1962).

16. M. Možíšek, Kaut. Gummi Kunstst., 24:53 (1971).

17. C. Matano, Jap. J. Phys., 8:109 (1932).

18. R. E. Pattle, P. J. A. Smith, and R. W. Hill, Trans. Faraday Soc., 63:2389 (1967).

19. G. V. Kidson, Phil. Mag., 13:247 (1966).

20. D. S. Glass and G. McGregor, R and D Report, I.C.I. Heavy Organic Chemicals Division, 1970.

21. M. C. Inman, D. Johnston, W. L. Mercer, and R. Shuttleworth, in Radioisotope Conference 1954, Vol. II (J. E. Johnston, R. J. Millet, eds.) Butterworths, London, 1954, p. 85.

22. T. S. Lundy, in Techniques in Metals Research (R. A. Rapp, ed.) Wiley-Interscience, New York, 1970, Chap. 9A.

23. J. P. Pemsler and E. J. Rapperport, Trans. Met. Soc. AIME, 230:90 (1964).

24. J. Crank, Mathematics of Diffusion, Clarendon, London, 1956, p. 15.

25. H. J. deBruin and R. L. Clark, Rev. Sci. Instrum., 35:227 (1964). S. J. Rothman and F. J. Sobachi, Rev. Sci. Instrum., 30:201 (1959).

26. D. K. Dawson and L. W. Barr, Proc. Brit. Ceram. Soc., 9:171 (1967).

27. J. Anderson and J. Richards, J. Chem. Soc., 537 (1946).

28. J. B. Holt, Mater. Sci. Res., 3:13 (1966).

29. G. v. Hevesey and A. Obrutscheva, Nature, 115:674 (1925).

30. G. v. Hevesey and W. Seith, Z. Phys., 56:790 (1929).

31. J. Steigman, W. Shockley, and F. C. Nix, Phys. Rev., 56:13 (1939).

32. F. E. Jaumont and R. L. Smith, Trans. AIME, 206:137 (1956).

33. F. Bueche, W. M. Cashin, and P. Debye, J. Chem. Phys., 20:313 (1952).

34. S. E. Bresler, G. M. Zakharov, and S. V. Kirillov, Vysokomol. Soedin, 3:1072 (1961). [Polym. Sci. USSR, 3:832 (1962)].

35. R. A. Jackson, S. R. D. Oldland, and A. Pajaczkowski, J. Appl. Polym. Sci., 12:1297 (1968).

36. R. S. Moore and J. D. Ferry, J. Phys. Chem., 66:2699 (1962). S. P. Chen, Ph.D. Thesis, Univ. of Wisconsin, 1968.

37. I. Auerbach and S. D. Gehman, Anal. Chem., 26:658 (1954).

38. R. B. Jenkins and G. S. Park, Paint Research Institute Report (1972).

39. I. Auerbach, W. R. Miller, W. C. Kuryla, and S. D. Gehman, J. Polym. Sci., 28:129 (1958).

40. K. E. Zimen, G. Johansson, and H. Hillert, J. Chem. Soc., S392 (1939).

41. K. D. Timmerhaus and H. G. Drickammer, J. Chem. Phys., 19:1242 (1951).

42. G. S. Park in Radioisotope Conference 1954, Vol. II (J. E. Johnston, R. A. Faires and R. J. Millett, eds.) Butterworths, London, 1954, p. 11.

43. L. A. Walker, Science, 112:757 (1950).

44. J. S. Anderson and K. Saddington, J. Chem. Soc., S381 (1939).

45. H. J. Wang, J. Am. Chem. Soc., 74:1182 (1952).

46. W. Bambynek and V. Freise, Z. Phys. Chem. Frankfurt Main, 7:317 (1956).

47. G. Chantrey and I. D. Rattee, J. Appl. Polym. Sci., 18:105 (1974).

48. A. Talbot and J. A. Kitchener, Brit. J. Appl. Phys., 7:96 (1956).

49. J. H. Northrop and M. L. Anson, J. Gen. Physiol., 12:543 (1929).

50. P. A. Johnson and A. L. Babb, Chem. Rev., 56:387 (1956).

51. R. Mills, Trans. Faraday Soc., 67:1654 (1971).

52. R. Mills, L. A. Woolf, and R. O. Watts, Am. Inst. Chem. Eng. J., 14:671 (1968).

53. R. Mills and L. A. Woolf, The Diaphragm Cell, A.N.U. Press, Canberra, 1968.

54. R. Jeanneret and F. Grün, Helv. Chim. Acta, 41:2156 (1958).

55. R. C. Koeller and H. G. Drickammer, J. Chem. Phys., 21:267, 575 (1953). R. B. Cuddebach, R. C. Koeller, and H. G. Drickammer, J. Chem. Phys., 21:589 (1953).

56. D. S. Brown and D. G. Tuck, Trans. Faraday Soc., 60:1230 (1964).

57. N. H. Nachtrieb and J. Petit, J. Chem. Phys., 24:746 and 1027 (1956).

58. D. W. Morgan and J. A. Kitchener, Trans. Faraday Soc., 50:51 (1954).

59. R. L. Saxton and H. G. Drickammer, J. Chem. Phys., 21:1362 (1955).

60. L. H. Jung and H. G. Drickammer, J. Chem. Phys., 20:13 (1952).

61. J. A. A. Ketelaar and E. P. Honig, J. Phys. Chem., 68:1596 (1964).

62. T. Williams and C. B. Monk, Trans. Faraday Soc., 57:447 (1961). J. R. Jones, D. L. G. Rowlands, and C. B. Monk, Trans. Faraday Soc., 61:1384 (1965).

63. S. Djordjevic and G. J. Hille, Trans. Faraday Soc., 56:269 (1960).

64. H. C. Thomas and J. C. Ku, J. Phys. Chem., 77:2233 (1973).

65. G. F. Allen, H. Schurig, L. Slade, and H. C. Thomas, J. Phys. Chem., 67:1402 (1963).

66. J. M. Nielsen, A. W. Adamson, and J. W. Cobble, J. Amer. Chem. Soc., 74:449 (1952).

67. G. E. Spalding, J. Phys. Chem., 73:3380 (1969).

68. R. Mills, J. Amer. Chem. Soc., 77:6116 (1955).

69. F. S. Nakayama and R. D. Jackson, J. Phys. Chem., 67:9232 (1963).

70. E. F. Broome and H. A. Walls, Trans. AIME, 242:2177 (1968).

71. A. G. Langdon and H. C. Thomas, J. Phys. Chem., 75:1821 (1971).

72. F. Hutchinson, Phys. Rev., 72:1256 (1947).

73. E. P. Ney and F. C. Armistead, Phys. Rev., 71:14 (1947).

74. L. Miller and C. Carman, Trans. Faraday Soc., 57:2143 (1961).

75. L. Miller and C. Carman, Trans. Faraday Soc., 60:33 (1964).

76. M. Brown and E. G. Murphy, Trans. Faraday Soc., 61:2444 (1965).

77. L. S. Darken, Trans. AIME, 175:184 (1948).

78. G. S. Hartley and J. Crank, Trans. Faraday Soc., 45:801 (1949).

79. O. Lamm, Acta Chem. Scand., 6:1331 (1952); 8:1120 (1954).

80. R. W. Laity, J. Phys. Chem., 63:80 (1959); J. Chem. Phys., 30:682 (1959).

81. W. A. Johnson, Trans. AIME, 147:331 (1942).

82. L. Miller and P. C. Carman, Trans. Faraday Soc., 58:1529 (1962).

83. R. A. Johnson and A. L. Babb, J. Phys. Chem., 60:14, 1671 (1956).

84. P. C. Carman and L. Miller, Trans. Faraday Soc., 55:1831, 1838 (1959).

85. K. R. Harris, C. K. N. Pua and P. J. Dunlop, J. Phys. Chem., 74:3518 (1970).

86. R. J. Bearman, J. Phys. Chem., 65:1961 (1961).

87. G. S. Park, Trans. Faraday Soc., 53:107 (1957).

88. L. Onsager, Ann. N.Y. Acad. Sci., 46:209 (1945).

89. R. Mills and R. W. Godbole, Aust. J. Chem., 11:1 (1958); 12:102 (1959).

90. G. S. Park, Macromolecular Symposium Wiesbaden 1959, short communication II A8.

91. F. C. Nix and F. E. Jaumot, Phys. Rev., 82:72 (1951).

92. T. S. Elleman, L. D. Meares and R. P. Christman, J. Am. Ceram. Soc., 51:560 (1968).

93. F. Schmitz and R. Lindner, J. Nucl. Mater., 17:259 (1965). R. Lindner, D. Reimann and F. Schmitz, in Plutonium as a Reactor Fuel Symp. Brussels, 1967, p. 265.

94. A. V. Savitskii, Fiz. Metal. Metalloved, 16:886 (1963).

95. R. H. Stokes, J. Am. Chem. Soc., 72:764 (1950).

96. J. Petit and N. H. Nachtrieb, J. Chem. Phys., 24:1027 (1956).

Chapter 13

INTERFACIAL PHENOMENA

John A. Spink

Division of Tribophysics
CSIRO
University of Melbourne
Parkville, Victoria
Australia

David E. Yates*

Colloid and Surface Chemistry Group
Department of Physical Chemistry
University of Melbourne
Parkville, Victoria
Australia

---

*Current affiliation: School of Chemistry, University of Bristol, Bristol, England.

I. INTRODUCTION

Interfaces, being two-dimensional or extremely thin three-dimensional systems, involve very small quantities of material, i.e., of the order of 10^{-7}g or 10^{-9} mol cm^{-2}. The extreme sensitivity of the radiotracer technique, whereby quantities of matter ranging from 10^{-7} to 10^{-17}g can be readily explored, as well as the near-absolute specificity, makes the method very attractive and in some instances essential for the localization in space and time of the atomic or molecular species concerned [1].

A wide range of radioactive isotopes and isotopically labeled compounds has been available now for three decades. The specific activities originally high enough to give thousands of counts per minute from the number of atoms or molecules in a monolayer covering 1 cm^2 of surface has been steadily increased to such a degree that today as little as 10^{-6} of a monolayer might now be detectable with a sufficiently short-lived isotope (e.g., ^{32}P, $t_{1/2}$ = 14.3d) and high percentage labeling. An example of this improved detectability is the steady increase in the specific activity available for [1-^{14}C]stearic acid, the classical long-chain polar molecule used in surface studies. From 1950 to the present day the specific activity available has increased from about 0.8 mCi $mmol^{-1}$ to >50 mCi $mmol^{-1}$.

The percentage labeling has thus increased from just over 1% to more than 80%. Under these circumstances a monolayer of [^{14}C]stearic acid on 1 cm^2 of surface can give rise to about 37,000 dpm. If this is counted with a thin end-window GM counter (window weight <2 mg cm^{-2}) so that the actual count is about 10% of the gross dpm and with a background of about 20 cpm we still have a count rate some 200 times background. Assuming that meaningful results can be obtained from a count rate equal to background we can expect significant detection of material equivalent to about 1/200 of a monolayer and representing about 10^{-9} g stearic acid.

It should be remembered, however, that for a given concentration of isotopic atoms, the specific activity decreases as the half-life $t_{1/2}$ increases. With ^{36}Cl, for example, with $t_{1/2} = 3.03 \times 10^5$y there would be 2.62×10^{12} dpm in 1 g atom of pure ^{36}Cl. If we assume that this isotope is incorporated in a molecule that is adsorbed at an area/molecule of 20 Å^2, then the number of molecules covering 1 cm^2 of surface will be 5×10^{14}. With an isotopic labeling of say, 1%, the decay rate from this 1 cm^2 of surface can be shown to be ≃22 dpm. If this radioactivity is counted with a thin end-window Geiger-Müller counter having a detection coefficient of ∿10% then the count rate will be reduced to about 2 cpm. This is only 10-20% of a normal background for such GM tubes and means that for ^{36}Cl to be of practical use in studies on adsorbed films of monolayer thickness either the percentage labeling must be dramatically increased or the counting efficiency raised.

A major problem in studying surface phenomena is that trace impurities that would have an insignificant influence on bulk properties are frequently concentrated at interfaces. This problem is even more acute in radiotracer studies since radioactive decay produces new molecular species which may be surface active. Fortunately, the radioisotope is usually only a small fraction of the total amount of the species present and thus the decay product forms an infinitesimal part of the total system. On the other hand, radiotracer methods give only the total number of radioactive atoms. If the labeled compound undergoes a change or reaction when it is adsorbed then the adsorption density Γ_{*atom} may not be equal to $\Gamma_{*compound}$. This lack of sensitivity to certain chemical changes in the adsorbed species could be overcome by dual labeling, e.g., with ^{14}C and ^{3}H, and dual counting.

In addition care should be taken to see that the incorporation of the radioactive atom into the surface-active species does not cause the latter to behave differently from the untagged compound. Neuman [153] has shown, for example, that the stability of [1-^{14}C]stearic acid monolayers (specific activity 14.3 mCi

$mmol^{-1}$) at a surface pressure of 31 dyn cm^{-1} differs from that of the inactive compound--being greater on pure water substrates (pH 2.0-6.0) but less on substrates containing calcium at pH $\geq$ 8.

Despite these difficulties, the value of the radiotracer technique in the localization of extremely small quantities of material is quite obvious. It remains to consider in some detail the special techniques that have been used to solve the problems peculiar to interfacial systems. In the limited space available we must, of necessity, be selective and consider only those studies in which the method of approach has advanced significantly the technique itself and resulted in a major contribution to an understanding of the behavior of material at interfaces. The areas to be discussed will include the gas/solution interface, the solid/solution interface, films of long-chain polar molecules deposited on solids either by the Langmuir/Blodgett process or from solution by simple adsorption or by retraction, and, finally, the gas/solid and solid/solid interfaces.

II. THE GAS/SOLUTION INTERFACE

Radioactive tracers have been employed at the gas/solution interface in basically two different ways. First, labeled atoms may be incorporated in the surface-active material to detect its presence at the interface or to study transformations that it may undergo. Second, a species in the bulk phase may be labeled to detect various kinds of interaction occurring at the surface. For instance, the ^{131}I exchange between α-iodostearic acid and KI in aqueous solution has been studied as a function of the surface pressure of the α-iodostearic acid monolayer [2]. The number of such radiometric applications is large and the techniques and data have been extensively reviewed by Muramatsu [1,3].

Radiotracers provide a direct means of measuring the amount of a substance at the gas/solution interface and some of the methods that can be employed are discussed below.

A. The Stable Film Method

In this technique, first used by Hutchinson [4], a platinum wire ring is used to draw a stable liquid film from the surface of the solution. The activity of the film is measured by the appropriate technique and the adsorption density determined from this measurement, the surface area of the film, and the activity of an equivalent amount of bulk solution. The volume of the film is usually determined by weighing. The value of this technique depends on whether adsorption equilibrium is maintained during the withdrawal of the loop, and, to a certain extent, this can be tested by studying the effect of withdrawal rate. The stable film method is not necessarily restricted to radiometric methods of analysis, but the latter are usually applied because of their high sensitivity.

A modification of this technique has been used to study the structure of soap films [5]. In the apparatus shown in Fig. 1 [6] a GM counter, shielded with an aluminum sheet containing a

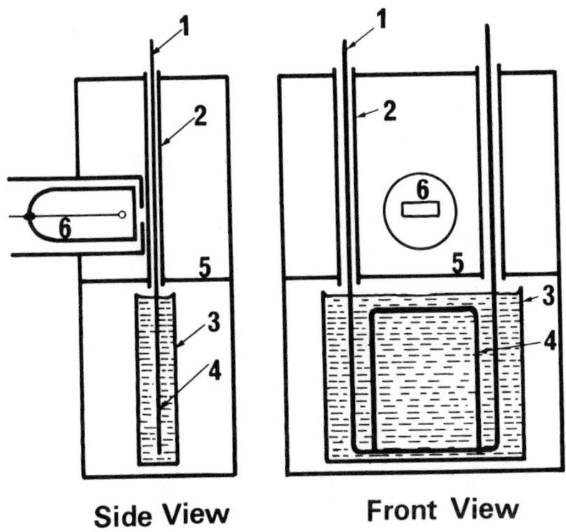

FIG. 1. Apparatus for measuring the radioactivity of thin detergent films. 1, glass rods; 2, guide for glass rods; 3, glass tank containing solution; 4, glass frame for films; 5, metal shield; 6, Geiger-Müller counter. (Reproduced from Ref. 6, by courtesy of the Chemical Society, London.)

narrow horizontal slit, was used to measure the activity from the thin film. The glass frame was withdrawn from the solution and fixed so that the film was adjacent to the counter slit. Radiotracers were incorporated into the electrolyte ions within the aqueous core and into the soap molecules that form the surface monolayers. In this way the symmetrical sandwich structure of monolayer-aqueous core-monolayer was directly confirmed.

The stable film method is limited to systems that can form very thin films; otherwise the contributions to the measured activity from the bulk species swamps that from the adsorbed species thus causing large errors. An advantage, however, is that a wide range of radiotracers is suitable for this technique.

B. The Anianssson-Salley Method

The technique developed by Salley *et al.* [7] and Aniansson [8] provides a direct method for measuring adsorption of any species at the solution-gas interface. The principle of this method is shown in Fig. 2. Since α or β particles are strongly absorbed by the solution, a counter placed just above the surface will detect only those particles from the surface region and a thin layer of the bulk region. Therefore the adsorption density Γ (mol cm^{-2}) can be calculated from

$$\Gamma = \frac{R - R_{sol}}{QA} \qquad (1)$$

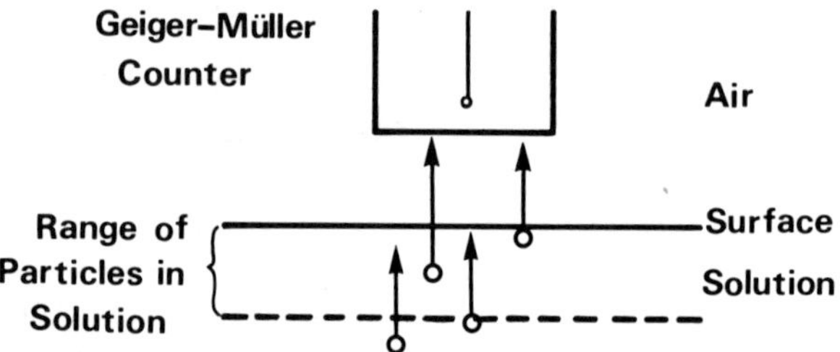

FIG. 2. Principle of the Aniansson-Salley method. Radiation emitted from below the dotted line is absorbed and does not reach the counter.

where R (cpm) is the measured activity from both surface and bulk, R_{sol} (cpm) is the activity from the bulk solution, Q is the proportionality constant (cpm/mol) between the activity and the amount present at the surface, and A (cm^2) is the surface area.

The shorter the range of the emitted particles in the solution, the smaller will be the activity (R_{sol}) from the bulk and therefore the more accurate the measurement. For this reason the Aniansson-Salley method is only useful for isotopes that emit low energy particles (e.g., ^3H, ^{35}S, and ^{14}C). For Eq. (1), the activity (R_{sol}) from the thin layer of solution below the interface must be determined. This can be done in a number of ways. First, the activity from the interior of a solution of a labeled surface-active species will be equal to that from the solution of a surface inactive species labeled with the same isotope of the same concentration. Therefore R_{sol} can be determined by substituting a radioactively equivalent surface-inactive species in place of the labeled surface-active one. For example, use has been made of $H_2{}^{35}SO_4$, $CH_3{}^{14}COONa$, and $NH_2CH_2{}^{14}COOH$ in aqueous solution, or [^3H]dodecanol in nonradioactive dodecanol. It is sometimes necessary to make corrections for the difference in self-absorption and self-scattering factors between the different media. Alternatively, the background activity R_{sol} can be calculated as shown by Salley *et al.* [7].

The relationship [Q in Eq. (1)] between the observed activity and the adsorption density is estimated by determining the count rate from either a standard spot source of known amount and specific activity or from an insoluble monolayer spread on a subphase under the same conditions as in the adsorption experiment. For further details of the estimation of the background activity, R_{sol}, and calculation of adsorption densities from observed activities the reader is referred to the recent detailed reviews by Muramatsu [1,3].

Nilsson [9] has shown for the Aniansson-Salley method that the experimental accuracy is given approximately by

$$\frac{R - R_{sol}}{R} \simeq \frac{1}{1 + (C/\mu\Gamma)} \tag{2}$$

where C (mol g^{-1}) is the solution concentration of the labeled species, and μ ($cm^2\ g^{-1}$) is the mass absorption coefficient of the radiation particles in the solution. The error will decrease as $(R - R_{sol})/R$ approaches unity so that the technique is most convenient for highly surface-active compounds (i.e., high Γ/C) or very low-range radiations (high μ).

A list of the relevant properties of most of the isotopes that have been used successfully is given in Table 1 [8,10,11]. The last column gives the ratio of the surface activity to the total activity for typical experimental conditions of $\Gamma = 1 \times 10^{-10}$ mol cm^{-2} and $C = 1 \times 10^{-3}$ M. Even though the soft β emitters such as ^{14}C and ^{35}S have been frequently used for measuring adsorption, the results obtained are not particularly accurate, especially for higher concentrations and weak adsorption.

Table 1 shows also that tritium offers considerably higher accuracy because of its lower range. In addition, it can be incorporated into a wide range of surface-active species. However, the use of ^{3}H-labeled compounds requires very sensitive counting methods. Although 2π-windowless gas-flow counters [9,12] have been used, they subject the gas/solution interface to a very large electric field which may affect the extent of adsorption. Also, as the solution surface is always exposed to flowing gas, surface enrichment of the solute due to evaporation of the solvent will occur unless extensive precautions are taken to humidify the gas. On the other hand, thin-windowed GM counters are not always successful because with a humidified atmosphere intrusion of water vapor through pin holes in the window causes interference. The most promising method is the sheet scintillation technique developed by

TABLE 1

Suitability of Various Radionuclides for Determination of Adsorption at Air/Solution Interface [8,10,11]

Isotope	Radiation used	E_{max} (MeV)	Half-life ($t_{1/2}$)	Range in water (μm)	R_{ads}/R_{total}
^{32}P	β^-	1.71	14.3d	10,000	0.001
^{22}Na	β^+	0.542	2.6y	1,960	0.006
^{45}Ca	β^-	0.260	164d	650	0.018
^{14}C	β^-	0.156	5730y	300	0.036
^{35}S	β^-	0.169	87.4d	340	0.036
^{63}Ni	β^-	0.065	85y	67	0.12
^{210}Po	α	5.30	138.4d	32	0.058
^{3}H	β^-	0.0186	12.35y	6	0.64*
^{212}Bi	$^{208}T\ell$ (α-recoil)	0.117	60.5m	0.08	0.95*

*Davies and Rideal [11] give a value of 1.8 for ^{3}H and 20 for ^{212}Bi.

Muramatsu and collaborators [13,14] which has been successfully used to measure accurately adsorbed amounts of tritiated surfactants at nitrogen/solution interfaces [1,3,15]. This technique is shown schematically in Fig. 3. The tritiated surfactant solution in a Teflon tray is attached to a mechanism so that it can be raised to a fixed distance (about 0.8 mm) from the scintillation sheet for determining the surface radioactivity. The whole apparatus, apart from the electronics, is kept in a dark thermostated glove box and the atmosphere humidified.

The sheet scintillation method for tritiated compounds is precise enough to test accurately the Gibbs adsorption isotherm for the first time. Muramatsu *et al.* [15] measured adsorbed

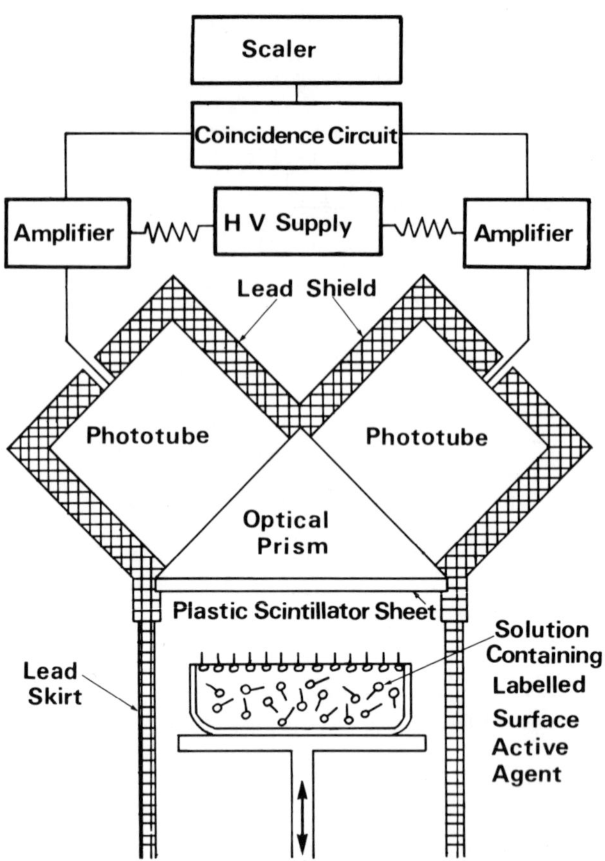

FIG. 3. Schematic diagram of the sheet scintillation counting method. (Reproduced from Ref. 13, by courtesy of North-Holland Publishing Company.)

amounts for solutions of two different surfactants in the presence and absence of strong electrolyte, and found in all cases good agreement between experimental and theoretical isotherms (see Fig. 4).

As seen in Table 1, the Aniansson-Salley method can be rendered even more accurate by using α-disintegration recoils which have an extremely short range. An example of its use is the measurement of $^{208}T\ell$ ions recoiling from the surface of a

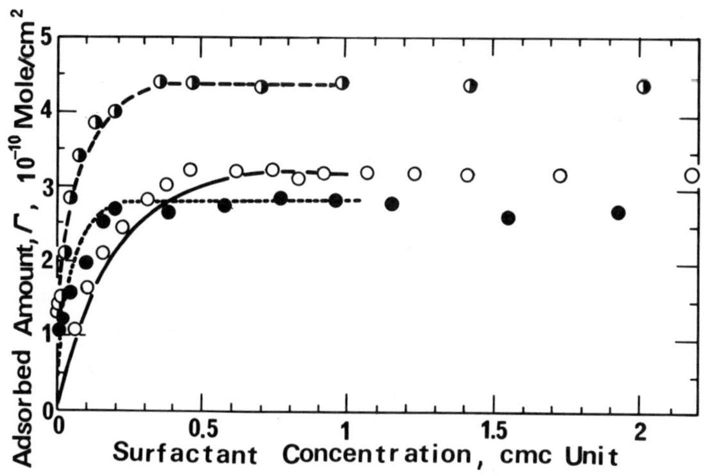

FIG. 4. Adsorption of sodium [2,3-^{3}H]dodecyl sulfate in the presence (◑) and absence (○) of 0.115 mol NaCl at 25.0 ± 0.2°C, and for hexaethyleneglycol mono[2,3-^{3}H]dodecyl ether at 30.0 ± 0.2°C (●) at the surfaces of the solutions exposed to 98 ± 3% humidified nitrogen. Experimental points were obtained by the sheet scintillation method and the curves represent derived Gibbs isotherms. The adsorbed amount is plotted against the ratio of the surfactant concentration to the critical micelle concentration (cmc). (Reproduced from Ref. 3.)

solution of ^{212}Bi ions [10,16,17]. The positive ^{208}Tℓ ions emitted from the surface are collected with a negatively charged electrode and the amount of ^{212}Bi in the surface region (about 800 Å thick) is determined from the activity of the collected ^{208}Tℓ (a strong β emitter). The recoil atom technique is the most sensitive tool for studying ion adsorption at ionized surface layers at the air/solution interface. However, the available nuclides for this method are not common and, being heavy metals, may complicate the chemistry of the system. For these reasons the recoil technique has so far been little used in studies of interfacial phenomena.

It is readily seen, then, that radiotracer techniques provide a powerful tool for studying the gas/solution interface and for measuring surface concentrations. This is particularly so when there is more than one surface-active species present where the determination of surface concentrations from surface tension measurements alone can in certain instances be quite ambiguous.

C. Properties of Insoluble Monolayers

Many amphipathic insoluble substances containing hydrophilic groups such as -COOH or -OH will form monomolecular films, or monolayers, at the air-water interface with the hydrophobic ends directed away from the water surface. This "surface-active" material is normally predissolved in a volatile solvent, such as hexane. The spreading solution is ejected onto the water surface so that the solvent evaporates leaving the insoluble monolayer. The film is usually contained between barriers on a Langmuir surface balance (or trough) so that the area per molecule (A) and the surface pressure (Π) can be varied. It is assumed that (a) the spreading solvent evaporates completely, (b) there is no leakage of the film around or beneath the barriers, and (c) the insoluble material does not evaporate or dissolve.

The validity of these assumptions can be tested with radiotracer techniques such as the stable film method (Section II A) and the Anianssson-Salley method (Section II B). In the following examples of the use of radiotracers, the Aniansson-Salley method has been employed unless stated otherwise. Barnes and coworkers [18] have shown that [^{14}C]decane evaporates completely from octadecanol monolayers so that the usual, more volatile, spreading solvents should also evaporate completely. By measuring the radioactivity from the "free" surface on the opposite side of the barrier from the monolayer, Muramatsu *et al.* [19] have shown directly that [^{14}C]methyl palmitate and [^{14}C]chloramphenicol palmitate leak past paraffin-coated barriers, which explains the hysteresis in their Π-A curves. On the other hand, these materials do not leak past barriers coated with polyethylene or FEP Teflon (a copolymer of fluoroethylene and fluoropropylene), nor does [^{14}C]stearic acid leak beyond even paraffined barriers.

These results suggest that leakage depends on the particular system and should be routinely checked for during monolayer experiments. Monolayers of alkyl esters of fatty acids show time dependence of surface pressure (Π) which was initially attributed to hydrolysis of the ester and dissolution of the resultant alcohol in the subphase. However, the direct determination of the hydrolysis rate from measurements of the time dependence of the surface radioactivity (R) for [^{14}C]methyl palmitate monolayers [20] has shown that the decrease in Π is due mainly to the evaporation of the ester and not to hydrolysis. The rate of decrease in R was too low to be due to hydrolysis with resulting loss of $^{14}CH_3OH$ and, besides, the radioactive ester was detected on the window of the GM tube which was kept 3 mm above the monolayer surface. In the same way, [^{14}C]myristic acid has been shown to evaporate quite rapidly (about 0.4% per min) at 30 to 40 $Å^2$ per molecule [21]. During measurement of evaporation rates, contamination of the counting equipment by the evaporating radioactive material can be minimized by covering the monolayer with moist paper between periods of counting. In this way the background radiation does not increase excessively during the experiment, while the accumulation of radioactive material in the paper can be measured.

Radioactivity-area measurements of [^{14}C]stearic acid monolayers at low pressures and large areas per molecule [22,23] depend on the lateral position of the counter, thus confirming an island structure already indicated by other techniques. The size, shape, and distribution of island structures appears to be affected by the nature of the spreading solvent.

Another aspect of monolayer properties studied with radiotracers is surface diffusion, i.e., the two-dimensional migration of molecules within the plane of the interface [21,24]. Two monolayers identical in all respects except that one is radioactive, are spread on opposite sides of a very thin barrier. The barrier is removed and the radioactivity from the monolayer is measured as a function of position and time. For this technique the counting tube is usually mounted on two horizontal rails to allow traversal of the diffusion path. To reduce the field of the counting tube a mask with a long narrow slit is placed over the end window so that the slit is perpendicular to the direction of diffusion. Surface diffusion coefficients have been shown in this way to be approximately equal to the corresponding bulk diffusion coefficients [21].

The detection and measurement of chemical reactions and other interactions in monolayers has usually been carried out by following changes in properties such as area per molecule, surface pressure, and surface potential. However, such properties are not always directly related to the nature and number of molecules present in the film. Radioactive tracers provide a direct method for following the *number* of molecules of a particular species in a monolayer. The application of this direct approach is sometimes very useful. For example, the large rate of decrease in area/molecule of [^{14}C]methyl palmitate has been shown to be due mainly to evaporation and not to hydrolysis of the ester as discussed above [20]. Other examples are the *cis*→*trans* reorientation [25] and enzymatic hydrolysis [26] of [^{14}C]chloramphenicol palmitate monolayers, the exchange between [^{131}I]iodostearic acid and I^- in the subphase [2,27], enzymatic hydrolysis of [^{32}P]lecithin monolayers [28], and the alkaline hydrolysis of [^{14}C]tripalmitin monolayers [29].

For chemical reactions of monolayers in which the radioactive component is released and diffuses into the aqueous substrate, the observed activity must be corrected for the increasing contribution from the active material in the subphase. Mason and Rabinovitch [30] developed a mathematical treatment to account for this varying background contribution by using conventional diffusion theory and assuming a simple exponential law for absorption of the radiation. When this treatment is applied to the results of their α-iodostearic acid exchange experiments [2], the observed rate constants were increased by 3-15%. This is only a small correction and for even weaker β emitters such as ^{14}C and ^{35}S it can probably be neglected. In the hydrolysis of [^{14}C]-tripalmitin [29], a mixture of reaction products is produced which remains within the monolayer so that the Aniansson-Salley method is inappropriate. Thus, the stable film method was used in which samples of the surface were taken after different intervals of time. Each sample was treated with acid to arrest the hydrolysis, the products separated by paper chromatography and the activity of each product measured. The accuracy of this *chromatography-radioassay* method is satisfactory but it is limited to materials which are easily separated by chromatography and which react only slowly.

The adsorption and interaction of subphase species, such as metal ions, with monolayers have been studied extensively by radiotracer techniques. Early experiments to determine the extent of binding of ^{90}Sr and $^{60}Co(II)$ to stearic acid monolayers [31] utilized the Langmuir/Blodgett technique (see Section IV) to transfer up to 20 monolayers onto a metal slide from a subphase containing the particular metal ion. In this case a nonradiometric method could have been used to determine the amount of metal transferred but the greater sensitivity of the radiotracer method requires fewer layers to be transferred and so decreases any transfer errors. Binding ratios of up to one ion per monolayer molecule were obtained. Later workers have preferred the more direct Aniansson-Salley method, although it is restricted to soft β emitters of which ^{45}Ca (E_{max} = 0.26 MeV) has been the most widely used. Some examples of general and biological interest are the interaction of ^{45}Ca ions with monolayers of stearic acid [32], with phospholipid monolayers in the presence of Na^+, K^+, and H^+ [33], and with lipid films in the presence of K^+ and Na^+ and chelating agents such as ATP and EDTA which were shown to desorb the Ca^{2+} ions [34]. Examples of the use of radiotracers in the study of interactions between monolayers and certain soluble species in the aqueous substrate are the interaction of steroid hormones and lipid films [35], and the penetration of sodium alkyl sulfates into monolayers of cholesterol, hexadecyl alcohol, and octadecylamine [36].

It is obvious from the above that radiotracers have become an important tool in the study of insoluble monolayers at the air/water interface.

III. THE SOLID/SOLUTION INTERFACE

Radiotracers have been used extensively in studying phenomena at the solid/solution interface such as surface concentration, or adsorption densities of the various species present, as well as the nature of the binding between adsorbate and adsorbent.

The *rate* of adsorption or desorption can indicate the nature of the adsorption binding. Thus, for ions that are adsorbed on a charged surface by electrostatic attraction, the rate will generally be rapid because it will be diffusion controlled. However, if the species is covalently bound to the surface then the rate of adsorption or desorption may be a chemically controlled process and occur much more slowly. Radiotracer techniques combined with electrochemical methods can also be used to measure directly exchange current densities for redox reactions at electrodes [37].

A. Isotopic Exchange

In determinations of surface areas or adsorbate surface concentrations, the rate and extent of exchange between a species in solution and the same species in the surface (or adsorbed at the surface) can be measured by radioactively labeling the species. If the specific surface area of the solid is known, then the area that the radioactively specified atoms occupy at the surface can be calculated from the amount of exchange. An example is the determination of the areas per PO_4^{3-} and Ca^{2+} ion at the surface of calcium hydroxyapatite, ($Ca_{10}(OH)_2(PO_4)_6$), using ^{32}P and ^{45}Ca radiotracers [38]. Conversely, if the area per species is known from previous studies or from crystallographic radii, then the specific surface area of the solid can be determined. This technique for measuring surface areas was developed by Paneth and coworkers using $^{212}PbSO_4$ and $^{212}PbCrO_4$ and has since been used for many solids [1].

Surface exchange is measured by first allowing the solid to come to equilibrium with the supernatent liquid, and then adding a given volume of the solution, this time containing the labeled spaces *X. This must not change the solution concentration of X which could lead to precipitation reactions. After surface equilibrium is reached the specific activity of the solution or the surface is measured and the number of moles of exchangeable labeled species in the surface is calculated using Eq. (145) of Chapter 10. Then the surface area or surface concentration can be calculated.

It is necessary to ensure that the exchange occurs only between the solution and the surface and does not involve the interior of the solid. Provided that the surface is not significantly porous, the surface exchange generally occurs rapidly (within several hours) whereas exchange with the interior occurs very slowly. An effective method for separating surface exchange from internal exchange is to measure the exchange as a function of time and extrapolate the portion of the curve due to the slow process back to zero time. This gives the amount of exchange in the surface layer alone.

The surface exchange method has been used mostly with sparsely soluble solids such as $BaSO_4$ for which the concentrations of the species in solution are large enough to be measured accurately and easily. For extremely insoluble solids such as the sulfides and oxides of certain metals the concentration and hence the specific activity of the soluble species is often too low to be accurately determined. However, metal sulfides and oxides have been studied by using $H_2{}^{35}S$ solutions for sulfides [39] and by labeling the OH groups at oxide surfaces with ^3H, removing the physically adsorbed water by evacuation, returning the solid to solution, and following the back-exchange from the surface into the solution [40]. The importance of the isotopic exchange method is that it enables one to measure adsorbate surface concentrations by a simple, nondestructive technique that does not involve removal of the solid from the solution. This advantage also applies to measurement of specific surface areas. In contrast, methods such as the BET gas adsorption technique require vigorous degassing of the solid, which can change the area and nature of the surface. However, the solid must be composed of elements of which radioactive isotopes are available, and the cross-sectional area of each exchangeable species in the surface must be reliably known.

B. Adsorption

The two basic methods by which radioactive tracers can be used to study adsorption at the solid/solution interface are by measuring the changes in the radioactivity of either the solution or the solid. The first approach is sometimes known as the differential method, since one measures the difference between the activity (and hence the concentration) before and after the solid is added. In the second approach the activity of the solid is measured either after the sample has been removed from the solution or *in situ* by the Joliot method (known also as the dip-counting method) in which the counting device is placed behind the solid sample. All these methods are useful for studying adsorption and exchange phenomena and involve measurements of adsorption densities, kinetics of adsorption or desorption, surface exchange, and finally the reversibility of adsorption or exchange.

1. The Differential Method

Radioactivities of R_0 and R of a solution of radiolabeled adsorbate are determined before and after, respectively, its contact with the solid sample. Provided that the relation between the measured activity and the solution concentration of the labeled species is linear, the amount Γ (mol m^{-2}) adsorbed or exchanged may then be calculated from the initial concentration, C (mol liter^{-1}), of the radioactive adsorbing species, the volume, V (liters), of the solution and the available surface area A (m^2) of the solid by

$$\Gamma = \frac{R_0 - R}{R_0} \frac{CV}{A} \qquad (3)$$

The feasibility and accuracy of this method depends on the relative change in the activity of the solution, $(R_0 - R)/R_0$. Normally the minimum change in the activity that can be measured easily and accurately would be about 5% and it follows from Eq. (3) that for a fixed Γ value the relative change in activity and hence the sensitivity is proportional to the area:volume ratio and inversely to the solution concentrations. Therefore one should arrange as high an area:volume ratio as possible, and even then there is an upper limit to the concentration that can be used.

There are a number of ways in which high area:volume ratios can be obtained. For particulate solids with high specific areas, such as clays, colloidal silver halides, metal oxides, sulfides, and polymer latexes, adsorption densities of or greater than 10^{-12} mol cm^{-2} can be easily measured in solution concentrations of up to at least 10^{-2} M. To determine the solution activity the solid must be separated from it, usually by centrifugation or filtration with Micropore filters.

For solid metal surfaces such as electrodes where pulverization is undesirable the differential technique can be used with reaction cells designed so that the volume is minimized. This is possible for radiotracers because the necessary volume of the solution sample removed for analysis can be very small. In other versions of the method the radiation counter is connected directly to the cell so that the solution can be transferred back and forth and measurements made without changing the volume of the solution [41,42]. Also, the radiation counter can be placed in the cell but separated from the solution by a thin membrane [41, 42]. However, the energy of the radiation must be high enough to pass through the membrane and low enough so that only the radiation from the solution immediately adjacent to the membrane is recorded.

Another method which has been used successfully to measure the adsorption of ^{131}I on iron in up to 10^{-3} M KI solutions [43] involved an increase in the metal surface area by electrolytic deposition of the metal on the inside of a thin platinized glass cylinder which served as the reaction vessel.

The differential method for studying adsorption can be and often is carried out with nonradiometric methods of analysis. The advantages of using radiometric methods are that they allow the volume of the aliquot removed for analysis to be reduced to a minimum, provide the high sensitivity necessary for measurements at low concentrations, and often involve the simplest procedures. Further, when this method is used for studying isotopic exchange, radiotracers usually provide the only feasible approach.

The usual problem encountered in using radiotracers with the differential method are the normal problems of radiotracer analysis and are discussed in Chapters 4 and 6. Difficulties associated with the nature of the system, such as standardization of interfacial conditions, confirmation of adsorption equilibrium, adsorption on the reaction vessel walls, and interference from surface active impurities, are beyond the scope of this article. For studying adsorption from extremely dilute solutions where the radiotracer must be almost carrier-free, the peculiar behavior due to the formation of so-called "radiocolloids" must be considered. This problem is discussed in Chapter 8.

In conclusion, the differential method is probably the easiest and most widely used method for studying adsorption at the solid/liquid interface; but it is only useful for solids with large surface areas (i.e., are highly dispersed), otherwise the following techniques are necessary.

2. The Sample-Removal Method

In this technique the adsorbate is estimated by measuring the radioactivity of the sample after it is removed from solution. It has been used often because it is simple and usually does not require elaborate equipment. An important application has been to specimens whose available surface areas are too small for use of the differential method described above. However, since removal may disturb the adsorption equilibrium it may not be a particularly reliable or precise method except for chemisorption or surface exchange where the adsorbed or surface species are kinetically stable.

With this method the solid is placed in a solution of labeled adsorbate. After the particular process under study has taken place the solid is removed, washed if necessary to remove the adhering solution, and its activity, R, measured. The amount Γ (mol m^{-2}) adsorbed or exchanged on a flat surface may be calculated from

$$\Gamma = \frac{R - R_{\text{ret}}}{Q\,A} \tag{4}$$

where R_{ret} is the contribution from the labeled species in the retained or adhering solution, A (m^2) is the surface area of the sample, and Q is the proportionality constant (cpm/mol) between the activity and the amount present at the surface. Q may be determined by measuring, under the same geometric conditions, the activity of known amounts of the radioactive species deposited on the sample. This technique for determining Q also allows for any back-scattering or self-absorption. The correction term, R_{ret}, for the activity of the adhering solution may be determined from the radioactive concentration (cpm/mℓ) of the solution and volume of the adhering layer which is determined by simply measuring the weight increase of the sample or by measuring the decrease on evaporation to dryness [44]. For samples with smooth surfaces the amount of active species in the adhering layer is often greater than the amount in the adsorbed layer, so that for acceptable accuracy it is often necessary to remove the solution layer (commonly done by rinsing [45,46] and/or mopping with filter paper [47]). This procedure increases the possibility of changing the adsorption equilibrium and thus decreasing the reliability of the results. Alternatively, "carry out" can be minimized by working with very dilute solutions. Successful applications of the sample-removal method include adsorption of sulfate ions on iron [48] and oleic acid on mineral surfaces [49], the effect of organic additives on metal-metal ion exchange reactions [50], and investigations of the electrical double-layer structure on platinum [51].

A modification of the sample-removal technique is to deduce the amount of adsorbate from the activity of the material desorbed from an equilibrated surface. The adsorbed layer is removed usually by dipping the equilibrated sample in a hot acid solution. An example of this procedure is the adsorption of $^{204}T\ell$ on platinum black catalysts [52]. In this case the amount of $^{204}T\ell$ in the adhering layer was very small relative to the adsorbed $^{204}T\ell$ because of the high surface area of platinum black so that the correction for the retained layer was not significant.

Another modification is the "*tape*" method described by Green *et al.* [53], which is really automation of the general removal technique (see Fig. 5). The solid used is a continuous metal tape which is equilibrated in a solution, then drawn out through narrow slits so that no back-diffusion occurs. In the adsorption chamber the tape is polarized against a Pt gauze auxiliary electrode. The amount of solution in the adhering layer is measured by a capacitance method and the activity of the adsorbed species is measured as the tape passes between two counters. Green *et al.* used the tape method to measure the adsorption of [^{14}C]thiourea on nickel as a function of potential, and it has

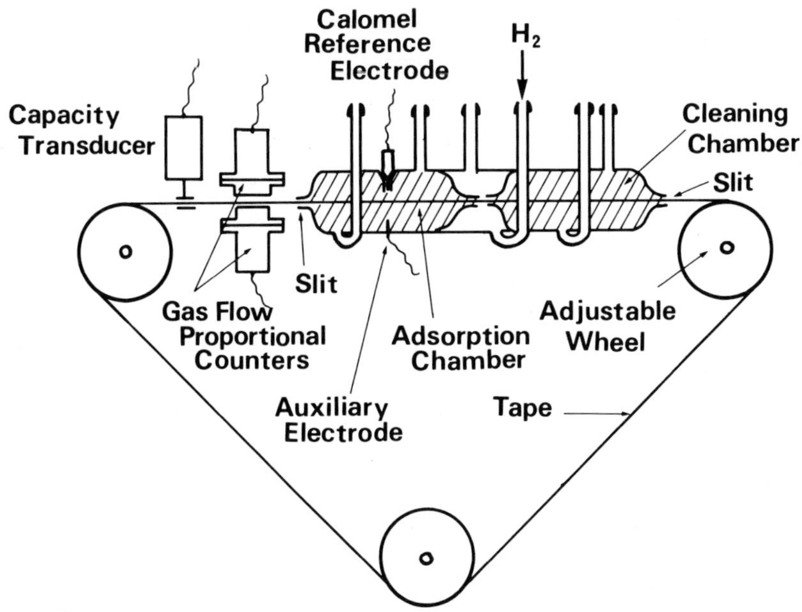

FIG. 5. The tape method of measuring adsorption of organic molecules at the metal/solution interface. (Reproduced from Ref. 53, by courtesy of American Institute of Physics.)

since been used to study various organic adsorbates on a number of metals [54].

One advantage of the sample-removal technique is that a restricted part of a surface may be studied; for example, Langdon *et al.* [55] placed a sheet of mica over a vessel so that the labeled solution contacted only the central portion of the sheet. In this way the adsorption of $^{59}Fe(III)$, $^{35}SO_4$, and $^{32}PO_4$ was measured for the {001} face of mica without any contribution from the crystal edges.

The greatest difficulty with the removal method is determining whether removing the sample from solution affects the amount adsorbed. In some cases for high-area surfaces and low bulk concentration it can be checked by measuring the adsorbed amount by the differential method. However, in most cases it is only possible to show that even if desorption does take place it is slow relative to the rate of removal of the solid from the solution so that there is not time for significant desorption to occur. Any effect of the washing technique can be estimated by determining the activity of the washings. In studies on metal-metal ion exchange, it has been observed that hot washing increases the exchange significantly [56].

3. The Dip-Counting Method

This is a direct method for measuring adsorption without removing the solid from solution. It was first used by Joliot [57] who showed that electrolytic deposition of metals could be followed by using thin electrodes which were at the same time, end windows of ionization counters. Dip-counting techniques based on the Joliot method for studying adsorption were first suggested by Aniansson [8] and have since been extensively used for studying the electrical double layer [41] as well as the adsorption of organic substances [58a] on metal electrodes, and adsorption on other window materials such as polymers [59], paraffin [60], and glass [61].

The solid to be studied is attached to or made into the counter window and the counter is then dipped into the solution. The basic arrangement is shown schematically in Fig. 6. Very thin layers of solid must be used in order to reduce absorption of the radiation in the solid. This has usually been accomplished by vacuum condensation of a thin (of the order of 1000 Å) film of metal onto a thin (about 2 mg cm^{-2}) mica sheet or some other material which is transparent to the radiation and sufficiently robust to act as a window for a radiation counter. Alternatively, higher counting efficiencies may be obtained by depositing the metal directly onto scintillation counters [58b].

The radioactivity detected comes from the labeled species adsorbed at the surface, R_{ads}, and also from the labeled species in the thin layer of solution, R_{sol}, that is within the maximum range of the particles emitted by the decay of the particular species. In order to obtain the radioactivity of the adsorbed

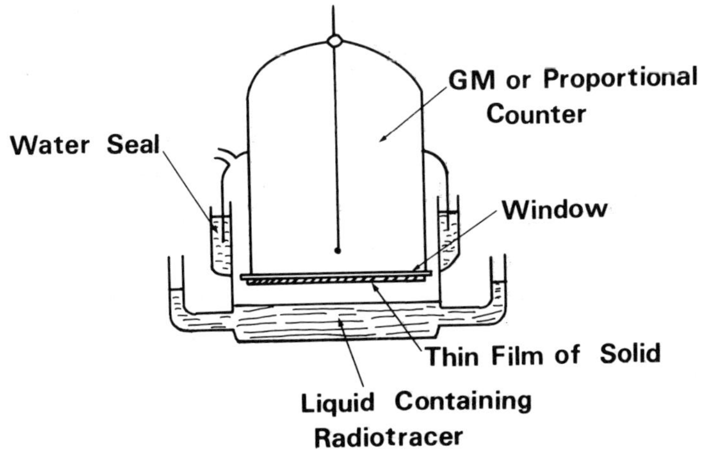

FIG. 6. Schematic diagram of the dip-counting method of measuring adsorption at the solid/solution interface.

layer, the background activity from the solution, R_{sol}, must be obtained and subtracted from the measured activity, $(R_{ads} + R_{sol})$. The most commonly used method of obtaining R_{sol} is by measuring, under the same conditions, the activity of a reference solution of a nonadsorbing species, labeled with the same radioisotope. Then the required activity can be calculated from the specific activities of the two species. An example of this is the use of $Na_2{}^{14}CO_3$ for the calibration of ^{14}C-labeled organic compounds [62]. If a matching nonadsorbing species is not available, a solution of the labeled species can be prepared in a solid matrix such as an epoxy resin in which surface enrichment cannot occur. The solid matrix should have the same absorption coefficient as the test solution. The activity can then be determined by placing the active solid on the end window of the counter [63]. A second method [41] involves the measurement of the activity for the same conditions as in the experiment but with a large excess of the unlabeled adsorbing species. Thus the fraction of active species in the adsorbed layer is decreased so that the activity of the adsorbed species is insignificant and the measured activity is effectively equal to the solution activity. In the third technique [64], the counter is mounted so that the end window is above and parallel to the surface of the solution. The distance between the end window and the solution is decreased in stages, and the count rate is determined as a function of distance (see Fig. 7). When the end window contacts the solution the count rate increases by an amount corresponding to the formation of the adsorbed layer. The activity from the solution is equal to the activity just before contact. This technique is not applicable to volatile compounds which might adsorb on the end window before it contacts the solution.

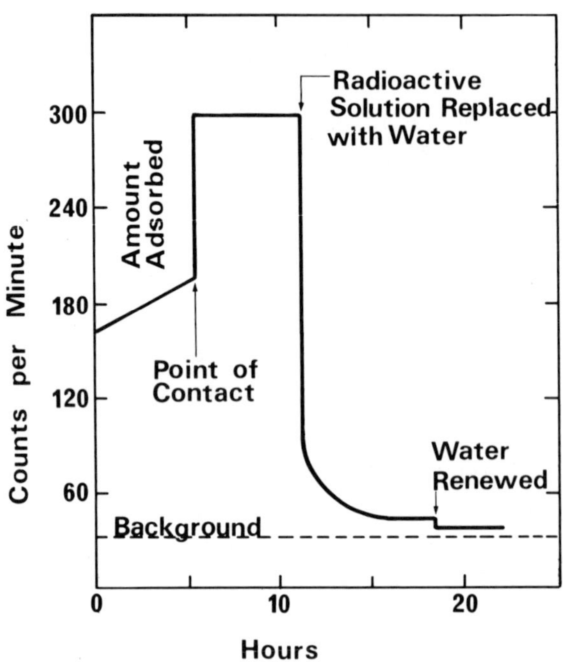

FIG. 7. Radiotracer adsorption measurement. Adsorbent: 12,500 Å gold film. Solution: 10^{-4} M [^{14}C]thiourea (specific activity 1 Ci mol^{-1}) in water. The overall shape of the count rate-vs.-time curve indicates that adsorption is not only very rapid but also highly reversible. The abscissa indicates the time elapsed under a constant rate of counter lowering. (Reproduced from Ref. 64, by courtesy of Macmillan Journals Ltd.)

The relationship between the activity from the adsorbed layer and the adsorbed amount (mol cm^{-2}) can be determined from the activity of a known quantity of the adsorbed species deposited on the end window under the same geometric conditions as in the adsorption experiment. This method of calibration is limited to nonvolatile compounds and errors may arise from nonuniform distribution of the compound after evaporation of the solvent. Alternatively calibration can be carried out by calculating the activity for a certain coverage in the following way [64]. The activity R_{ads}(cps) originating from the adsorbed layer is given by

$$R_{ads} = k\ 3.7 \times 10^{10}\ \alpha A Y \Gamma \tag{5}$$

where k is the detection coefficient, α the specific activity of the adsorbed species (Ci mol^{-1}), A the geometric surface area (cm^2), Y the roughness factor, and Γ the adsorption density (mol cm^{-2}).

The activity R_{sol}(cps) originating from the solution is given approximately by

$$\begin{aligned} R_{sol} &\simeq k\ 3.7 \times 10^{10}\ \alpha A C \int_0^{\infty} \exp(-\mu x)\ dx \\ &\simeq k\ 3.7 \times 10^{10}\ \frac{\alpha A C}{\mu} \end{aligned} \tag{6}$$

where C is the concentration of the active species (mol cm^{-3}), μ the absorption coefficient for the radiation in that solution (cm^{-1}), and x the perpendicular distance from the surface. From Eqs. (5) and (6) we obtain

$$\Gamma = R_{ads} \frac{C}{R_{sol}} Y \mu \tag{7}$$

so that the adsorption densities can be calculated from known and measurable quantities. Wroblowa and Green [65] found that both methods agreed for ^{14}C radiation and gold-foil coated windows thus confirming the validity of the calibration techniques. The absolute accuracy was estimated to be about 5%.

It is also apparent from Eq. (7) that

$$\frac{R_{ads}}{R_{sol}} = \frac{Y\Gamma\mu}{C} \tag{8}$$

and that this ratio must be greater than about 0.01 for acceptable accuracy. Therefore for flat surfaces, where $Y \simeq 1$, only isotopes that emit radiation with large absorption coefficients (short ranges in solution) are usable in reasonably concentrated solutions. Even for soft β emitters such as ^{14}C, ^{35}S, and ^{45}Ca, the highest concentration can only be approximately 10^{-3} M. Tritium has also been used [59,60], but the range is so short (∿0.6 mg cm^{-2} or about 6 μm of water) that ultrathin windows (less than about 1 μm) are required and this adds to the experimental difficulties. This is best overcome by using solid scintillator discs as reported by Wieckowski [58b].

The dip-counting method for determing the adsorption of organic compounds on platinized platinum electrodes has been found to be in good agreement with electrochemical methods [66,58b]. Several modifications have been introduced that facilitate the use of harder β, or γ radiation or higher concentrations. The first was proposed by Kafalas and Gatos [67] who designed a cell to measure the activity of the adsorbed layer from the solution side. The activity measured from the solution layer is decreased by making the layer of solution between the surface being studied and the counter extremely thin (<1 mm). A version of this apparatus, shown in Fig. 8, has been used by Schwabe and Schwenke [68] for measuring adsorption of $^{131}I^-$ (E_{max}, 0.61 MeV) on Pt and Au. A similar apparatus has been described by Kazarinov [69] in which, after adsorption equilibrium is established, the electrode is lowered down onto a counter which makes up the floor of the cell. Again the measured activity is the sum of the activities

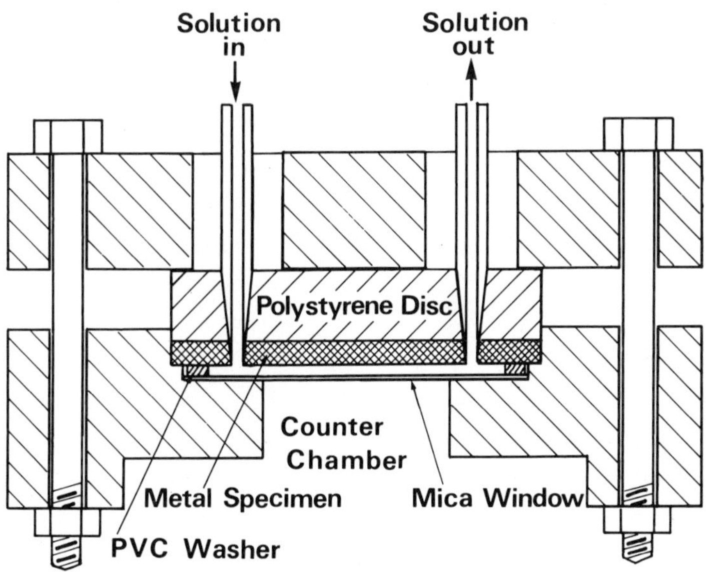

FIG. 8. Cross-sectional view of cell for measurement (from solution side) of adsorption of $^{131}I^-$ at the metal/solution interface. (Reproduced from Ref. 68, by courtesy of Microforms International Marketing Corporation.)

from the adsorbed species, the layer of solution between the end window and electrode, and the background activity of the solution. All these quantities can be readily determined experimentally by methods described above. One of the advantages of these two modifications is that they can be used for solids that cannot be made into thin films. The basic limitation is that they are only applicable to active solute concentrations of less than about 10^{-4} M unless large-area surfaces are used.

A third modification is based on increasing the roughness factor (γ) in Eq. (5) and is useful for moderately high concentrations as well as a wider range of β emitters. Horanyi *et al.* [70] used this approach by platinizing a gold-plated plastic foil which served as the "*window*" for a scintillation counter and the bottom of their reaction vessel. They also showed that covering the surface of the end window with Pt powder was as successful for studying adsorption as platinizing it to roughness factors of 100 ∿ 1000. Therefore, this technique should be useful for all powders (as an alternative to the differential method) and has been used for studying surface exchange of ^{45}Ca on calcite muds [71]. Horanyi's technique with platinized platinum electrodes has been successfully used to investigate the adsorption and exchange of species labeled with ^{32}P (E_{max}, 1.71 MeV), ^{36}Cl (0.714 MeV) as well as with the weaker β emitters ^{14}C, ^{35}S and ^{45}Ca in concentrations up to 10^{-1} M [72].

In contrast to the sample removal method the dip-counting techniques have the advantage that the measurements are made *in situ* without disturbing the adsorption equilibrium. Another advantage is that adsorption can be followed continuously, thus facilitating studies of the kinetics of surface processes. The main disadvantage is the difficulty in the preparation of clean and reproducible surfaces attached to the end windows of radiation counters. Metal surfaces may, of course, be prepared by vapor condensation onto a mica window supported within a high vacuum apparatus, followed by introduction of the labeled solution to one side of the window and the counter to the other without exposing the metal film to the atmosphere [63].

C. Adsorbate Distribution

The distribution of the adsorbed species can be studied by applying autoradiography to a solid surface after it has been removed from solution. For example, this technique has been used to show increased activity, and therefore preferential adsorption, along grain boundaries of metals [73] and different amounts of adsorption of [^{35}S]xanthates on different crystal faces of sulfide materials [74]. However, the resolution is generally limited to ∿5 μm.

Autoradiography at the electron microscope level, frequently applied to biological samples (see Chaps. 4 and 17), has also been used to study the distribution of ^{55}Fe, ^{85}Sr, and ^{125}I over clay particles [75]. Examples are given in Fig. 9 which shows the distribution of the photographic silver grains which denote the presence of radioactivity and appear as black spots or ribbons over clay particles containing adsorbed $^{55}Fe(III)$ ions. There is a preference for edge sites in adsorption from solutions at pH $\leq$ 4.0 (Fig. 9a, b), but uniform coverage for adsorption from solutions at pH 4.5 (Fig.9c) where Fe(III) is hydrolyzed. It is apparent that with this technique a resolution of about 0.2 µm can be obtained with isotopes that emit low-energy radiation and for fine-grain emulsions.

Autoradiography is, of course, just a variation of the sample-removal method discussed above. Therefore it suffers from the same disadvantages, i.e., adsorption equilibrium can be disturbed by the removal from solution unless the adsorbed species is kinetically stable.

IV. LANGMUIR/BLODGETT MONOLAYERS ON SOLIDS

A. General

Monomolecular films of long-chain polar molecules may be transferred from a water surface where the area/molecule is accurately known to that of a solid by a process developed in the 1930s by Langmuir and Blodgett [76]. Figure 10 illustrates schematically the process by which the transfer is effected. Usually the clean hydrophilic solid is placed beneath the water surface, the monolayer spread, and then compressed by means of a mechanical device or a "piston" oil which maintains a constant pressure on the spread monolayer, while the solid is slowly drawn vertically through the air/water interface. Such transferred films, have provided ideal models for studies not only of the behavior of solid surfaces on which they are deposited but also of monomolecular films adsorbed from solution onto solids (see Section V). Such adsorbed films are of prime importance in

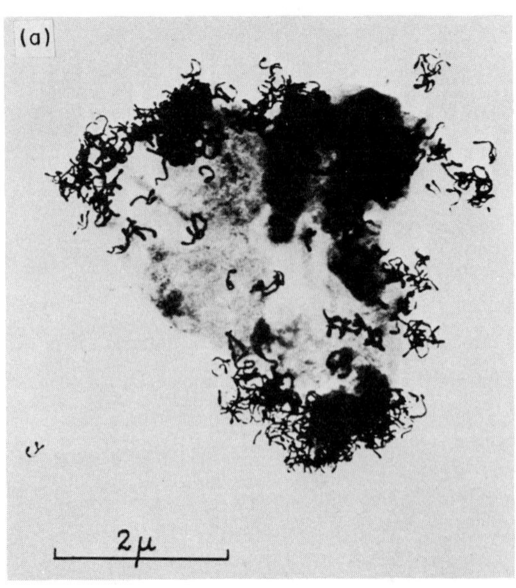
(a)
2μ

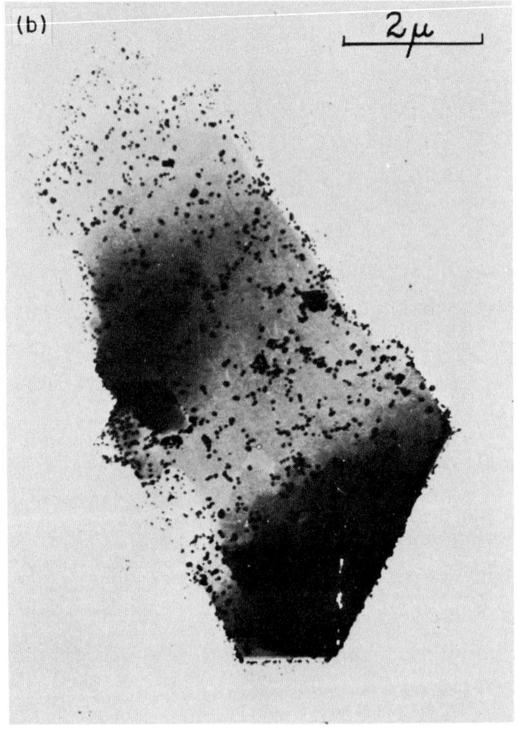
(b)
2μ

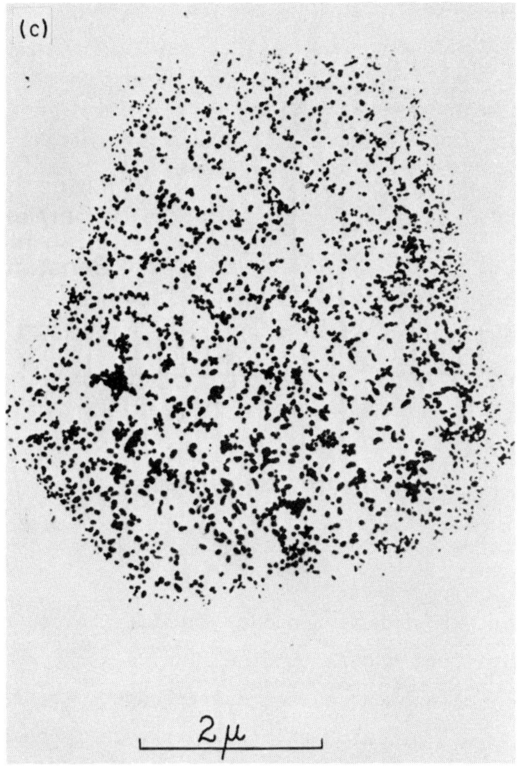

FIG. 9. Autoradiograph of $^{55}Fe(III)$ sorbed from 0.15 M NaCl by (a), South American kaolinite after 7 days at pH 3.7; (b), dickite after 3 days at pH 4.0; and (c), dickite after 3 days at pH 4.5. (Reproduced from Ref. 75, by courtesy of Pergamon Press.)

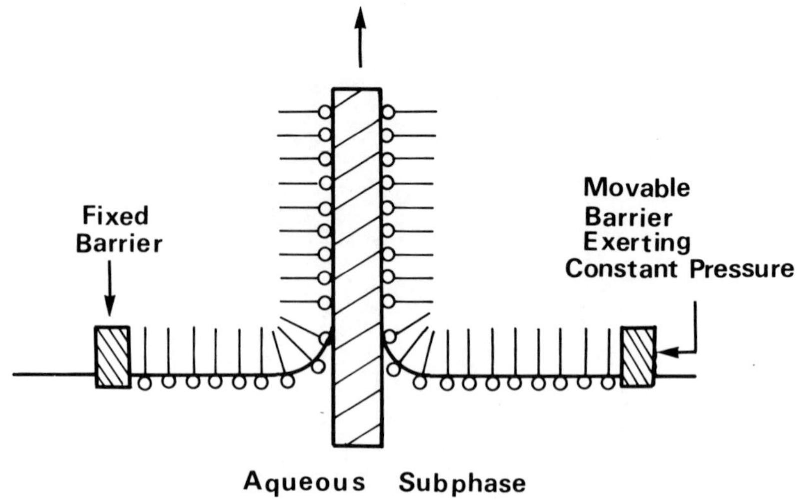

FIG. 10. Transfer of Langmuir/Blodgett monolayers to solids.

technical processes such as boundary lubrication, adhesion, corrosion inhibition, and flotation. Thus a study of the nature and strength of attachment of Langmuir/Blodgett monolayers on solids can give useful information relevant to processes occurring in real systems. The amphipathic substances most widely used as deposited monolayers have been the long-chain fatty acids and their soaps.

The transfer process may in favorable circumstances be repeated a number of times so that multilayers can be built up. Such built-up films have found wide use as step wedges (thickness gauges) by virtue of their light interference colors [77] as well as diffracting crystals in x-ray spectroscopy [78] and also in electronic devices [79]. Films which transfer as the solid both enters and leaves the monolayered water surface are termed Y films, those which transfer only on immersion are termed X films, while Z films are considered to deposit only on withdrawal. Conflicting views about the nature and mechanism of the deposition process have arisen especially since x-ray diffraction has shown that both X and Y films give a lattice constant equal to 2ℓ, viz., twice the length of a single molecule. This suggested that in X deposition entire molecular layers turn over--presumably during the transfer process itself.

An ingenious method was devised by Berlovich, Kutsentov, and Fleischer [154] to determine the actual positions of the polar groups of the deposited molecules i.e. the orientation of the molecules relative to the solid to which they had been transferred. Monolayers of cetyl[^{32}P]phosphate [$C_{16}H_{33}O^{32}PO(OH)_2$], having a specific activity of ∿1 μCi cm^{-2} were transferred to aluminum foil. The ^{32}P β decay is accompanied by the emission of very slow electrons from the outer electron shells as a result of a sudden change of the nuclear charge. As the mean path of such "jolted" electrons is comparable with the length of a molecule their emergence will depend significantly on the orientation of the molecules. Thus when the molecules are oriented with the polar groups adjacent to the solid the electrons must pass through a layer whose thickness is equal to the length of the hydrocarbon chain. If the second monolayer of unlabeled cetyl phosphate is deposited and the orientation of the first layer remains unchanged the distance the "jolted" electron must pass through will have increased at least twofold, resulting in an appropriate decrease in the yield. Kutsentov and Fleischer [155] observed no change in yield under these conditions and concluded that the first (radiolabeled) layer had turned over during transfer of the second layer. Success in this work depended on the separation of the "jolted" electrons from the background of secondary electrons produced as the β particles passed through the monolayers. This was achieved by counting the number of coincidences between β particles and the slow electrons at various angles relative to the substrate normal.

Because of their specificity and high sensitivity radiotracers have been used extensively in following the transfer process itself as well as the structure and behavior of the monolayers after deposition. The isotopes useful for incorporation in the organic molecules forming the monolayer are principally ^{3}H, ^{14}C, ^{32}P, and ^{35}S, all being weak β emitters. Should soap monolayers be required cations such as ^{45}Ca (pure β emitter of E_{max}, 0.26 MeV, $t_{1/2}$ = 164d) may be incorporated in the monolayer by reaction with labeled calcium cations in the aqueous subphase prior to transfer.

The use of radiolabeled monolayers was pioneered by Beischer [80]. Since it is generally accepted that monolayers may be transferred from a water surface to a solid support without noticeable change in their structure they have the following important properties: (i) the homogeneous distribution gives a uniform activity per unit area; (ii) the thickness of a monolayer is such that even films some thousands of monolayers thick may be treated as weightless sources, making corrections for self-absorption unnecessary for such weak β emitters as ^{14}C and ^{35}S; (iii) the monolayers or multilayers can be transferred with good reproducibility and are relatively permanent. For example, barium stearate monolayers deposited on copper plates showed no change in activity after storage at 2°C for as long as half a year (Beischer [81]).

Beischer showed that radioactive monolayers provide better reference sources for quantitative absolute autoradiography than films prepared by other means. He prepared step wedges of [^{14}C]-stearic acid on mica and obtained approximately linear density-activity relationships following exposure to medical no-screen x-ray films. By exposing simultaneously to the same photographic material the stepped wedge and a specimen carrying a ^{14}C-labeled film of unknown concentration, a precise estimate of the activity of the latter may be made. Further, knowing the disintegration rate of the active isotope the method may be extended to absolute measurements of activity by autoradiography. Sobotka and co-workers [82,83] using [^{14}C]stearic acid studied the behavior of built-up films, such as skeletonization[1] of mixed acid/soap multilayers [77] as well as the relative stability of acid and soap molecules within monolayers and multilayers. They confirmed that there is no measurable self-absorption of ^{14}C β particles in built-up films up to at least 100 layers (or 2500 Å). In fact, calculations show that the built-up films would need to contain at least 2500 double layers before the count rate would be reduced by 10% by self-absorption.

Sobotka *et al.* [83] also studied the actual process of soap formation within the monolayer at the water surface by transferring films to chromium-plated slides from $^{45}CaCl_2$ solutions adjusted to different pH values. The activity within built-up films 100 layers thick gave an indication of the degree of saponification for the particular pH of the aqueous substrate. The results are shown in Table 2. For purposes of calibration the amount of ^{45}Ca in the film for 100% soap formation was estimated and this, in the form of the chloride, was deposited on a planchet and counted with precisely the same geometry as the built-up films.

TABLE 2

^{45}Ca Content of Built-Up Films as a Function of pH [83]

pH	Count rate (cpm)	Stearic acid neutralized (%)
6.04	43.8	34.6
6.77	64.8	51.9
7.15	87.6	70.2
7.68	117.0	93.9
7.90	133.8	107.9
Control	124.8	100.0

[1]Preferential removal of one of the constituents of a mixed system without collapse of the overall structure.

B. The Transfer Process

In his early studies in this field Langmuir had already recognized the importance of the chemical nature of the solid substrate with regard to the ease of transfer and subsequent strength of attachment. Later, however, Bikerman [84] stressed the role played by the contact angle that the water made with the solid during the passage of the latter through the air/water interface. He concluded that a monolayer on a water surface could only touch and attach itself to the solid if the direction of movement of the latter and that of the water surface adjacent to the solid formed an obtuse angle. Subsequent work, discussed below, indicates that both of these factors are important since the chemical nature and even the detailed atomic structure of the solid surface can influence the contact angle in question. The long-held view that under a wide variety of conditions fatty acid and soap monolayers are transferred unchanged, i.e., as a carpet, to the solid with a transfer ratio ρ of unity, was called into question by the radiotracer studies of Gaines [85] and Spink [86]. Both authors used thin end-window GM counters to determine the activity and hence amount of [^{14}C]stearic acid within the transferred film. In addition, autoradiography and electron microscopy were employed to check the uniformity of the film material.

Not having precise values for geometric efficiency and specific activity factors Gaines [85] had to convert count rate to concentration within the transferred monolayer by assuming that at 30 dyn cm^{-1} the film on the water surface (whose area/molecule is well established) was transferred unchanged to the solid. This assumption was partly justified by parallel experiments in which, for a wide range of surface pressures, count rates of the film at the water surface were compared with those of the material transferred to the mica plate of a Wilhelmy balance. On this basis Gaines found that at low surface pressures ($\sim$10 dyn cm^{-1}) the transfer ratio for stearic acid never reached unity for glass, mica, and platinum and even for reactive metals such as copper and aluminum. On the other hand, for films transferred at a surface pressure of 17.5 dyn cm^{-1} ρ equaled unity for glass, mica, Pt, Cr, Cu, and Al.

In later studies Spink [86] used a calibration procedure which permitted an estimation of the absolute concentration of molecules in the monolayers transferred to a variety of solids. As described below the counter constant was determined for a monolayer of [^{14}C]stearic acid on the water surface at the particular surface pressures used later in the deposition process and thus at specific values of area/molecule.

If R_s is the count rate from a monolayer on a specimen, then the surface concentration Γ_s of long-chain polar molecules is given by

$$\Gamma_s = \eta_s R_s \tag{9}$$

where η_s is the detection coefficient that contains a back-scattering factor η_{b_s} which will depend on the nature of the substrate. Spink obtained η_s values for labeled monolayers at various surface pressures when the water surface was brought closer to the counter aperture (Fig. 11). Extrapolation gives the count rate R_s for the close contact of the counter window to the monolayer surface. It was thus possible to calculate Γ_s for any solid if its back-scattering value and that for water were known. It is convenient to use the ratio of the area/molecule on the solid to that on water. This has been introduced already as the transfer ratio ρ given by

$$\rho = \frac{\Gamma_s}{\Gamma_w} = \frac{R_s \eta_{b_s}}{R_w \eta_{b_w}} \tag{10}$$

where the subscript w refers to a water surface. An evaluation of ρ still requires the back-scattering factors η_{b_w} for water and η_{b_s} for the solid to which the monolayer is transferred. According to Faires and Parks [87] η_b is proportional to the

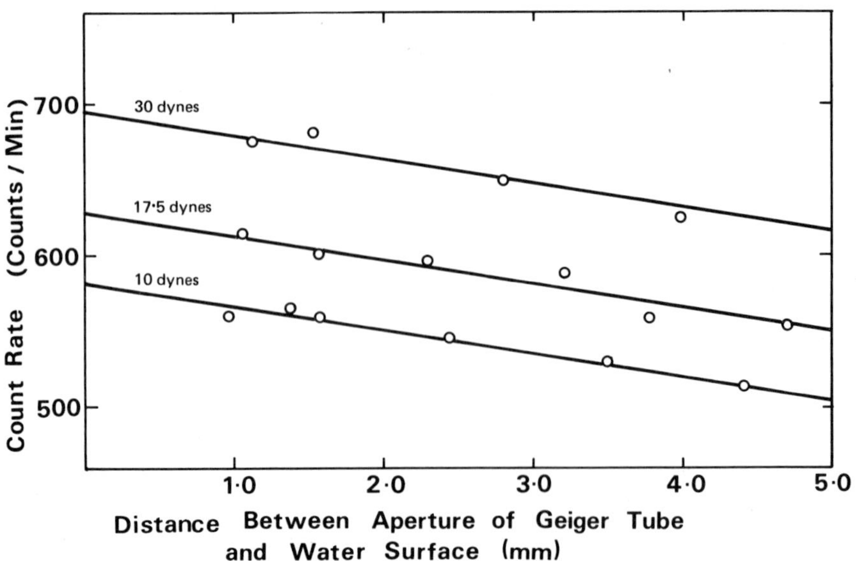

FIG. 11. Determination of counter constant for monolayers of stearic acid on a water surface. (Reproduced from Ref. 86, by courtesy of Academic Press.)

square root of the atomic number of the back-scatterer. Fig. 12 shows selected published values of the saturation percentage increase in radioactivity for ^{14}C β radiation for a variety of substances [85,86,88] as a function of the square root of the atomic number.

For weak β radiation, back-scattering corrections are particularly sensitive to the counting geometry [89]. For the work under discussion [83] the geometry and equivalent thickness of counter window and air path were the same as those used by Calvin *et al.* [88] to obtain the results shown in Fig. 12. In order to avoid assumptions about the transfer ratios, data for Langmuir/Blodgett films of ^{14}C-labeled material have not been included in Fig. 12. Instead the values refer to thin films deposited from solution on to the different substrates. A straight line drawn through these points enables saturation values

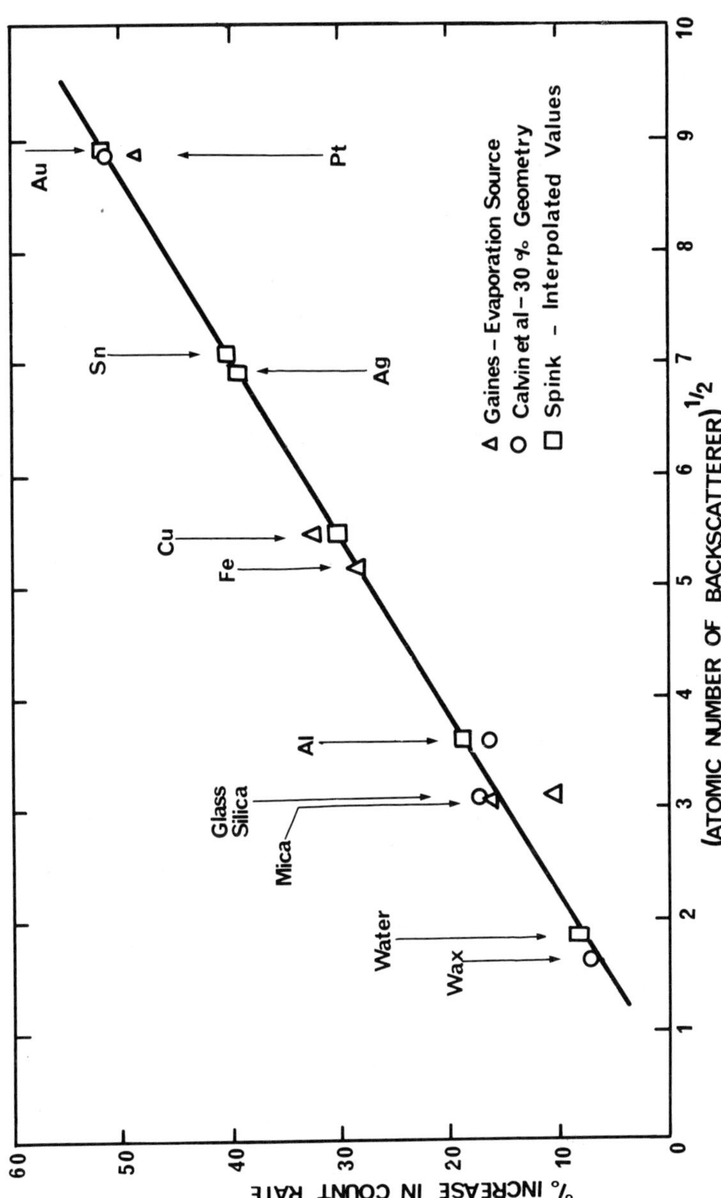

FIG. 12. Saturated back-scattering of ^{14}C β rays as a function of atomic number of the back-scatterer. Data based on the reports by Gaines [85], Spink [86], and Calvin et al. [88]. (Reproduced from Ref. 86, by courtesy of Academic Press.)

for η_{b_s} to be obtained by interpolation for substances for which no satisfactory data are available [86].

On the basis of this work Spink [86] found that the transfer ratio was unity for mica, silica, and glass as well as for abraded silver, copper and iron, and flamed platinum. The ratio, ρ, was not influenced by the pH of the aqueous substrate between 2 and 6, the piston oil pressure between 10 and 30 dyn cm^{-1} and the rate of transfer between 2.5 and 12.5 mm/min. Due to the hydrophilic nature of these solids the films were deposited as a monolayer/water carpet and irrespective of the reactivity of the solid a transfer ratio of one was obtained. Large-grained polycrystals of silver, the grains of which were large enough to be counted separately did, however, show a dependence of the transfer ratio on the orientation of the individual crystal faces. Figure 13 shows an autoradiograph of one such specimen following transfer of a monolayer of $[^{14}C]$stearic acid (Π = 30 dyn cm^{-1}) together with the orientation of the grains and their values of ρ. It is clear that ρ is lowest for those crystals having macroscopic surfaces closest to the {111} face which for the face-centered cubic lattice is smoothest in atomic models. On the other hand ρ is highest for crystals with surfaces close to the {210} face which is atomically the roughest. These experiments were extended to very flat epitaxed films of silver condensed on mica where, again the deposition ratios depended inversely on the contribution of the atomically smoothest plane, {111}, to the overall surface. For a withdrawal rate of 2.5 mm min^{-1} the ratio was just over 0.8 for 10-20% {111} contribution and between 0.5 and 0.6 for 100% {111}. Gold showed a similar behavior, whereas tin films showed $\rho = 1$ even for surface pressures of 10 dyn cm^{-1} [90]. This difference in behavior between metals such as silver and gold on the one hand and copper, tin, and iron on the other may be explained in terms of the ready formation of thick oxide films on the latter but not on the former. The relative weakness of dipole-metal image forces proposed by Fowkes [91] suggests that

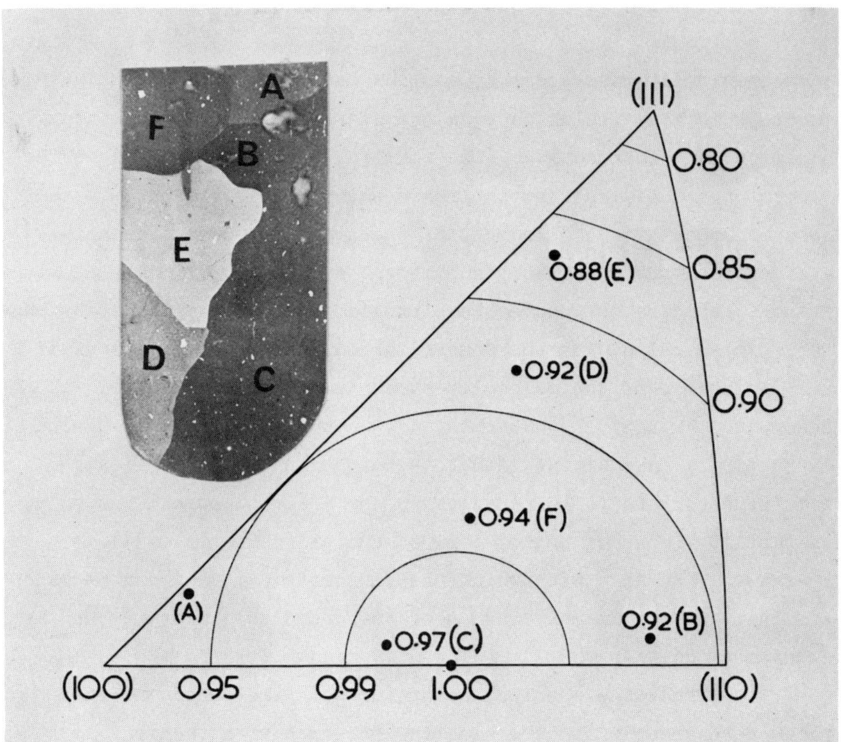

FIG. 13. Autoradiograph of [^{14}C]stearic acid monolayer deposited on large-grained silver polycrystal. The stereographic triangle shows the orientations of the grains together with the transfer ratios. Contour lines indicate number of broken nearest-neighbor bonds in a face-centered cubic crystal normalized to 1.00 at {210}. (Reproduced from Ref. 86, by courtesy of Academic Press.)

oxide-free metal surfaces are not wetted by water. The resultant restricted wettability must then affect the efficiency (at least from the purely geometrical point of view advanced by Bikerman) with which monolayers can be transferred to solid surfaces of the noble metals.

C. The Stability of Monolayers on Solids

The behavior of monolayers at various interfaces is of continuing interest in many fields such as transfer processes, detergency, adhesion, and lubrication. The chemical nature and atomic perfection of the solid substrate already shown to be important in the case of deposition of Langmuir/Blodgett monolayers must also play a significant role in their subsequent stability. In many instances radiotracer studies have provided valuable information on their interaction with the underlying solids from estimations of the resistance of the molecules to surface diffusion as well as to desorption brought about by heating, vacuum exposure, or solvent extraction. In addition, other techniques have been employed simultaneously to follow changes in surface concentration such as wettability (contact angle) [85,92] and ellipsometry [93]. Because of the ease with which the radiometric method can be made absolute, it has in many instances been used to calibrate the other methods.

Whereas the chemical nature of both the solid and the polar head group will determine the strength of attachment or "anchoring" of the molecule to the solid, the atomic flatness of the substrate would ensure a strict parallel arrangement of the molecules (standing normal to the surface) resulting in an efficient close packing which should give a maximum cohesive energy (van der Waals). The importance of the polar head group/solid surface interaction has been the subject of many investigations but a complete analysis of the sometimes conflicting data has yet to be made.

The simplest experiment has been to deposit a film on a restricted part of a surface and measure the rate of diffusion into the film-free area as a function of temperature. A similar experiment has involved placing in contact a clean surface and one carrying a monolayer of radiolabeled material, and observing the rate of transfer from one to the other either at room or elevated temperatures. Measurements of residual activities in the above system have been relatively straightforward. Although most workers have used thin end-window Geiger counters, Beicher [81], in particular, used quantitative autoradiography which possesses the added advantage that the amount of substance and its distribution are well defined. Depletions or accumulations or other artifacts can be immediately visualized.

Studies of the complete loss of molecules from close-packed monolayers have involved either vacuum or solvent desorption. The monitoring of residual activity in such cases can be much more involved since the equilibrium at any point might be upset when the specimen is removed either from the vacuum or the solvent for counting. Consequently the advances in this field of study have depended on the development of techniques permitting continuous monitoring of the radioactive material either remaining on or desorbed from the solid substrate. We will now consider briefly the methods used as well as the significance of the results.

Using quantitative autoradiography Beischer [81] investigated the lateral movement of stearic acid on mica, glass, platinum, copper, aluminum, and lead. At temperatures up to 100°C, he found that, whereas the molecules moved freely over the first three substrates, no migration from the originally covered areas was observed for the last three. The ease with which the surface-active molecules could move over the solid surfaces reflected the degree of stearate formation.

Using a GM counter Rideal and Tadayon [94] showed that transfer of radiolabeled stearic acid from one metal surface to another brought into contact again reflected their ability to form stearates. Table 3 shows the equilibrium established after about 80 hr at 36°C. Considerably lower transfers of the order of 5% were observed when barium stearate on mica was contacted with mica. The authors concluded from these as well as further experiments on surface diffusion of stearic acid on mica that surface migration as well as overturning of the molecules were involved in the transport process.

TABLE 3

Equilibrium Transfer (%) of [^{14}C]Stearic Acid [94]

From / To	Mica	Ag	Sn	Ni	Cu	Pt
Mica	50-58	--	5	--	--	--
Ag	--	45	30	--	4	15
Sn	85	65	50	8	6	--
Ni	--	--	65	6	--	--
Cu	--	80	85	10	4	8
Pt	90	90	--	--	8	--

The importance of surface diffusion was later questioned by Young [95] who carried out detailed studies on the conditions necessary for long-range surface diffusion to be energetically more likely than complete desorption. By ventilating specimens (mica, copper, and platinum) carrying stepped films of [^{14}C]-stearic acid *toward* the region of higher surface concentration of the acid Young found no surface transport at all, the behavior of all layers being confined to progressive desorption.

Conditions favorable to vapor phase transport were then used by Young. This involved the mounting of a specimen carrying monolayers of [^{14}C]stearic acid at various distances from a film-free specimen in a glass tube which was sealed off and then immersed in a thermostat. After an hour at about 90°C vapor phase transfer was considerable for the mica and platinum and slightly so for the copper pairs. Although this order of loss confirmed earlier conclusions on the importance of chemical reactivity, Young deduced that in the systems under discussion "probability favors desorption and the maximum movement anywhere of a molecule by true surface diffusion before being desorbed is below the limit of the observational technique." In similar radiotracer experiments Gaines [96] confirmed that vapor phase transfer between mica surfaces occurred more readily from stearic acid than from barium stearate monolayers.

Other radiotracer studies involving thermal loss from mixed acid/soap multilayers [82], solvent extraction of monolayers on various substrates [85] and finally stability in vacuum of acid and soap monolayers on a variety of solids [97] have all confirmed the greater stability of the soap layers. In addition, the *in situ* formation of soap films from free acids on reactive substrates give surface films of even greater stability.

Finally, Spink [98] examined continuously the rate of loss in vacuum of [^{14}C]stearic acid from Langmuir/Blodgett monolayers deposited on mica and condensed silver films on mica. As already mentioned, the main problem to be overcome in monitoring gain or loss of material at an interface is that the equilibrium not be disturbed during the measurement of activity at any particular time during the process. This was overcome in the case of thermal desorption (in vacuum) by use of the specially designed apparatus shown in Fig. 14. The specimen (attached to the small block furnace J) and a thin end-window GM tube H were both mounted within the vacuum chamber A on a 30° sector plate fixed to a post rotatable through 30° intervals by means of a Wilson seal G and a pin F and graduated disk E. The specimen and GM tube window were close to (∿1 mm) and parallel to the refrigerated steel plate C. Any increment of loss from the monolayered specimen would strike the cold plate and become firmly anchored. Subsequent rotation of the specimen (and the GM tube) through a 30° angle would then permit the activity of the desorbed material to be counted during which time a further increment would be desorbed and captured by a fresh area of the cold plate. This procedure could be repeated until the desorption process was complete. For more details of the operation of this equipment reference should be made to the original publication.

The considerable difference in activation energies, 43.3 and ∿10 kcal mol^{-1}, found for desorption from the two substrates mica and {111} faces of epitaxed Ag films on mica was related principally to the perfection of the lateral arrangement within

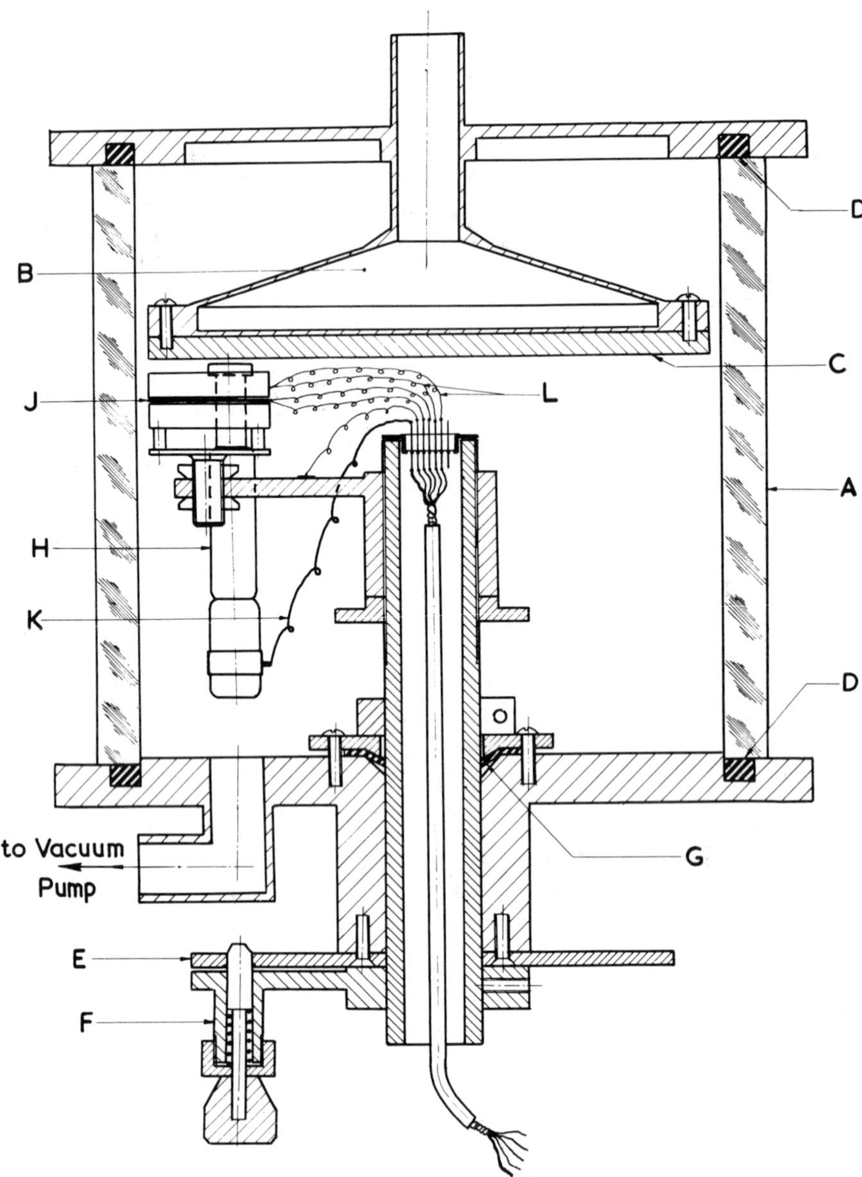

FIG. 14. Cross-sectional view of vacuum desorption apparatus. A, Pyrex cylinder; B, refrigerating chamber; C, mild steel cold plate; D, O-ring gaskets; E, graduated disk; F, spring-loaded locating pin; G, Wilson seal; H, Geiger-Müller counting tube; J, block furnace; K, high-tension lead to GM counter; L, thermocouple leads. (Reproduced from Ref. 98, by courtesy of Academic Press.)

the Langmuir/Blodgett monolayers. Only with a complete ($\rho = 1$) close-packed monolayer on an atomically flat surface can the van der Waals intermolecular attraction reach its maximum and this was considered to be the situation for mica substrates. For the epitaxed silver films the monolayers were not close packed ($\rho < 1$) so that the principal contribution to binding was the head group/metal interaction. Thus the relative importance of these two contributions (lateral and polar attractions) to molecular stability within monolayers was clearly demonstrated in these experiments.

V. NATURE OF ADSORBED FILMS ON SOLIDS

In parallel with the above studies on the stability of deposited Langmuir/Blodgett monolayers, a considerable amount of attention has been paid to the mechanism of adsorption of long-chain molecules from solution as well as their subsequent behavior. A general treatment of the solid/liquid interface is given in Section III of this chapter and we will mention briefly only those studies with long chain polar adsorbates. Again, the use of radiotracers has played an essential role, both reactive adsorbates and adsorbents being utilized. For example, in studies on fundamentals of boundary lubrication Bowden and Moore [99] immersed radioactive foils of metals in solutions of inactive octadecyl alcohol, ethyl stearate, and stearic acid. After washing the foils in boiling benzene these authors deduced chemisorbal amount from the activity of the washings. Gold and platinum showed little or no activity in contrast to zinc, cadmium, and copper. Most other work in this field, however, has involved the use of radiolabeled organic adsorbates.

In order to avoid possible errors associated with interrupting the adsorption process for purposes of making a measurement, Walker and Ries [63] monitored continuously the adsorption of $[^{14}C]$stearic acid from hexadecane solutions by using as their adsorbent a metal condensed as a thin film on the outer face of the mica window of a GM tube. The principle of this technique is shown schematically in Fig. 6. The detection coefficient [k in Eq. (5)] was determined by reference to the radioactivity of a close-packed monolayer of $[^{14}C]$stearic acid on a water surface, with assumption of identical back-scattering factors for water and hexadecane. The results showed that adsorption varied with substrate, but in no instance exceeded about one monolayer. Exposure of the condensed metal films to dry air rather than dry helium caused slightly greater adsorption, the saturation values being $\sim$1.1 monolayers for iron, 0.6 for gold, 0.4 for copper, and no detectable adsorption for mica.

In similar work Seimiya [100] studied the uptake of sodium dodecyl $[^{35}S]$sulfate from aqueous solutions by various metals condensed on GM counter windows (1.3 mg cm^{-2}) of polyethylene terephthalate. Iron, silver, and gold proved to be so nonreactive as to give less than or equal to monolayer thickness. Copper and lead, on the other hand, showed thick film formation.

A novel method was introduced by Patrick and Payne [101] to measure continuously the amount of $[^{14}C]$stearic acid desorbed by toluene from a film preadsorbed onto ferrotype plates from hexane solutions. The monitoring of the activity of the desorbed material was carried out by incorporating a scintillator in the desorbing solvent and using a liquid scintillation spectrometer. Although reproducible desorption isotherms were obtained, a lack of knowledge of the number of molecular layers present in the original films as well as the mutual disposition of the molecules within the layers did not permit a complete elucidation of the desorption mechanism, nor the fraction of stearic acid removed at any stage of the process.

Finally, let us consider those studies where the specimen is removed from the solution periodically for monitoring of the activity of the adsorbed species. In most systems where the specimen emerges wet some rinsing off of the solution carried out is necessary to avoid errors. Parallel experiments need to be performed to check that this rinsing does not disturb the equilibrium. One method of avoiding this difficulty has been by the use of conditions whereby the adsorbed films are either hydrophobic, oleophobic, or even autophobic: when a slide is withdrawn from a solution or the melt of a pure surface-active substance the liquid peels back from the solid which emerges "dry." Thus carry-out is eliminated. The formation of such nonwetted surfaces, by what has been termed the retraction method (Zisman *et al.* [102]), requires a solvent with high surface tension and low volatility as well as a dissimilar chemical structure from and low solubility for the solute. In work related to studies on adhesion Shepard and Ryan [103] utilized this technique to develop a simple reproducible method of preparation of standard adsorbed monolayers on various solids. Knowing that long-chain molecules containing outermost CF_3 groups conferred the greatest nonwetting property on adsorbed monolayers, they were able to produce satisfactory films (with negligible carry-out) on platinum, quartz, glass, and aluminum slides by adsorption from solutions of $C_7F_{15}{}^{14}COOH$ in *n*-decane. The results enabled these authors to determine apparent surface areas and hence roughness factors provided that the area occupied by the long chain molecule, the specific activity of the adsorbate, the detection coefficient of the counter, and the area of the sample counted were all known. Timmons [92] extended this work by preparing retracted monomolecular films of $[^{14}C]$stearic acid on a variety of metals from its solutions in nitrobenzene. Calibration of the counting rate was accomplished by depositing in a restricted area on each of the metals a known aliquot of the working solution of stearic acid and evaporating to dryness. Since count rates were determined for each of the metal substrates back-scattering corrections did not have to be made. By making assumptions regarding the roughness factors, areas/molecule within the monolayers for the various metals were deduced (see Table 4.) A relationship between contact angle with a drop of methylene iodide and monolayer coverage was achieved by taking a complete monolayer and successively reducing the amount of stearic acid present by solvent desorption (CCl_4 at 25°C). For all metals, iron, chromium, nickel, and two types of stainless steel, the contact angle fell from about 70° at 100% coverage to about 35° at 30% coverage, the slope being about 0.5°/percent coverage. In similar work Gottlieb [45] found a linear relationship between surface potential and coverage for [^{14}C]octadecyl acetate adsorbed from solution on to chrome plate, platinum and stainless steel.

A direct comparison of the radiotracer method with other physical methods in the measurement of adsorption on solids has been made by several workers. Bartell and Betts [93], for example, used ellipsometric and radiometric techniques to study [^{14}C]octadecylamine Langmuir/Blodgett monolayers and monolayers adsorbed from *n*-hexadecane solution onto chromium-plated slides at various stages of solvent depletion. Although the optical method showed equal results for complete layers the tracer technique showed a smaller activity for the adsorbed film. This was attributed to coadsorption of significant amounts of the inactive solvent in the adsorbed film. The possibility of coadsorption of solvent with solute was confirmed in later studies by Doyle and Ellison [46]. Again the retraction method was used but in this case both the solute and the solvent were radiolabeled. Freshly prepared 1000 Å films of silver, platinum, copper, and iron condensed on polished slides of the same metals were treated with *n*-[1,2-^{3}H]octadecane solutions of [^{14}C]stearic acid. The amounts of each substance in the adsorbed film were counted using a gas flow GM counter. The ^{14}C β radiation was detected through a thin Mylar window which absorbed completely the much weaker ^{3}H β radiation. Total radiation was then counted with the window removed. Count rates were converted to equivalent monolayers of

TABLE 4

Some Properties of Adsorbed Stearic Acid Monolayers*

Metal	Contact angle (deg; ±1 deg)	Acid adsorbed (μg/real cm^2)**	Surface concentration (× 10^{14} mol/real cm^2)**	Area/molecule ($Å^2$)
Iron	71	0.215	4.55	22
Nickel	69	0.205	4.34	23
Chromium	67	0.194	4.10	24
Stainless steel, Type 304	70	0.215	4.55	22
Stainless steel, Type 416	70	0.215	4.55	22

*Reproduced from Ref. 92 by courtesy of Academic Press.

**Calculated using a roughness factor of 1.33.

stearic acid by comparison with Langmuir/Blodgett monolayers of the radiolabeled acid deposited on the relevant metals. From the initial counts and from counts after several successive rinses of the specimens in cyclohexane it was deduced that, although coadsorption does occur, especially in long-term adsorption, nearly all of the stearic acid is adsorbed at the metal surface, the solvent being weakly retained in the spaces between the solute chains and in layers extending beyond the first.

VI. THE GAS/SOLID INTERFACE

A. Adsorption and Surface Area

In investigations of chemical reactions between a solid and a gas the surface area of the solid is an important factor. The traditional method for establishing the surface area has been to measure either volumetrically or gravimetrically the amount of an inert gas adsorbed at low temperature as a function of the gas pressure. This adsorption can usually be represented by the BET equation

$$\frac{p}{x(p_0 - p)} = \frac{k}{x_m} + \frac{(1 - k)p}{x_m p_0} \qquad (11)$$

where x is the mass of adsorbate in g/kg adsorbent, p the partial gas pressure, p_0 the saturation vapor pressure of the adsorbate, x_m the amount of gas adsorbed per gram adsorbent at monolayer coverage, and k is a constant. A plot of $p/x(p_0 - p)$ against p/p_0 usually gives a straight line from which x_m (and k) may be calculated. The radiometric technique has found a ready application in this field since the traditional approach, usually with nitrogen, is slow and tedious and requires extensive calculations. The pressures are usually measured with a McLeod gauge and a total surface area of at least several square meters is necessary. The substitution of radiolabeled krypton for nitrogen has several advantages [104,105]. First, at the temperature of boiling N_2 (78°K) the saturation vapor pressure of krypton is only 2 mm so that the proportion of gas unadsorbed relative to that adsorbed is small. In order to obtain a desired accuracy it is necessary to use samples with large surface areas and minimize the volume in the sample cell. Second, krypton labeled with ^{85}Kr ($t_{1/2}$ 10.6y; 0.67 MeV β, 0.51 MeV γ) permits not only a higher precision in measuring the residual gas in the system but also the amount adsorbed on the solid itself. Aylmore and Jepsen [104] for example, incorporated a thin end-window GM counter into their apparatus and used the β activity R of the unadsorbed gas as a measure of its pressure, the relationship being:

$$P = R\alpha T \tag{12}$$

where α is a constant for the apparatus and T the temperature. Using iron oxide as adsorbent these authors showed that areas from 0.025 to 1 m^2 could be measured with precisions from 5 to 1.5%, respectively. This method still suffers from the disadvantage peculiar to small surface area measurements, that adsorbed amounts have to be estimated by subtracting the quantity of gas present at equilibrium from that initially present. Clarke [105], on the other hand, measured directly the amount of ^{85}Kr adsorbed on graphite by counting its γ radiation. The equilibrium Kr pressure was determined with a thermistor gauge. The activity of the sample was counted using a scintillation crystal and a photomultiplier placed in close proximity to the adsorbent. Small samples (0.5-1.0 g) having widely different surface areas (0.001-10.00 $m^2\ g^{-1}$) were successfully used in this study. The method was found to give the same results as the standard BET but could be used on samples of one-tenth the weight. Since the amount of Kr adsorbed is directly proportional to the count rate, the much simpler procedure and calculations meant that the time for a surface area measurement was reduced to about one-third.

Houtman and Medema [106] developed a rapid and accurate BET method (see Fig. 15) for measurement of a wide range (from 0.003 to 1000 $m^2\ g^{-1}$) of surface areas using two GM counters for

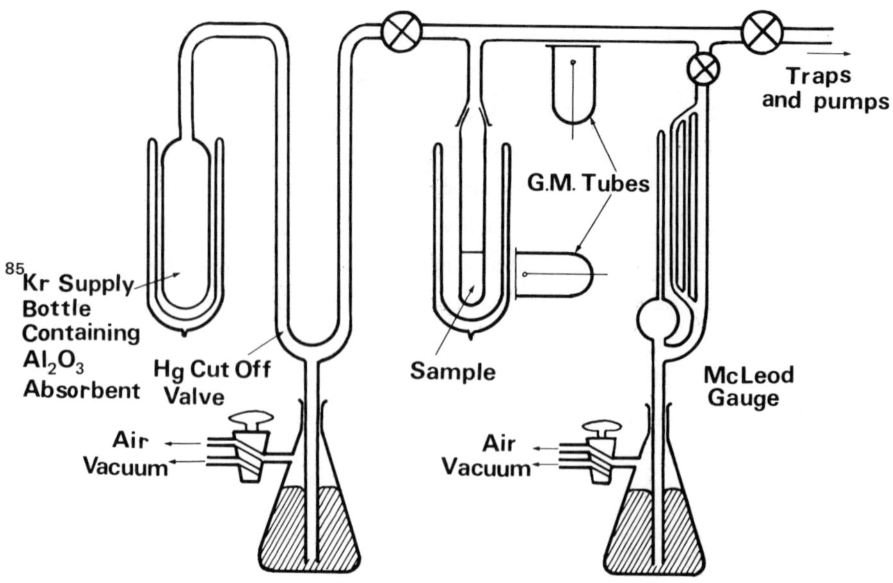

FIG. 15. Houtman and Medema's ^{85}Kr BET apparatus for determining surface areas of solids. (Reproduced from Ref. 106, by courtesy of Verlag Chemie.)

determining radioactivities of ^{85}Kr in adsorbed and gaseous phases. For each surface area determination the amount of gas required depends on the amount of sample, its specific surface area, and the volume of the apparatus. For 100 mg samples and an apparatus volume of 100 ml the amount of gas at NTP is 0.2 ml for a surface area of 1 $m^2\ g^{-1}$ and 8 ml for an area of 300 $m^2\ g^{-1}$. The weak β activity of ^{85}Kr means that the measured activities must be attributable mostly to the γ radiation from the adsorbed and gaseous phases. The upper GM counter (Fig. 15) measures the latter and gave about 3×10^5 counts $sec^{-1}\ bar^{-1}$ while the lower counter measures the former and gave about 80 counts $sec^{-1}\ m^{-2}$. The optimum values for the specific activity of the krypton were estimated for various surface areas and these are listed in Table 5. The small amount of krypton (1 mCi) needed for areas of 100 $m^2\ g^{-1}$ could give insufficient counts by the upper GM counter if γ rays only are detected. However, if a thin glass wall is used

TABLE 5

Necessary Krypton Activities for Surface Area Determinations [106]

Adsorbent surface area (m^2 g^{-1})	Radioactive concentration (mCi ml^{-1})	Radioactivities in Adsorbed phase (mCi)	Radioactivities in Gaseous phase (mCi)	Total activity required (mCi)
100	0.42	0.84	0.08	1
10	4.2	0.84	0.8	1.80
1	42.0	0.84	8.0	10.00
0.1	420.0*	0.84	80.0	90.00

*Maximum attainable specific activity if separated from Xe isotopes.

or the counter window forms part of the wall of the apparatus the β rays might also be counted. For example, an 0.4 mm-thick glass wall will absorb only 50% of the β particles so that this approach becomes a real possibility for the measurement of high surface areas.

The measurement of adsorption isotherms at low surface coverages is important for a proper understanding of the nature of the relatively small number of more active sites known to exist on solids of large specific areas such as carbon blacks. The lower pressures required ($<10^{-3}$ torr) preclude the use of McLeod gauges and one must resort either to ultrahigh vacuum systems containing ion gauges or to radioactive methods. Cochrane *et al.* [107] used ^{133}Xe as a tracer to study surface coverage on graphitized carbon blacks in the pressure range of 10^{-9} to 2 torr corresponding to surface coverages of 10^{-11} to 10^{-2}. The isotherms were extended to a surface coverage of 0.9 by the use of a conventional volumetric system. A unique feature of this study was the method of measuring the ^{133}Xe concentration either for calibration purposes or after adsorption equilibrium had been attained. Helium carrying the ^{133}Xe to be assayed was passed at a known rate through a coil of thin-walled aluminum tubing wound around a 5 cm × 5 cm NaI scintillator crystal and the 81 keV γ radiation counted. The detector was calibrated by supplying to the aluminum coil a calculated amount of ^{133}Xe released from a melted crystal of uranyl nitrate hexahydrate containing a known amount of the xenon isotope.

A number of important catalysts used today in petroleum processing contain small amounts of metal supported on microporous oxides of large specific areas. It is generally believed that catalytic activity is directly related to the effective area of the supported metal which is highly dispersed over the surface of the support as very small separate particles. In contrast to surface area measurements by the BET method, determination of this surface area depends upon the use of an adsorbate which is strongly chemisorbed by the catalytically active metal and adsorbed weakly or not at all by the oxide supports.

Hughes *et al.* [108] have developed a flow-adsorption method for determining metal surfaces of catalysts by measurement of carbon monoxide chemisorption. An important aspect of this approach is that it avoids the necessity of evacuating the catalyst chamber after prereduction of the catalyst. The ^{14}CO diluted with an inert carrier gas, helium, was passed over the reduced metal catalyst and the concentration of unadsorbed CO in the exit gas monitored in a counting chamber sealed to a GM tube. Disregarding a small correction for dead space, the volume V_{ads} (ml) of adsorbed gas is given by

$$V_{ads} = \frac{273.2}{760T} pfn_0\left(t - \frac{c}{c_0}\right) \tag{13}$$

where p is the pressure in torr, T the absolute temperature, f the flow rate in ml sec^{-1}, n_0 the mole fraction of CO in the inlet gas, t run time in sec, c the cumulative count at time t from the gas stream following adsorption, and c_0 the count rate for the incoming gas. The total CO concentration was, as usual, assumed to be proportional to the count rate. The ^{14}CO

concentration chosen gave 35-40 cps which was more than 100 times background but still sufficiently low to keep the coincidence error down to 0.3%. Standard deviations in the measurements for typical platinum-reforming catalysts was about 4%. The method was successfully applied to Pt, Ni, Rh, and reduced chromium oxide catalysts and was shown to be useful for studying dispersion and sintering of supported metals and for correlation of catalytic activity with available area of metal surface.

B. Surface Heterogeneity

That all parts of a catalyst surface are not equally effective in promoting a particular chemical reaction has been known for some time. In addition, heterogeneity has been held largely responsible for the well-known reduction in adsorption affinity with increasing surface coverage. Much of the effort expended in elucidating the origins of this variation in adsorption sites has made use of radiotracers and the studies have involved largely exchange and differential isotope methods.

An early example of the exchange method was due to Eischens [109] who examined the exchange process between inactive gaseous CO and ^{14}CO preadsorbed on iron powder. The thin mica end window of a GM tube formed part of the counting chamber which was in direct contact with the adsorption cell. Time-dependence of the activity in the gas phase thus showed the rate of exchange. The fact that exchange occurred in several stages and was temperature-dependent indicated a heterogeneity which was found to be due to differences between a relatively few homogeneous regions rather than to a wide spectrum of chemisorption bond strengths.

A simple form of the differential isotope method involves the exposure of a surface successively to two samples of gas, the first radiolabeled and the second unlabeled. Desorption should then indicate whether or not the adsorbent surface is uniform. If the desorbate contains no radioactive molecules then desorption will have occurred solely from the second sample and it may be concluded that the surface is heterogeneous. If, on the other hand, the whole of the adsorbate comes off with uniform activity it may be concluded either that the surface is homogeneous or that rapid exchange has occurred between active and inactive species at the surface. Kummer and Emmett [110] examined Fischer-Tropsch catalysts by the differential isotope method but the results were inconclusive, especially since the extent of exchange on the surface was unknown. Other catalysts have been similarly examined [111] and nickel, in particular, has been shown to be heterogeneous from successive adsorption batch studies with labeled and unlabeled acetylene [112].

The poisoning of catalyst surfaces by extremely small quantities of adsorbate is an indication not only of the heterogeneous nature of catalyst surfaces but also of the limited areas that participate in promoting the chemical reaction. For example, it is well established that traces of mercury vapor can poison nickel catalysts in certain hydrogenation reactions but not in others. For a better understanding of this problem as well as of the general one involving the adsorption and/or displacement of various species in the presence of one another, Campbell and Thomson [113] followed the displacement by radioactive mercury of $^{3}H_2$ preadsorbed on nickel films. The displaced tritium was counted in a gas counter and the adsorbed tracer, ^{203}Hg, simultaneously monitored with a gamma counter placed next to the adsorption vessel. The results showed that mercury readily displaces hydrogen at 20°C but not completely, about 16.5% of the original monolayer being retained. This indicates some type of heterogeneity in bonding sites for the hydrogen.

A further use of tracers to study the nature of catalyst poisoning was reported by Minachev and Isagulyants [114]. Thiophene, an impurity in cyclohexane, acts as a poison in the conversion of the latter to benzene on a platinum-alumina catalyst. The use of [^{35}S]thiophene led to the discovery that the yield of benzene falls off linearly with the amount of thiophene adsorbed on the catalyst.

Single crystal faces are known to exhibit different reactivities and this is yet another aspect of heterogeneity. The exposure of large-grained polycrystalline specimens of iron to ^{35}S from the vapor phase, for example, gave rise to greatly differing amounts of activity on each crystal face [115]. This variation was clearly indicated by autoradiography and showed adsorption on the principal faces to be in the order {111} > {110} >> {100}.

A more quantitative study of this adsorption anisotropy was carried out by Dillon and Farnsworth [116] who measured *in situ* the adsorption of $^{14}CO_2$ on the {110} and {100} faces of a nickel single crystal by incorporating in the adsorption chamber a thin end-window GM counter. This avoided errors usually associated with removal of the sample from the adsorption apparatus. The counter had a 2 mg cm^{-2} mica window sealed to soft glass and tests showed that CO_2 does not adsorb significantly on clean mica at room temperature and at pressures below 10^{-1} torr.

C. The Catalytic Process

In order to elucidate the nature of events taking place on the surface of a catalyst *during* the catalytic process Thomson and Wishlade [117] made a direct observation of radiolabeled reactants on the catalyst surface itself. This involved the development of an apparatus in which the catalyst film could be formed on a baked and degassed substrate and then moved into a position where the metal surface could be monitored by a GM counter capable of counting ^{14}C β particles. A simplified version of the apparatus is shown in Fig. 16. Once the nickel film had been condensed on the inside of the substrate tube the latter could be slid down to surround the GM counter by rotating the catalyst vessel. A thin-walled counter was produced by drilling a copper tube which acted as the cathode with many 3-mm holes. The tube was then covered with a sheet of Melinex (4.54 mg cm^{-2}) which rendered the tube vacuum-tight. Taking into account absorption of the β particles by the Melinex and the limited area of the perforations in the copper

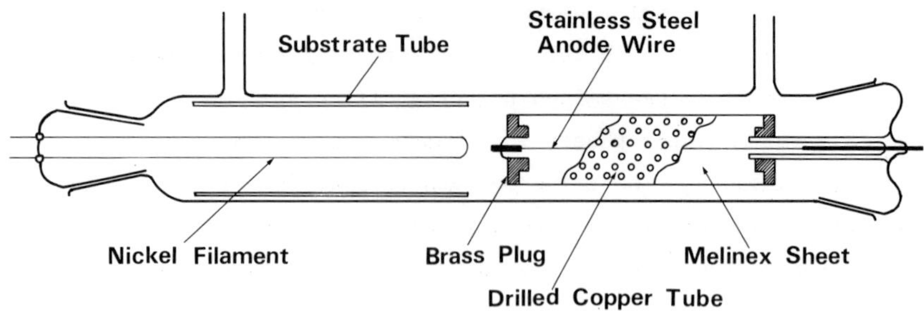

FIG. 16. Catalyst vessel showing catalyst substrate and cylindrical thin-walled GM counter. (Reproduced from Ref. 117, by courtesy of the Chemical Society, London.)

tube, a counter efficiency of 8% was established. In the experiment, hydrogen and unlabeled ethylene were admitted to the Ni films on which [^{14}C]ethylene had been preadsorbed. As hydrogenation proceeded any change in concentration of the chemisorbed molecules could be monitored directly by the thin-walled counter. Very little of the preadsorbed ethylene was removed during the reaction, the remainder being firmly held during extensive hydrogenation of the unlabeled ethylene. Thus negligible exchange between gas phase and chemisorbed ethylene occurred and it was concluded that only a fraction of the chemisorbed species participated in the hydrogenation. However, it was suggested that reaction might occur through hydrogen adsorbed on residual sites left on the surface.

The decomposition of formic acid catalyzed by various metals has provided a model system for extensive studies of the catalytic process. An important aspect of this work is a knowledge of the degree of coverage of the catalyst surface by the decomposing molecules. With powders or polycrystalline materials it was observed that under reaction conditions coverage of formic acid on nickel and copper was high but on the more noble metals such as silver it was low--only one formic acid molecule per 10 silver atoms. With the aim of learning more about the coverage for a noble metal such as silver and of investigating the influence of crystal orientation on adsorption Lawson [118] devised an apparatus for measuring the adsorption of [^{14}C]formic acid on poly- and monocrystalline silver films at low coverage (10^{-13} mol cm^{-2}) on surfaces of a few square centimeters. The apparatus shown in Fig. 17 permitted *in situ* preparation of silver films on mica substrates as well as counting of the activity of the

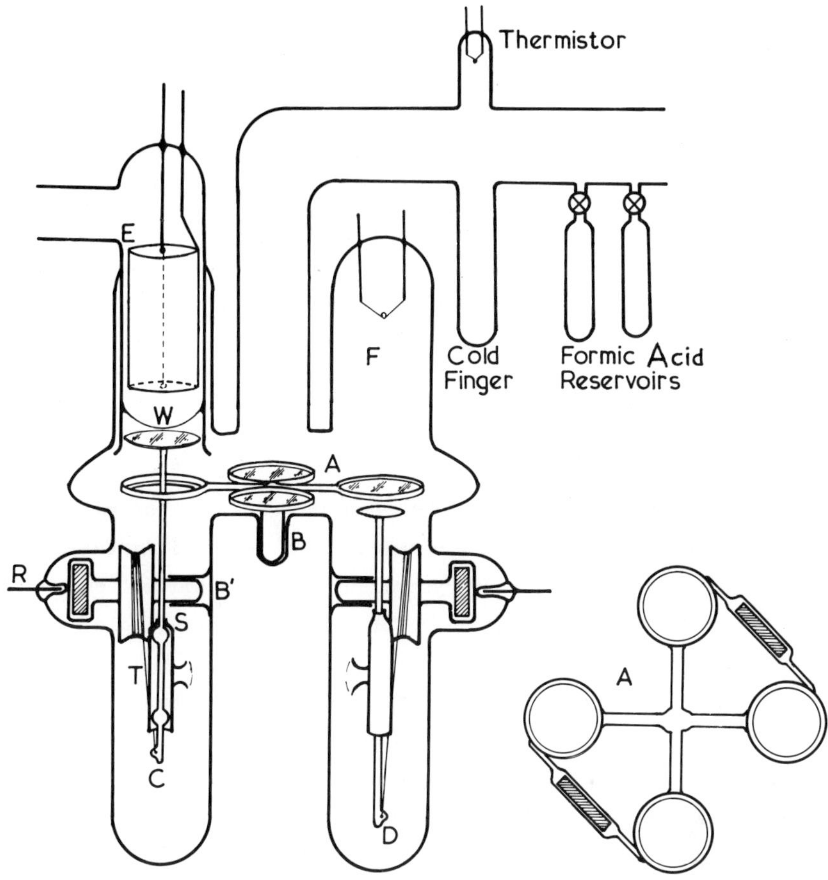

FIG. 17. Lawson's adsorption vessel. A, rotatable tray carrying four specimen holders; B, B', bearings; C, D, pistons and magnetically operated winches; E, GM counter; F, evaporation source; R, tungsten rod pivot; S, stop; T, piston guide; W, GM counter window. (Reproduced from Ref. 118, by courtesy of Academic Press.)

adsorbed [^{14}C]formic acid. The rotatable tray A carrying four specimen holders together with two magnetically operated pistons C and D enabled the specimens to be raised up either into tube F containing the source for evaporating silver or into tube E containing the counter. The counter window W was made from thin Pyrex glass about 10 μm thick by the following procedure. One end of a 27-mm internal diameter tube was closed to form a hemisphere which was heated to melting point and the molten glass sucked into the tube over a length of about 3 cm until the thin bubble came into contact with a shaped carbon cylinder which molded it into a curved surface. The carbon cylinder was removed and the tube containing the window W joined onto the tube E. The total count rate R_t contained contributions from the adsorbent R_{sub}, the counter window and walls of the adsorption vessel R_w, and the gas phase R_g; thus

$$R_t = R_{sub} + R_w + R_g \tag{14}$$

By making a series of corrections the components R_w and R_g, both of which are pressure-dependent, were allowed for and true adsorption isotherms obtained. The results were in close agreement with earlier work on noble metals, the formic acid/metal atom ratios being 1:9 for polycrystalline films of silver (one-third {111} faces exposed) to 1:17 for fully epitaxed single crystals on mica (100% {111}).

The reaction of xenon with fluorine to form xenon difluoride is known to occur readily on heated palladium and nickel. Earlier experiments indicated that under reaction conditions large concentrations of chemically bound xenon were expected to be present on the catalyst surface. Using a thin-window GM counter similar to that used earlier by Lawson [118], Baker and Lawson [119] examined the uptake of ^{133}Xe on nickel and palladium foils in the presence of fluorine. Again, gas-phase contribution to the total count rate was carefully estimated and the corrected isotherms indicated that at 20°C xenon did not adsorb on nickel unless fluorine was present. For both substrates it was concluded that xenon is chemically bound to fluorine in the adsorbed state and that this constitutes the intermediate in the catalytic formation of xenon difluoride.

A further application of tracers to fundamental problems of the catalytic process has been the extensive studies on the Fischer-Tropsch synthesis of hydrocarbons from carbon monoxide and hydrogen over metal catalysts. The nature of the process has been in doubt from the very beginning. One theory postulated a surface metal carbide as an intermediate while another suggested that carbon monoxide and hydrogen formed initially an oxygen complex which acted as a nucleus for a chain-like building up of higher hydrocarbons as successive monoxide molecules were added. Emmett using ^{14}C-labeled reactants disproved the first theory and found the second more tenable [120]. He concluded that primary and secondary alcohols combined were most probably involved as principal intermediates in the formation of the higher hydrocarbons. In later work with ketene ($^{14}CH_2=CO$) as an additive to the Fischer-Tropsch synthesis gas Blyholder and Emmett [121] indicated that the C_2 intermediate complex probably resembles adsorbed ketene.

Surfaces involved in catalytic reactions are known to be heterogeneous and hence it is reasonable to assume that not all adsorbed complexes will be equally active in catalytic processes. In an attempt to measure the active fraction, Cormack, Thomson, and Webb [122] used [^{14}C]ethylene in studies of the adsorption and hydrogenation of ethylene at 20°C on 5% alumina-supported Ni, Rh, Pd, Ir, and Pt catalysts. The catalyst holder could be moved directly beneath a GM counter within the adsorption chamber to measure the activity of the adsorbate and then moved away so that the gas phase contribution could be obtained. The surface count rate was obtained by difference. Equimolecular mixtures of inactive ethylene and hydrogen were interacted with [^{14}C]ethylene preadsorbed on the catalysts. For all catalysts it was found that removal of the preadsorbed ethylene during the hydrogenation reaction paralleled the loss by simple evacuation. Retention percentages were Pd, 63.5; Ni, 24; Rh, 22.5; Ir, 16; and Pt, 6.5. The results indicated two modes of ethylene adsorption: one involved in hydrogenation and molecular exchange and considered to be an associatively bonded complex, and the other a dissociative complex which was responsible for the observed ^{14}C retentions. It was proposed further that the reactive adsorbate was olefinic in nature.

In catalytic reactions promoted by supported metal catalysts the respective contributions of metal and support are not fully understood. Some studies have indicated a surface or gas-phase migration of hydrocarbon species from the metal to the support. Thus, it is suggested that with bifunctional catalysts hydrogenation-dehydrogenation takes place on the metal and rearrangement or isomerization of the intermediate occurs on the acidic sites of the support. The use of [^{14}C]ethylene and ^{14}CO in a specially developed adsorption vessel enabled Reid *et al.* [123] to examine the possibility of hydrocarbon migration over the catalyst surface. The specimen could be moved by means of a magnet from one part of the vessel, where it could be heat-treated in hydrogen and then in vacuum, to the other end so as to be positioned beneath one of two adjacent GM tubes. The use of the second tube permitted simultaneous monitoring of the gas-phase contribution to the total radioactivity. This contribution was found to be nearly proportional to the gas pressure over the range used in the work. Comparison of the separate coverages by ethylene and carbon monoxide as well as suppression of ethylene adsorption by preadsorbed carbon monoxide led to the conclusion that ethylene is adsorbed in two modes--a primary and, more extensively, a secondary. The former was attributed to adsorption on the metal itself and the latter to adsorption on the support following migration from the metal.

The influence of crystal defects on the catalytic behavior of solids has been studied widely over the past few years. Although for certain systems the catalytic activity is shown to depend on the nature and concentration of the defects [148], Jaeger [149] has shown for the catalytic decomposition of formic acid by silver crystals that the activity does *not* depend on concentration of crystal defects but is influenced by surface orientation. The application of radiotracers to this field is reported recently by Yasumori, Kabe, and Inoue [150] who examined the catalytically active sites on deformed and annealed palladium foils. [^{14}C]Acetylene was preadsorbed on the foil and the surface density of effective sites was deduced from the amount desorbed during hydrogenation and the specific activity per site evaluated. In an apparatus similar to that used by Thompson and coworkers [122] the count rate of radioactive species on the surface was estimated from the difference between the total and the gas phase count rates. The reaction sites on the deformed catalysts had specific activities of the order of 10 times greater than those on catalysts annealed above 200°C. A close correlation between these more active sites and surface vacancies was established.

VII. SOLID/SOLID INTERFACE

A. Friction and Wear

The efficient operation of sliding mechanisms and machines in general will depend sensitively on physical quantities such as the coefficient of friction between the sliding surfaces and the surface damage or metallic transfer. Because of its high sensitivity and resolution the radiotracer technique has proved to be invaluable in the rapid estimation of the extremely small quantities of material transferred or lost as a result of frictional and wear processes. Above all, it has been possible to measure the transfer between chemically identical surfaces sliding against one another.

As early as 1944, Sakmann, Burwell, and Irvine [124] showed in experiments on the transfer from a radioactive Be-Cu surface to an inactive hemispherical slider that the minimum detectable transfer (or pickup) was of the order of 10^{-10} g. In this work the use of GM counters meant that only the *total* amount of activity transferred was measured, whereas the use of autoradiography enabled Gregory [125] to indicate the detailed distribution (and particulate nature) of the transferred material when a radioactive lead (ThB) slider was moved over flat surfaces of steel, copper, and lead.

The use of autoradiography in this field was developed further by Rabinowicz and Tabor [126] who examined the relative importance of the coefficient of friction and metallic transfer (or surface damage) when metals slide on one another. Hemispherical sliders of Cu, Cd, Zn, Pt, Ag, and various steels made radioactive by irradiation in the BEPO Nuclear Reactor at Harwell, U.K., were pressed down on to a flat metal plate which was moved at a slow uniform speed, the sliders responding to the frictional force at the interface. The numbers and distribution of metallic particles of the slider transferred to the plate were determined by autoradiography. Calibration of the autoradiographs involved the use of step wedges, consisting of different thicknesses of the metal used in the sliding experiment plated onto aluminum (which is only weakly activated in the reactor) and irradiated together with the friction specimen. From the photographic darkening produced by this wedge, exposed on a similar photographic film under standard conditions, the masses of radioactive fragments transferred (and hence the wear) could be estimated. Rabinowicz and Tabor concluded that for both clean and lubricated conditions metallic transfer is immensely more sensitive to changes in surface conditions than is the coefficient of friction. The results also supported the view [127] that friction between metals is due principally to the formation and shearing of *metallic junctions* and that the main function of a lubricant is to reduce the amount of metallic interaction at these junctions rather than to reduce their number.

In recent studies on the relationship between material transfer and wear Talke [128] used autoradiography to measure the transfer from a radioactive ferrite pin sliding at high speed and low load on a steel disk, a nickel-plated disk, and a disk coated with a polymer containing finely dispersed abrasive particles. Following neutron activation of the ferrite probe and a brief cooling down period, the activity (x, γ, and β rays) was due to ^{55}Fe, ^{59}Fe. ^{63}Ni, ^{65}Ni, and ^{65}Zn. Calibration of the autoradiographic method was carried out as follows. A radioactive ferrite probe was dissolved in hot acid and following neutralization and subsequent slight acidification (to keep the iron chloride in solution) a standard solution was prepared. From this a range of concentrations of the radioactive material was obtained by dilution and aliquots were distributed evenly over areas one square inch each. When these had dried autoradiographs were obtained of the various deposits. To allow for nonuniform optical density across individual squares a microdensitometer was used in such a way (see Fig. 18a) as to perform an optical integration over each square to give the desired relationship between calibrated optical density and the amount of radioactive material per unit area (see Fig. 18b). This relationship was then used to obtain a quantitative estimate of the amount of radioactive material contained in a wear track. The close similarity in the results for all three disks indicated that the observed large amounts of transfer are characteristic not only of adhesive wear (Fig. 19a) but also of abrasive wear caused by hard particles dispersed in a soft matrix (Fig. 19b).

Using radiotracers Golden and Rowe [129] studied the influence of the relative hardness of the two sliding surfaces on the wear mechanism. Examination of the wear of cobalt-bonded tungsten carbide sliding against soft metals such as duralumin and copper showed that not only was the soft metal transferred to the hard slider, but significant quantities of the latter were

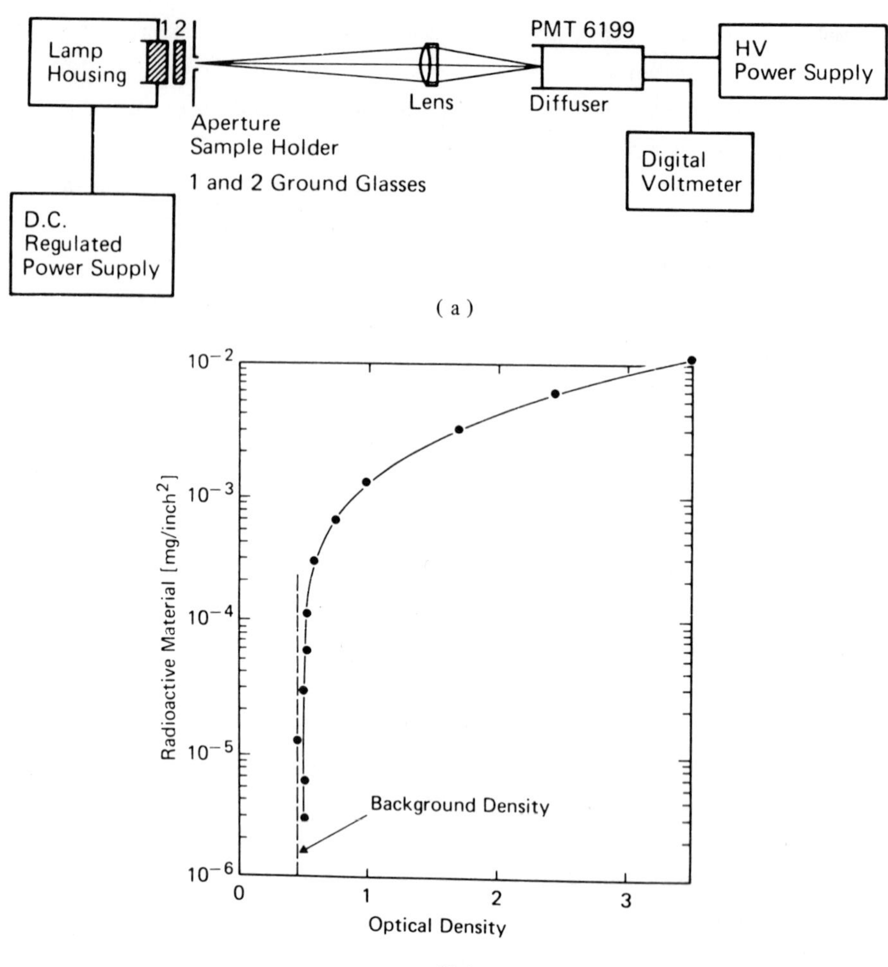

FIG. 18. Calibration of autoradiographic technique in wear studies. (a), Schematic diagram of optical integration set-up. (b), Calibration curve: radioactive material per unit area as a function of optical density. (Reproduced from Ref. 128, by courtesy of Elsevier Sequoia S.A.)

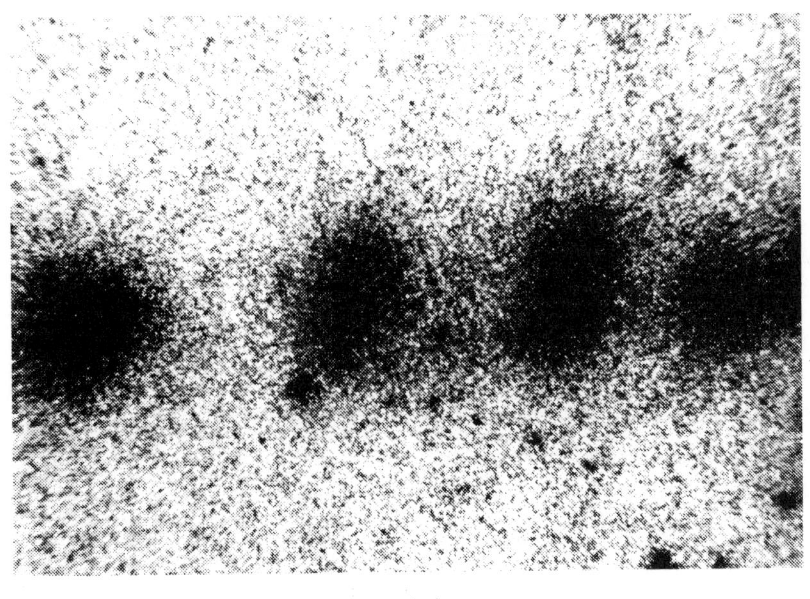

(a)

(b)

FIG. 19. Photomicrographs of autoradiograph along the wear track of a radioactivated ferrite pin sliding on (a) a steel disk and showing the discrete nature of adhesive wear and (b) an abrasive disk showing the uniform distribution of abrasive wear. Magnification 100x. (Reproduced from Ref. 128 by courtesy of Elsevier Sequoia S.A.)

lost to the much softer metal plates. An important observation in this work was that, within a week of activation of the carbide, ^{187}W was the dominant activity but that after about 1 year the activity was primarily ^{60}Co. Thus one could, in principle, carry out experiments at different times following activation in order to assess the different roles of the carbide and the cobalt matrix in the wear process.

The problems associated with the use of highly active metal sliders in boundary friction machines mentioned above have been discussed by Campbell and Harriden [130] who described modifications and special techniques which would facilitate their use. If a transfer in the region of 10^{-10} g is to be measured the radioactivity of a 1 g slider irradiated in an atomic reactor would need to be of the order of 100 mCi. The authors showed that by the use of special jigs and lead shielding a copper slider having the above activity could be safely and easily used on a typical boundary friction machine.

Rabinowicz [131] showed earlier that radioactivation analysis could eliminate completely the use of radioactive specimens on friction machines. In the example used, the author moved a copper surface over a steel one and then activated both in an atomic reactor. An hour after the steel plate had been irradiated, the activities of the iron radioisotopes produced were such that an exposure of 10 sec to the autoradiographic plate caused only a light fogging, whereas adhering copper fragments showed up as intensely black spots. For success this method requires a large difference in the specific activities of the two metals and in this case the difference was >4000 (Cu: 50 mCi/g; Fe: 11 μCi/g). Because of the great difference in the half-lives ($t_{1/2}$) between the radioisotopes produced from copper and iron, the transfer of steel to copper could also be monitored. By allowing a decay of several weeks following activation of the copper plate, by which time the activity of the copper isotope ^{64}Cu ($t_{1/2}$ = 12.8h) had practically disappeared, the transferred steel fragments still showed up strongly on the autoradiograph, $t_{1/2}$ for ^{55}Fe and ^{59}Fe being 2.7y and 49d respectively. Here, an exposure of 3 days was necessary, but it should be emphasized that for this type of experiment to be successful impurities in the copper that are strongly activated in the reactor must be eliminated in order to avoid a high background.

More recently Bush *et al.* [132] showed that radioactivation analysis can be successfully applied to measurement of the wear of automobile braking materials. Resin-asbestos pads were pressed against revolving cast-iron disks and the wear dust in milligram quantities collected and subjected to neutron activation analysis. The quantity of wear debris required for analysis need be only one-hundredth that required for analysis by the usual method of measurement of reduction in thickness of the pad. In addition, the activation analysis is not affected by changes in the moisture content which would reduce greatly the accuracy of methods involving direct weighing of the pad.

A novel method of studying wear was recently proposed by Zemskova *et al.* [133]. The scintillation properties of mineral oils form the basis of the method. Activated by additives and with a ^{60}Co gamma source providing ionizing radiation the mineral oil is placed in a 4-ball friction machine. Contaminants such as metals are effective as suppressors and so wear debris in the form of a fine powder will disperse and absorb the glow of scintillation. Thus we have a continuous and direct measure of the amount of wear in terms of the loss in light intensity. Excellent correlation was found between this estimate of the amount of wear and the diameter of the wear spot on the lower balls.

B. Metal Working and Transfer

In metal working operations where the work piece is brought into contact with a cutting, grinding, or deforming tool some loss of material from the latter is inevitable, and radiotracers have been widely used in estimating wear of the cutting tools and dies. A standard test procedure first introduced by Marchant *et al.* [134], and now used in many laboratories, has been the use of radioactive tools with measurement of the loss of activity as a rapid assessment of tool life. At least 90% of the tool wear adheres to the metal chips removed from the work piece so that their activity is a direct measure of tool wear. Following neutron irradiation, cutting tools give materials of high specific activity--isotopes such as ^{60}Co, ^{182}Ta, and ^{187}W for the carbides and ^{51}Cr, ^{60}Co, ^{185}W, and ^{59}Fe for the high-speed steels. In a similar manner the active debris on drawn wires can also give an indication of the wear of a radioactivated wire-drawing die [135,136].

Of considerable importance in the field of orthopedic surgery is the transfer of metal from the screwdrivers and spanners to screws and nuts attached to splints etc., that must remain for considerable periods of time within the human body. Such "foreign" materials might interact chemically with the surrounding tissue and seriously interfere with healing processes. This led Bowden and coworkers [137] to examine such transfer to vitallium products by using normal tool steel made radioactive by neutron irradiation--only the activity acquired by tungsten being significant. Quantitative estimates of total transfer were made by Geiger or scintillation counting of γ rays, self-shielding thus being negligible. The distribution of the transferred material over the surface of the vitallium was checked by autoradiography in order to learn more about the actual mechanism of transfer.

C. Lubrication

Monomolecular films of amphipathic molecules have been used extensively in fundamental studies of lubrication. The application of radiotracer techniques in their preparation and properties has already been discussed in Sections IV and V and will not be considered here.

Present-day technology has increased the demands on lubricants to the point where they must perform satisfactorily under extreme conditions of pressure, temperature, and atmosphere (or lack of it). Solid lubricants can meet some of these demands and the two most widely used are graphite and molybdenum disulfide. Unfortunately, graphite cannot be made radioactive by neutron irradiation so that other materials have to be incorporated into it. One such material is silver and transfer of graphite has been studied by following the movement of ^{110}Ag [138]. Even impurities normally present in graphite could be used for the same purpose. Molybdenum disulfide, on the other hand, is readily activated by neutron irradiation giving ^{99}Mo and, to a lesser degree, ^{35}S. Thus not only can movement of the lubricant be traced but also the stability of the material by following loss of ^{35}S. However, there are limitations to the use of radiolabeled MoS_2 in lubrication studies and some of these are discussed by Lancaster [139].

D. Oxidation

A full understanding of the processes involved in the growth of oxide films on metals has depended critically on a knowledge of the nature of the diffusing entities. One successful approach has involved a study of the behavior of the metal/metal oxide interface by depositing a thin layer of a radioactive material on the metal surface as a marker and observing its position as oxidation proceeds. An early example of this technique is given by Davies, Simnad, and Birchendall [140] who used ^{110}Ag as a marker in studies of the oxidation of iron. During the formation of wüstite the silver marker layer remained precisely at the metal/oxide interface which indicated that this oxide grows almost exclusively by diffusion of iron ions. In the same way it was concluded that diffusion of both iron and oxygen ions contributes to the building of magnetite layers, while hematite is built up solely by diffusion of oxygen ions.

Despite the success of this and later work using the marker technique any observation of the movement and final position of the oxygen itself within the oxide layer by radiotracer techniques has been hampered by the lack of sufficiently long-lived oxygen radioisotopes. Nevertheless, nuclear techniques have been developed which do permit the determination of ^{18}O concentrations at various depths within an oxide layer [141]. Thus Brückman *et al.* [142] oxidized an Fe-3.5% Cr specimen *sequentially* in ^{16}O and ^{18}O at 1000°C and determined the resulting ^{18}O distributions in the approximately 1 mm-thick scale which consisted of an outer and an inner layer. Negligible oxygen transport should give a sharp boundary between ^{16}O and ^{18}O distributions, whereas if bulk oxygen diffusion is appreciable there will be uniform distribution of both isotopes. The ^{18}O distribution was measured using a (neutron) counting and sectioning technique involving the $^{18}O(\alpha,n)^{21}Ne$ reaction by α particles from a ^{210}Po source. In addition an autoradiographic method was used in which the ^{18}O was revealed by irradiation with protons to give $^{18}O(p,n)^{18}F$, the product ^{18}F being an 0.64 MeV positron-emitting radioisotope with $t_{1/2}$ = 110 min. These authors established that the outer wüstite layer grew by cation transport and the inner layer by movement of the oxygen through microcracks. Barnes and coworkers [143] used a similar approach to study the oxidation of a series of Fe-Cr alloys. The ^{18}O distribution in the scale was determined using the yield of α particles from a resonance in the reaction $^{18}O(p,\alpha)^{15}N$, following irradiation of the scale with a beam of protons from a van de Graaff generator. These authors varied the depth within the target at which resonance occurred by varying the incident proton energy. Thus the yield of α particles from the resonance as a function of proton energy gives a measure of the ^{18}O concentration profile within the surface. Accurate quantitative concentration profiles could then be computed for depths up to 3 μm. Again, it was found that bulk oxygen diffusion plays a negligible role in the oxidation reaction, any apparent diffusion being due to movement of the oxygen through cracks in the scale.

The above resonance technique is, however, restricted to cases where there is a strong narrow resonance in a suitable nuclear reaction and, in addition, demands prolonged times for the recording and analyzing of a separate spectrum for each increment in the incident particle energy. These disadvantages have been overcome by Wise *et al.* [151] by utilizing nonresonant nuclear reactions in which the actual shape of the energy spectrum of the emitted particles is used rather than the areas of successive resonance peaks. Of more general application, this technique can be applied to any reaction that has a reasonable crosssection over the useful energy range. In addition, when using the $^{18}O(p,\alpha)^{15}N$ reaction both accelerator and computing times can be reduced by a factor of three. The theoretically expected charged-particle/ energy spectrum resulting from bombardment of the specimen is calculated and compared with the experimentally observed spectrum to obtain the concentration profile. For a thin $W^{18}O$ target a depth range of analysis of ∿6 μm was achieved. Successful results were also obtained with a uniform $Cr_2{}^{18}O_3$ target, the assumed oxide thickness of 3.5 μm for best fit α-particle spectrum agreeing well with the thickness calculated from a weight gain measurement. This technique offers great promise for the estimation not only of oxidation rates but also of oxygen diffusion coefficients [152] and concentration profiles of atomic species in thin surface films [151].

E. Diffusion

The use of radiotracers in diffusion studies is treated fully in Chapter 12. In this section we will consider only a small fraction of those studies related to the interfacial processes dealt with in the present chapter.

In studies on the diffusion of one solid into another, the technique usually applied is to deposit either electrolytically or by vapor condensation onto a specimen surface a thin (∿0.1 μm) layer of radiolabeled metal. A second specimen is then placed on this active surface, the whole clamped together and subjected to the required thermal treatment. This sandwich arrangement minimizes loss of the active material by evaporation. Penetration of the radioactive element into the inactive metal can be determined by various methods. Most frequently the activities of layers on either side of the original interface are measured by etching off successive layers in a suitable acid and recovering the dissolved metal by electrolysis or precipitation. For certain metals it is often possible to remove successive layers of the specimen by means of a microtome or a lathe. The autoradiographic technique has also been used to examine diffusion (see, for example, Ref. 144). Here, the sandwich-type specimen is sectioned at a very small angle to the plane of separation of the two metals and the sections autoradiographed. The variation in photographic blackening along a line crossing the original interface gives a measure of the concentration profile of the radioactive element. It should be pointed out, however, that in certain types of bulk diffusion experiments the transport of material by surface diffusion could lead to erroneous conclusions unless such a contribution was carefully taken into account.

In studies on surface diffusion, where single crystal substrates are involved the diffusion itself might be anisotropic as has been reported by Hackerman and Simpson [145] who examined the surface self-diffusion of copper on copper single crystal faces. Not only did the diffusion coefficients differ measurably on the {100}, {110}, and {111} faces but also along different axes of the {100} face. Problems associated with identification of the precise mechanisms by which material is transported in "surface" diffusion experiments have been discussed by Choi and Shewman [146] in their work on surface diffusion of gold (^{198}Au) and copper (^{64}Cu) on {100} and {111} surfaces of copper. Attempts to measure surface diffusion coefficients of silver (^{110}Ag) on copper and silver surfaces were frustrated because of indications that vapor phase transport was a dominant factor.

This section on solid/solid interactions has been concerned almost exclusively with metals and mention should be made of the excellent treatment by Houseman [147] of the application of radiotracers to this field of study. Although Houseman's report appeared a number of years ago it contains much useful information about the properties of relevant isotopes as well as the design of experiments.

ACKNOWLEDGMENT

The authors wish to thank Dr. T. W. Healy for valuable comments. One of us (DEY) acknowledges receipt of a Commonwealth Postgraduate Research Award, during the tenure of which this chapter was prepared.

REFERENCES

1. M. Muramatsu, in Surface and Colloid Science, Vol. 6 (E. Matijevic, ed.), Wiley-Interscience, New York, 1973, p. 101.
2. W. Rabinovitch, R. F. Robertson, and S. G. Mason, J. Colloid Sci., 13:600 (1958).
3. M. Muramatsu, in Surface and Colloid Techniques (R. J. Good and R. R. Stromberg, eds.), Plenum Press, New York, to be published.
4. E. Hutchinson, J. Colloid Sci., 4:599 (1949).
5. J. S. Clunie, J. F. Goodman, and B. T. Ingram, in Surface and Colloid Science, Vol. 3 (E. Matijevic, ed.), Wiley-Interscience, New York, 1971, p. 171.
6. J. M. Corkill, J. F. Goodman, D. R. Haisman, and S. P. Harrold, Trans. Faraday Soc., 57:821 (1961).
7. D. J. Salley, A. J. Weith, Jr., A. A. Argyle, and J. K. Dixon, Proc. Roy. Soc. (Lond.), A203:42 (1950).
8. G. Aniansson, J. Phys. Colloid Chem., 55:1286 (1951).
9. G. Nilsson, J. Phys. Chem., 61:1135 (1957).

10. M. Avrahami and N. H. Steiger, J. Colloid Sci., 18:863 (1963).

11. J. T. Davies and E. K. Rideal, Interfacial Phenomena, Academic Press, New York, 1961, p. 207.

12. M. Muramatsu, K. Tajima, and T. Sasaki, Bull. Chem. Soc. Japan, 41:1279 (1968).

13. M. Muramatsu, N. Tokunaga, and A. Koyano, Nuclear Instr. Methods, 52:148 (1967).

14. M. Muramatsu, A. Shigematsu, and N. Tokunaga, Nuclear Instr. Methods, 55:249 (1967).

15. M. Muramatsu, K. Tajima, M. Iwahashi, and K. Nukina, J. Colloid Interface Sci., 43:499 (1973).

16. G. Aniansson and N. H. Steiger, J. Chem. Phys., 21:1299 (1953).

17. N. H. Steiger and G. Aniansson, J. Phys. Chem., 58:228 (1954).

18. G. T. Barnes, A. J. Elliott, and E. C. M. Grigg, J. Colloid Interface Sci., 26:230 (1968).

19. M. Muramatsu, O. Hirata, and M. Iwahashi, J. Colloid Interface Sci., 39:424 (1972).

20. M. Muramatsu and T. Ohno, J. Colloid Interface Sci., 35:469 (1971).

21. P. A. Good and R. S. Schechter, J. Colloid Interface Sci., 40:99 (1972).

22. H. D. Cook and H. E. Ries, Jr., J. Phys. Chem., 60:1533 (1956).

23. E. R. Moss, D. M. Himmelblau, R. S. Schechter, and R. L. Pitzer, Nature, 217:349 (1968).

24. E. K. Sakata and J. C. Berg, IEC Fundamentals, 8:570 (1969).

25. M. Iwahashi, K. Aruga, O. Hirata, T. Horiuchi, and M. Muramatsu, J. Colloid Interface Sci., 42:349 (1973).

26. M. Iwahashi, J. Colloid Interface Sci., 50:572 (1975).

27. R. F. Robertson, C. A. Winkler, and S. G. Mason, Can. J. Chem., 34:716 (1956).

28. A. D. Bangham and R. M. C. Dawson, Biochem. J., 75:133 (1960).

29. J. E. Charbonneau and G. N. Kowkabany, J. Colloid Interface Sci., 33:183 (1970).

30. S. G. Mason and W. Rabinovitch, Proc. Roy. Soc. (Lond.), A249:90 (1959).

31. T. Sasaki and M. Muramatsu, Bull. Chem. Soc. Japan, 29:35 (1956).

32. A. Matsubara, R. Matuura, and H. Kimizuka, Bull. Chem. Soc. Japan, 38:369 (1965).

33. E. Rojas and J. M. Tobias, Biochem. Biophys. Acta, 94:394 (1965).

34. H. Kimizuka, T. Nakahara, H. Uejo, and A. Yamauchi, Biochem. Biophys. Acta, 137:549 (1967).

35. N. L. Gershfeld and M. Muramatsu, J. Gen. Physiol., 58:650 (1971).

36. A. Matsubara, Bull. Chem. Soc. Japan, 38:1254 (1965).

37. For a review of this topic see V. V. Losev, in Modern Aspects of Electrochemistry, No. 7 (B. E. Conway and J. O'M Bockris, eds.), Plenum Press, New York, 1972, Chap. 5, p. 336.

38. M. Kukura, L. C. Bell, A. M. Posner, and J. P. Quirk, J. Phys. Chem., 76:900 (1972).

39. H. R. Luckens, Jr., R. G. Meisenheimer, and J. N. Wilson, J. Phys. Chem., 66:469 (1962).

40. Y. G. Bérubé, G. Y. Onoda, Jr., and P. L. de Bruyn, Surface Sci., 7:448 (1967). D. E. Yates and T. W. Healy, J. Colloid Interface Sci., 55:9 (1976).

41. For a review of radiotracers and metal electrodes see N. A. Balashova and V. E. Kazarinov, in Electroanalytical Chemistry, Vol. 3 (A. J. Bard, ed.), Marcel Dekker, New York, 1969, p. 135.

42. N. A. Balashova and N. S. Merkulova, in New Methods of Physico-Chemical Investigations, Trans. Inst. Physical Chem. No. VI, p. 12, 1957, Acad. Sci. USSR (in Russian).

43. K. E. Heusler and G. H. Cartledge, J. Electrochem. Soc., 108:732 (1961).

44. M. Muramatsu and T. Sasaki, Isotopes and Radiation, 5:451 (1959).

45. M. H. Gottlieb, J. Phys. Chem., 64:427 (1960).

46. W. P. Doyle and A. H. Ellison, Adv. Chem. Ser., 43:268 (1963).

47. C. V. King and B. Levy, J. Phys. Chem., 59:910 (1955). H. A. Smith and R. M. McGill, J. Phys. Chem., 61:1025 (1957). A. Block and B. B. Simms, J. Colloid Interface Sci., 25:514 (1967).

48. N. Hackerman and S. J. Stephens, J. Phys. Chem., 58:904 (1954).

49. J. Vance Batty, W. W. Agey, and B. F. Andrews, U. S. Bureau of Mines, Report of Investigations 7094, U. S. Department of

the Interior, Bureau of Mines, Salt Lake City, Utah, 1968.

50. D. S. Newman, J. McCarthy, and M. Heckaman, J. Electrochem. Soc., 118:541 (1971).

51. N. A. Balashova and V. E. Kazarinov, Russian Chem. Rev. (Uspekhi Khim.), 34:730 (1965).

52. I. R. Jonasson and D. R. Stranks, Electrochem. Acta, 13:1147 (1968).

53. M. Green, D. A. J. Swinkels, and J. O'M. Bockris, Rev. Sci. Instr., 33:18 (1962).

54. J. O'M. Bockris and D. A. J. Swinkels, J. Electrochem. Soc., 111:736 (1964); J. O'M. Bockris, M. Green, and D. A. J. Swinkels, J. Electrochem. Soc., 111:743 (1964).

55. A. G. Langdon, K. W. Perrott, and A. T. Wilson, J. Colloid Interface Sci., 44:486 (1973).

56. C. V. King and N. E. McKinney, Can. J. Chem., 37:205 (1959).

57. F. Joliot, J. Chim. Phys., 27:119 (1930).

58a. For references to this work see B. B. Damaskin, O. A. Petrii, and V. V. Batrokov, Adsorption of Organic Compounds on Electrodes, Plenum Press, New York, 1971, p. 211.

58b. J. Sobkowski and A. Wieckowski, J. Electroanal. Chem., 34:185 (1972); A. Wieckowski, J. Sobkowski, and A. Jablonska, ibid., 55:383 (1974); A. Wieckowski, ibid., 122:252 (1975).

59. S. J. Rehfeld, J. Colloid Interface Sci., 31:46 (1969).

60. T. Seimiya, S. Saito, and T. Sasaki, J. Colloid Interface Sci., 30:153 (1969).

61. E. L. Mark, R. A. Porter, and R. N. Chanda, J. Colloid Interface Sci., 35:133 (1971).

62. H. Dahms and M. Green, J. Electrochem. Soc., 110:1075 (1963).

63. D. C. Walker and H. E. Ries, Jr., J. Colloid Sci., 17:789 (1962).

64. E. A. Blomgren and J. O'M. Bockris, Nature, 186:305 (1960).

65. H. Wroblowa and M. Green, Electrochim. Acta, 8:679 (1963).

66. E. Gileadi, L. Duic, and J. O'M. Bockris, Electrochim. Acta, 13:1915 (1968).

67. J. A. Kafalas and H. C. Gatos, Rev. Sci. Instr., 29:47 (1958).

68. K. Schwabe and W. Schwenke, Electrochim. Acta, 9:1003 (1964).

69. V. E. Kazarinov, Elektrokhimiya, 2:1170 (1966).

70. G. Horanyi, J. Solt, and F. Nagy, J. Electroanal. Chem., 31:87 (1971).

71. P. Möller, Radiochimica Acta, 18:144 (1972).

72. G. Horanyi and E. M. Rizmayer, J. Electroanal. Chem., 36:496 (1972); G. Horanyi, J. Solt, and G. Vertes, ibid., 32:271 (1971); G. Horanyi and F. Nagy, ibid., 32:275 (1971); G. Horanyi, ibid., 36:247 (1972).

73. K. Schwabe, Chem. Tech., (Berlin), 13:275 (1961).

74. A. S. Joy and A. J. Robinson, Recent Prog. Surface Sci., 2: 198 (1964).

75. A. W. Fordham, Clays and Clay Minerals, 21:175 (1973).

76. K. B. Blodgett, J. Am. Chem. Soc., 57:1007 (1935).

77. K. B. Blodgett and I. Langmuir, Phys. Rev., 51:964 (1937); K. B. Blodgett, J. Phys. Chem., 41:975 (1937).

78. B. L. Henke, Adv. X-ray Anal., 7:460 (1964); R. C. Ehlert and R. A. Mattson, ibid., 10:389 (1967); M. W. Charles, J. Appl. Phys., 42:3329 (1971).

79. R. M. Handy and L. C. Scala, J. Electrochem. Soc., 113:109 (1966); B. Mann and H. Kuhn, J. Appl. Phys., 42:4398 (1971); M. H. Nathoo, Thin Solid Films, 16:215 (1973).

80. D. E. Beischer, J. Phys. Chem., 57:134 (1953).

81. D. E. Beischer, in Monomolecular Layers (H. Sobotka, ed.), Amer. Assoc. Adv. Sci., Washington, 1954, p. 107.

82. H. Sobotka, J. Phys. Chem., 62:527 (1958).

83. H. Sobotka, M. Demeny, and J. D. Chanley, J. Colloid Sci., 13:565 (1958).

84. J. J. Bikerman, Proc. Roy. Soc. (Lond.), A170:130 (1939).

85. G. L. Gaines, J. Colloid Sci., 15:321 (1960).

86. J. A. Spink, J. Colloid Interface Sci., 23:9 (1967).

87. R. A. Faires and B. H. Parks, Radioisotope Laboratory Techniques, Newnes, London, 1958, p. 223.

88. M. Calvin, C. Heidelberger, J. C. Reid, B. M. Tolbert, and P. F. Yankwich, Isotopic Carbon, Wiley, New York, 1949, p. 311.

89. G. L. Gaines, J. Appl. Phys., 31:741 (1960).

90. J. A. Spink, J. Electrochem. Soc., 114:646 (1967).

91. F. M. Fowkes, Ind. Eng. Chem., 56:40 (12) (1964).

92. C. O. Timmons, J. Colloid Interface Sci., 43:1 (1973).

93. L. S. Bartell and J. F. Betts, J. Phys. Chem., 64:1075 (1960); J. R. Miller and J. E. Berger, J. Phys. Chem., 70: 3070 (1966).

94. E. K. Rideal and J. Tadayon, Proc. Roy. Soc. (Lond.), A225: 346, 357 (1954).

95. J. E. Young, Austral. J. Chem., 8:173 (1955).

96. G. L. Gaines, Nature, 186:384 (1960).

97. R. W. Roberts and G. L. Gaines, Jr., Trans. 9th Nat. Vacuum Symp., Amer. Vacuum Soc., 1962, p. 515; G. L. Gaines, Jr. and R. W. Roberts, Nature, 197:787 (1963).

98. J. A. Spink, J. Colloid Interface Sci., 24:61 (1967).

99. F. P. Bowden and A. C. Moore, Research, 2:585 (1949); Trans. Faraday Soc., 47:900 (1951).

100. T. Seimiya, Bull. Chem. Soc. Japan, 38:745 (1965).

101. R. L. Patrick and G. O. Payne, Jr., J. Colloid Interface Sci., 16:93 (1961).

102. W. C. Bigelow, D. L. Pickett, and W. A. Zisman, J. Colloid Sci., 1:513 (1946).

103. J. W. Shepard and J. P. Ryan, J. Phys. Chem., 63:1729 (1959).

104. D. W. Aylmore and W. B. Jepson, J. Sci. Instr., 38:156 (1961).

105. J. T. Clarke, J. Phys. Chem., 68:884 (1964).

106. J. P. W. Houtman and J. Medema, Ber. Bunsenges, Phys. Chem., 70:489 (1966).

107. H. Cochrane, P. L. Walker, Jr., W. S. Diethorn, and H. C. Friedman, J. Colloid Interface Sci., 24:405 (1967).

108. T. R. Hughes, R. J. Houston, and R. P. Sieg, Ind. Eng. Chem. Process Design Dev., 1:96 (1962).

109. R. P. Eischens, J. Am. Chem. Soc., 74:6167 (1952).

110. J. T. Kummer and P. H. Emmett, J. Am. Chem. Soc., 73:2886 (1951).

111. S. Z. Roginskii, Theoretical Principles of Isotope Methods for Investigating Chemical Reactions (Russ.) Moscow, 1956, English Transl., New York, 1957; Zh. Fiz. Khim., 32:737 (1958); N. P. Keier, Probl. Kinet. Katal., 8:224 (1955); Dokl. Akad. Nauk, SSSR, 111:1274 (1956).

112. N. P. Keier, Izv. Akad. Nauk SSSR, Otd. Khim. Nauk, 616 (1952); 48 (1953).

113. K. C. Campbell and S. J. Thomson, Trans. Faraday Soc., 55:306 (1959).

114. Kh. M. Minachev and G. V. Isagulyants, Third Congress on Catalysis, Vol. I, p. 204, 1965.

115. B. Le Boucher, C. Libanati, and P. Lacombe, Acad. Sci., Compt. Rend. (Paris), 248:2578 (1959).

116. J. A. Dillon, Jr. and H. E. Farnsworth, Rev. Sci. Instr., 25:96 (1954); J. Chem. Phys., 22:1601 (1954).

117. S. J. Thomson and J. L. Wishlade, Trans. Faraday Soc., 58: 1170 (1962); J. Sci. Instr., 39:570 (1962).

118. A. Lawson, J. Catalysis, 11:283 (1968).

119. B. G. Baker and A. Lawson, J. Catalysis, 16:108 (1970).

120. P. H. Emmett, Adv. Catalysis, 9:645 (1957).

121. G. Blyholder and P. H. Emmett, J. Phys. Chem., 63:962 (1959).

122. D. Cormack, S. J. Thomson, and G. Webb, J. Catalysis, 5:224 (1966).

123. J. U. Reid, S. J. Thomson, and G. Webb, J. Catalysis, 29:421 (1973).

124. B. Sakmann, J. T. Burwell, and J. W. Irvine, J. Appl. Phys., 15:459 (1944). See also J. T. Burwell, Nucleonics, 1:38 (1947).

125. J. N. Gregory, Nature, 157:443 (1946).

126. E. Rabinowicz and D. Tabor, Proc. Roy. Soc. (Lond.), A208:455 (1951).

127. F. P. Bowden and D. Tabor, The Friction and Lubrication of Solids, Part I, Oxford University Press, Oxford, 1950; Part II, 1964.

128. F. E. Talke, Wear, 22:69 (1972).

129. J. Golden and G. W. Rowe, Brit. J. Appl. Phys., 9:120 (1958); 10:367 (1959).

130. R. B. Campbell and G. Harriden, Wear, 1:173 (1957/58).

131. E. Rabinowicz, Proc. Phys. Soc., A64:939 (1951).

132. H. D. Bush, D. M. Rowson, and S. E. Warren, Wear, 20:211 (1972).

133. I. I. Zemskova, R. M. Matveevsky, and M. M. Khruschov, Wear, 23:225 (1973).

134. M. E. Merchant, H. Ernst, and E. J. Krabacher, Trans. Am. Soc. Mech. Eng., 75:549 (1953).

135. J. C. E. Button, A. J. Davies, and R. Tourret, Nucleonics, 9:34 (1951).

136. W. Dahl and W. Leug, Stahl Eisen, 76:257 (1957); W. Lueg and P. Funke, ibid., 79:996 (1959).

137. F. P. Bowden, J. P. B. Williamson, and P. G. Laing, J. Bone Joint Surgery, 37B:676 (1955).

138. D. G. Flom, J. Appl. Phys., 28:850 (1957).

139. J. K. Lancaster in "A Review of Radioactive Tracer Applications in Friction Lubrication and Wear", R. A. E. Tech. Note No. CPM 64, March 1964.

140. M. H. Davies, M. T. Simnad, and C. E. Birchendall, J. Metals, 3:889 (1951).

141. D. J. Neild, P. J. Wise, and D. G. Barnes, J. Phys. D: Appl. Phys., 5:2292 (1972).

142. A. Brückman, R. Emmerich, and S. Mrowec, Oxid. Metals, 5:137 (1972).

143. D. G. Barnes, J. M. Calvert, K. A. Hay, and D. G. Lees, Phil. Mag., 28:1303 (1973).

144. H. C. Gatos and A. Azzam, Trans. Amer. Inst. Min. Met. Eng., 194:407 (1952).

145. N. Hackerman and N. H. Simpson, Trans. Faraday Soc., 52:628 (1956).

146. J. Y. Choi and P. G. Shewmon, Trans. Met. Soc. AIME, 230:123 (1964).

147. D. H. Houseman, in The Physical Examination of Metals, 2nd ed. (B. Chalmers and A. G. Quarrell, eds.), Edward Arnold, London, 1960, Chap. XVI, p. 737.

148. H. M. C. Sosnovsky, J. Phys. Chem. Solids, 10:304 (1959); J. Tuul and H. E. Farnsworth, J. Am. Chem. Soc., 83:2247, 2253 (1961); I. Uhara, S. Kishimoto, Y. Yoshida, and T. Hikino, J. Phys. Chem., 69:880 (1965); M. A. Bhakta and H. A. Taylor, J. Chem. Phys., 44:1264 (1966).

149. H. Jaeger, J. Catalysis, 9:237 (1967).

150. I. Yasumori, T. Kabe, and Y. Inoue, J. Phys. Chem., 78:583 (1974).

151. P. J. Wise, D. G. Barnes, and D. J. Neild, J. Phys. D: Appl. Phys., 7:1475 (1974).

152. J. M. Calvert, D. J. Derry and D. G. Lees, J. Phys. D: Appl. Phys., 7:940 (1974).

153. R. D. Neuman, Nature, 250:725 (1974).

154. E. E. Berlovich, L. M. Kutsentov, and V. G. Fleischer, Soviet Phys.-JETP, 21:675 (1965).

155. L. M. Kutsentov and V. G. Fleischer, Russ. J. Phys. Chem., 40:753 (1966).

Chapter 14

RADIONUCLIDES IN ENVIRONMENTAL STUDIES

A. A. Moghissi and M. W. Carter

Georgia Institute of Technology
Atlanta, Georgia

I. INTRODUCTION

Due to the complexity and the interactive nature of various elements and subelements of the environment, tracer methods described in other chapters of this book can be directly or indirectly related to environmental studies. Biomedical studies relative to the kinetics or effects of trace metals and organic compounds are of fundamental importance to the understanding of environmental effects of these materials and thus can be rightfully included in environmental studies. Similarly, kinetics of pollutants in air and water can be effectively studied in the laboratory using radioactive tracers. As some of these topics have been discussed earlier in this book, this chapter will be concerned with environmental field studies or laboratory studies which, due to their peculiarity, have not been adequately described elsewhere.

Radiotracers have been extensively used in environmental studies. Numerous symposia have been partially or totally devoted to this subject [1-33]. In addition, several reviews and monographs have been published covering various aspects of the environment [34-38].

One characteristic of many environmental studies is caused by the large masses of air, water, soil, and other environmental elements involved and thus tracer studies in the field are usually associated with a substantial dilution. As radiological safety and other environmental considerations usually place an upper limit on the quantity of radioactive tracer which may be released, these studies require particular attention to the optimization of experimental design, sampling, and analytical manipulation. If one considers that dilution factors of 10^{10} or more are not uncommon in environmental tracer studies, the importance of optimization becomes apparent.

II. MEASUREMENT SYSTEMS FOR ENVIRONMENTAL STUDIES

In many environmental experiments, samples contain sufficient quantity of the tracer and thus generally applicable counting systems and statistics may be used. However, in many cases, the activity of the sample in terms of counts per minute is not significantly above the background of the counting equipment and thus chemical or other enrichment processes must be applied. In these cases, the rules of low-level counting must be carefully considered.

A. Low-level Counting Systems

There are many factors which affect low-level counting systems. The foremost factor is the stability and reproducibility of the system. Although the literature on statistical and many other parameters of the low-level counting systems is voluminous, a simpler and more practical approach is treated here and the reader is referred to the two excellent reports [39,40] for further details.

It is generally recognized that in a <u>stable</u> counting system there are three major factors affecting the processing of samples containing low levels of activity; these are background B (cpm), counting efficiency E (cpm/dpm), and the quantity of the sample M which is used to obtain B and E. In the majority of environmental tracer studies, the sample size can be chosen <u>ad libitum</u>. If the flow of a river is measured, for example, relatively large volumes of river water are available. In most hydrological studies, any quantity of water, within limits, can be collected for analysis. Even in metabolism studies in which a high degree of dilution occurs, such as metabolism of pollutants in dairy cows, a sufficient volume of milk is available.

It has been well established [41] that the relationship among the three above-mentioned factors follows Eq. (1).

$$Y = \frac{\sqrt{B}}{KEM} \tag{1}$$

Where K is a proportionality factor relating dpm to Ci, e.g., 2.22 if Y is the minimum limit of detection at 1 sigma confidence level and 1 min counting time and is expressed in pCi/unit M.

As in background dominant counting, the sample count rate is comparable to the background count rate and counting times for the background and sample measurements are generally chosen at the same level.

B. Sample Preparation and Laboratory Counting Systems

Environmental studies often require the processing of a large number of samples. Therefore, it is advisable to design a proper identification and coding system for their processing.

1. Sample Preparation

The preparation of samples for counting of γ-emitting radionuclides is simple and consists of attempting to maintain a reasonable and reproducible geometry. Occasionally, the sample is ground and/or slurried in water or other liquids. Sometimes, the sample is ashed (see below) to obtain a smaller volume and thus increase the sensitivity of the system.

Hard β-emitting radionuclides can also be counted using unprocessed samples. If liquid scintillation counting is used, the sample generally must be processed to remove "quenching" materials (see further details in Chaps. 4 and 6).

Soft β-emitters and α-emitting radionuclides require sample processing in almost every case.

a. Ashing. Wet ashing and dry ashing are common and well-established techniques. The selection of a proper wet ashing procedure relates to the requirements of each experiment and the preference of the investigator. The most common wet ashing agents are nitric acid, perchloric acid, and H_2O_2 in the presence of ferrous ions [42].

Dry ashing is extensively used in environmental studies primarily because of its simplicity. The sample is placed in a porcelain or borosilicate glass dish and heated at 150°-200°C for several hours to remove the water. Subsequently, it is heated at 350°-400°C for several hours resulting in oxidation of organic materials. Halogens are retained in the sample by the addition of sodium hydroxide.

Dry ashing results in loss of carbon, hydrogen, part of sulfur, part of cesium and certain other elements. If these elements are used as tracers, other techniques must be used.

b. Combustion. Conventional combustion methods described previously in this book apply to many environmental studies. For many low-level measurements more elaborate combustion methods must be used primarily because of the requirements of an increased sample size. The oxygen flask combustion, commonly known as the Schöniger combustion, can process samples up to approximately 100 mg. Kaartinen [43] partially automated this technique by simplifying the sample feeding and oxygen flow systems and thus larger samples may be processed on a repetitive basis. The classical Liebig apparatus can be readily scaled up for processing of larger samples [43]. Samples up to 10 g can be processed in an oxygen bomb [44] shown in Fig. 1. Although oxygen bomb combustion is associated with high pressures and certain safety requirements, it is simple and can be easily used for repetitive measurements.

c. Separation of water from environmental samples. In certain cases, it is necessary to separate water from environmental samples either to use the water for tritium analysis or to reduce the volume of the sample. The choice of the methods depends to a great extent upon the conditions of the experiment and the preference of the investigator.

Distillation under normal or reduced pressure, freeze drying, and azeotropic distillation [45] using benzene have been repeatedly applied. Distillation is usually used for purification of water and aqueous samples, whereas freeze drying and particularly azeotropic distillation are used for biological samples. Essential features of azeotropic distillation are shown in Fig. 2. The sample is refluxed in the presence of benzene. As the distillation is advanced, benzene flows back and water is collected in the receiving container. Subsequent to the completion of the

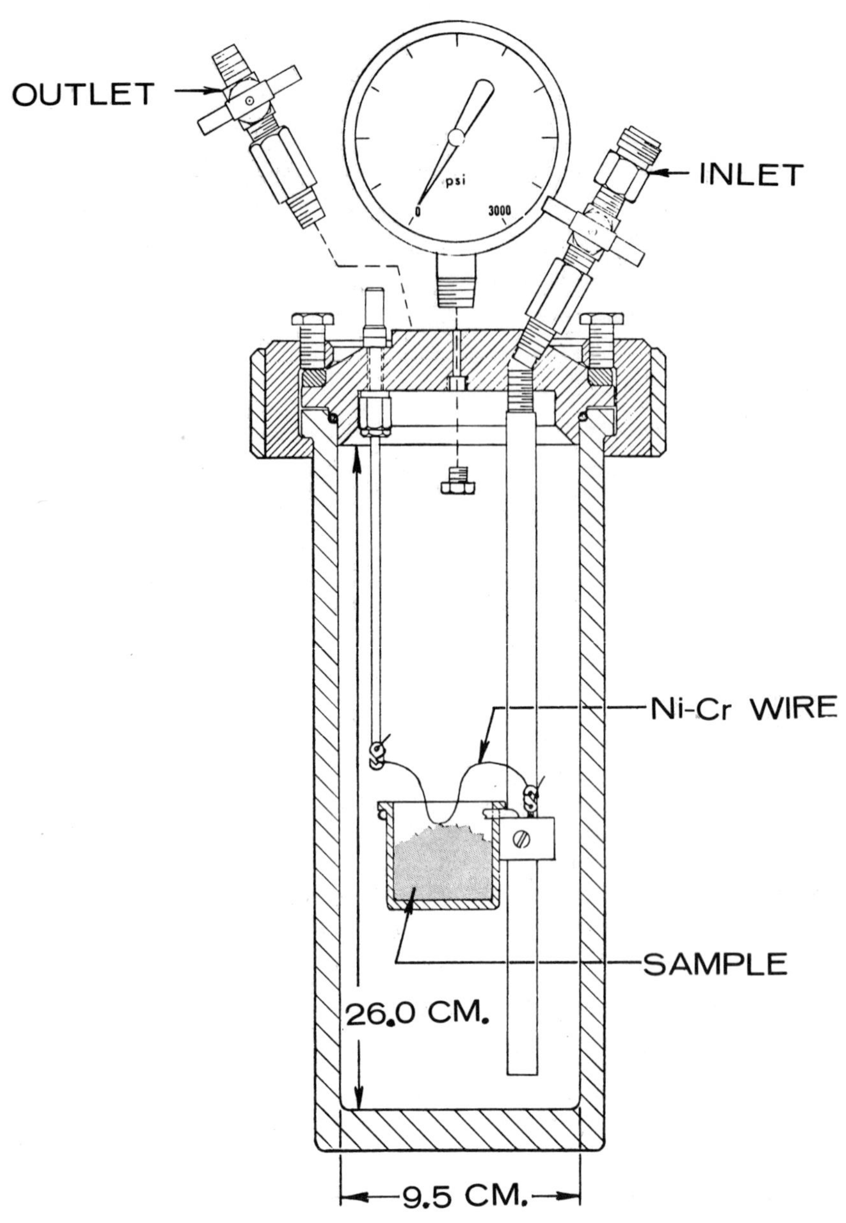

FIG. 1. Oxygen bomb combustion.

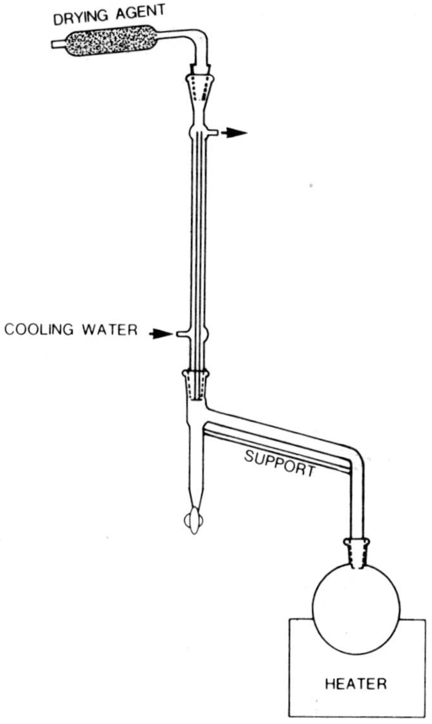

FIG. 2. Azeotropic distillation for removal of water from environmental samples.

removal of water, benzene is distilled and the organic and other materials are left behind.

2. Laboratory Counting Systems for α-Emitting Radionuclides

The difficulties in detection and the biological hazards of α-emitting radionuclides are the main reasons for the unpopularity of these radionuclides as tracers in environmental studies.

Windowless proportional counters can be effectively used for gross measurement and to some extent for spectroscopy of α-emitting radionuclides. Solid state detectors are somewhat simpler to operate than proportional counters and thus have gained in popularity.

Alpha spectroscopy requires separation of the tracer from its bulk material and preparation of a thin and evenly distributed sample preferably in a monomolecular form. Electroplating procedures may be used to prepare monomolecular samples suitable for high-resolution spectroscopy. However, electroplating of plutonium [46] exemplifies the complexity of the sample preparation if this process is desired. If α-emitting radionuclides must be used, they should be chosen at such a level to permit gross counting by proportional counters, solid state counters, or liquid scintillation counters (see below). If the levels are low enough to require alpha spectroscopy, a coprecipitation technique may be more rapid. This well-known phenomenon makes it possible to separate minute quantities of a radionuclide as a result of precipitation of certain compounds. The only requirement for alpha spectroscopy is to maintain a small mass, not exceeding a few milligrams. Coprecipitation techniques using quantities of 1 mg or less on a 2.5 cm microfilter resulting in resolutions of better than 100 keV for alpha spectroscopy have been reported [47].

3. Laboratory Counting Systems for β-Emitting Radionuclides

Techniques used for measurement of hard β emitters are somewhat different from those used for soft β emitters such as tritium and ^{14}C. As these two radionuclides have certain significance in studies related to water management they are discussed in greater detail than other radionuclides (see also Chaps. 4 and 6).

a. Low-level measurement of tritium and carbon-14. Procedures for low-level measurement of tritium and ^{14}C go back to Libby [48]. He used electrolytic enrichment of tritium in association with internal gas counting for tritium measurement.

The literature related to this subject is voluminous and several symposia have been devoted to low-level counting of tritium and ^{14}C and their applications [1,2,4].

(1) Tritium enrichment. For many applications, the specific activity of tritium in the environmental samples is too

low to be measured with a reasonable degree of accuracy with the presently available instruments. Therefore, an enrichment method must be applied. As the majority of samples to be processed are either already in the form of water or easily convertible to water, enrichment processes are generally based on water.

Many techniques have been proposed for tritium enrichment [49]. Based on differences of vapor pressures of tritiated and ordinary water, fractional distillation methods have been proposed. Gas chromatographic separation of hydrogen isotopes on a preparative scale may also be used for tritium enrichment. Also, thermal diffusion in a setup similar to that used by Clusius has been proposed. The only practical method based on a related concept is proposed by Hayes and Hoy [50], who use the differences of absorption abilities of hydrogen isotopes in a palladium sponge. Some of these techniques require the conversion of water to hydrogen and/or exhibit inadequate separation factors. For these reasons, the electrolysis process has remained the most popular method.

Many variations to this simple method have been proposed, although only two procedures have gained acceptance in routine applications. These are described by Bainbridge [51] and Oestlund [52]. The cell described by Bainbridge is an improved Libby cell and is shown in Fig. 3. This cell consists of an iron tube constituting the cathode, whereas a nickel tube is used as an anode. A more popular cell was designed by Oestlund. Although it results in a somewhat higher tritium loss than the previous cell, it is more extensively used because of the simplicity of its operation. The Oestlund cell is shown in Fig. 4. As it is made of glass, problems associated with electrical insulation are small or nonexistent. Water is added in 50 ml increments and thus the cell size remains small. After a 100-fold reduction in volume, less than 25% of the tritium is lost. Typically, the cell is run at 3A during the first 10-fold volume reduction and at 0.3A at the last 10-fold reduction. Sodium hydroxide is used as the electrolyte at 0.8% initial concentration. The degree of enrichment is determined

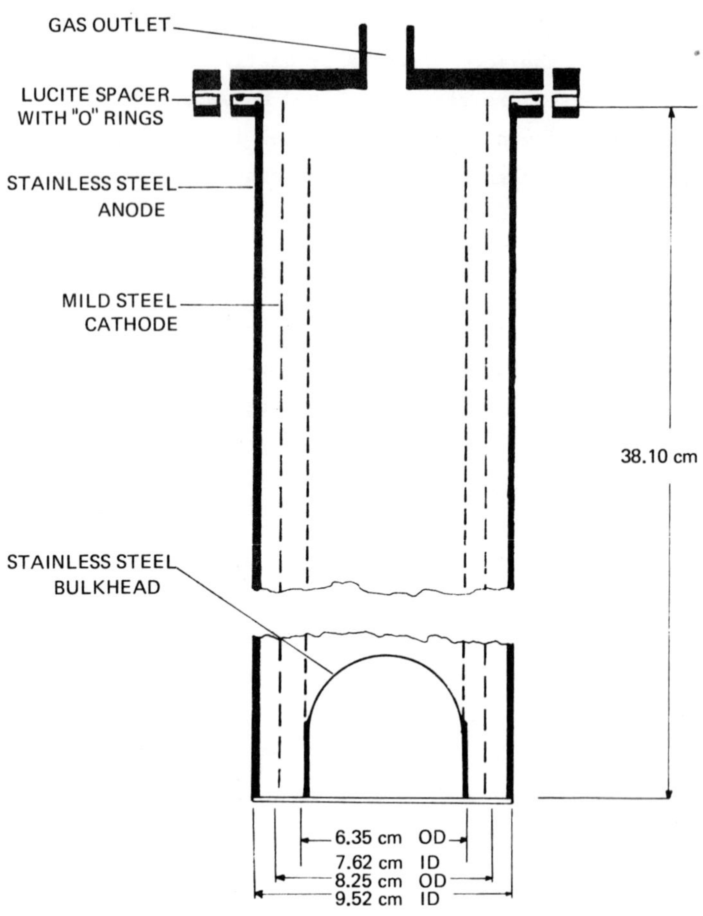

FIG. 3. Electrolytic cell for enrichment of tritium.

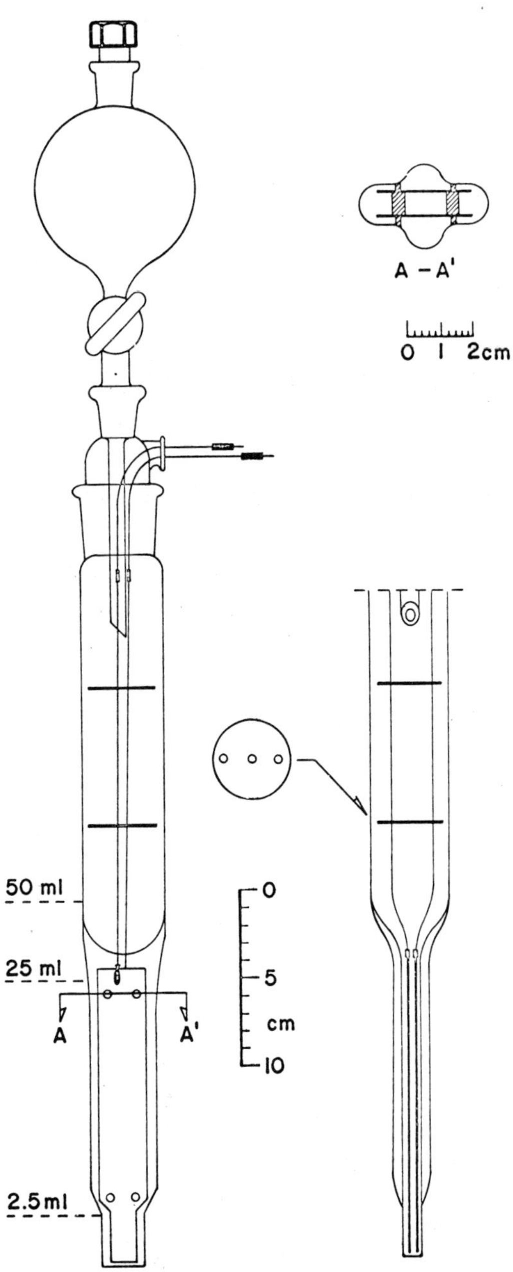

FIG. 4. Oestlund cell for electrolytic enrichment of tritium.

in both cells by running a standard cell with a known tritium concentration and comparing its enrichment factor to the unknown samples. If a higher degree of accuracy is desired, deuterium enrichment may be used as an indicator for tritium measurement according to the Oestlund equation [52]. The expected accuracy for 100-fold reduction using a standard cell is 2-4%.

(2) Internal gas counting. An excellent review on internal gas counting has been reported by Schell [53]. According to Schell, internal proportional counting represents the most sensitive method for measurement of low-levels of tritium and ^{14}C. If the system is properly designed and operated, the detection efficiency for both radionuclides is essentially 100% with background of 1 cpm or less. The principle difficulty of operation of an internal gas counter is the necessity of conversion of the radionuclide to a countable gas. Both CO_2 and water are unsuitable for that purpose; thus, they require chemical processing. The most suitable reactions for ^{14}C are (2)-(4).

$$CO_2 + H_2 \xrightarrow{\text{catalyst}} CH_4 + 2H_2O \tag{2}$$

$$2CO_2 + 9Li \longrightarrow LiC_2 + 4Li_2O \tag{3}$$

$$Li + LiC_2 + 2H_2O \longrightarrow 2LiOH + C_2H_2 \tag{4}$$

Both methane and acetylene are suitable counting gases. Lithium must be used in excess resulting in production of hydrogen as a result of reaction of lithium and water. However, the separation of hydrogen and acetylene does not constitute any problem. The reaction of water and carbide can be enhanced by the addition of small concentrations of sulfuric acid. The conversion of water to a suitable gas is done by the following reactions:

$$CaC_2 + 2H_2O \longrightarrow C_2H_2 + Ca(OH)_2 \tag{5}$$

$$H_2O + Zn \xrightarrow{400°C} H_2 + ZnO \tag{6}$$

Reaction (5) is associated with an isotope effect which, however, can be made reproducible. Many other metals, such as magnesium, can be used in reaction (6). Oestlund [54] uses hydrogen whereas the conversion of hydrogen to ethane by the following reaction is preferred by others [53].

$$C_2H_2 + 2H_2 \xrightarrow{\text{catalyst}} C_2H_6 \qquad (7)$$

The reaction of an acid with carbide for tritium counting has not been used, although it holds a great deal of promise, as it avoids the isotope effect associated with Eq. (5). Internal gas counters are preferentially operated in the proportional region. Due to the importance of low background the counter is not only shielded with lead or iron but also with an anticoincidence ring. There have been numerous designs for anticoincidence shields. Many investigators use an arrangement shown in Fig. 5. The center ring contains the sample, while outer rings are sealed and negate any event which simultaneously is recorded by either one of them and the inner ring. Oeschger [55] has used the arrangement shown in Fig. 6. In Oeschger's system the inner and outer rings are separated by a thin wall and filled with the same gas. Although the Oeschger counter exhibits a lower background, the requirement of a larger sample has somewhat reduced its popularity.

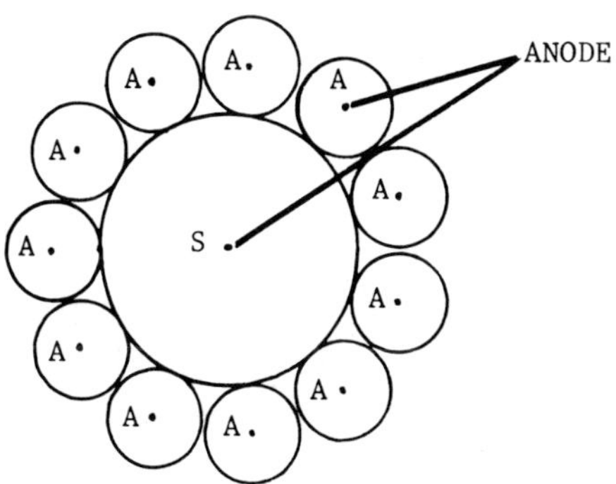

FIG. 5. Conventional anticoincidence shield.

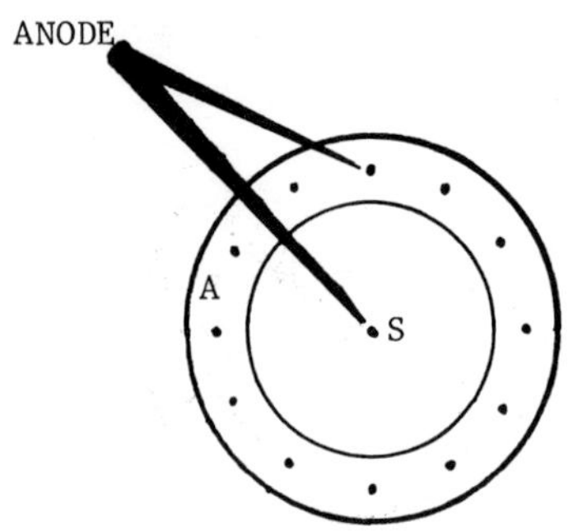

FIG. 6. Anticoincidence shield after Oeschger.

(3) Liquid Scintillation Counting. Techniques for low-level liquid scintillation counting have improved significantly during the last few years. This is due to development of multiplier phototubes with quantum efficiencies of approximately 30%, whereas previously quantum efficiencies had a range of 11-18%. If one considers that, because liquid scintillation counters operate in coincidence, the efficiency of the system improves with the square of the quantum efficiency, the importance of this improvement becomes apparent. The second reason is a better understanding of emulsion systems which has made it possible to incorporate a significant quantity of water into the scintillation liquid.

There are many brand named emulsion systems available which are also useful for low-level counting. These are all based on alkylphenol detergents mixed with toluene or xylene in ratios of 1:2-3. Triton X-100, Triton N101 (Rohm and Haas brand names), and Instagel, PCS, and Aquasol are examples of these mixtures. The lowest Y value (see Section II A for definition of Y value) has been reported for Triton N101-xylene mixture in a volume ratio of 1:2 to 1:2.5 [56]. Xylene contains in this case 7 g of 2,5-diphenyloxazol (PPO) and 1.5 g p-bis-(o-methylstyryl)benzene (bis-MSB). Such a system can incorporate 40% water at optimum conditions.

The introduction of CO_2 into the system for low-level counting occurs rapidly using 0.5 N sodium hydroxide which is subsequently mixed with a scintillation liquid in volume ratio of 1:1. The latter consists of *p*-xylene containing 7 g PPO and 1.5 g bis-MSB per liter in detergent and volume ratio of 2:1. The detergent consists of 28% of Triton QS 44 in Triton N101 (both manufactured by Rohm and Haas). This mixture can incorporate easily 150 mg of carbon in the form of CO_2 [57]. A somewhat slower but more efficient method [58] consists of mixing toluene (45%), methanol (27.5%) and phenethylamine (27.5%) containing 7 g PPO and 1.5 g bis-MSB per liter of the mixture. This mixture can incorporate 650 mg of carbon in the form of CO_2.

The best liquid scintillation method for ^{14}C consists of the conversion of CO_2 to acetylene and subsequently to benzene [59] by the sequence of reactions (2)-(5) and (8).

$$3C_2H_2 \xrightarrow{\text{catalyst}} C_6H_6 \tag{8}$$

The conversion efficiency is better than 90% and thus this technique has a lower Y value than the internal gas counting, provided sufficient quantity of sample is available.

The same technique can be used for tritium measurements. In this case, because reaction (4) is associated with an isotope effect, C_2H_2 is somewhat depleted of tritium. In addition, the Y values obtained using this technique are only slightly better than those obtained using emulsion counting. Therefore, benzene synthesis for tritium is seldom used.

b. <u>Measurement of other β emitters</u>. Low-level measurement of other beta emitters is usually carried out in proportional counters or by liquid scintillation. Proportional counters exhibit generally a lower Y value (see Section II A for definition of Y value) than liquid scintillation counters. However, the latter are more convenient to operate, and are easier to automate. In addition, the sample preparation for liquid scintillation is both simpler and more reproducible as compared to proportional counters.

4. Measurement of γ-Emitting Radionuclides

These radionuclides are preferred because of no or little sample preparation. In many environmental studies several radionuclides are used simultaneously. In these cases gamma spectroscopy rather than gross counting is carried out. Scintillation using NaI(Tl) or other crystals is the most widely applied counting method. In order to increase the sensitivity of the system, the sample is placed in a beaker designed by Marinelli as shown in Fig. 7. This improves the geometry and thus the performance of the system.

III. APPLICATION OF TRACERS IN WATER STUDIES

Radiotracers have been extensively used in evaluating various aspects of water resources and water pollution. Many symposia have been partially or totally dedicated to this subject [8-17]. Also, several reviews have appeared among them, the excellent bibliographies by Schultz [35] and Rhodehamel et al. [60] being examples.

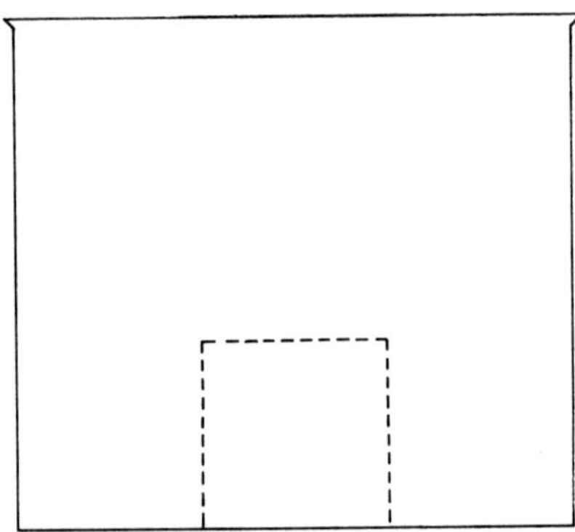

FIG. 7. Marinelli beaker used for measurement of γ-emitting radionuclides. Typically 1- or 4-liter beakers are used.

A. Dating of Underground Waters

Libby, who first described the properties of atmospheric tritium, indicated its potential as a tracer in hydrological studies [61]. Tritium is produced in the upper atmosphere and reaches the surface of the earth primarily as a result of precipitation. Prior to the atmospheric testing of nuclear devices, the concentration of tritium in rain was in the order of a few tritium units (1 TU = 10^{-18} T:H): during the spring and in the fall this concentration was increased by 1-2 orders of magnitude. Figure 8 shows the concentration of tritium in rain, including concentration of tritium in surface waters indicating that spring and fall peaks are smoothed out in the dilution process. Underground waters containing tritium can, therefore, be related to the time of precipitation and thus the underground water recharge rate can be

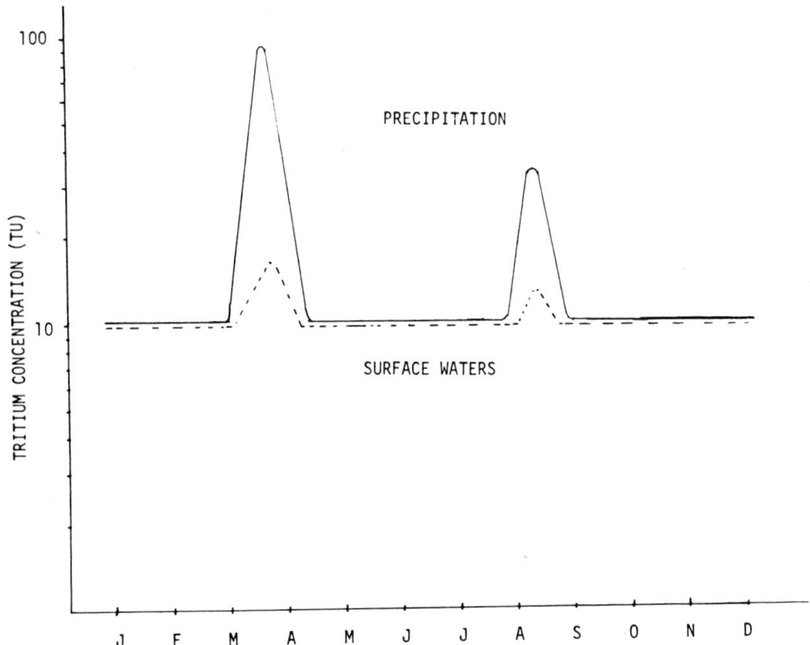

FIG. 8. Tritium concentration in precipitation and in surface waters.

estimated. Nuclear weapons testing has greatly disturbed the natural equilibrium of tritium by substantially increasing the total concentration immediately following a particular test during which it was formed. These peak levels can and have been effectively used as tracers in a technique similar to the methods described in Section III B.

Application of ^{14}C dating to studies of underground waters was described by Muennich [62]. This method is based on the ^{14}C produced in the atmosphere which is converted to CO_2 at an established constant level of ^{14}C in terms of activity ($^{14}C/gC$). Rain, therefore, contains CO_2 with a ^{14}C content at the same level as natural ^{14}C. During passage through limestone, this CO_2 contained in groundwater reacts with $CaCO_3$ according to Eq. (9).

$$CaCO_3 + CO_2 + H_2O \longrightarrow Ca^{2+} + 2HCO_3^- \qquad (9)$$

Bicarbonate is soluble and migrates with water. It can be easily seen that one half of the carbon in bicarbonate originates from limestone and due to its age reduces the specific activity of ^{14}C by 50%. There are several other reactions which complicate this equilibrium. However, age determinations using this technique and those using tritium, if done simultaneously, could be useful in establishing the validity of either one.

Other methods used for determination of various properties of underground water are generally similar to those for surface water described under Section III B.

B. Water Velocity Measurements

River flow measurements are of significant value in establishing basic parameters of surface water. In addition, they are important in establishing criteria for pollution abatement as they determine the dispersion rate of pollutants. The choice of a tracer is usually based on many factors such as availability, ease of detection, desired accuracy, and public health considerations. Unfortunately, there is no isotope which satisfactorily meets all these requirements. In many cases, absorption of radiotracers by bottom sediments is a problem. Tritiated water, which is ideally suited for these kinds of studies, is hard to measure and has a relatively long half-life of 12.35 y. Depending upon specific application, ^{22}Na ($T_{1/2}$ = 2.6y), ^{131}I ($T_{1/2}$ = 8.05d), and ^{82}Br ($T_{1/2}$ = 35.4h) are used. In every case, results from use of these radionuclides have been compared with HTO and found to deviate only slightly from values obtained using HTO, indicating the validity of application of these radionuclides. There have been several techniques used [63] for flow measurements and these are discussed below.

1. Pulse Measurement Method

This method represents a simple technique. The radiotracer is introduced at some point in the river and a detector is placed sufficiently downstream from the injection point to allow proper mixing. The time elapsed between the injection point and the appearance of the peak divided by the distance is the desired value.

As the appearance of the peak is measured, absorption of the radiotracer does not influence the results. This technique gives only limited information and thus other techniques are generally preferred.

2. Dilution Method

This method is based on the comparison of the concentration of tracer (C_0) introduced into the stream at a constant rate with the concentration of samples (C_d) collected from a point downstream. The distance must be selected far enough downstream to assure complete mixing and thus reasonable accuracy in measurement. As the mass flow of the tracer is equal to the mass flow of the river, the ratio of C_d to C_0 can be used to determine the stream flow.

Dilution is one of the most accurate methods available. Its major disadvantage is the necessity of continuous injection of radiotracer and the associated large quantities of radioactive material to be transported and used.

3. Continuous Sample or Total Count Method

This method constitutes the opposite approach to the dilution method. In this technique a single injection is made and, at a position downstream, the concentration of the tracer is continuously measured and numerically integrated.

In this case, the ratio of the initial activity and the integral of the concentration downstream over time relates to the flow rate.

If the integral is done by inserting a detector into the stream and measuring the total activity over time, the method is called total count method. This procedure is simple to perform and results in an accurate measurement provided the calibration is done properly. It also lends itself readily to automation.

C. Measurement of Reaeration of Water

Tsivoglou et al. [64] have introduced an interesting and ingenious method for the measurement of reaeration of streams at any point of the river. This method is based on the dissolution of injected krypton, an inert gas, in water. The key to the success of the method is the linear relationship between the transfer of oxygen and krypton in the gas-liquid interface. The practical application consists of injection into the stream of a dye, tritiated water and ^{85}Kr, all mixed in a single solution. Dye is used to indicate the sampling points. Krypton loss, measured as reduction in $^{85}Kr/HTO$ ratio, is directly related to reaeration at the sampling point. This method is particularly useful for measurement of reaeration ability of polluted waters.

D. Measurement of Movement of Sand and Silt

The description of details of studies relative to dynamics of rivers and other bodies of water including excellent descriptions of various methods and their limitations are available [8-17]. This description shall be limited to the movement of silt and sand as related to pollution control.

In polluted waters the bottom sediments act as a sink and thus by various physical, chemical, and biological processes concentrate certain pollutants. If these sediments move from their original location, they contribute to the pollution of other areas.

The single most important problem in labeling silt and sand is producing a labeled material reasonably similar to the material whose movement is being measured. The choice of radionuclide is therefore arbitrary and depends to a great extent upon the ease of labeling and detection. Isotopes of gold and iridium, ^{198}Au and ^{192}Ir, respectively, have been extensively used as they are easy to precipitate on sand and silt and are not subject to removal by the action of water. The mathematical evaluation of the data is relatively simple and follows basically the same rules as those for river flow.

E. Dispersion of Pollutants in Water

Although numerous papers report studies related to specific pollutants [65-67], the methodology for these studies is identical to those in Sections III A-C. With the increased emphasis on water pollution control it is expected these methods will find increased applications.

IV. APPLICATION OF RADIOTRACERS IN STUDIES RELATED TO THE MIGRATION OF LIVING THINGS

An understanding of movement, reproduction rate, and other ecological factors related to terrestrial and aquatic animals as well as insects is essential in assessing environmental impact of pollution. Since the availability of inexpensive radiotracers, they have been extensively used to investigate a number of parameters related to living things [28-31,35-38].

A. Vertebrates

Tester [68] has reviewed techniques for studying movement of vertebrates. In a more recent paper by Gentry et al. [69] the literature for small animals has been reviewed.

The basic technique in all cases is the same. A certain quantity of a suitable radionuclide is placed on or in an animal in such a way that the animal is not injured and its behavior is not significantly altered. The animal is subsequently checked periodically or continuously by monitoring its "*tag*."

The major advantage of application of radiotracers is that, if done properly, no interference occurs during the labeling or detection. The label is undetectable by the labeled organisms or by predators which may prey differentially on animals marked with conspicuous labels such as fluorescent or other easily detectable dyes.

The choice of radionuclides depends on the environmental and public health considerations, the desired length of the study, the availability of the tracer, the ease of labeling and the ease of detection. The labeling techniques vary somewhat and can be separated into metabolized and external labeling.

External labeling consists of rings or wires attached to the animal or inserting a wire under the skin [70]. Although these techniques are relatively simple, the metabolized labels have been more frequently applied as they can be done with virtually no interference with the animal's natural behavior. An example of this kind of labeling consists of application of peanut butter containing ^{131}I, ^{65}Zn, and ^{59}Fe to label small animals.

One disadvantage of this kind of labeling is that it is difficult to distinguish between labeled individuals when there are many in the same area. This difficulty could be partially overcome by trapping the animals and feeding each of them the same quantity of radionuclides. In this case, the level of radioactivity is indicative of the number of labeled animals.

Study of fish movements using radiotracers is somewhat more difficult because of the possibility of human consumption. Again here the review by Tester is recommended. An elegant method of fish migration study was proposed by Loeffel who studied salmon, stemming from the Columbia River, contaminated with ^{65}Zn as a result of Hanford reactor operations [71]. His results showed that salmon from the Oregon-Washington coasts make a circular journey around the Gulf of Alaska during their first year of ocean life.

B. Invertebrates

Application of radiotracers in entomology has been both extensive and useful. The first reported use of a radiotracer in tracing movements of an animal [72] dealt with beetles. Tomes and Brian used radium labeling, enabling them to detect beetles 10 cm underground with a portable GM counter.

Jenkins [73] reviewed radiotracer techniques in entomology. Although that report is somewhat outdated, it contains all the significant methods related to this subject. More recent bibliographies have been prepared by the International Atomic Energy Agency [36-38]. Virtually all available radionuclides with a reasonable half-life have been used in labeling insects. The criteria for selection of a suitable radionuclide for invertebrates are essentially the same as those for vertebrates, except for the additional problem associated with the sensitivity of certain insects to the trace element which often accompanies a radionuclide as an isotopic or nonisotopic carrier.

Contrary to the vertebrates, invertebrates can be advantageously labeled with β emitters such as ^{32}P and ^{89}Sr with maximum β energies of 1.7 and 1.5 MeV, respectively. These isotopes are easily detected in insects with relatively simple radioactive detection instruments.

Insects are labeled using a variety of techniques. These include contamination of food, growing contaminated medium, external contamination of surfaces of the insect, and injections.

The insects are released subsequent to the labeling and collected at a suspected location. Also, the longevity of the insects can be easily studied using the radiotracer technique.

V. APPLICATION OF RADIOTRACERS IN OTHER ENVIRONMENTAL STUDIES

Application of radiotracers in global mixing of air is of significant environmental interest. As these studies are primarily related to meteorology, they will not be discussed here in any detail. It should be mentioned that both weapons test and reactor produced radionuclides have been used for this purpose [6].

As the major constituents of air, oxygen and nitrogen, have no isotopes with a half-life longer than 10 min, no radiotracer studies have been carried out which would relate to the long-term environmental reactions of these elements. For the same reason, studies related to nitrogen oxide pollution applying radiotracer techniques are unknown. Fortunately, ^{35}S is radioactive with a half-life of 87.4 days making it well suited for environmental studies. Bergstrom et al. used $^{35}SO_2$ to study the movement and chemical behavior of SO_2 over 400 km [74]. They showed the ability of this technique to establish the contribution of background SO_2 to the measured SO_2 levels around a source of release. This method was based on the change of specific activity of SO_2 measured at particular sampling points. Smith et al. used $^{35}SO_2$ to study the effect of aerosols on the transport of SO_2 [75].

Transport properties of trace metals through environmental elements (air, water, soil, plants, animals, and humans) and subelements (rivers, lakes, fish, wildlife, insects, etc.) are advantageously studied using radiotracers. The impressive advances in understanding the pollution caused by mercury would have been impossible without the application of radiotracers. The description of the environmental mercury problem is beyond the scope of this chapter and is the subject of several books [76-79]. However, mercury analysis shall be used as an example. As the problem of mercury contamination became apparent during the early 1970s large numbers of samples were analyzed for mercury. Schulert et al. [80] showed, however, that essentially all analytical mercury methods were subject to errors of up to two orders of magnitude. This investigation was carried out using ^{203}Hg. Subsequently, the development of a mercury method using ^{203}Hg as tracer was pursued which made it possible to correct for losses throughout the process [81].

REFERENCES

1. Tritium in the Physical and Biological Sciences, Vols. I and II, IAEA, Vienna, 1962.
2. Radiocarbon and Tritium Dating, Proc. 6th Int. Conf., Pullman, Wash., 1965 (Conf. 650652, TID 4500).
3. Assessment of Airborne Radioactivity, IAEA, Vienna, 1967.
4. Radioactive Dating and Methods of Low-Level Counting, IAEA, Vienna, 1967.
5. Radioisotope Tracers in Industry and Geophysics, IAEA, Vienna, 1967.
6. Nuclear Techniques in Environmental Pollution, IAEA, Vienna, 1971.

7. A. A. Moghissi and M. W. Carter (Eds.), Tritium, Messenger Graphics, Phoenix, Arizona, 1973.

8. Application of Isotope Techniques in Hydrology, IAEA, Vienna, 1962.

9. Radioisotopes in Hydrology, IAEA, Vienna, 1963.

10. Isotope Techniques for Hydrology, IAEA, Vienna, 1964.

11. E. Eriksson, Y. Gustafsson, K. Nilsson (eds.), Ground Water Problems, Pergamon Press, New York, 1966.

12. G. E. Stout (ed.), Isotope Techniques in the Hydrologic Cycle, American Geophysical Union, Washington, D.C., 1967.

13. Isotopes in Hydrology, IAEA, Vienna, 1967.

14. Tritium and Other Environmental Isotopes in the Hydrological Cycle, IAEA, Vienna, 1967.

15. Isotope Hydrology, IAEA, Vienna, 1970.

16. E. Gaspar and M. Oncescu, Radioactive Tracers in Hydrology, Elsevier, New York, 1972.

17. Ven Te Chow (ed.), Adv. Hydrosci., 8 (1972).

18. Radioisotopes in Tropical Medicine, IAEA, Vienna, 1962.

19. Radioisotopes in Soil-Plant Nutrition Studies, IAEA, Vienna, 1962.

20. Radioisotopes in Animal Nutrition and Physiology, IAEA, Vienna, 1965.

21. Radioisotope Sample Measurement Techniques in Medicine and Biology, IAEA, Vienna, 1965.

22. Isotopes and Radiation in Soil-Plant Nutrition Studies, IAEA, Vienna, 1965.

23. Isotope and Radiation Techniques in Soil Physics and Irrigation Studies, IAEA, Vienna, 1967.

24. Isotopes and Radiation in Soil Organic-Matter Studies, IAEA, Vienna, 1968.

25. Radiation and Radioisotopes for Industrial Microorganisms, IAEA, Vienna, 1971.

26. Dynamic Studies with Radioisotopes in Medicine, IAEA, Vienna, 1971.

27. Isotopes and Radiation in Soil-Plant Relationships Including Forestry, IAEA, Vienna, 1972.

28. C. L. Comar (ed.), Radioisotopes in Entomology, American Association for the Advancement of Science, Washington, D.C., 1949.

29. Radioisotopes and Radiation in Entomology, IAEA, Vienna, 1962.

30. Radiation and Radioisotopes Applied to Insects of Agricultural Importance, IAEA, Vienna, 1963.

31. Isotopes and Radiation in Entomology, IAEA, Vienna, 1968.

32. Sterility Principle for Insect Control or Eradication, IAEA, Vienna, 1971.

33. D. J. Nelson (ed.), Radionuclides in Ecosystems, U.S. Atomic Energy Commission, Oak Ridge, Tenn., 1971.

34. A. W. Klement, Jr. and V. Schultz, Terrestrial and Freshwater Radioecology: A Selected Bibliography, U.S. Atomic Energy Commission, Washington, D.C., Report TID-3910, 1962, TID-3910, Suppl. 1, 1963, TID-3910, Suppl. 2, 1964, TID-3910, Suppl. 3, 1965, TID-3910, Suppl. 4, 1966, TID-3910, Suppl. 5, 1968, TID-3910, Suppl. 6, 1970.

35. V. Schultz, Ecological Techniques Utilizing Radionuclides and Ionizing Radiation: A Selected Bibliography, U.S. Atomic Energy Commission, Washington, D.C., Report RLO-2213, 1961, RLO-2213, Suppl. 1, 1972.

36. M. Binggeli, Radioisotopes and Ionizing Radiation in Entomology, Vol. I, IAEA, Vienna, 1963.

37. M. Binggeli, Radioisotopes and Ionizing Radiation in Entomology, Vol. II, IAEA, Vienna, 1965.

38. Radioisotopes and Ionizing Radiations in Entomology, Vol. III, IAEA, Vienna, 1964-65.

39. Measurement of Low-Level Radioactivity, ICRU Report 22, International Commission on Radiation Unit, and Measurements, Washington, D.C., 1974.

40. J. M. R. Hutchinson and W. B. Mann, Nucl. Instr. Methods, 112:305 (1973).

41. A. A. Moghissi, H. L. Kelley, J. E. Regnier, and M. W. Carter, Int. J. Appl. Radiat. Isotopes, 20:145 (1969).

42. B. Sansoni and W. Kracke, Z. Analyt. Chem., 243:209 (1968).

43. N. Kaartinen, Described by J. D. Davidson, V. T. Oliverio, and J. I. Paterson, in The Current Status of Liquid Scintillation Counting (E. D. Bransome, Jr., ed.), Grune & Stratton, New York, 1970.

44. A. A. Moghissi, E. W. Bretthauer, E. L. Whittaker, and D. M. McNelis, Int. J. Appl. Radiat. Isotopes 26:339 (1975).

45. A. A. Moghissi, E. W. Bretthauer, and E. H. Compton, Anal. Chem., 45:1565 (1973).

46. N. A. Talvitie, Anal. Chem., 44:280 (1972).

47. R. Lieberman and A. A. Moghissi, Health Phys., 15:359 (1968).

48. S. Kaufmann and W. F. Libby, Phys. Rev., 93:1337 (1954).

49. D. G. Jacobs, Sources of Tritium and Its Behavior Upon Release to the Environment, U.S. Atomic Energy Commission, Oak Ridge, Tenn., 1968.

50. D. W. Hayes and J. E. Hoy, in Tritium (A. A. Moghissi and M. W. Carter, eds.), Messenger Graphics, Phoenix, Arizona, pp. 127-133, 1973.

51. A. E. Bainbridge, Rev. Sci. Instr., 36:1779 (1965).

52. H. G. Oestlund and E. Werner, in Tritium in Physical and Biological Sciences, IAEA, Vienna, pp. 95-105, 1962.

53. W. R. Schell, in Tritium (A. A. Moghissi and M. W. Carter, eds.), Messenger Graphics, Phoenix, Arizona, pp. 113-127, 1973.

54. H. G. Oestlund, in Tritium in the Physical and Biological Sciences, Vol. I, IAEA, 1962.

55. H. Oeschger, in Low-Level Counting Methods, IAEA, pp. 13-34, 1963.

56. R. Lieberman and A. A. Moghissi, Int. J. Appl. Radiat. Isotopes, 21:319 (1970).

57. A. A. Moghissi, D. N. McNelis, W. F. Plott, and M. W. Carter, in Rapid Methods for Measuring Radioactivity in the Environment, IAEA, Vienna, pp. 391-394, 1971.

58. F. H. Woeller, Anal. Biochem., 2:508 (1961).

59. M. A. Tamers, in Organic Scintillators D. L. Horrocks, ed.), Gordon and Breach, New York, pp. 261-276, 1968.

60. E. C. Rhodehamel, V. B. Kron, and V. M. Doughery, Bibliography of Tritium Studies Related to Hydrology through 1966, U.S. Geological Survey Paper 1900, U.S. Government Printing Office, Washington, D.C., 1971.

61. W. F. Libby, Proc. Nat. Acad. Sci., 39:245 (1953).

62. K. O. Muennich, Naturwissenschaften, 4432 (1957).

63. C. G. Clayton and D. B. Smith, in Radioisotopes in Hydrology, IAEA, Vienna, pp. 1-24, 1963.

64. E. C. Tsivoglou, J. B. Cohen, S. D. Shearer, and P. J. Godsil, J. Water Poll. Cont. Fed., 40:285 (1968).

65. J. L. Putnam, A. M. Wildblood, and J. E. Robson, Water Sanit. Engr., 6:99 (1956).

66. E. Sons, Atompraxis, 3:443 (1957).

67. P. Harremoes, J. Water Poll. Cont. Fed., 38:1323 (1966).

68. J. R. Tester, in Radioecology (V. Schultz and A. W. Klement, eds.), Reinhold, New York, pp. 445-451, 1963.

69. J. B. Gentry, M. H. Smith, and R. J. Beyers, in Radionuclides in Ecosystems (D. J. Nelson, ed.), U.S. Atomic Energy Commission, Oak Ridge, Tenn., pp. 253-260, 1971.

70. S. V. Kaye, Science, 131:824 (1960).

71. R. E. Loeffel and W. O. Forster, Oregon Fish Com. Res. Rep. 2 (No. 1), 15-27, 1970.

72. G. A. R. Tomes and M. V. Brian, Nature, 158:551 (1946).

73. D. W. Jenkins, in Radioisotopes and Radiation in Entomology, IAEA, Vienna, pp. 3-22, 1962.

74. S. O. W. Bergstrom, L. Devell, C. Gyllander, R. Hesbol, and L. Bergstrom, in Peaceful Uses of Atomic Energy, Vol. 14, United Nations and IAEA, pp. 481-493, 1972.

75. B. M. Smith, J. Wagman, and B. R. Fish, Environ. Sci. Technol., 3:558 (1969).

76. H. R. Jones, Mercury Pollution Control, Noyes Data Corporation, Park Ridge, N.J., 1971.

77. F. M. D'Itri, The Environmental Mercury Problem, CRC Press, Cleveland, Ohio, 1970.

78. L. Friberg and J. Vostal (eds.), Mercury in the Environment, CRC Press, Cleveland, Ohio, 1971.

79. R. Hartung and B. D. Dinman (eds.), Environmental Mercury Contamination, Ann Arbor Science Publishers, Ann Arbor, Mich., 1972.

80. A. R. Schulert, J. T. Davis, and D. G. Nicholson, in Environmental Mercury Contamination (R. Hartung and B. D. Dinman, eds.), Ann Arbor Science Publishers, Ann Arbor, Mich., pp. 153-154, 1972.

81. E. W. Bretthauer, A. A. Moghissi, S. S. Snyder, and N. W. Mathews, Anal. Chem., 45:1565 (1973).